VOLUME ONE HUNDRED AND TWENTY

CURRENT TOPICS IN DEVELOPMENTAL BIOLOGY

Mammalian Preimplantation Development

CURRENT TOPICS IN DEVELOPMENTAL BIOLOGY

"A meeting-ground for critical review and discussion of developmental processes"
A.A. Moscona and Alberto Monroy (Volume 1, 1966)

VOLUME ONE HUNDRED AND TWENTY

CURRENT TOPICS IN DEVELOPMENTAL BIOLOGY

Mammalian Preimplantation Development

Edited by

MELVIN L. DePAMPHILIS

Eunice Kennedy Shriver National Institute of Child Health and Human Development, National Institutes of Health, Bethesda, MD, United States

ELSEVIER

AMSTERDAM • BOSTON • HEIDELBERG • LONDON
NEW YORK • OXFORD • PARIS • SAN DIEGO
SAN FRANCISCO • SINGAPORE • SYDNEY • TOKYO
Academic Press is an imprint of Elsevier

AP

Academic Press is an imprint of Elsevier
50 Hampshire Street, 5th Floor, Cambridge, MA 02139, United States
525 B Street, Suite 1800, San Diego, CA 92101-4495, United States
The Boulevard, Langford Lane, Kidlington, Oxford OX5 1GB, United Kingdom
125 London Wall, London, EC2Y 5AS, United Kingdom

First edition 2016

ISBN: 978-0-12-801428-8
ISSN: 0070-2153

For information on all Academic Press publications visit our website at https://www.elsevier.com/

Publisher: Zoe Kruze
Acquisition Editor: Zoe Kruze
Editorial Project Manager: Shellie Bryant
Production Project Manager: Radhakrishnan Lakshmanan
Cover Designer: Greg Harris

Typeset by SPi Global, India

CONTENTS

CONTRIBUTORS

Y.D. Alvarez
Facultad de Ciencias Exactas y Naturales, Universidad de Buenos Aires, CONICET, Buenos Aires, Argentina

S. Bissiere
Institute of Molecular and Cell Biology, A*STAR, Singapore, Singapore

L. Bury*
University of Cambridge, Cambridge, United Kingdom

A. Cervero
Igenomix, Parc Científic Universitat de Valéncia, Valencia, Spain

P.A. Coelho
University of Cambridge, Cambridge, United Kingdom

M.L. DePamphilis
Eunice Kennedy Shriver National Institute of Child Health and Human Development, National Institutes of Health, Bethesda, MD, United States

S. Garcia-Herrero
Igenomix, Parc Científic Universitat de Valéncia, Valencia, Spain

V. Garg
Developmental Biology Program, Sloan Kettering Institute, Memorial Sloan Kettering Cancer Center; Biochemistry, Cell and Molecular Biology Program, Weill Cornell Graduate School of Medical Sciences, New York, NY, United States

D.M. Glover
University of Cambridge, Cambridge, United Kingdom

A.-K. Hadjantonakis
Developmental Biology Program, Sloan Kettering Institute, Memorial Sloan Kettering Cancer Center; Biochemistry, Cell and Molecular Biology Program, Weill Cornell Graduate School of Medical Sciences, New York, NY, United States

P. Kaldis
Institute of Molecular and Cell Biology (IMCB), A*STAR (Agency for Science, Technology and Research); National University of Singapore (NUS), Singapore, Republic of Singapore

K.J. Kaneko
Center for Food Safety and Applied Nutrition, U.S. Food and Drug Administration, College Park, MD, United States

*Present address: Whitehead Institute and Department of Biology, MIT, Nine Cambridge Center, Cambridge, Massachusetts 02142, USA.

B.B. Knowles
Emerita, The Jackson Laboratory, Bar Harbor, ME, United States; Siriraj Center of Excellence for Stem Cell Research, Mahidol University, Bangkok, Thailand

M.S.H. Ko
Keio University School of Medicine, Tokyo, Japan

C.Y. Leung†
University of Cambridge, Cambridge, United Kingdom

C.Y. Lim
Institute of Medical Biology, A*STAR, Singapore, Singapore

A.A. Lokken
Michigan State University, East Lansing, MI, United States

E. Mateu
Igenomix, Parc Científic Universitat de Valéncia, Valencia, Spain

D.M. Messerschmidt
Institute of Molecular and Cell Biology, A*STAR, Singapore, Singapore

P. Mir
Igenomix, Parc Científic Universitat de Valéncia, Valencia, Spain

S. Morgani
Developmental Biology Program, Sloan Kettering Institute, Memorial Sloan Kettering Cancer Center, New York, NY, United States

N. Palmer
Institute of Molecular and Cell Biology (IMCB), A*STAR (Agency for Science, Technology and Research); National University of Singapore (NUS), Singapore, Republic of Singapore

N. Plachta
Institute of Molecular and Cell Biology, A*STAR, Singapore, Singapore

M.E. Póo
Igenomix, Parc Científic Universitat de Valéncia, Valencia, Spain

L. Prezzoto
Agricultural Research Centers, Montana State University, Bozeman, MT, United States

A. Ralston
Michigan State University, East Lansing, MI, United States

R.A. Reijo Pera
Montana State University, Bozeman, MT, United States

L. Rodrigo
Igenomix, Parc Científic Universitat de Valéncia, Valencia, Spain

† Current address: Integrated Cardio Metabolic Centre, Novum, Hälsovägen 7, Karolinska Institutet, Huddinge, 141 57 Sweden.

C. Rubio
Igenomix, Parc Científic Universitat de Valéncia, Valencia, Spain

D. Solter
Siriraj Center of Excellence for Stem Cell Research, Mahidol University, Bangkok, Thailand; Emeritus, Max-Planck Institute of Immunobiology and Epigenetics, Freiburg, Germany

M. Vera
Igenomix, Parc Científic Universitat de Valéncia, Valencia, Spain

M.D. White
Institute of Molecular and Cell Biology, A*STAR, Singapore, Singapore

M. Zernicka-Goetz
University of Cambridge, Cambridge, United Kingdom

M. Zhu
University of Cambridge, Cambridge, United Kingdom

PREFACE

One critical difference between an embryo and a machine is that a machine is never required to function until after it is built, whereas every animal has to function as it builds itself.

Scott Gilbert, in "Developmental Biology," 6th ed., 2000

1. THE FOUNDATION OF MAMMALIAN DEVELOPMENT

Development of a fertilized human egg into an adult of 75 kg requires the production of 29 trillion cells or about 60 trillion meters of DNA. That's enough DNA to stretch from the earth to the sun and back 200 times! Include the approximately 1 trillion hematopoietic cells produced each day in adults, together with the turnover in epithelial cell populations, and the number of cell divisions in a human life span is truly astronomical!

Remarkably, the first 7–9 rounds of cell division that follow fertilization, the period termed preimplantation to peri-implantation development, establish the machinery that regulates cell proliferation and differentiation throughout these trillions of cell divisions. Each time a cell divides, it must duplicate its nuclear genome accurately once, but only once. Genes in the fertilized egg must soon be expressed to compensate for the loss of maternally inherited RNA and proteins. A developmental program must be established that determines which genes will become active, in which cells, and at what stage in development. The three primary cell lineages that are established in the blastocyst provide the progenitor cells for the fetus, placenta and adult. They also are the source of stem cells that can be derived, propagated, manipulated and differentiated *ex utero*. Thus, preimplantation development is the foundation on which the adult is built. Unfortunately, mistakes appear to occur frequently at the beginning of human development, because miscarriage during early pregnancy is as high as 31% (Regan & Rai, 2000; Wilcox et al., 1988). This book is dedicated to the proposition that understanding mammalian preimplantation development is critical to understanding fertility, embryonic development, and the birth of healthy offspring.

2. THE UNIQUE GOALS OF PREIMPLANTATION DEVELOPMENT

The mouse has been a popular experimental model, because it is cost-effective, amenable to the tools of genetics, biochemistry, and molecular biology, and has a short gestation period. For example, implantation occurs 4 days postcoitum in mice, whereas humans average 9 days and cows require 30 days until implantation (Lee & DeMayo, 2004). These attributes permitted an extraordinary understanding of preimplantation development in mice that is now being extended to humans. Although the time elapsed between events can vary markedly, the sequence of events from fertilization to implantation is essentially the same in mice, rats, guinea pigs, rabbits, pigs, sheep, cows, and primates. What is striking is that the developmental goals of mammals are unique when compared with animals of other phyla.

In sexual reproduction, the onset of development is triggered by fertilization. Since nuclear DNA replication is dormant in the sperm, oocytes and eggs of mollusks, birds, fish, reptiles, insects, amphibians, and mammals, nuclear DNA replication begins in the *fertilized egg* (also termed *1-cell embryo* or *zygote*). In the well-characterized examples of insects, amphibians, and mammals, fertilization produces a zygote containing two *pronuclei*, one for the paternal chromosomes and one for the maternal chromosomes, and both the maternal and paternal pronuclei replicate their DNA concurrently. The pronuclei then fuse during the first mitosis and cytokinesis to form a *2-cell embryo* in which each cell contains a single diploid nucleus with one set of paternal and one set of maternal chromosomes (Fig. 1). Mitochondrial DNA replication occurs independently of nuclear DNA replication and is linked to the energy requirements of the cell.

The events that occur in a *1-cell embryo* are essentially the same in most, if not all, metazoa, but then they change dramatically following the first mitosis. In sexual reproduction, DNA transcription does not occur in sperm, and it stops in oocytes when they undergo meiotic maturation to form eggs (Zhang et al., 2013). Transcription of the zygote's genes follows DNA replication, although the time when it occurs varies widely among animals. For example, the zygotes of insects develop by repeated rounds of nuclear DNA replication, each separated by a mitosis, but without DNA transcription and cytokinesis. The result is a multinucleated cell termed a syncytial blastoderm that then undergoes cellularization and zygotic gene expression. Amphibian embryos undergo repeated cell cleavage events (cell division in the absence of cell growth), and nuclear DNA replication is followed by mitosis and cytokinesis in the absence of DNA transcription. The result is a multicellular blastula that then activates zygotic gene expression and continues cell proliferation using the mitotic cell cycle, which includes a period of cell growth termed the G1 phase. Mammalian zygotes, however, continue development by cell cleavage in the presence of zygotic gene expression. In the

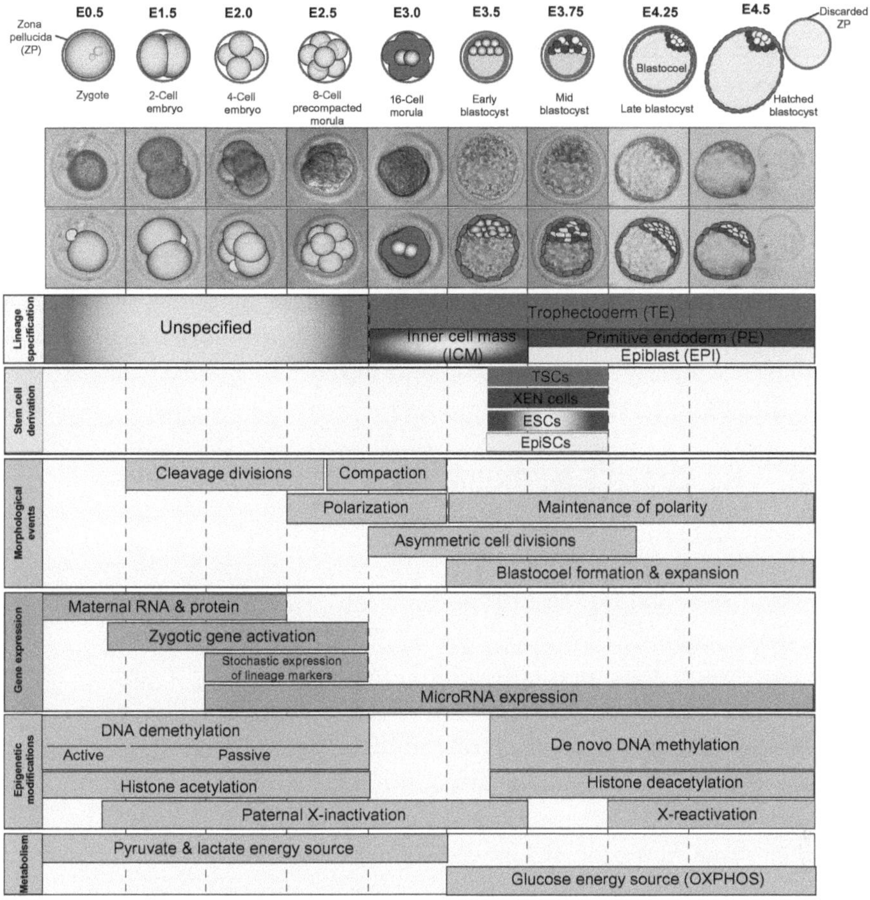

Fig. 1 Mouse preimplantation development. The fertilized egg, also called the 1-cell embryo or zygote, encapsulated by a glycoprotein layer termed the zona pellucida (ZP), undergoes three rounds of cell cleavage divisions to form the 8-cell embryo. At this stage, the totipotent blastomeres, yet unspecified for any defined cell lineage, become flattened, polar and compacted together. Over the next two divisions ($8 \rightarrow 16$ and $16 \rightarrow 32$ cells), the first cell lineage decision generates two cell populations. The outer layer of the embryo consists of a monolayer of epithelial trophoblast cells that constitutes the trophectoderm (TE), which give rise to the placenta. The TE (*blue* (*gray* in the print version)) expresses the *Cdx2* gene, but not the *Pou5f1/Oct4* gene. The inner cell mass (ICM) enclosed by the trophectoderm consists of pluripotent blastomeres, which give rise to both the embryo and the extraembryonic endoderm. The ICM expresses the *Oct4* gene, but not the *Cdx2* gene. This step is followed by formation of the early blastocyst, which is characterized by a fluid-filled cavity (blastocoel) generated by water pumped into the blastocyst by the TE. Blastocoel formation requires a major increase in oxidative phosphorylation (OXPHOS) by the trophectoderm that is reflected

(Continued)

mouse, DNA replication in the zygote is followed by a burst of nuclear DNA transcription prior to mitosis and cytokinesis. A second, larger burst of transcription occurs at the 2-cell stage and then continues throughout preimplantation development. The normal mitotic cell cycle begins during the formation of a compacted morula (Fig. 1). This change in strategy allows the "placental mammals" (eutherians) to begin cell differentiation prior to implantation, which ensures that placenta and fetus develop concurrently (O'Farrell, Stumpff, & Su, 2004). Development is essentially the same in the early human embryo, although compaction occurs at the 16-cell stage and the trophectoderm and inner cell mass (ICM) are not established until the late blastocyst stage at 6 days postcoitum (Cockburn & Rossant, 2010).

3. THE EVENTS IN PREIMPLANTATION AND PERI-IMPLANTATION DEVELOPMENT

Preimplantation development begins with formation of a zygote and ends when the developing embryo implants into the inner mucous membrane of the uterus (endometrium) (Zhang et al., 2013). As the zygote travels along the oviduct toward the uterus, the cell undergoes a series of sequential cleavages to form a 2-cell, 4-cell, and then 8-cell embryo in which each of the blastomeres remains separate from one another (Fig. 1). During the transition from an 8-cell to a 16-cell embryo (morula), the blastomeres undergo increased cell-to-cell adhesion, and the embryo adopts a more spherical, smoother shape in a process known as *compaction* that marks the beginning of cell differentiation. When development reaches the morula stage, the outer blastomeres differentiate into a monolayer of epithelial cells (trophectoderm) that is essential for the formation of the fluid-filled cavity (blastocoel)

Fig. 1—Cont'd in a switch in energy sources from pyruvate and lactate to glucose. Concomitant with blastocoel formation, the second lineage decision results in the segregation of the ICM into primitive endoderm (PE) cells that express *Gata6* (*red* (*dark gray* in the print version)) and epiblast cells (Epi) that express *Nanog* (*yellow* (*light gray* in the print version)). *Nanog*- and *Gata6*-expressing cells are arranged in a "salt and pepper" pattern in the early blastocyst, but by the late blastocyst stage, are sorted into a layer of primitive endoderm that separates the epiblast from the blastocoel. Prior to implantation, the late blastocyst hatches out of the zona pellucida. Different stem cell populations are typically isolated in vitro from blastocyst-stage outgrowths. These include trophoblast stem cells (TSCs), embryonic stem cells (ESCs), epiblast stem cells (EpiSCs), and extraembryonic endoderm stem cells (XEN cells). Numbers above the phase contrast images indicate the number of days postcoitum at the indicated embryonic stage. *Images and figures were provided by Alyson A. Lokken and Amy Ralston.*

characteristic of a *blastocyst*. The preimplantation embryo now consists of two distinct cell lineages: the trophectoderm that surrounds the blastocoel, and a cluster of cells within the blastocoel termed the *inner cell mass* (ICM), which remains in contact with the trophectoderm. The trophectoderm is essential for the subsequent implantation of the embryo, and it contributes exclusively to formation of the placenta, whereas the fetus is derived from the ICM, with minor contributions from the endoderm (Sills et al., 2016).

A protein shell termed the *zona pellucida* surrounds the zygote prior to implantation and prevents it from attaching to the oviduct. The cells of the blastocyst continue to proliferate until the pressure produced by the expanding blastocoel allows the embryo to hatch out of its zona pellucida and implant. The formation and expansion of the blastocoel requires dramatic metabolic changes in order to maintain the embryo's *energy homeostasis*. By the late blastocyst stage (100–200 cells), three distinct tissue lineages are present: the trophectoderm that gives rise exclusively to the placenta, the epiblast that gives rise to the fetus, and the primitive endoderm, which emerges as a polarized epithelium adjacent to the blastocoel with a basement membrane separating it from the epiblast. From the time the blastocyst is free in the uterus, through the process of attachment, until the beginning of trophoblast differentiation is known as *peri-implantation*. A peri-implantation blastocyst results from the ICM segregating into two discrete layers: the epiblast, which gives rise to the fetus, and the primitive endoderm that, along with the trophectoderm, produces extraembryonic tissues, such as the placenta (Fig. 2).

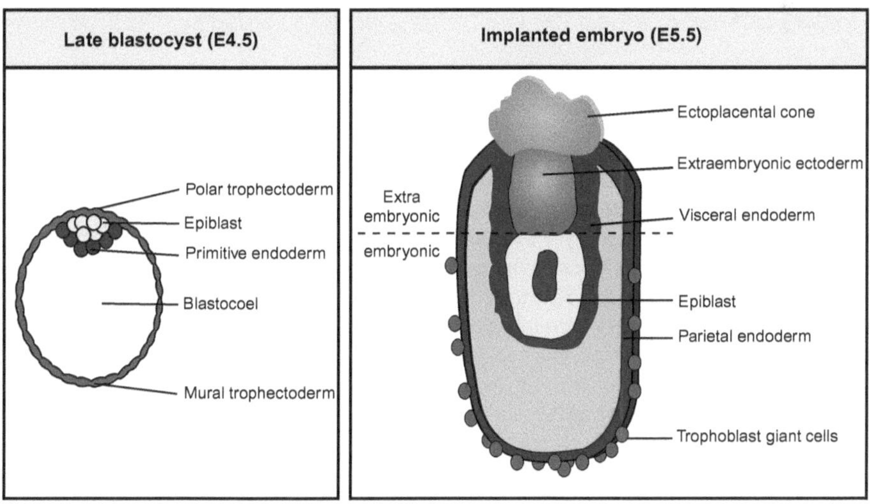

Fig. 2 See legend on next page.

4. RECENT ADVANCES

The purpose of this book is to highlight the fundamental importance of preimplantation development, and the recent breakthroughs in our understanding of the processes that govern it. Taken together, the 12 chapters provide a comprehensive picture of mammalian development from the 1-cell embryo to the period encompassing implantation of the blastocyst. The chapters are grouped into three sections.

Chapters 1–4 address cell proliferation. They address the problems and solutions encountered in triggering DNA replication in the zygote, ensuring that the entire genome is accurately duplicated only once each time a cell divides, but allowing special developmentally programmed cells to overcome this restriction and produce multiple copies of the genome in response to environmental signals. Fertilization also results in transcription and translation of genes in preimplantation embryos. Which genes are activated, and when are they activated? Are they activated in specialized cells at different times during preimplantation development? How are oocytes and eggs that are programmed to undergo meiosis reprogrammed into cells that undergo mitosis? How do cells transit from cell cleavage cycles, when DNA replication and mitosis occur in the absence of cell growth, to mitotic cell cycles when cell growth precedes DNA replication and mitosis?

Chapters 5–10 address cell differentiation. They address the transition from totipotent cells that produce lineages that will form both the embryo and the placenta, to pluripotent cells that give rise to either the embryo, or to

Fig. 2 Mouse peri-implantation development. A late-stage blastocyst hatches out of its zona pellucida in order to implant into the uterine endometrium. In the first 2 days postimplantation, the trophoblast cells not in contact with the ICM (the mural trophectoderm) give rise to polyploid trophoblast giant cells that are essential for implantation. The remaining trophectoderm becomes the ectoplacental cone and the extraembryonic ectoderm that give rise to the placenta. Migration of cells into the primitive endoderm produces the parietal endoderm that covers the inner surface of the mural trophectoderm, and the visceral endoderm that covers the egg cylinder and the epiblast. The egg cylinder quickly undergoes gastrulation to form the three germ layers: ectoderm, endoderm, and mesoderm. The outer layer of ectoderm gives rise to the epidermis, neural crest, and tissues that later form the nervous system. The inner layer of endoderm gives rise to the epithelium of the digestive and respiratory systems, and organs associated with the digestive system, such as liver and pancreas. The middle layer of mesoderm gives rise to the somites that form muscle, cartilage, bone, and other connective tissues. *Figures were provided by Alyson A. Lokken and Amy Ralston.*

multipotent stem cells that produce placenta and other extraembryonic tissues. Obvious signs of differentiation include compaction, wherein a loose collection of totipotent cells suddenly form a compact ball of tightly adhering cells. Embryo compaction is accompanied by a major transition in metabolic energy pathways concomitant with the first example of cell differentiation, specification of the trophectoderm lineage that drives blastocoel formation. Changes in gene expression correlate with epigenetic changes such as histone acetylation and methylation, DNA methylation, and inactivation of the paternal X chromosome. Cell fate acquisition involves changes in gene expression, cell polarity, and cell signaling. Cell fate acquisition not only includes separation of the trophectoderm from ICM, but it also involves the specification of progenitors of embryonic, trophoblast, and extraembryonic endoderm stem cells, all three of which can self-renew and are capable of differentiating into multiple lineage-specific cell types.

Chapters 11 and 12 address the differences that exist in preimplantation development among various mammalian species, and how genetic analysis of preimplantation embryos can overcome human infertility and facilitate the development of healthy offspring.

5. THE FUTURE OF PREIMPLANTATION DEVELOPMENT

5.1 Genetic Diagnosis

Preimplantation genetic screening has the power to improve successful pregnancy rates, minimize miscarriage risk, and limit multiple gestations. Nevertheless, background aneuploidy remains a major factor driving implantation failure and miscarriage for patients experiencing infertility, suggesting that genetic screening of preimplantation embryos should become a routine procedure for patients requesting in vitro fertilization (Sills et al., 2016). This raises ethical questions for parents using technologies to select genetic traits of their offspring. Should we select for genetic traits such as gender and giftedness, as well as against genetic traits such as Huntington's, Parkinson's, and Alzheimer's disease? Whole-genome sequencing of embryos is technically possible (Winand et al., 2014). Many loss-of-function mutations exist in the general population without serious effects on the phenotype of the individual, but more than 40% of individuals who can be considered healthy have mutations that are predicted to be damaging in genes associated with severe Mendelian disorders or are annotated as disease causing. Thus, preimplantation development is already a window into genetic diagnosis and embryo selection.

5.2 Genetic Intervention

How long will it be before preimplantation development is also a window into genetic modification of mammals? Embryonic stem cells, as well as induced pluripotent stem cells, are widely used to model early embryonic events, provide a vehicle for introducing new or altered genes into specific mammalian tissues, and form the basis of cell replacement therapies. Embryonic stem cells have been reverted to a totipotent state that can differentiate in the direction of trophectoderm as well as ICM. Stem cells have been derived from trophectoderm, endoderm, and epiblast tissues. Such stem cells could soon be used to correct genetic traits that are harmful to development, as well as to enhance those traits we deem helpful. It is all a question of ethics (Ethics Committee of the American Society for Reproductive Medicine, 2015), a discussion in which the entire society should be taking part. Just because genetic modifications in the germ line are possible, it does not mean that they are desirable.

6. CANCER STEM CELLS

Preimplantation development not only provides the foundation for construction of a healthy adult, but it might also sow the seeds for its demise. Consider the possibility that some pluripotent embryonic cells do not differentiate during postimplantation development, but enter a quiescent state that allows them to survive into the adult where they eventually resurface as the progenitors of cancers. If this hypothesis is correct, then great caution must be exercised in the application of pluripotent cells to gene therapy.

MELVIN L. DEPAMPHILIS

REFERENCES

Cockburn, K., & Rossant, J. (2010). Making the blastocyst: Lessons from the mouse. *The Journal of Clinical Investigation, 120*, 995–1003.

Ethics Committee of the American Society for Reproductive Medicine. (2015). Use of reproductive technology for sex selection for nonmedical reasons. *Fertility and Sterility, 103*, 1418–1422.

Lee, K. Y., & DeMayo, F. J. (2004). Animal models of implantation. *Reproduction, 128*, 679–695.

O'Farrell, P. H., Stumpff, J., & Su, T. T. (2004). Embryonic cleavage cycles: How is a mouse like a fly? *Current Biology, 14*, R35–R45.

Regan, L., & Rai, R. (2000). Epidemiology and the medical causes of miscarriage. *Baillière's Best Practice & Research. Clinical Obstetrics & Gynaecology, 14*, 839–854.

Sills, E. S., Anderson, R. E., McCaffrey, M., Li, X., Arrach, N., & Wood, S. H. (2016). Gestational surrogacy and the role of routine embryo screening: Current challenges and future directions for preimplantation genetic testing. *Birth Defects Research. Part C, Embryo Today: Reviews*, *108*(1), 98–102.

Wilcox, A. J., Weinberg, C. R., O'Connor, J. F., Baird, D. D., Schlatterer, J. P., Canfield, R. E., et al. (1988). Incidence of early loss of pregnancy. *The New England Journal of Medicine*, *319*, 189–194.

Winand, R., Hens, K., Dondorp, W., de Wert, G., Moreau, Y., Vermeesch, J. R., et al. (2014). In vitro screening of embryos by whole-genome sequencing: Now, in the future or never? *Human Reproduction*, *29*, 842–851.

Zhang, S., Lin, H., Kong, S., Wang, S., Wang, H., Wang, H., et al. (2013). Physiological and molecular determinants of embryo implantation. *Molecular Aspects of Medicine*, *34*, 939–980.

CHAPTER ONE

Regulation of the Embryonic Cell Cycle During Mammalian Preimplantation Development

N. Palmer*,†, P. Kaldis*,†,1
*Institute of Molecular and Cell Biology (IMCB), A*STAR (Agency for Science, Technology and Research), Singapore, Republic of Singapore
†National University of Singapore (NUS), Singapore, Republic of Singapore
[1]Corresponding author: e-mail address: kaldis@imcb.a-star.edu.sg

Contents

Current Topics in Developmental Biology, Volume 120
ISSN 0070-2153
http://dx.doi.org/10.1016/bs.ctdb.2016.05.001

Abstract

The preimplantation development stage of mammalian embryogenesis consists of a series of highly conserved, regulated, and predictable cell divisions. This process is essential to allow the rapid expansion and differentiation of a single-cell zygote into a multicellular blastocyst containing cells of multiple developmental lineages.

This period of development, also known as the germinal stage, encompasses several important developmental transitions, which are accompanied by dramatic changes in cell cycle profiles and dynamics. These changes are driven primarily by differences in the establishment and enforcement of cell cycle checkpoints, which must be bypassed to facilitate the completion of essential cell cycle events.

Much of the current knowledge in this area has been amassed through the study of knockout models in mice. These mouse models are powerful experimental tools, which have allowed us to dissect the relative dependence of the early embryonic cell cycles on various aspects of the cell cycle machinery and highlight the extent of functional redundancy between members of the same gene family.

This chapter will explore the ways in which the cell cycle machinery, their accessory proteins, and their stimuli operate during mammalian preimplantation using mouse models as a reference and how this allows for the usually well-defined stages of the cell cycle to be shaped and transformed during this unique and critical stage of development.

1. INTRODUCTION

1.1 The Somatic Cell Cycle

The word mitosis derives from the Greek word "mitos" meaning to warp or thread. This term was first applied to the cell cycle following the pioneering observations of condensed chromatin threads within cell nuclei during prophase, by the noted anatomist and cell biologist Walther Fleming in 1882.

The mitotic cell cycle is an exquisitely complex process but can be broadly split into two phases known as the mitotic phase, also known as M phase and a much longer growth phase, known as interphase, which comprises 95% of the total cell cycle length (for a comprehensive review, see Morgan, 2007).

Interphase can be viewed as the preparatory period, which is required by a cell to allow the initial entry into M phase. This is the longest and most complex stage of the cell cycle and like the proceeding M phase can be further subdivided into stages, which are known as G1, S, and G2. Unlike M phase these substages are not discernable by physical, observable events occurring within the cell but can be determined indirectly through the measurement of DNA content or the quantification of specific proteins within the cell.

For the entry of a cell into M phase to become possible, the cell must first grow in size and promote the transcription of mRNA and translation of proteins required for cell division. This occurs during the G1 and G2 stages of interphase. These are known as gap or growth phases due to their segregation of S phase and M phase by periods of cellular growth and allow for the maintenance of cellular size despite the occurrence of reductional divisions. By definition, the cell cycle is a cyclical process and cells can either exist in a state of active cycling whereby they will reenter the cell cycle at the G1 phase following a successful mitotic division or alternatively enter a state of cellular stasis known as G0.

The decision by a cell to enter either G1 or G0 following a successful cell division is governed mainly by the availability of factors such as growth factors, nucleotides, and amino acids. If sufficient factors are available, the cell will proceed through the restriction point. At this point an irreversible all or nothing chain of cellular events is initiated which drives cell entry into G1/S. From this point onward the cell is committed to completing one complete cell cycle. If the restriction point cannot be bypassed due to unfavorable cellular conditions, the cell will enter G0. Nondividing cells such as hepatocytes and differentiated neurons will arrest in this phase.

G1 entry sets the stage for the following S phase in which the initial diploid cell (2N) undergoes high fidelity semiconservative replication of its chromosomes to achieve 4N status. In addition, the centrosome of the cell undergoes duplication as an essential preparatory step for the proceeding M phase. Following successful S phase progression, the cell is able to enter G2 phase and undergoes a second round of cell growth. Before entry into M phase the cell is halted at the G2/M checkpoint. This checkpoint responds to many inputs from upstream DNA damage response pathways and arrests the cell until any DNA damage accumulated in interphase is repaired.

The mitotic phase can be viewed as the physical division of a cell into two equal mother and daughter cells. M phase follows a series of observable cellular events which like interphase can be simplistically broken down into well-defined substages each of which is crucial for the progression and successful completion of a single-cell division.

M phase is initiated at prophase where replicated chromosomes undergo condensation of their chromatin and associate with their homologous counterparts (sister chromatids) via their centromeres. In addition, newly replicated centrosomes segregate to opposite poles of the nucleus (centrosomal poles). Each centrosome acts as a microtubule-organizing center and drives the formation of the mitotic spindle by catalyzing the dynamic polymerization and depolymerization of microtubules, which actively seek out and bind a centromeric structure known as a kinetochore on each sister chromatid. During metaphase the pull and push forces exerted by the mitotic spindle cause the chromosomes to align along the metaphase plate approximately equidistant from each centrosomal pole. At this time the sister chromatids are held in a state of high tension by the mitotic spindle due to a cohesin ring structure, which prevents their separation. Upon complete bipolar attachment of all chromosomal kinetochores to the mitotic spindle a third checkpoint known as the spindle assembly checkpoint (SAC) is silenced and the cohesin ring is destroyed. This allows the initiation of anaphase and the release of sister chromatids to opposite poles, simultaneously the nuclear envelope rapidly breaks down resulting in the bilateral division of the nucleus into two daughter nuclei. During telophase these nuclei are quickly enveloped by a new nuclear membrane and their chromosomes decondense reversing the events of prophase and metaphase. Subsequently the cell forms a contractile ring composed of actin and myosin II filaments, which surrounds the midline of the cell. Inward constriction of this contractile ring driven by myosin ATPases causes an abscission cleaving the cell into two daughter cells, which segregates organelles, cytoplasm, and the cell membrane into two new daughter cells completing the final stage of cell division cytokinesis.

2. THE PREIMPLANTATION CELL CYCLE

Although in general terms, the embryonic cell cycles, which occur prior to implantation, follow the same broad patterns of events as somatic cell divisions, this complex period of development cannot be considered as a single event due to intrinsic differences in both the length and composition of cell cycles during preimplantation.

Instead the preimplantation cell cycle can be broken down into several smaller periods of development each with distinct characteristics. They consists of the initial embryonic cell divisions, zygotic genome activation, blastulation and compaction, epiblast formation, and finally the formation of an implantation-competent blastocyst (reviewed in Wang & Dey, 2006).

2.1 Preparation for the First Cell Cycle

Preparation for the first mammalian cell division is arguably initiated approximately 2 weeks prior to the ovulation of unfertilized secondary oocytes from the ovary. During this stage of intraovarian growth, the immature oocyte increases in size (De Leon, Johnson, & Bachvarova, 1983), there is an upregulation of maternal effect genes, and there is a subsequent accumulation of mRNA (Christians, Davis, Thomas, & Benjamin, 2000; Tong et al., 2000). These events prime the oocyte for potential fertilization by facilitating the stockpiling of macromolecules and organelles, which will be required for the first cell division immediately following fertilization.

The relative contribution of the paternal spermatozoon toward the resultant single-cell zygote following successful fertilization is small with the donation of a male pronucleus and a functional centrosome (Sathananthan, 1997). This is speculated to be because unlike secondary oocytes, spermatozoa are readily and continually remade in large numbers during spermatogenesis and need only maintain their DNA for a short period until fertilization. This is in stark contrast to oocytes, which require a large investment of time and energy to ensure that their DNA properly is maintained and repaired from initial formation of primary oocytes until fertilization since arrested oocytes are stored for the entire fertility time of the organism.

2.2 Initial Embryonic Cell Cycles

Although the cell division kinetics between different cell types can vary dramatically, the typical somatic eukaryotic cell division lasts approximately 24 h. This consists 11 h spent in G1, 8 h in S phase, 4 h in G2 phase, and finally 1 h taken to complete M phase and cytokinesis.

In terms of length the initial embryonic cell divisions are very similar to these somatic cell divisions but exhibit specialized differences in their cell growth dynamics and composition. This period of early preimplantation is referred to as the cleavage stage of division due to the lack of cellular growth, which occurs between divisions in somatic cells. The cells within the embryo during this stage, therefore, increase in number but reduce in size with each successive division (Aiken, Swoboda, Skepper, & Johnson, 2004). This is due to the fact that at this time the developing embryo remains surrounded by a glycolipid membrane barrier known as the zona pellucida, which restricts access to outside sources of nutrition required to sustain growth.

The first cell cycle is initiated when a mature spermatozoon fertilizes an ovulated secondary oocyte. This event triggers the oocyte to exit its prolonged metaphase II arrest and complete its second meiotic division resulting in the formation of a haploid nucleus and the extrusion of the second polar body.

The events which follow lead to the formation of a female and a male pronucleus through the decondensation of the chromatin of the oocyte and spermatozoon nuclei (Wright & Longo, 1988). These pronuclei enter S phase and replicate their DNA independently and do not interact until the breakdown of the nuclear membranes of both pronuclei during the first prometaphase which allows the intermixing and combination of maternal and paternal genetic DNA (Donahue, 1972; Mayer, Smith, Fundele, & Haaf, 2000). This is followed by the first metaphase of the single-cell zygote at which time homologous chromosome pairs align on the metaphase plate for segregation (Adenot, Szollosi, Geze, Renard, & Debey, 1991; Mayer, Smith, et al., 2000).

The first two embryonic cell divisions are the longest during mammalian preimplantation development lasting approximately 18–20 h in both murine and human zygotes (Ciemerych, Maro, & Kubiak, 1999). Although these cycles have been extensively studied and reviewed in the mouse, various studies have continually reported varying lengths for each of the substages of interphase, possibly due to differences in the genetic background of mice used in each study. Approximate lengths for the first cell division duration are represented in Fig. 1 and have been reported as: 5–12 h for G1, 4–7 h for S phase, and 1–8 h for the G2/M transition with the resultant mitotic phase

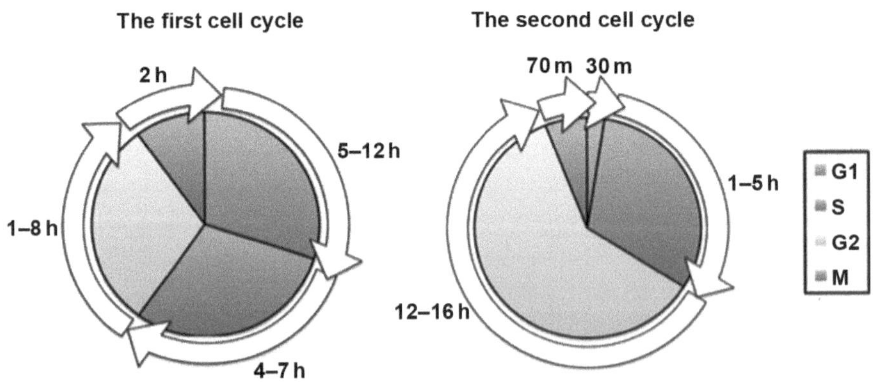

Fig. 1 The estimated duration of each of the substages of interphase and mitosis during the first and second murine embryonic cell cycles. (See the color plate.)

lasting approximately 120 min (Abramczuk & Sawicki, 1975; Domon, 1980; Howlett & Bolton, 1985; Krishna & Generoso, 1977; Molls, Zamboglou, & Streffer, 1983).

The second cell cycle starts approximately 1.5 days postfertilization and is similarly long, lasting approximately 20 h. In contrast to the first cell cycle, the second one consists of an extremely short G1 phase of 0.5–1.3 h (Gamow & Prescott, 1970) followed by 1–5 h for S phase (Bolton, Oades, & Johnson, 1984) and an extremely long G2/M transition making up the bulk of the cell cycle (12–16 h) (Bolton et al., 1984; Luthardt & Donahue, 1975; Molls, Pon, Streffer, van Beuningen, & Zamboglou, 1983; Molls, Zamboglou, et al., 1983; Sawicki, Abramczuk, & Blaton, 1978; Siracusa, Coletta, & Monesi, 1975; Streffer, van Beuningen, Molls, Zamboglou, & Schulz, 1980). In contrast to the first cell cycle the mitotic phase is much quicker during the second cell cycle culminating after just 70 min.

2.3 Maternal–Zygotic Transition, Genome Activation, and Epigenetic Reprogramming

Although de novo transcription of zygotic mRNA can be seen as early as in the G2 phase of the first embryonic cell division (Bouniol, Nguyen, & Debey, 1995), full zygotic genome activation does not occur until the second G2/M transition in mice and the 4–8 cell stage in humans (Flach, Johnson, Braude, Taylor, & Bolton, 1982). This period is known as the maternal–zygotic transition and marks the handover of the transcriptional responsibility, from stored maternal mRNAs to de novo synthesis directly from the zygotic genome (Bolton et al., 1984; Braude, 1979; De Leon et al., 1983; Flach et al., 1982; Reik & Walter, 2001). The transcriptional burst associated with zygotic genome activation facilitates this transitional period by driving the proteolytic destruction of maternally inherited mRNA and proteins which are required for maintenance of the maternal oocyte and by promoting de novo synthesis of proteins required for the maintenance of embryonic cell divisions (Lee et al., 2013).

At this time there is also a wave of epigenetic reprogramming to allow the cells of the early embryo to assume a pluripotent state and prevent premature differentiation. This is propagated primarily by the downregulation of the DNA methyltransferase family of methylases which passively promotes the global demethylation of maternally inherited DNA over several cycles of DNA replication (Mayer, Niveleau, Walter, Fundele, & Haaf, 2000; Oswald et al., 2000; Santos, Hendrich, Reik, & Dean, 2002). This is in contrast to paternally inherited DNA which is actively demethylated through a currently unknown mechanism even before the onset of DNA replication in the male pronucleus (Mayer, Niveleau, et al., 2000; Oswald et al., 2000; Santos et al., 2002; reviewed by Messerschmidt, Knowles, & Solter, 2014). This demethylation of the zygotic DNA is not fully penetrant however and methylation remains within imprinted gene regions to facilitate the maintenance of "parent-of-origin-specific" gene expression in the developing embryo.

2.4 Blastocyst Formation and Compaction

Following zygotic genome activation, a "mulberry-like" structure known as the morula is formed at the 8-cell stage of development. Unlike the initial two embryonic cell divisions further divisions from the 8-cell stage until the formation of the blastocyst at approximately the 32- to 64-cell stage of development are shorter and more homogenous in length lasting approximately 12–14 h in total (Smith & Johnson, 1986) with an approximate composition of (G1:1–2 h, S:7 h, G2/M:1–5 h) (Molls, Zamboglou, et al., 1983; Sawicki et al., 1978; Smith & Johnson, 1986; Streffer et al., 1980).

The formation of the blastocyst is initiated during the late 8-cell stage at approximately E2.75 in mice and promotes the polarization of developing blastomeres within the morula to either an inner or an outer layer (Plusa et al., 2005; Thomas et al., 2004). This marks the earliest instance of cellular fate determination in the early embryo as the outer layer will develop into the embryonic epithelial tissue, collectively known as the trophoblast, whereas blastomeres remaining within the inner layer will become the pluripotent inner cell mass (ICM) of the embryo (Chen, Wang, Wu, Ma, & Daley, 2010).

Following blastomere polarization, the embryo undergoes a process known as compaction (Tarkowski & Wroblewska, 1967) in which cell–cell contacts are maximized and trophoblast cells flatten and form a watertight seal via the interlocking of desmosomes and gap junctions (Fernandez & Izquierdo, 1980). This newly formed trophoblast layer is rich in ion channels and actively transports sodium and chloride ions to alter the osmotic gradient and water flux within the embryo resulting in the formation of a fluid-filled cavity, the blastocoel (Manejwala, Cragoe, & Schultz, 1989).

The resultant structure is known in mammals as the blastocyst, which is formed approximately 5–6 days postfertilization in humans (Hardy, Handyside, & Winston, 1989) and 3.5 days postfertilization in mice. The blastocyst consists of a hollow sphere of cells containing an ICM and a blastocoel, encapsulated by a trophoblast layer.

When compared to human adult fibroblasts, pluripotent blastomeres isolated from the 8-cell stage of human embryogenesis show downregulation of important cell cycle checkpoint mRNA transcripts such as Rb and their enforcers. This is accompanied by a significant upregulation of key cell cycle drivers including Cdc25A, cyclin A2, and cyclin E1 (Kiessling et al., 2009, 2010). This increased cell cycle drive coupled with weakened cell cycle checkpoints is mirrored by a decreased time spent by the cell in the G1 and G2 phases and a reduction in the total cell cycle length.

2.5 Late Preimplantation

During late preimplantation, epiblast and hypoblast lineages are formed from the ICM of the blastocyst (Najm et al., 2011). At this time the epiblast cells, which will eventually form the endoderm, ectoderm, and mesoderm cell lineages of the embryo proper, undergo a proliferative burst and expand through an explosive series of unusually rapid cell divisions. The rate of division within the epiblast far exceeds those of typically dividing cells approximately equating to a single-cell division once every 10 h (Snow, 1977; Solter, Skreb, & Damjanov, 1971). These rapid cell cycles also exhibit a globally elevated transcriptional output (Power & Tam, 1993) and characteristically display truncated G1 and G2 phases enabling them to enter S phase and replicate their DNA immediately following exit from the preceding M phase (Guzman-Ayala et al., 2015; Hogan, Beddington, Constantini, & Lacy, 1994).

The mechanism behind the rapid cycling of these cells was studied in detail (Stead et al., 2002), indicating that cells of the embryonic epiblast origin lack usual cell cycle control mechanisms leading to the loss of periodicity. This was primarily driven by heightened expression of Cdk2/cyclin A and Cdk2/cyclin E complexes allowing for precocious G1/S entry and DNA replication. This was also coupled with high levels of Cdk1/cyclin B complexes allowing the immediate progression through the G2/M checkpoint, essentially truncating all gap phases of the cell cycle. The speed of these divisions during this proliferative bust is thought to be a protective mechanism, which prevents the premature or ectopic differentiation of pluripotent cells within the embryo (Calegari & Huttner, 2003). Similar phenomena have also been seen in embryonic stem cells (ESCs) (Hyslop et al., 2005).

In terms of length and composition these epiblast divisions are comparable to much earlier embryonic cell divisions in other organisms. In this respect the mammalian embryonic cell cycles have been described as an exception to those of many other phyla (O'Farrell, Stumpff, & Su, 2004). Indeed many model organisms including leeches (Bissen & Weisblat, 1989), sea urchins (Parisi, Filosa, De Petrocellis, & Monroy, 1978), *Xenopus* (Newport & Kirschner, 1982), and *Drosophila* (Milan, Campuzano, & Garcia-Bellido, 1996) display rapid cell divisions lacking gap phases throughout early embryonic development. In addition to the lack of gap phases, the preimplantation cycles of other organisms are also uncoupled from normal cell cycle DNA damage checkpoints as has been demonstrated in *Drosophila* by the lack of DNA damage-mediated checkpoint arrest upon administration of genotoxic agents such as aphidicolin (Raff & Glover, 1988).

In contrast, mouse embryos are acutely sensitive to aphidicolin-mediated G2 arrest as early as the 4-cell stage but progressively lose this sensitivity through preimplantation development (Spindle, Nagano, & Pedersen, 1985). The loss of checkpoint control by late preimplantation is supported by the fact that pluripotent mammalian ESCs isolated from epiblast also do not arrest following gamma irradiation (Fluckiger et al., 2006). This suggests that checkpoint activity is established in the early embryo but is lost by the onset of this proliferative burst. It is currently unknown why these gap phase-independent cycles establish themselves during a later stage of preimplantation in mammals although the establishment of cell cycle checkpoints for the repair of DNA damage during embryogenesis seems to be an essential event for early embryogenesis and will be discussed in Section 5.

3. REGULATION OF THE PREIMPLANTATION CELL CYCLE

3.1 Cdks and Cyclins

To complete mitosis, a cell must pass several major quality control checkpoints during both interphase and M phase, which ensures that the cell has adequate resources required for growth and enables the cell to monitor for DNA double-strand breaks improper chromosome segregation and other abnormalities to safeguard against the heritance of damaged DNA. The cell is equipped with an arsenal of cell cycle machinery, which consists of a highly conserved array of regulatory proteins, which can drive, halt, or even stop cell division to enforce these checkpoints. The most notable of these proteins during interphase are arguably the cyclin-dependent kinases (Cdks) and their binding partner proteins known as cyclins. These complexes have been aptly described as the engines, which provide the driving force for cell cycle progression through their phosphorylation of crucial substrates (for a comprehensive review, see Morgan, 2007).

The Cdks are a family of low molecular weight serine–threonine kinases, ranging from 32 to 40 kDa, which have been conserved throughout eukaryotic evolution. Indeed, this is highlighted by the fact that human Cdk1 can functionally replace the homologous protein cdc2 in the fission yeast *Schizosaccharomyces pombe* harboring an inactivating mutation in the native cdc2 (Lee & Nurse, 1987). Based on sequence similarity studies, there are currently 21 described *Cdk* genes in humans and 5 further, more distantly related, *Cdk-like kinase* (*CdkL*) genes. Of these genes, only six (Cdk1, 2, 3, 4, 6, 7) have confirmed roles in cell cycle progression (reviewed in Sherr & Roberts, 2004).

Alongside the Cdks, at least 30 cyclin genes have also been described, with 11 of these having currently known roles in cell cycle progression (reviewed by Malumbres et al., 2009). Despite the tight association of Cdks and cyclins with transcriptional regulation and ubiquitin-mediated degradation, the subsequent stimulation of the cell cycle by the different Cdk/cyclin complexes is highly diverse.

The structure of numerous Cdks as determined by X-ray crystallography (Brown et al., 2015; De Bondt et al., 1993; Echalier, Endicott, & Noble, 2010; Endicott & Noble, 2013; Jeffrey et al., 1995; Jeffrey, Tong, & Pavletich, 2000; Krzywda et al., 2004; Lu, Chang, Baratte, Meijer, & Schulze-Gahmen, 2005; Pavletich, 1999; Russo, Jeffrey, & Pavletich, 1996; Russo, Tong, Lee, Jeffrey, & Pavletich, 1998; Takaki et al., 2009) is broadly similar to other protein kinases, which share a tertiary structure comprising a large carboxy-terminal α-helical lobe and a smaller amino-terminal β-sheet-rich lobe. The space between these two lobes forms a cleft known as the active site, which allows both the binding of ATP within the cleft and the binding of protein substrates at the mouth of the cleft with additional interactions occurring with the surface of the carboxy-terminal lobe. The proper orientation of ATP with its phosphates facing toward the outside of the cleft and the positioning of the substrate within the mouth of the active site are important to allow the gamma-terminal phosphate of the ATP to be efficiently transferred to a protein substrate bound within the active site. Typical substrates of Cdks contain a highly specific [S/T*]PX [K/R] motif with a serine or threonine followed by a proline residue, which is specifically targeted for the transfer of the terminal gamma phosphate group from ATP (Holmes & Solomon, 1996).

3.2 Cellular Regulation of Cdk/Cyclin Complexes

Inactive Cdks contain two main structural features, which interfere with the optimal interaction between ATP and substrate and thus prevent premature activation of their enzymatic kinase activity and can be utilized as on–off switch allowing them to be regulated by other proteins.

First, a structure known as a T-loop or activation loop extends from the carboxy-terminus to block the binding of substrates within the mouth of the active site. Second, several side chains are present which interfere with the correct positioning of ATP within the active site. To overcome these inactivating structural features, an alpha-helical structure on the Cdk known as the PSTAIRE domain (named after the target amino acid sequence of the cyclin for the Cdk) or the α1-helix is able to interact directly with an activating partner protein known as a cyclin. The binding of a cyclin to the PSTAIRE helix causes several structural alterations to the Cdk, including the restructuring of a second alpha-helix known as the small L12 helix, which directly precedes the T-loop, and changes it to a beta-strand conformation. The subsequent reorganization of the T-loop exposes several activating phosphorylatable residues and primes the Cdk to bind both ATP and substrates. The most crucial phosphorylatable residues exposed on the T-loop after cyclin binding (Thr161, Thr160, or Thr172 in Cdk1, Cdk2, and Cdk4, respectively) are targets of the Cdk-activating kinase (CAK) which will be discussed later in Section 3. Phosphorylation of this residue by the CAK further contorts the T-loop allowing a more favorable interaction of the active site with a substrate.

A second phosphorylatable tyrosine residue (Tyr15, Tyr15, or Tyr17 in Cdk1, Cdk2, and Cdk4, respectively) alongside a neighboring threonine residue is found within the ATP-binding site of the Cdk. In contrast to the activating effect of T-loop phosphorylation, phosphorylation of this "TY" motif is inhibitory and is mediated by negative regulators of the cell cycle including the Wee1, Myt1, and Mik1 dual-specificity kinases. This is a bidirectional reaction and removal of this tyrosine phosphorylation by the activating dual-specificity phosphatases Cdc25 must occur to allow the full activation of Cdk/cyclin complexes (Johnson & Lewis, 2001; Morgan, 1997).

Additional layers of regulation are imposed by the Cdk inhibitor family of molecules which includes the CIP/KIP family of proteins $p21^{CIP1/WAF1}$ (Chen, Jackson, Kirschner, & Dutta, 1995; el-Deiry et al., 1993; Harper, Adami, Wei, Keyomarsi, & Elledge, 1993), $p27^{KIP1}$ (Polyak et al., 1994;

Slingerland et al., 1994; Toyoshima & Hunter, 1994), and p57^{KIP2} (Lee, Reynisdottir, & Massague, 1995; Matsuoka et al., 1995), which inhibit the function of cyclin D-, E-, and A-dependent kinases and also the INK4 family of proteins; p15^{INK4b} (Hannon & Beach, 1994), p16^{INK4a} (Serrano, Hannon, & Beach, 1993), p18^{INK4c} (Guan et al., 1994), and p19^{INK4d} (Chan, Zhang, Cheng, Shapiro, & Winoto, 1995) which specifically inhibit the cyclin D-dependent kinases Cdk4/6. These proteins are able to physically associate with either the Cdk directly or its partner cyclin and block their kinase function preventing cell cycle progression.

4. MOUSE MODELS OF CELL CYCLE REGULATORS AS DEVELOPMENTAL TOOLS

4.1 Cell Cycle-Related Cdk/Cyclin Complexes

In mammals, the activation of different Cdk/cyclin complexes has been well documented to promote the timely advancement between the different stages of interphase. Typically the cell utilizes Cdk4/6–cyclin D1 and Cdk2/cyclin E1 complexes for G1/S transition, Cdk2/cyclin A2 complexes for S phase progression, and Cdk1/cyclin A2 and Cdk1/cyclin B1 complexes for the G2/M transition. This is not always the case however, as different Cdks are able to exhibit cross-specificity for the same cyclin allowing for a large degree of compensation between the Cdk family members (for review, see Bisteau, Caldez, & Kaldis, 2014; Gopinathan, Ratnacaram, & Kaldis, 2011).

The function and relative importance of Cdks and cyclins during development has been thoroughly scrutinized through the generation of knockout mouse models, which have allowed researchers to determine the minimal amount of cell cycle machinery needed for normal development to occur. As a result the majority of the interphase-related Cdks and cyclins have been comprehensively shown to be dispensable for both the somatic cell cycle and early embryonic development (reviewed by Sherr & Roberts, 2004; Bisteau et al., 2014).

In fact, very few genes have been shown to be absolutely essential for mammalian preimplantation development. These include Cdk1, Cdk8, Wee1, polo-like kinase 1 (Plk1), and the CAK comprising Cdk7, ménage-à-trois 1 (Mat1), and cyclin H.

Although convincing, some caveats remain when interpreting these knockout studies, including the fact that maternal transcripts of some

Cdks and cyclins persist throughout preimplantation in knockout mice born to heterozygous parents. Much of the available literature has neglected to report the effect of the depletion or interference of these transcripts within the early embryo, and without this information, it is difficult to draw definitive conclusions as to whether these transcripts extend embryonic development until their eventual degradation.

4.1.1 Cdk4/Cdk6

Mice which lack either Cdk4 or Cdk6 are viable and show no major phenotype due to a highly overlapping range of substrates including Rb which leads to complete functional redundancy (Hu et al., 2009; Malumbres et al., 2004; Rane et al., 1999; Tsutsui et al., 1999). Concomitant Cdk4/Cdk6 deletion resulted in lethality during midgestation at approximately E14.5. This lethality, however, was demonstrated to be due to severe anemia caused by defects in erythroid progenitor cells rather than cell cycle progression (Malumbres et al., 2004). Further studies showed that Cdk2/Cdk6 double-knockout mice develop normally and show abnormalities no more severe than have been previously described each of the single-knockout models (Malumbres et al., 2004). In contrast, Cdk2 and Cdk4 double-knockout mice died at approximately E15 due to defects during heart development (Berthet et al., 2006). This series of experiments suggested that embryonic preimplantation development is unaffected by the loss of Cdk2, 4, 6, or any combination of the above.

4.1.2 D-Type Cyclins

Deletion of any one of the three D-type cyclins in mice, cyclin D1 (Fantl, Stamp, Andrews, Rosewell, & Dickson, 1995; Sicinski et al., 1995), cyclin D2 (Sicinski et al., 1996), or cyclin D3 (Sicinska et al., 2003) causes mild and tissue-specific phenotypes. These phenotypes arise due to the fact that some tissues specifically express only a single isoform of cyclin D and therefore cannot be compensated by the remaining two. Cyclin D1, for example, is the only D-type expressed in retinal and breast tissue, and deletion of this isoform in isolation leads to defects in embryonic retina development and insensitivity to steroid-induced proliferation of mammary epithelium during pregnancy (Sicinski et al., 1995).

Intercrossing of these knockout mice to generate mice expressing only a single D-type cyclin revealed that expression of any single isoform of cyclin D is sufficient for full completion of embryogenesis. These single isoform expressing embryos were described as indistinguishable from wild-type embryos during preimplantation and displayed normal cellular proliferation. Interestingly these mice lost all tissue-specific cyclin D expression regardless of the two D-type cyclins which were ablated and ubiquitously expressed the remaining D-type cyclin in all tissues, suggesting that cells can utilize each D-type cyclin interchangeably during early development (Ciemerych et al., 2002).

Deletion of all three D-type cyclins in the mouse led to embryonic lethality at approximately E13.5. This lethality, however, was not due to a cell cycle defect but instead was caused by a failure of hematopoietic stem cell expansion, which caused hypoplastic anemia (Kozar et al., 2004). The D-type cyclins are therefore also considered dispensable for cell cycle progression in the early embryo.

The fact that observed embryonic lethality in the Cdk4/6 double knockout was slightly later than that of the cyclin D triple-knockout embryos is still unexplained but may be due to the fact that Cdk2 and/or Cdk1 is able to bind D-type cyclins allowing for suboptimal kinase activity which ultimately is not sufficient for the survival of the embryo to term. This is supported by observations that Cdk2 in Cdk4/6 double-knockout mouse embryonic fibroblasts is able to form active kinase complexes with cyclin D2 and the fact that recombinant Cdk2 with either cyclin D2 or D3 is able to form an active complex in insect Sf9 cells (Ewen et al., 1993; Padmakumar, Aleem, Berthet, Hilton, & Kaldis, 2009; Sweeney, Sarcevic, Sutherland, & Musgrove, 1997; Xiong, Zhang, & Beach, 1992).

4.1.3 Cdk2

Cdk2 was originally thought to be essential for mitotic cell cycle progression by driving G1/S entry in complex with E-type cyclins. This was supported by early studies which had shown that overexpression of a kinase dead dominant-negative version of Cdk2 (van den Heuvel & Harlow, 1993) or direct inhibition of Cdk2 activity using either antibodies against Cdk2 (Pagano et al., 1993; Tsai, Lees, Faha, Harlow, & Riabowol, 1993) or p21/p27 inhibitors could induce potent G1 arrest in human cell lines (Sherr & Roberts, 1995).

The perceived essential nature of Cdk2, however, was comprehensively disproven by the observation that cells could proliferate in the absence of Cdk2 (Tetsu & McCormick, 2003) and by the generation of germ line Cdk2 knockout mice by two independent groups (Berthet, Aleem, Coppola, Tessarollo, & Kaldis, 2003; Ortega et al., 2003). In these studies, Cdk2 knockout mice were shown to be viable but infertile due to defects in homologous recombination during meiosis. In contrast to meiosis, mitotic cell cycle progression was grossly normal apart from a slight delay in S phase entry (Berthet et al., 2003). The viability of these mice has since been shown to be possible the formation of Cdk1/cyclin E complexes which allow Cdk1 to drive G1/S transition in the absence of Cdk2 (Aleem, Kiyokawa, & Kaldis, 2005).

The observation of a Cdk1/cyclin E complex in the absence of Cdk2 also reconciled the fact that the p27 knockout phenotype which causes multiple organ hyperplasia and spontaneous tumorigenesis caused by the loss of cyclin E-complex repression (Fero et al., 1996; Kiyokawa et al., 1996; Nakayama et al., 1996) is not attenuated in Cdk2/p27 double-knockout mice (Aleem et al., 2005; Martin et al., 2005). This allowed for the explanation that p27 regulates a separate E-type cyclin controlled complex, namely Cdk1/cyclin E, which becomes dysregulated in the Cdk2/p27 double-knockout mice.

4.1.4 The E-Type Cyclins

Following the confirmation that Cdk2 was nonessential for mitotic progression, two further studies demonstrated that cyclin E1 and E2 knockout mice (Geng et al., 2003; Parisi et al., 2003) were also viable with no observable cell cycle defects. Interestingly the combined E1, E2 double knockout resulted in embryonic lethality with no live double-knockout embryos observed after E11.5. The lethality in this compound knockout, however, was determined to be due to a failure of the endoreplication of the giant trophoblast cells of the placenta rather than a defect of cell cycle progression. Consequently this embryonic lethality could be rescued through the injection of cyclin E1/E2 double-knockout ES cells into tetraploid blastocysts to create chimeric cyclin E double-knockout embryos with wild-type trophoblast. These mice subsequently developed to term but died postnatally due to defects in cardiovascular development. Both E-type cyclins therefore are considered dispensable for cell cycle progression during embryonic development.

4.1.5 The A-Type Cyclins

Like Cdk2, cyclin A1 was found to be essential for male meiosis and knockout mice are viable but infertile (Liu et al., 1998; van der Meer et al., 2004). Deletion of cyclin A2 however results in early embryonic lethality directly postimplantation at approximately E5.5 (Murphy et al., 1997).

Survival to blastocyst stage in this model was unexpected, as cyclin A2 in complex with Cdk2 had previously been thought to be vital for DNA replication during S phase. In support of this idea earlier experiments had shown that the injection of both antisense oligonucleotide constructs targeting cyclin A2 or antibodies against cyclin A2 could completely stall DNA synthesis in both mouse embryonic fibroblasts and epithelial cells (Girard, Strausfeld, Fernandez, & Lamb, 1991; Pagano, Pepperkok, Verde, Ansorge, & Draetta, 1992; Zindy et al., 1992).

One possible explanation for the survival of these cells to blastocyst stage was the presence of inherited maternal pools of cyclin A2 protein during preimplantation development. This has since been suggested to be unlikely due to the observation of the degradation of maternal cyclin A2 mRNA as early as the 2-cell stage (Winston et al., 2000). Another explanation was that S phase could be driven by other Cdk/cyclin complexes. This was supported by the observation that Cdk2/cyclin E complexes have been shown to drive replication initiation in the absence of cyclin A (Furstenthal, Swanson, Kaiser, Eldridge, & Jackson, 2001; Jackson, Chevalier, Philippe, & Kirschner, 1995; Strausfeld et al., 1996) with cyclin A becoming an absolute requirement for DNA replication in typical somatic cells (Girard et al., 1991; Pagano et al., 1992).

Later an in-depth analysis of cyclin A2 single-knockout and A1/A2 double-knockout mouse embryonic fibroblasts revealed that levels of cyclin E1 protein were markedly increased in both cases and took on a broad pattern of cell cycle expression which encompassed the combined expression patterns of both A- and E-type cyclins observed in wild-type MEFs. In addition, the level and pattern of expression of other cyclins or Cdks remained unchanged and cyclin A1/A2 double-knockout fibroblasts were able to enter S phase from G0 with no observable delay. These observations suggested that E-type cyclin–Cdk complexes were able to drive S phase in the absence of cyclin A2. This theory was tested in an elegant experiment in the same paper by generating quadruple-knockout MEFs lacking all A- and E-type cyclins (Kalaszczynska et al., 2009). These cells showed absolute cessation of cell proliferation and an inability to complete S phase, which, when taken into consideration with all previous results, confirmed that at least one of E-type or A-type cyclin is required for DNA replication during early embryonic development.

4.1.6 Cdk1

The role of Cdk1 in cell cycle progression in mammals is absolutely essential as has been shown by germ line knockout of Cdk1 (Diril et al., 2012) and gene trap mutation studies (Santamaria et al., 2007), which lead to a failure to develop to blastocyst stage and early embryonic lethality after E3.5.

Cdk1 is frequently referred to as the master regulator of the mitotic cell cycle and has been shown to form atypical complexes with D- and E-type cyclins (Aleem et al., 2005; Koseoglu, Graves, & Marzluff, 2008; Padmakumar et al., 2009; Satyanarayana & Kaldis, 2009), in addition to its major partner proteins the A- and B-type cyclins. The functional redundancy exhibited within the Cdk family described in this section is most dramatically supported by the ability of Cdk1 to compensate for the loss of any interphase-related Cdk (2, 3, 4, 6) by binding to their respective cyclin partner proteins. Even more impressive is the fact that although quadruple Cdk2, Cdk3, Cdk4, Cdk6 knockout mice die during embryogenesis, early development seems to be largely unaffected. The blastocysts of these embryos are able to implant, undergo organogenesis, and do not arrest until midgestation (Santamaria et al., 2007).

This observation is in accordance with the conservation of Cdk1 throughout evolutionary development. Now classical experiments performed in the 1980s and 1990s convincingly demonstrated that the genomes of many lower organisms such as fission yeast *S. pombe* and baker's yeast *Saccharomyces cerevisiae* encode only a single Cdk gene, cdc2, and Cdc28, respectively, which is highly homologous to mammalian Cdk1. These Cdk1 homologs were shown to be sufficient to drive the cell cycle in isolation of any other Cdk which has arisen throughout eukaryotic evolution (Richardson, Wittenberg, Cross, & Reed, 1989). Further deletion studies in yeast also confirmed that in yeast the cyclin B1 homologue (clb5/6) is the only cyclin required for cell cycle progression cementing Cdk1/cyclin B complex as the evolutionary master cell cycle regulator (Fisher & Nurse, 1996).

4.1.7 B-Type Cyclins

The essential role of Cdk1 in mammalian embryogenesis is supported by the fact that mice lacking cyclin B1 die during early development before E10.5 (Brandeis et al., 1998). Currently the later onset embryonic lethality of the cyclin B1 knockout as compared to Cdk1 knockout is unexplained. Survival to this stage this may be due to the maternal inheritance of cyclin B1 transcripts or crosscompensation by other cyclins although this has never been conclusively proven.

In contrast to cyclin B1, cyclin B2 knockout mice are viable and show no observable cell cycle defects indicating a dispensable role in mitotic cell cycle progression.

Although a knockout model for cyclin B3 is yet to be reported, it is presumed that this is also nonessential for cell cycle progression due to the confinement of its expression exclusively to male germ cells (Nguyen et al., 2002).

4.1.8 Summary of Cell Cycle-Driving Cdks/Cyclins

When considered as a whole, a large body of research covering the effects of the ablation of the interphase Cdks and cyclins indicates that at least until development to the blastocyst stage of development, the requirement of any Cdk/cyclin complex save from Cdk1/cyclin B1 is surplus to requirement. The major role which Cdk1/cyclin B1 complexes seem unable to replace however is the initiation of DNA replication by cyclin A2 or cyclin E1 complexes, revealing that at least one of these cyclins is required for the completion of DNA replication in preimplantation development.

Interestingly, in many of the compound knockout models a recurring pattern of postimplantation lethality is observed, which emerges often during gastrulation and organogenesis. This manifests itself in the form of tissue-specific phenotypes such as defects in cardiovascular development and hematopoiesis, which cannot be compensated for even by the presence of Cdk1/cyclin B complexes. This however is presumably due to the establishment of new roles of Cdks and cyclins throughout evolutionary development. It has been proposed therefore that within the mammalian cell cycle, there exist two types of cyclin proteins. The A- and B-type cyclins, which under normal circumstances, are an absolute requirement for the cell cycle machinery and the remaining interphase cyclins which are required for highly cell type-specific roles later in development (Hochegger, Takeda, & Hunt, 2008; Murphy et al., 1997). Nevertheless, the situation is likely to be cell-type specific and context dependent. Therefore, a general rule cannot be made (Fig. 2).

4.1.9 Cdk-Activating Kinase

Cdk7 was originally identified as a Cdk-activating kinase (CAK) following observations that it was able to phosphorylate threonine residues within the T-loop of other kinases including Cdk1 and Cdk2 (Thr161 and Thr160, respectively) promoting their activation (Fesquet et al., 1993; Poon, Yamashita, Adamczewski, Hunt, & Shuttleworth, 1993; Solomon, Harper, & Shuttleworth, 1993; reviewed in Fisher, 2005; Kaldis, 1999).

As is true for all Cdks, Cdk7 has no functional kinase activity in isolation and requires association with cyclin H for activation (Fisher & Morgan,

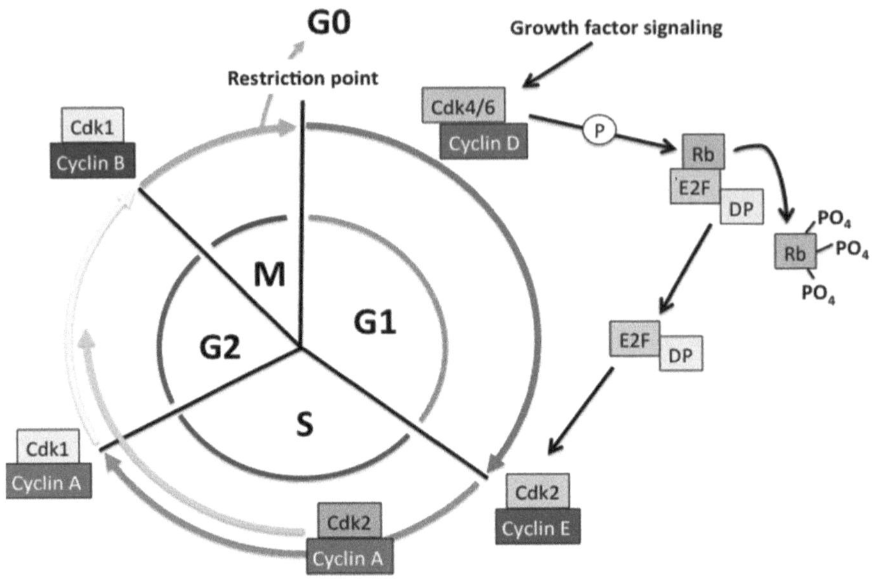

Fig. 2 The role of the interphase cyclin-dependent kinases in a typical mammalian cell cycle. (See the color plate.)

1994; Labbe et al., 1994). Cdk7/cyclin H complexes then bind an additional accessory protein called Mat1 in a trimeric fashion (Adamczewski et al., 1996; Fisher, Jin, Chamberlin, & Morgan, 1995; Tassan et al., 1995). Mat1 association has been proposed to convey substrate specificity to the active complex (Yankulov & Bentley, 1997).

In addition to its Cdk-activating role, the activated CAK complex has also been described to associate with the general transcription factor TFIIH (Akoulitchev, Makela, Weinberg, & Reinberg, 1995; Palancade & Bensaude, 2003; Serizawa et al., 1995; Shiekhattar et al., 1995). The resultant protein complex is then able to phosphorylate Ser5 of RNA polymerase II to regulate gene transcription, although this has not been shown to be essential during preimplantation and Ser5 can be phosphorylated independently of the CAK complex.

Although no cyclin H knockout mice are described to date, it has been shown that cyclin H transcripts are present at the 4-cell stage of the mouse embryo and remain high through blastocyst stage. Ablation of cyclin H via lentiviral shRNA infection in mouse ES cells or in a developing single-cell mouse embryo was shown to increase the levels of lineage promoting gene transcripts and drive the loss of stem cell identity in the ICM (Patel & Simon, 2010). As the ICM of the blastocyst is a transient population of pluripotent stem cells, the maintenance of stemness is essential to facilitate their continued division in the early embryo. Indeed, silencing of cyclin H corresponds to a loss of ICM expansion and lethality at blastocyst stage.

In support of this concept, Mat1 knockout mice die before E7.5, shortly after implantation and this coincides with the loss of detectable maternally inherited Mat1 transcripts (Rossi et al., 2001). The lethality of this genotype was also shown to be due to an essential role of Mat1 in proliferating mitotic ICM, which causes a loss of ICM expansion in the knockout.

In Cdk7 knockout mice, evidence of ICM-specific apoptosis induction is also present as early as E4.5, and by approximately E6.5–7.5, there is a complete loss of ICM cells leading to lethality (Ganuza et al., 2012). Interestingly in this study it is reported that unlike Mat1 and cyclin H, maternal copies of Cdk7 are not inherited by the embryo and development until E4.5 is grossly normal, suggesting that progression to this stage is Cdk7 independent.

The phenotypical similarity of ICM apoptosis and embryonic lethality following depletion of any one of the proteins in the CAK complex suggests that the role of the active trimeric complex is required for proper ICM lineage maintenance or proliferation during preimplantation and that loss of any one of these factors may cause the differentiation of the ICM leading to a loss of expansion and early embryonic lethality. Another possibility is that the CAK complex is required for the phosphorylation of Cdks and stimulation of cell cycle genes during early development although this is yet to be tested.

4.2 The Retinoblastoma and Related Pocket Proteins

The major mechanism whereby active Cdk/cyclin complexes are able to activate the transcription of genes essential for cell cycle progression during the G1/S and G2/M transitions is through the hyperphosphorylation and inhibition of the retinoblastoma pocket proteins Rb/p105, p107, and p130. This inhibitory phosphorylation is usually driven by Cdk/cyclin complexes (Grafstrom, Pan, & Hoess, 1999) but can also be driven by MAPK signaling (Chen et al., 2003) both of which become activated as a result of upstream mitogenic signaling (for review, see Dyson, 1998).

These pocket proteins serve as important negative regulators of the cell cycle, which under normal conditions bind to and sequester heterodimeric complexes formed by the association of the E2F and DP families of transcription factors (Hiebert, Chellappan, Horowitz, & Nevins, 1992). Upon pocket protein hyperphosphorylation, E2F–DP complexes are released (Knudsen & Wang, 1997) and translocate to the nucleus binding strongly to E-box motifs located within gene promoters. Many of these E-boxes are found on genes essential for cell cycle progression, such as the A- and E-type cyclins, and also on proteins involved in DNA replication (Muller et al., 1997).

During preimplantation the expression of Rb follows a highly distinctive pattern. Initially in unfertilized oocytes, Rb is highly expressed but this decreases dramatically at the 2-cell stage and remains at low until the embryo develops to the blastocyst stage when Rb expression is restored (Iwamori, Naito, Sugiura, & Tojo, 2002; Xie et al., 2005). The reduced level of Rb expression during this time is correlated with a reduction in G1- and G2-phase length after the 2-cell stage leading to progressively shorter cycling times as described previously. It seems therefore that the preimplantation embryo sacrifices some checkpoint control in able to increase the speed of embryonic cell division. Gap phases and checkpoint control are then restored postimplantation with the resurgence of Rb, and the cell cycle length adopts a profile similar to somatic cell divisions. The dispensable nature of the major pocket protein Rb during preimplantation was highlighted by knockout studies, which show that Rb ablation in mice causes postimplantation embryonic lethality at E14.5 (Clarke et al., 1992; Jacks et al., 1992; Lee et al., 1992).

At present the expression pattern for p107 and p130 during preimplantation has not been described, but the requirement for these proteins also seems minimal as knockout mouse models for each of these genes also result in viable mice, which do not display an observable phenotype (Cobrinik et al., 1996). Interestingly late embryonic lethality can be observed in p107-null mice when crossed with mice heterozygous for Rb (Lee et al., 1996). Together these results indicate a central role for Rb during embryogenesis, which can only be partially compensated by p107 and p130, but ultimately all pocket proteins seem wholly dispensable for preimplantation development. The lack of Rb expression during preimplantation is particularly striking, suggesting that checkpoint control is less essential in the early embryo or is deliberately downregulated to serve a developmental function.

4.3 The E2F Family as a Downstream Target of Cdk/Cyclin Complexes

In total there are eight E2F genes (E2F1–8), with E2F3 comprising two functionally distinct isoforms E2F3a and E2F3b and also two DP genes

(DP1–2) (reviewed in DeGregori & Johnson, 2006; Dyson, 1998; Trimarchi & Lees, 2002; Zhang & Chellappan, 1995).

The resultant E2F proteins can be subdivided into two distinctive classes, activators (E2F1–3a) or repressors (E2F3b–8) of transcription, based on their transcriptional activity and sequence homology. Although all E2F proteins bar E2F7 and E2F8 are able to heterodimerize with a DP family protein to form a functional E2F complex, only E2F1–3 have sufficiently strong activity to drive active transcription and S phase entry in quiescent cells (DeGregori, Leone, Miron, Jakoi, & Nevins, 1997). E2F4–5/DP complexes are also able to bind to E-boxes, but their transactivation activity is low, and as a result, they act as transcriptional repressors by blocking the normal binding E2F1–3.

E2F6 retains a dimerization domain but lacks a transactivation domain. E2F6 has been shown to repress transcription through the recruitment of repressive Polycomb group proteins (Trimarchi, Fairchild, Wen, & Lees, 2001).

E2F3b, 7, and 8 are the most recently characterized members of the E2F family. E2F3b is produced as the result of a unique mRNA transcribed from the E2F3 gene and has been described as acting to maintain transcriptional repression in quiescent cells through association with Rb which is the main pocket protein expressed in quiescent cells (He et al., 2000; Leone et al., 2000). E2F7 and E2F8 are also described as repressors of transcription, but the mechanism is poorly understood but mediated independently of DP heterodimerization due to a unique structure which lacks a DP dimerization domain but contains a duplicated DNA-binding domain (de Bruin et al., 2003; Di Stefano, Jensen, & Helin, 2003).

A large body of data has detailed knockout mouse models for each of the E2F family members. Like the Cdks, cyclins, and pocket proteins, this revealed a significant degree of functional redundancy between family members. Deletion of any one of the activating E2F members E2F1 (Field et al., 1996), E2F2 (Leone et al., 2001; Murga et al., 2001), or E2F3 (Cloud et al., 2002) led to viable births although some prenatal lethality was detected in the E2F3 knockout mice from approximately E13.5 as a result of congestive heart failure. Although mice with ablation of both E2F1 and E2F2 were viable, combined deletion of E2F3 with either E2F1 or E2F2 led to embryonic lethality at E9.5 or earlier in all embryos generated. Total ablation of E2F1–3 using the conditional cre–lox system in embryonic fibroblasts caused absolute cessation of cell proliferation and arrest of cell cycle progression, regardless of the position of the cell cycle the deletion occurred suggesting that E2F activity is essential for all stages of the cell cycle (Wu et al., 2001). This arrest correlated with a sharp rise in p21 expression and the hypophosphorylation of Rb. Supplementation of E2F1–3 triple-knockout MEFs with exogenous E2F3a or E2F3b was able to partially restore cell cycle progression and supplementation with both almost nullified the colony-forming deficit shown by these cells.

Ablation of the repressive E2F family members, E2F4 (Gaubatz et al., 2000; Rempel et al., 2000), E2F5 (Gaubatz et al., 2000), E2F6 (Storre et al., 2002), E2F7, or E2F8 (Li et al., 2008) in isolation, also showed no effect upon the completion of embryogenesis. Interestingly however compound E2F7-/E2F8-null mice exhibit embryonic lethality at E11.5 due to widespread embryonic apoptosis and vascular defects (Li et al., 2008).

In contrast to the redundant nature of the E2F family, DP1 knockout mice are not viable (Kohn, Bronson, Harlow, Dyson, & Yamasaki, 2003). The phenotype of these mice indicated that DP1 is required for endoreduplication in trophoblast giant cells and development of extraembryonic tissues required for postimplantation survival. A DP2 knockout mouse is yet to be described, but the late embryonic lethality phenotype demonstrated by the DP1 knockout suggests that E2F–DP2 complexes may have nonoverlapping functions with E2F–DP1 complexes postimplantation. The relative importance of these complexes during preimplantation will hopefully be addressed in the future by the generation of a DP2 knockout mouse and a compound DP1 DP2 knockout mouse.

4.4 Cdk3/Cyclin C

The Cdk3/cyclin C complex was first identified as regulator of G0 exit in quiescent cells through its ability to phosphorylate Rb-driving G1 entry (Ren & Rollins, 2004; Sage, 2004). Although no knockout mouse model is described for Cdk3, a serendipitous nonsense mutation rendering Cdk3 nonfunctional was described in most laboratory mouse strains, indicating that the role of Cdk3 is nonessential during development (Ye, Zhu, & Harper, 2001). In mice, cyclin C was found to associate with Cdk1 and Cdk2 indicating functional compensation (Li et al., 2014).

Conditional cyclin C knockout mice were generated and were shown to die at E10.5 due to defects in placental development (Li et al., 2014). Interestingly this same study identified the association of cyclin C with an "orphan" Cdk, Cdk19 which shares high structural homology with the previously described cyclin C partner Cdk8. In addition, all three known cyclin

C complexes (Cdk3, 8, 19) were shown to target the Notch1 intracellular domain (ICN1) to promote its proteolytic degradation, although the functional implications of this are still unclear and seem to have no role in the preimplantation embryo.

4.5 Cdks and Transcription

4.5.1 Cdk7/Cyclin H and Cdk9/Cyclin T

Although not directly required for interphase, the action of several Cdks is an essential requirement for the initiation of transcription. The Cdk7/cyclin H (previously discussed as part of the CAK complex) and Cdk9/cyclin T1 complexes are able to phosphorylate the C-terminal domain (CTD) of the RNA polymerase II holoenzyme on Ser5 and Ser2 (Price, 2000), respectively. In its nonphosphorylated state, the CTD of RNA polymerase II associates with and assembles transcriptional machinery proteins, general transcription factors, and other cofactors collectively known as the preinitiation complex (PIC) onto the promoters of DNA (Myers & Kornberg, 2000; Usheva et al., 1992). Phosphorylation of Ser5 of the CTD by Cdk7 however causes RNA polymerase II to become transcriptionally active and supports its dissociation from DNA leaving a subset of PIC proteins bound to the promoter. Ser5 phosphorylation however is not sufficient for the DNA transcription as it is unable to elongate transcribed mRNAs. For RNA polymerase II to become elongation competent, it must first be recognized by and associate with elongation factors such as the elongator. This is promoted by an additional phosphorylation on Ser2 of the CTD by Cdk9/cyclin T1 (Otero et al., 1999).

Although no knockout mouse model has currently been described for Cdk9, depletion of this Cdk in lower organisms such as *Caenorhabditis elegans* (Shim, Walker, Shi, & Blackwell, 2002) and *Drosophila* (Eissenberg, Shilatifard, Dorokhov, & Michener, 2007) led to embryonic lethality. A knockout model for cyclin T1 has yet to be described, but deletion of its highly homologous family member cyclin T2 also led to early preimplantation lethality prior to the observation of blastocyst-stage embryos (Kohoutek et al., 2009).

4.5.2 Cdk8/Cyclin C

Cdk8 in complex with its activating cyclin, cyclin C, has been shown to mediate transcription through its association with the mediator complex. The core mediator complex, which can also form in isolation of Cdk8/cyclin C, is a large collection of 25–30 proteins, which arranges into structurally interacting modules. It is thought that mediator acts as a scaffold whereby general transcription factors can be stabilized at gene promoters to form the PIC in association with RNA polymerase II (Baek, Kang, & Roeder, 2006; Cantin, Stevens, & Berk, 2003; Hahn, 2004). For association with the mediator, Cdk8/cyclin C first forms a multiprotein complex with med12 and med13 to create a further mediator submodule (Hoeppner, Baumli, & Cramer, 2005). Upon interaction with this submodule, the RNA polymerase II binding site within mediator undergoes a conformational change. This structural modification is associated with an increase in transcriptional repression at promoter regions. This repression has already been convincingly shown in yeast models (Carlson, 1997) and in mouse models (Wang, Piotrowska, et al., 2004). It is assumed that the Cdk8/cyclin C submodule of mediator is able to act as a negative regulator of mediator–RNA polymerase II interaction leading to a loss of recruitment of RNA Pol II to DNA.

Inactivation of Cdk8 in mice ES cell lines using a gene trap mutation in the Cdk8 gene locus results in early embryonic lethality. These cells arrest specifically at E2.5–3 and Cdk8-null embryos were found to have fragmented blastomeres resulting in a total lack of progression to the compaction stage. Interestingly, the essential role of Cdk8 seems to be specific to preimplantation with conditional knockout in the adult mouse causing no observable phenotype (Westerling, Kuuluvainen, & Makela, 2007). This early embryonic phenotype of the Cdk8 knockout could possibly be due to the repression of crucial genes during preimplantation. In mice, a broad repression in transcription can be observed at the 2-cell and 8-cell stages and is correlated with an induction of Cdk8 transcription just prior to compaction (Wang, Piotrowska, et al., 2004). The importance of this repression in maintaining early embryonic cell cycles however remains unclear.

5. DNA DAMAGE AND CHECKPOINT SIGNALING IN THE PREIMPLANTATION EMBRYO

In somatic cells, three major cell cycle checkpoints are utilized to ensure the timely and robust progression of the cell cycle; these include the interphase checkpoints G1/S (for review, see Shaltiel, Krenning, Bruinsma, & Medema, 2015), G2/M (for review, see Taylor & Stark, 2001), and the SAC (for review, see Musacchio, 2015). Under deleterious circumstances such as DNA damage or low growth factor availability these checkpoints allow the implementation of cell cycle arrest (see Fig. 3).

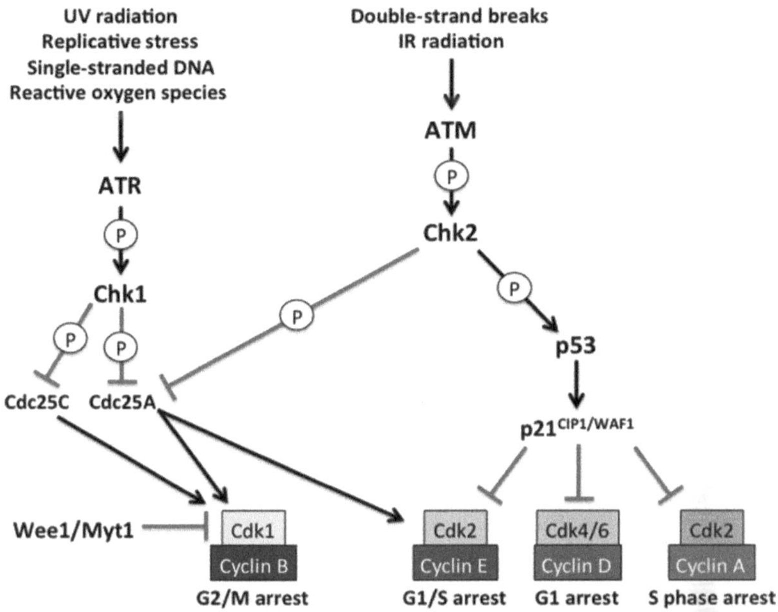

Fig. 3 Initiation and effectors of the mammalian DNA damage checkpoint response.

DNA damage signaling pathways are crucial in both adult somatic cells and the preimplantation embryo for the activation of downstream effectors in response to noxious stimuli, which threaten the integrity of genomic DNA. Typically these stimuli result from ionizing radiation and environmental stresses such as hypoxia and reactive oxygen species production and which contribute toward stalling of the replication fork, double-strand break formation, or single-stranded DNA production.

5.1 G1/S Checkpoint

The first cell cycle checkpoint is the G1/S checkpoint. As previously discussed, S phase entry from G1 is driven primarily by the Cdk4/6–cyclin D complex-mediated inhibition of the retinoblastoma-related pocket proteins and the ensuing activation of E2F–DP complexes which then drive the transcription of Cdk2/cyclin E complexes. In turn, the Cdk2/cyclin E complexes initiate DNA replication through the further phosphorylation of E2F–DP complexes which drive the transcription of replication machinery genes and cyclin A, which localize to replication origins on DNA and promote DNA replication.

The G1/S checkpoint is extremely sensitive to DNA damage response pathways, which lead to the upregulation of p21$^{CIP1/WAF1}$ in a p53-dependent manner. p21$^{CIP1/WAF1}$ can subsequently inhibit the activation of Cdk2, or alternatively in the absence of Cdk2, Cdk1 (Satyanarayana, Hilton, & Kaldis, 2008) to prevent S phase entry and DNA replication (Harper et al., 1995).

5.1.1 Embryonic Dependence on Cell Cycle Checkpoint Proteins

The primary response of a cell to DNA damage is implemented during the G1/S, intra-S, and G2/M transitions and is mediated by two major serine/threonine kinases of the phosphatidylinositol-3-kinase (PI3K)-related protein kinase superfamily, ataxia telangiectasia and Rad3 related (ATR) and ataxia telangiectasia mutated (ATM). These proteins usually exist as inactive monomers but undergo oligomerization and activation upon receipt of appropriate stimuli (Bakkenist & Kastan, 2003). In the case of ATR, activation is mediated by the presence of single-stranded DNA formed at stalled replication forks during replication stress. On the other hand, ATM is activated by genotoxic agents, which cause DNA double-strand breaks such as ionizing radiation (Smith, Tho, Xu, & Gillespie, 2010). These upstream DNA damage response genes activate a multitude of downstream effectors and, depending on the severity of the damage, may range from stalling of the cell cycle at cell cycle checkpoints until the DNA damage is resolved or even result in the induction of apoptosis (Zhou & Elledge, 2000).

Unsurprisingly, many of the major target proteins of ATR and ATM act to inhibit cell cycle progression such as Chk1, Chk2, p53, and BRCA1 and following DNA damage cells typically stall at G1/S, S, or G2/M phases of the cell cycle (Hartwell & Kastan, 1994).

5.1.2 ATR

ATR knockout mice embryos are able to successfully reach blastocyst stage, but postimplantation embryonic lethality is observed before E7.5 (Brown & Baltimore, 2000; Cortez, Guntuku, Qin, & Elledge, 2001; de Klein et al., 2000). Cultured blastocysts from ATR-null embryos show an absolute failure of dividing cells to proliferate postimplantation, and by E6.0, there is a complete loss of the ICM, leaving only the nondividing giant trophoblast cells. The embryonic lethality observed in ATR-null embryos is correlated with caspase-independent chromosome fragmentation, indicating that ATR is essential for preventing the loss of genomic integrity in the early embryo. The phenotype of ATR-null embryos also mirrors the loss of ATR- and

ATM-mediated downstream effectors including the DNA repair genes BRCA1, BRCA2, RAD51 with all three knockout models also leading to early embryonic lethality.

The importance of these DNA repair mechanisms in the early embryo is underlined by the fact that deletion of many of these genes in the adult mouse is not deleterious. Global deletion of ATR in the adult mouse using inducible cre–lox methodology did not lead to premature death, but mice demonstrated a loss of regenerative tissue capacity and a depletion of stem and progenitor cells (Ruzankina et al., 2007).

5.1.3 ATM

Unlike ATR-null mice, ATM knockout mice are viable but show postnatal growth retardation and infertility (Barlow et al., 1996; Elson et al., 1996; Xu et al., 1996). Interestingly, point mutations introduced into the kinase domain of ATM cause embryonic lethality before E9.5 (Daniel et al., 2012) and cells expressing this kinase-dead form of ATM exhibit chromatid breaks at a much higher rate than ATM-null cells (Yamamoto et al., 2012). It was proposed that the complete loss of ATM can activate compensatory mechanisms within the embryo, which are not stimulated by the inactivating mutation alone. This is supported by the fact that ATM compound knockout models show embryonic lethality when combined with mutations in other DNA damage response genes such as PARP1 (Menisser-de Murcia, Mark, Wendling, Wynshaw-Boris, & de Murcia, 2001) and DNA-dependent protein kinase (Gurley & Kemp, 2001). This indicates that several DNA damage response pathways are active and essential in the embryo and may have functional overlap allowing for the loss of one or more components.

5.1.4 Chk1/2

Upon activation ATR and ATM phosphorylate Chk1 and Chk2, respectively. These proteins are regulatory checkpoint kinases, which play a crucial role in the G2/M checkpoint control.

Deletion of Chk1 in the mouse leads to embryonic lethality between E3.5 and E7.5 (Liu et al., 2000). It was shown that the early embryonic divisions in this mouse model are sustained by maternally inherited Chk1 protein, but directly following depletion, there is apoptosis leading to the loss of the ICM. It was suggested that this embryonic lethality was due to a failure of DNA-damaged cells to enter the G2/M checkpoint in a Chk1-dependent manner.

Preimplantation mouse embryos have been demonstrated to be acutely sensitive to DNA damage induced by radiation (Goldstein, Spindle, & Pedersen, 1975; Heyer, MacAuley, Behrendtsen, & Werb, 2000). This sensitivity may be contributed by the rapid cycling time and truncation of gap phases typical of cells during late preimplantation. When coupled with the early lethality of the Chk1-null mice, this suggests a reliance of the embryo on a functional G2/M checkpoint during preimplantation for the correction of DNA damage before M phase entry. This is further supported by the fact that improper removal of DNA-damaged cells during preimplantation severely impairs embryonic development (Kamjoo, Brison, & Kimber, 2002).

In contrast to Chk1 ablation, Chk2 knockout mice are viable and seemingly normal aside from increased resistance to IR-mediated apoptosis (Hirao et al., 2002; Takai et al., 2002). This suggests that although ATR and ATM are both important for DNA repair, only the ATR–Chk1 pathway is absolutely required during early embryogenesis and that the activation of cell cycle checkpoints through Chk1 but not Chk2 is essential for development to implantation.

5.2 Effectors of the G2/M Checkpoint

The G2/M checkpoint marks the final barrier before entry into M phase. This checkpoint is influenced by a multitude of upstream inputs such as the previously discussed DNA damage response pathways, which act to stall the cell in G2 stage if the cell detects DNA damage or chromosomal aberrations. The timely progression through G2 phase and subsequent entry into M phase are dictated by a large increase in the levels of active Cdk1/cyclin B complexes. Consequently to prevent premature or inappropriate M phase entry the cell exerts tight control of both Cdk1 and cyclin B via several inhibitory mechanisms (Kellogg, 2003). G2/M transition therefore occurs when this checkpoint inhibition is removed allowing the rapid initiation of prometaphase.

5.2.1 The Cdc25 Phosphatases

Three members of the Cdc25 gene family are described in mammals: Cdc25A, B, and C. These genes encode highly similar dual-specificity phosphatases which are able to dephosphorylate both phospho-threonine and -tyrosine residues (Dunphy & Kumagai, 1991). The different members of Cdc25 are responsible for influencing transition between specific phases of the cell cycle through their dephosphorylation of crucial tyrosine and threonine residues on inactivated Cdks (Pines, 1999).

Although Cdc25A has been shown to increase in expression during both the G1/S transition and the G2/M transition, its peak expression occurs during G1 phase. It is during this phase that Cdc25A fulfills its major role as an effector of G1/S entry through its activating dephosphorylation of Cdk2/cyclin E complexes. In the presence of DNA damage however, Cdc25A is phosphorylated by both Chk1 (Zhao, Watkins, & Piwnica-Worms, 2002) and Chk2 (Falck, Mailand, Syljuasen, Bartek, & Lukas, 2001) allowing it to become targeted for degradation (Donzelli et al., 2002; Molinari, Mercurio, Dominguez, Goubin, & Draetta, 2000). This facilitates rapid p53-independent G1 arrest due to the inability of the cell to remove Cdk2 from an inactive conformation, and as a result the cell is unable to complete DNA replication and S phase entry. Premature G1 arrest can also be initiated by the depletion of Cdc25A (Costanzo et al., 2000; Falck, Mailand, Syljuasen, Bartek, & Lukas, 2001; reviewed in Bartek & Lukas, 2001). Cdc25A-null mice survive to the blastocyst stage but die just prior to implantation due to defects in hatching from the zona pellucida (Ray et al., 2007). Preimplantation development in these mice however was determined to be grossly normal (Lee, White, Hurov, Stappenbeck, & Piwnica-Worms, 2009), suggesting that the role of Cdc25A in G1/S transition during this time may be dispensable due to the lack of checkpoint signaling in the early embryo. Alternatively functional redundancy could exist originating from the remaining Cdc25 family members or other cellular phosphatases.

Unlike Cdc25A, Cdc25B and Cdc25C have a more confined role in the cell cycle transition and are expressed specifically at the G2/M transition. Cdc25C, for example, has been well documented to become activated during the G2/M transition due to activating phosphorylation by the Plk1. Additional activating phosphorylation of Cdc25C can be mediated by Cdk1/cyclin B complexes in a positive feedback loop fashion to rapidly drive the massive upregulation of Cdk1/cyclin B activity (Izumi & Maller, 1993; Villa-Moruzzi, 1993).

Despite the fact that G2 arrest can be initiated following microinjection with antibodies against either Cdc25B or Cdc25C (Lammer et al., 1998; Millar et al., 1991), ablation of Cdc25B (Lincoln et al., 2002), Cd25C (Chen, Hurov, White, Woodford-Thomas, & Piwnica-Worms, 2001), or both in combination led to healthy, viable mice, which display no defects in cell cycle progression (Ferguson, White, Donovan, & Piwnica-Worms,

2005). This observation suggests compensation of Cdc25B and Cdc25C by another activator of Cdk1 in these mice, which enables the G2/M checkpoint to be bypassed. To date Cdc25A cannot be satisfactory ruled out as the phosphatase, which mediates G2/M transition in these mice as no Cdc25A/B/C triple-knockout mouse is currently described.

5.2.2 Wee1/Myt1 Kinases

The action of the Cdc25 phosphatases is directly opposed by the action of the Wee1 and Myt1 kinases, which are able to drive the inhibitory phosphorylation of Thr14, Tyr15 on Cdk1 to enforce cell cycle arrest at the G2/M checkpoint. Wee1 itself can be activated by Cdk1 leading to a positive feedback loop whereby Cdk1 drives a cascade of its own activation through Wee1 (Harvey, Charlet, Haas, Gygi, & Kellogg, 2005; Nigg, 1998; van de Weerdt & Medema, 2006).

A global knockout mouse model of Myt1 is currently unavailable, but ablation of Wee1 in mice leads to early embryonic lethality caused by major apoptosis and morphological defects, which prevent progression to the blastocyst stage of development. Wee1-null embryos also show defective G2/M checkpoint signaling fail to enter G2 arrest in response to γ-radiation (Tominaga, Li, Wang, & Deng, 2006). These results suggest that Wee1 enforcement of checkpoint signaling is essential during preimplantation to correct accumulated DNA damage. Similar to Cdc25A, Wee1 can also be phosphorylated by upstream DNA damage pathways such as Chk1 (O'Connell, Raleigh, Verkade, & Nurse, 1997). This phosphorylation however is stimulatory and contributes toward G2/M arrest.

5.2.3 Polo-Like Kinase 1

The Plks are a family of highly conserved serine–threonine kinases with divergent and diverse roles described between family members. Five Plks have been identified in mammals (Plk1–5) with knockout models available for Plk1–4 (Hudson et al., 2001; Inglis et al., 2009; Myer et al., 2011). Of these knockout studies only Plk1 has been shown to be essential for preimplantation development. Plk1 is a master of the G2/M transition but is involved in multiple roles including bipolar spindle formation chromosome segregation and M phase exit. For brevity this review will focus on the G2/M transition, but an extensive overview of Plk1's other functions is available in a review by Barr, Sillje, and Nigg (2004).

The first suggestion that Plk1 could influence Cdk1 activity was demonstrated in *Xenopus* whereby the homologue of Plk1, Plx1 was shown to indirectly activate Cdk1 through the phosphorylation and activation of Cdc25 phosphatase (Kumagai & Dunphy, 1996). This observation has since been confirmed in human cell lines, revealing that Plk1 could directly phosphorylate and activate Cdc25C (Roshak et al., 2000). In addition to activation of Cdk1 through stimulation of its activator Cdc25, Plk1 also is able to prevent the inactivation of Cdk1 through the phosphorylation and inhibition of the Myt1 (Nakajima, Toyoshima-Morimoto, Taniguchi, & Nishida, 2003) and Wee1 kinases (Watanabe et al., 2004). Evidence also suggests that Plk1 may also activate Cdk1 through the direct phosphorylation of cyclin B (Jackman, Lindon, Nigg, & Pines, 2003; Toyoshima-Morimoto, Taniguchi, Shinya, Iwamatsu, & Nishida, 2001).

The importance of Plk1 in preimplantation development is highlighted by the fact that Plk1 knockout causes embryonic lethality in mice at approximately the 8-cell stage. This however seems to arise as a result of loss of its roles in the SAC as these mice display the formation of multipolar or monopolar spindles leading to improper chromosome segregation and impaired cytokinesis (Lu et al., 2008).

5.3 The Spindle Assembly Checkpoint

The major checkpoint in M phase is the SAC, which is activated during the metaphase–anaphase transition. The SAC prevents the premature separation of sister chromatids during prometaphase, by delaying anaphase onset until all sister chromatid kinetochore pairs are correctly biorientated and attached to spindle poles. This helps to preserve the genomic integrity of daughter cells by preventing missegregation of sister chromatids which could lead to aneuploidy through the inheritance of an incomplete set of chromosomes (Lara-Gonzalez, Westhorpe, & Taylor, 2012; Musacchio, 2015; Sacristan & Kops, 2015).

A multiprotein E3 ubiquitin ligase complex known as the anaphase-promoting complex or cyclosome (APC/C) is the main effector which drives sister chromatid separation during anaphase. When active the APC targets mitotic cyclins and other substrates such as the anaphase inhibitor securin for ubiquitination and subsequent proteolytic degradation by the 26S proteasome. The degradation of securin during M phase is an essential step, which leads to the activation separase. Separase is a protease which cleaves the Ssc1 subunit of the cohesin ring complex releasing the tension of the spindle and allowing segregation of sister chromatids to opposite centrosomal poles. For full activity, the APC associates with the activating adaptor protein Cdc20 during the metaphase–anaphase transition, or alternatively with Cdh1 from telophase to G1 (Peters, 2002). These adaptors convey substrate specificity to the APC through their recognition of substrate D-box and also in the case of Cdh1, KEN box motifs (Burton & Solomon, 2001; Hilioti, Chung, Mochizuki, Hardy, & Cohen-Fix, 2001; Pfleger, Lee, & Kirschner, 2001). It is this association between the APC/C and its adaptor proteins, which is targeted by the SAC to regulate its activity during different phases of the cell cycle.

Under normal circumstances the SAC is activated during prometaphase by the presence of unattached kinetochores, which act as a "wait signal" by recruiting the core SAC mitotic arrest-deficient proteins Mad1 and Mad2 (Rieder, Cole, Khodjakov, & Sluder, 1995; Rieder et al., 1997). When bound to unattached kinetochores, Mad2 undergoes a conformational change to a closed state, which allows it to capture Cdc20 preventing its association with the APC/C (Han et al., 2013). Upon chromosomal biorientation Mad1 and Mad2 dissociate and Cdc20 is released allowing the activation of the APC/C. Knockout mice harboring a deletion in Cdc20 exhibit formation of two independent spindles within the male and female pronuclei resulting in permanent metaphase arrest at the 2-cell stage (Li, York, & Zhang, 2007). This suggests that the early metaphase–anaphase transition is still reliant on the activation of the APC/C.

An additional multiprotein complex known as the mitotic checkpoint complex (MCC) comprising Mad2, BUBR1 and BUB3, and Cdc20 (Chao, Kulkarni, Zhang, Kong, & Barford, 2012; Sudakin, Chan, & Yen, 2001) is also able to directly inhibit APC activity (Fang, Yu, & Kirschner, 1998; Morrow et al., 2005; Sudakin et al., 2001). The importance APC/C repressive control during early development is highlighted by the fact that deletion of the components of the MCC Bub3 (Kalitsis, Earle, Fowler, & Choo, 2000), BubR1 (Wang, Liu, et al., 2004), or Mad2 (Dobles, Liberal, Scott, Benezra, & Sorger, 2000) causes early embryonic lethality.

5.3.1 Emi1

Emi1 is an APC/C inhibitor, which allows for the tight control and activation of the APC/C only during the SAC (Reimann, Freed, et al., 2001; Reimann, Gardner, Margottin-Goguet, & Jackson, 2001). This regulation is imposed mainly by the interaction of Emi1 with Cdc20. Cdc20 is expressed at low levels during G1 and S phases but rises during G2 phase. Despite this rise APC activity remains at a low level until prophase. This is due to the inhibitory action of Emi1 which able to bind to the substrate-binding regions of Cdc20 blocking its interaction with APC/C. There is also some evidence, which suggests that Emi1 can also bind to and inhibit activated APC^{Cdc20} and APC^{Cdh1} complexes. During prophase Emi1 becomes degraded allowing the release of Cdc20 and Cdh1 and their interaction with APC/C during prometaphase. This degradation is mediated by the phosphorylation of Emi1 by Plk1, which causes it to become a target for the SCF ubiquitin ligases leading to its eventual destruction by the 26S proteasome (Hansen, Loktev, Ban, & Jackson, 2004). In contrast to the APC/C, which is active only during late mitosis to G1 phase, these SCF ligases are active throughout the cell cycle allowing the quick removal of Emi1 and thus APC/C activation immediately following Plk1 phosphorylation. The inhibitory action of Emi1 on the APC has been shown to be essential during normal preimplantation development and abnormal mitotic progression is seen by the 8- to 16-cell stage leading to defects in the developing blastocysts (Lee et al., 2006). Emi1-null blastocysts exhibited major cellular apoptosis leading to the lack of formation of ICM blastocoel and trophoblast.

6. CONCLUDING REMARKS

Although knockout mouse models have unquestionably advanced our current understanding of the role of various cell cycle proteins during early development, caution must be taken when inferring meaning from the resultant phenotypes.

First, throughout evolution multiple gene duplication events have increased the number of members of many of the cell cycle protein families in mammals as compared to lower organisms. Many of these gene duplication events convey functional redundancy, which effectively removes selective pressures from the gene in question. This allows for the accumulation of mutations in the original or newly duplicated gene copy, which over time can lead to additional functionality, despite displaying high homology to the original gene. This can result in altered tissue or cellular compartment-specific expression or cause interaction with novel proteins, be it substrates, inhibitors, or activating partner proteins.

An excellent example of this are the cyclins, many of which are shown to be absolutely dispensable during preimplantation development but are essential for more specialized compartment or tissue-specific processes such as hematopoiesis and organ development during later embryonic development, the mechanisms of which are still not fully understood.

Second, the knockout mouse methodology simplistically allows the organism to be viewed as a system lacking a single genetic component. This however does not allow the detection of compensatory mechanisms especially as many genes are highly homologous and carry out similar functions. Newer gene-editing technologies will undoubtedly pave the way for a new wave of complex multiple knockout models to help in this regard.

Although many of the major questions concerning the cell cycle in early development have been addressed, several important questions are still to be answered in this field including the mechanisms whereby checkpoint signaling is downregulated and cell cycle drivers are upregulated to allow the shortened cell division time and increased proliferation. The events, which dictate the removal of checkpoint control in these cells, may be of particular clinical interest due to the parallels of early embryonic cell cycles with those of cancer cells during oncogenic proliferation.

REFERENCES

Abramczuk, J., & Sawicki, W. (1975). Pronuclear synthesis of DNA in fertilized and parthenogenetically activated mouse eggs. *Experimental Cell Research*, *92*(2), 361–371.

Adamczewski, J. P., Rossignol, M., Tassan, J. P., Nigg, E. A., Moncollin, V., & Egly, J. M. (1996). MAT1, cdk7 and cyclin H form a kinase complex which is UV light-sensitive upon association with TFIIH. *The EMBO Journal*, *15*(8), 1877–1884.

Adenot, P. G., Szollosi, M. S., Geze, M., Renard, J. P., & Debey, P. (1991). Dynamics of paternal chromatin changes in live one-cell mouse embryo after natural fertilization. *Molecular Reproduction and Development*, *28*(1), 23–34. http://dx.doi.org/10.1002/mrd.1080280105.

Aiken, C. E., Swoboda, P. P., Skepper, J. N., & Johnson, M. H. (2004). The direct measurement of embryogenic volume and nucleo-cytoplasmic ratio during mouse preimplantation development. *Reproduction*, *128*(5), 527–535. http://dx.doi.org/10.1530/rep.1.00281.

Akoulitchev, S., Makela, T. P., Weinberg, R. A., & Reinberg, D. (1995). Requirement for TFIIH kinase activity in transcription by RNA polymerase II. *Nature*, *377*(6549), 557–560. http://dx.doi.org/10.1038/377557a0.

Aleem, E., Kiyokawa, H., & Kaldis, P. (2005). Cdc2-cyclin E complexes regulate the G1/S phase transition. *Nature Cell Biology*, 7(8), 831–836. http://dx.doi.org/10.1038/ncb1284.

Baek, H. J., Kang, Y. K., & Roeder, R. G. (2006). Human mediator enhances basal transcription by facilitating recruitment of transcription factor IIB during preinitiation complex assembly. *The Journal of Biological Chemistry*, *281*(22), 15172–15181. http://dx.doi.org/10.1074/jbc.M601983200.

Bakkenist, C. J., & Kastan, M. B. (2003). DNA damage activates ATM through intermolecular autophosphorylation and dimer dissociation. *Nature*, *421*(6922), 499–506. http://dx.doi.org/10.1038/nature01368.

Barlow, C., Hirotsune, S., Paylor, R., Liyanage, M., Eckhaus, M., Collins, F., ... Wynshaw-Boris, A. (1996). Atm-deficient mice: A paradigm of ataxia telangiectasia. *Cell*, *86*(1), 159–171.

Barr, F. A., Sillje, H. H., & Nigg, E. A. (2004). Polo-like kinases and the orchestration of cell division. *Nature Reviews. Molecular Cell Biology*, *5*(6), 429–440. http://dx.doi.org/10.1038/nrm1401.

Bartek, J., & Lukas, J. (2001). Mammalian G1- and S-phase checkpoints in response to DNA damage. *Current Opinion in Cell Biology*, *13*(6), 738–747.

Berthet, C., Aleem, E., Coppola, V., Tessarollo, L., & Kaldis, P. (2003). Cdk2 knockout mice are viable. *Current Biology*, *13*(20), 1775–1785.

Berthet, C., Klarmann, K. D., Hilton, M. B., Suh, H. C., Keller, J. R., Kiyokawa, H., & Kaldis, P. (2006). Combined loss of Cdk2 and Cdk4 results in embryonic lethality and Rb hypophosphorylation. *Developmental Cell*, *10*(5), 563–573. http://dx.doi.org/10.1016/j.devcel.2006.03.004.

Bissen, S. T., & Weisblat, D. A. (1989). The durations and compositions of cell cycles in embryos of the leech, Helobdella triserialis. *Development*, *106*(1), 105–118.

Bisteau, X., Caldez, M. J., & Kaldis, P. (2014). The complex relationship between liver cancer and the cell cycle: A story of multiple regulations. *Cancers (Basel)*, *6*(1), 79–111. http://dx.doi.org/10.3390/cancers6010079.

Bolton, V. N., Oades, P. J., & Johnson, M. H. (1984). The relationship between cleavage, DNA replication, and gene expression in the mouse 2-cell embryo. *Journal of Embryology and Experimental Morphology*, *79*, 139–163.

Bouniol, C., Nguyen, E., & Debey, P. (1995). Endogenous transcription occurs at the 1-cell stage in the mouse embryo. *Experimental Cell Research*, *218*(1), 57–62. http://dx.doi.org/10.1006/excr.1995.1130.

Brandeis, M., Rosewell, I., Carrington, M., Crompton, T., Jacobs, M. A., Kirk, J., … Hunt, T. (1998). Cyclin B2-null mice develop normally and are fertile whereas cyclin B1-null mice die in utero. *Proceedings of the National Academy of Sciences of the United States of America*, *95*(8), 4344–4349.

Braude, P. R. (1979). Control of protein synthesis during blastocyst formation in the mouse. *Developmental Biology*, *68*(2), 440–452.

Brown, E. J., & Baltimore, D. (2000). ATR disruption leads to chromosomal fragmentation and early embryonic lethality. *Genes & Development*, *14*(4), 397–402.

Brown, N. R., Korolchuk, S., Martin, M. P., Stanley, W. A., Moukhametzianov, R., Noble, M. E., & Endicott, J. A. (2015). CDK1 structures reveal conserved and unique features of the essential cell cycle CDK. *Nature Communications*, *6*, 6769. http://dx.doi.org/10.1038/ncomms7769.

Burton, J. L., & Solomon, M. J. (2001). D box and KEN box motifs in budding yeast Hsl1p are required for APC-mediated degradation and direct binding to Cdc20p and Cdh1p. *Genes & Development*, *15*(18), 2381–2395. http://dx.doi.org/10.1101/gad.917901.

Calegari, F., & Huttner, W. B. (2003). An inhibition of cyclin-dependent kinases that lengthens, but does not arrest, neuroepithelial cell cycle induces premature neurogenesis. *Journal of Cell Science*, *116*(Pt. 24), 4947–4955. http://dx.doi.org/10.1242/jcs.00825.

Cantin, G. T., Stevens, J. L., & Berk, A. J. (2003). Activation domain-mediator interactions promote transcription preinitiation complex assembly on promoter DNA. *Proceedings of the National Academy of Sciences of the United States of America*, *100*(21), 12003–12008. http://dx.doi.org/10.1073/pnas.2035253100.

Carlson, M. (1997). Genetics of transcriptional regulation in yeast: Connections to the RNA polymerase II CTD. *Annual Review of Cell and Developmental Biology*, *13*, 1–23. http://dx.doi.org/10.1146/annurev.cellbio.13.1.1.

Chan, F. K., Zhang, J., Cheng, L., Shapiro, D. N., & Winoto, A. (1995). Identification of human and mouse p19, a novel CDK4 and CDK6 inhibitor with homology to p16ink4. *Molecular and Cellular Biology*, *15*(5), 2682–2688.

Chao, W. C., Kulkarni, K., Zhang, Z., Kong, E. H., & Barford, D. (2012). Structure of the mitotic checkpoint complex. *Nature*, *484*(7393), 208–213. http://dx.doi.org/10.1038/nature10896.

Chen, Y. F., Chiu, H. H., Wu, C. H., Wang, J. Y., Chen, F. M., Tzou, W. H., … Lin, S. R. (2003). Retinoblastoma protein (pRB) was significantly phosphorylated through a Ras-to-MAPK pathway in mutant K-ras stably transfected human adrenocortical cells. *DNA and Cell Biology*, *22*(10), 657–664. http://dx.doi.org/10.1089/104454903770238139.

Chen, M. S., Hurov, J., White, L. S., Woodford-Thomas, T., & Piwnica-Worms, H. (2001). Absence of apparent phenotype in mice lacking Cdc25C protein phosphatase. *Molecular and Cellular Biology*, *21*(12), 3853–3861. http://dx.doi.org/10.1128/MCB.21.12.3853-3861.2001.

Chen, J., Jackson, P. K., Kirschner, M. W., & Dutta, A. (1995). Separate domains of p21 involved in the inhibition of Cdk kinase and PCNA. *Nature*, *374*(6520), 386–388. http://dx.doi.org/10.1038/374386a0.

Chen, L., Wang, D., Wu, Z., Ma, L., & Daley, G. Q. (2010). Molecular basis of the first cell fate determination in mouse embryogenesis. *Cell Research*, *20*(9), 982–993. http://dx.doi.org/10.1038/cr.2010.106.

Christians, E., Davis, A. A., Thomas, S. D., & Benjamin, I. J. (2000). Maternal effect of Hsf1 on reproductive success. *Nature*, *407*(6805), 693–694. http://dx.doi.org/10.1038/35037669.

Ciemerych, M. A., Kenney, A. M., Sicinska, E., Kalaszczynska, I., Bronson, R. T., Rowitch, D. H., … Sicinski, P. (2002). Development of mice expressing a single D-type cyclin. *Genes & Development*, *16*(24), 3277–3289. http://dx.doi.org/10.1101/gad.1023602.

Ciemerych, M. A., Maro, B., & Kubiak, J. Z. (1999). Control of duration of the first two mitoses in a mouse embryo. *Zygote*, 7(4), 293–300.

Clarke, A. R., Maandag, E. R., van Roon, M., van der Lugt, N. M., van der Valk, M., Hooper, M. L., … te Riele, H. (1992). Requirement for a functional Rb-1 gene in murine development. *Nature*, *359*(6393), 328–330. http://dx.doi.org/10.1038/359328a0.

Cloud, J. E., Rogers, C., Reza, T. L., Ziebold, U., Stone, J. R., Picard, M. H., … Lees, J. A. (2002). Mutant mouse models reveal the relative roles of E2F1 and E2F3 in vivo. *Molecular and Cellular Biology*, *22*(8), 2663–2672.

Cobrinik, D., Lee, M. H., Hannon, G., Mulligan, G., Bronson, R. T., Dyson, N., … Jacks, T. (1996). Shared role of the pRB-related p130 and p107 proteins in limb development. *Genes & Development*, *10*(13), 1633–1644.

Cortez, D., Guntuku, S., Qin, J., & Elledge, S. J. (2001). ATR and ATRIP: Partners in checkpoint signaling. *Science*, *294*(5547), 1713–1716. http://dx.doi.org/10.1126/science.1065521.

Costanzo, V., Robertson, K., Ying, C. Y., Kim, E., Avvedimento, E., Gottesman, M., … Gautier, J. (2000). Reconstitution of an ATM-dependent checkpoint that inhibits chromosomal DNA replication following DNA damage. *Molecular Cell*, *6*(3), 649–659.

Daniel, J. A., Pellegrini, M., Lee, B. S., Guo, Z., Filsuf, D., Belkina, N. V., … Nussenzweig, A. (2012). Loss of ATM kinase activity leads to embryonic lethality in mice. *The Journal of Cell Biology*, *198*(3), 295–304. http://dx.doi.org/10.1083/jcb.201204035.

De Bondt, H. L., Rosenblatt, J., Jancarik, J., Jones, H. D., Morgan, D. O., & Kim, S. H. (1993). Crystal structure of cyclin-dependent kinase 2. *Nature*, *363*(6430), 595–602. http://dx.doi.org/10.1038/363595a0.

de Bruin, A., Maiti, B., Jakoi, L., Timmers, C., Buerki, R., & Leone, G. (2003). Identification and characterization of E2F7, a novel mammalian E2F family member capable of blocking cellular proliferation. *The Journal of Biological Chemistry*, *278*(43), 42041–42049. http://dx.doi.org/10.1074/jbc.M308105200.

DeGregori, J., & Johnson, D. G. (2006). Distinct and overlapping roles for E2F family members in transcription, proliferation and apoptosis. *Current Molecular Medicine*, *6*(7), 739–748.

DeGregori, J., Leone, G., Miron, A., Jakoi, L., & Nevins, J. R. (1997). Distinct roles for E2F proteins in cell growth control and apoptosis. *Proceedings of the National Academy of Sciences of the United States of America, 94*(14), 7245–7250.

de Klein, A., Muijtjens, M., van Os, R., Verhoeven, Y., Smit, B., Carr, A. M., … Hoeijmakers, J. H. (2000). Targeted disruption of the cell-cycle checkpoint gene ATR leads to early embryonic lethality in mice. *Current Biology, 10*(8), 479–482.

De Leon, V., Johnson, A., & Bachvarova, R. (1983). Half-lives and relative amounts of stored and polysomal ribosomes and poly(A) + RNA in mouse oocytes. *Developmental Biology, 98*(2), 400–408.

Diril, M. K., Ratnacaram, C. K., Padmakumar, V. C., Du, T., Wasser, M., Coppola, V., … Kaldis, P. (2012). Cyclin-dependent kinase 1 (Cdk1) is essential for cell division and suppression of DNA re-replication but not for liver regeneration. *Proceedings of the National Academy of Sciences of the United States of America, 109*(10), 3826–3831. http://dx.doi.org/10.1073/pnas.1115201109.

Di Stefano, L., Jensen, M. R., & Helin, K. (2003). E2F7, a novel E2F featuring DP-independent repression of a subset of E2F-regulated genes. *The EMBO Journal, 22*(23), 6289–6298. http://dx.doi.org/10.1093/emboj/cdg613.

Dobles, M., Liberal, V., Scott, M. L., Benezra, R., & Sorger, P. K. (2000). Chromosome missegregation and apoptosis in mice lacking the mitotic checkpoint protein Mad2. *Cell, 101*(6), 635–645.

Domon, M. (1980). Cell cycle-dependent radiosensitivity in two-cell mouse embryos in culture. *Radiation Research, 81*(2), 236–245.

Donahue, R. P. (1972). Fertilization of the mouse oocyte: Sequence and timing of nuclear progression to the two-cell stage. *The Journal of Experimental Zoology, 180*(3), 305–318. http://dx.doi.org/10.1002/jez.1401800303.

Donzelli, M., Squatrito, M., Ganoth, D., Hershko, A., Pagano, M., & Draetta, G. F. (2002). Dual mode of degradation of Cdc25 A phosphatase. *The EMBO Journal, 21*(18), 4875–4884.

Dunphy, W. G., & Kumagai, A. (1991). The cdc25 protein contains an intrinsic phosphatase activity. *Cell, 67*(1), 189–196.

Dyson, N. (1998). The regulation of E2F by pRB-family proteins. *Genes & Development, 12*(15), 2245–2262.

Echalier, A., Endicott, J. A., & Noble, M. E. (2010). Recent developments in cyclin-dependent kinase biochemical and structural studies. *Biochimica et Biophysica Acta, 1804*(3), 511–519. http://dx.doi.org/10.1016/j.bbapap.2009.10.002.

Eissenberg, J. C., Shilatifard, A., Dorokhov, N., & Michener, D. E. (2007). Cdk9 is an essential kinase in Drosophila that is required for heat shock gene expression, histone methylation and elongation factor recruitment. *Molecular Genetics and Genomics, 277*(2), 101–114. http://dx.doi.org/10.1007/s00438-006-0164-2.

el-Deiry, W. S., Tokino, T., Velculescu, V. E., Levy, D. B., Parsons, R., Trent, J. M., … Vogelstein, B. (1993). WAF1, a potential mediator of p53 tumor suppression. *Cell, 75*(4), 817–825.

Elson, A., Wang, Y., Daugherty, C. J., Morton, C. C., Zhou, F., Campos-Torres, J., & Leder, P. (1996). Pleiotropic defects in ataxia-telangiectasia protein-deficient mice. *Proceedings of the National Academy of Sciences of the United States of America, 93*(23), 13084–13089.

Endicott, J. A., & Noble, M. E. (2013). Structural characterization of the cyclin-dependent protein kinase family. *Biochemical Society Transactions, 41*(4), 1008–1016. http://dx.doi.org/10.1042/BST20130097.

Ewen, M. E., Sluss, H. K., Sherr, C. J., Matsushime, H., Kato, J., & Livingston, D. M. (1993). Functional interactions of the retinoblastoma protein with mammalian D-type cyclins. *Cell, 73*(3), 487–497.

Falck, J., Mailand, N., Syljuasen, R. G., Bartek, J., & Lukas, J. (2001). The ATM-Chk2-Cdc25A checkpoint pathway guards against radioresistant DNA synthesis. *Nature*, *410*(6830), 842–847. http://dx.doi.org/10.1038/35071124.

Fang, G., Yu, H., & Kirschner, M. W. (1998). The checkpoint protein MAD2 and the mitotic regulator CDC20 form a ternary complex with the anaphase-promoting complex to control anaphase initiation. *Genes & Development*, *12*(12), 1871–1883.

Fantl, V., Stamp, G., Andrews, A., Rosewell, I., & Dickson, C. (1995). Mice lacking cyclin D1 are small and show defects in eye and mammary gland development. *Genes & Development*, *9*(19), 2364–2372.

Ferguson, A. M., White, L. S., Donovan, P. J., & Piwnica-Worms, H. (2005). Normal cell cycle and checkpoint responses in mice and cells lacking Cdc25B and Cdc25C protein phosphatases. *Molecular and Cellular Biology*, *25*(7), 2853–2860. http://dx.doi.org/10.1128/MCB.25.7.2853-2860.2005.

Fernandez, M. S., & Izquierdo, L. (1980). Blastocoel formation in half and double mouse embryos. *Anatomy and Embryology*, *160*(1), 77–81.

Fero, M. L., Rivkin, M., Tasch, M., Porter, P., Carow, C. E., Firpo, E., … Roberts, J. M. (1996). A syndrome of multiorgan hyperplasia with features of gigantism, tumorigenesis, and female sterility in p27(Kip1)-deficient mice. *Cell*, *85*(5), 733–744.

Fesquet, D., Labbe, J. C., Derancourt, J., Capony, J. P., Galas, S., Girard, F., … Cavadore, J. C. (1993). The MO15 gene encodes the catalytic subunit of a protein kinase that activates cdc2 and other cyclin-dependent kinases (CDKs) through phosphorylation of Thr161 and its homologues. *The EMBO Journal*, *12*(8), 3111–3121.

Field, S. J., Tsai, F. Y., Kuo, F., Zubiaga, A. M., Kaelin, W. G., Jr., Livingston, D. M., … Greenberg, M. E. (1996). E2F-1 functions in mice to promote apoptosis and suppress proliferation. *Cell*, *85*(4), 549–561.

Fisher, R. P. (2005). Secrets of a double agent: CDK7 in cell-cycle control and transcription. *Journal of Cell Science*, *118*(Pt. 22), 5171–5180. http://dx.doi.org/10.1242/jcs.02718.

Fisher, R. P., Jin, P., Chamberlin, H. M., & Morgan, D. O. (1995). Alternative mechanisms of CAK assembly require an assembly factor or an activating kinase. *Cell*, *83*(1), 47–57.

Fisher, R. P., & Morgan, D. O. (1994). A novel cyclin associates with MO15/CDK7 to form the CDK-activating kinase. *Cell*, *78*(4), 713–724.

Fisher, D. L., & Nurse, P. (1996). A single fission yeast mitotic cyclin B p34cdc2 kinase promotes both S-phase and mitosis in the absence of G1 cyclins. *The EMBO Journal*, *15*(4), 850–860.

Flach, G., Johnson, M. H., Braude, P. R., Taylor, R. A., & Bolton, V. N. (1982). The transition from maternal to embryonic control in the 2-cell mouse embryo. *The EMBO Journal*, *1*(6), 681–686.

Fluckiger, A. C., Marcy, G., Marchand, M., Negre, D., Cosset, F. L., Mitalipov, S., … Dehay, C. (2006). Cell cycle features of primate embryonic stem cells. *Stem Cells*, *24*(3), 547–556. http://dx.doi.org/10.1634/stemcells.2005-0194.

Furstenthal, L., Swanson, C., Kaiser, B. K., Eldridge, A. G., & Jackson, P. K. (2001). Triggering ubiquitination of a CDK inhibitor at origins of DNA replication. *Nature Cell Biology*, *3*(8), 715–722. http://dx.doi.org/10.1038/35087026.

Gamow, E. I., & Prescott, D. M. (1970). The cell life cycle during early embryogenesis of the mouse. *Experimental Cell Research*, *59*(1), 117–123.

Ganuza, M., Saiz-Ladera, C., Canamero, M., Gomez, G., Schneider, R., Blasco, M. A., … Barbacid, M. (2012). Genetic inactivation of Cdk7 leads to cell cycle arrest and induces premature aging due to adult stem cell exhaustion. *The EMBO Journal*, *31*(11), 2498–2510. http://dx.doi.org/10.1038/emboj.2012.94.

Gaubatz, S., Lindeman, G. J., Ishida, S., Jakoi, L., Nevins, J. R., Livingston, D. M., & Rempel, R. E. (2000). E2F4 and E2F5 play an essential role in pocket protein-mediated G1 control. *Molecular Cell*, *6*(3), 729–735.

Geng, Y., Yu, Q., Sicinska, E., Das, M., Schneider, J. E., Bhattacharya, S., … Sicinski, P. (2003). Cyclin E ablation in the mouse. *Cell*, *114*(4), 431–443.

Girard, F., Strausfeld, U., Fernandez, A., & Lamb, N. J. (1991). Cyclin A is required for the onset of DNA replication in mammalian fibroblasts. *Cell*, *67*(6), 1169–1179.

Goldstein, L. S., Spindle, A. I., & Pedersen, R. A. (1975). X-ray sensitivity of the preimplantation mouse embryo in vitro. *Radiation Research*, *62*(2), 276–287.

Gopinathan, L., Ratnacaram, C. K., & Kaldis, P. (2011). Established and novel Cdk/cyclin complexes regulating the cell cycle and development. *Results and Problems in Cell Differentiation*, *53*, 365–389. http://dx.doi.org/10.1007/978-3-642-19065-0_16.

Grafstrom, R. H., Pan, W., & Hoess, R. H. (1999). Defining the substrate specificity of cdk4 kinase-cyclin D1 complex. *Carcinogenesis*, *20*(2), 193–198.

Guan, K. L., Jenkins, C. W., Li, Y., Nichols, M. A., Wu, X., O'Keefe, C. L., … Xiong, Y. (1994). Growth suppression by p18, a p16INK4/MTS1- and p14INK4B/MTS2-related CDK6 inhibitor, correlates with wild-type pRb function. *Genes & Development*, *8*(24), 2939–2952.

Gurley, K. E., & Kemp, C. J. (2001). Synthetic lethality between mutation in Atm and DNA-PK(cs) during murine embryogenesis. *Current Biology*, *11*(3), 191–194.

Guzman-Ayala, M., Sachs, M., Koh, F. M., Onodera, C., Bulut-Karslioglu, A., Lin, C. J., … Ramalho-Santos, M. (2015). Chd1 is essential for the high transcriptional output and rapid growth of the mouse epiblast. *Development*, *142*(1), 118–127. http://dx.doi.org/10.1242/dev.114843.

Hahn, S. (2004). Structure and mechanism of the RNA polymerase II transcription machinery. *Nature Structural & Molecular Biology*, *11*(5), 394–403. http://dx.doi.org/10.1038/nsmb763.

Han, J. S., Holland, A. J., Fachinetti, D., Kulukian, A., Cetin, B., & Cleveland, D. W. (2013). Catalytic assembly of the mitotic checkpoint inhibitor BubR1-Cdc20 by a Mad2-induced functional switch in Cdc20. *Molecular Cell*, *51*(1), 92–104. http://dx.doi.org/10.1016/j.molcel.2013.05.019.

Hannon, G. J., & Beach, D. (1994). p15INK4B is a potential effector of TGF-beta-induced cell cycle arrest. *Nature*, *371*(6494), 257–261. http://dx.doi.org/10.1038/371257a0.

Hansen, D. V., Loktev, A. V., Ban, K. H., & Jackson, P. K. (2004). Plk1 regulates activation of the anaphase promoting complex by phosphorylating and triggering SCFbetaTrCP-dependent destruction of the APC Inhibitor Emi1. *Molecular Biology of the Cell*, *15*(12), 5623–5634. http://dx.doi.org/10.1091/mbc.E04-07-0598.

Hardy, K., Handyside, A. H., & Winston, R. M. (1989). The human blastocyst: Cell number, death and allocation during late preimplantation development in vitro. *Development*, *107*(3), 597–604.

Harper, J. W., Adami, G. R., Wei, N., Keyomarsi, K., & Elledge, S. J. (1993). The p21 Cdk-interacting protein Cip1 is a potent inhibitor of G1 cyclin-dependent kinases. *Cell*, *75*(4), 805–816.

Harper, J. W., Elledge, S. J., Keyomarsi, K., Dynlacht, B., Tsai, L. H., Zhang, P., et al. (1995). Inhibition of cyclin-dependent kinases by p21. *Molecular Biology of the Cell*, *6*(4), 387–400.

Hartwell, L. H., & Kastan, M. B. (1994). Cell cycle control and cancer. *Science*, *266*(5192), 1821–1828.

Harvey, S. L., Charlet, A., Haas, W., Gygi, S. P., & Kellogg, D. R. (2005). Cdk1-dependent regulation of the mitotic inhibitor Wee1. *Cell*, *122*(3), 407–420. http://dx.doi.org/10.1016/j.cell.2005.05.029.

He, S., Cook, B. L., Deverman, B. E., Weihe, U., Zhang, F., Prachand, V., … Weintraub, S. J. (2000). E2F is required to prevent inappropriate S-phase entry of mammalian cells. *Molecular and Cellular Biology*, *20*(1), 363–371.

Heyer, B. S., MacAuley, A., Behrendtsen, O., & Werb, Z. (2000). Hypersensitivity to DNA damage leads to increased apoptosis during early mouse development. *Genes & Development, 14*(16), 2072–2084.

Hiebert, S. W., Chellappan, S. P., Horowitz, J. M., & Nevins, J. R. (1992). The interaction of RB with E2F coincides with an inhibition of the transcriptional activity of E2F. *Genes & Development, 6*(2), 177–185.

Hilioti, Z., Chung, Y. S., Mochizuki, Y., Hardy, C. F., & Cohen-Fix, O. (2001). The anaphase inhibitor Pds1 binds to the APC/C-associated protein Cdc20 in a destruction box-dependent manner. *Current Biology, 11*(17), 1347–1352.

Hirao, A., Cheung, A., Duncan, G., Girard, P. M., Elia, A. J., Wakeham, A., ... Mak, T. W. (2002). Chk2 is a tumor suppressor that regulates apoptosis in both an ataxia telangiectasia mutated (ATM)-dependent and an ATM-independent manner. *Molecular and Cellular Biology, 22*(18), 6521–6532.

Hochegger, H., Takeda, S., & Hunt, T. (2008). Cyclin-dependent kinases and cell-cycle transitions: Does one fit all? *Nature Reviews. Molecular Cell Biology, 9*(11), 910–916. http://dx.doi.org/10.1038/nrm2510.

Hoeppner, S., Baumli, S., & Cramer, P. (2005). Structure of the mediator subunit cyclin C and its implications for CDK8 function. *Journal of Molecular Biology, 350*(5), 833–842. http://dx.doi.org/10.1016/j.jmb.2005.05.041.

Hogan, B. L. M., Beddington, R., Constantini, F., & Lacy, E. (1994). *Manipulating the mouse embryo.* Cold Spring Harbor, New York: Cold Spring Harbor Laboratory Press.

Holmes, J. K., & Solomon, M. J. (1996). A predictive scale for evaluating cyclin-dependent kinase substrates. A comparison of p34cdc2 and p33cdk2. *The Journal of Biological Chemistry, 271*(41), 25240–25246.

Howlett, S. K., & Bolton, V. N. (1985). Sequence and regulation of morphological and molecular events during the first cell cycle of mouse embryogenesis. *Journal of Embryology and Experimental Morphology, 87*, 175–206.

Hu, M. G., Deshpande, A., Enos, M., Mao, D., Hinds, E. A., Hu, G. F., ... Hinds, P. W. (2009). A requirement for cyclin-dependent kinase 6 in thymocyte development and tumorigenesis. *Cancer Research, 69*(3), 810–818. http://dx.doi.org/10.1158/0008-5472.CAN-08-2473.

Hudson, J. W., Kozarova, A., Cheung, P., Macmillan, J. C., Swallow, C. J., Cross, J. C., & Dennis, J. W. (2001). Late mitotic failure in mice lacking Sak, a polo-like kinase. *Current Biology, 11*(6), 441–446.

Hyslop, L., Stojkovic, M., Armstrong, L., Walter, T., Stojkovic, P., Przyborski, S., ... Lako, M. (2005). Downregulation of NANOG induces differentiation of human embryonic stem cells to extraembryonic lineages. *Stem Cells, 23*(8), 1035–1043. http://dx.doi.org/10.1634/stemcells.2005-0080.

Inglis, K. J., Chereau, D., Brigham, E. F., Chiou, S. S., Schobel, S., Frigon, N. L., ... Anderson, J. P. (2009). Polo-like kinase 2 (PLK2) phosphorylates alpha-synuclein at serine 129 in central nervous system. *The Journal of Biological Chemistry, 284*(5), 2598–2602. http://dx.doi.org/10.1074/jbc.C800206200.

Iwamori, N., Naito, K., Sugiura, K., & Tojo, H. (2002). Preimplantation-embryo-specific cell cycle regulation is attributed to the low expression level of retinoblastoma protein. *FEBS Letters, 526*(1–3), 119–123.

Izumi, T., & Maller, J. L. (1993). Elimination of cdc2 phosphorylation sites in the cdc25 phosphatase blocks initiation of M-phase. *Molecular Biology of the Cell, 4*(12), 1337–1350.

Jackman, M., Lindon, C., Nigg, E. A., & Pines, J. (2003). Active cyclin B1-Cdk1 first appears on centrosomes in prophase. *Nature Cell Biology, 5*(2), 143–148. http://dx.doi.org/10.1038/ncb918.

Jacks, T., Fazeli, A., Schmitt, E. M., Bronson, R. T., Goodell, M. A., & Weinberg, R. A. (1992). Effects of an Rb mutation in the mouse. *Nature, 359*(6393), 295–300. http://dx.doi.org/10.1038/359295a0.

Jackson, P. K., Chevalier, S., Philippe, M., & Kirschner, M. W. (1995). Early events in DNA replication require cyclin E and are blocked by p21CIP1. *The Journal of Cell Biology*, *130*(4), 755–769.

Jeffrey, P. D., Russo, A. A., Polyak, K., Gibbs, E., Hurwitz, J., Massague, J., & Pavletich, N. P. (1995). Mechanism of CDK activation revealed by the structure of a cyclinA-CDK2 complex. *Nature*, *376*(6538), 313–320. http://dx.doi.org/10.1038/376313a0.

Jeffrey, P. D., Tong, L., & Pavletich, N. P. (2000). Structural basis of inhibition of CDK-cyclin complexes by INK4 inhibitors. *Genes & Development*, *14*(24), 3115–3125.

Johnson, L. N., & Lewis, R. J. (2001). Structural basis for control by phosphorylation. *Chemical Reviews*, *101*(8), 2209–2242.

Kalaszczynska, I., Geng, Y., Iino, T., Mizuno, S., Choi, Y., Kondratiuk, I., … Sicinski, P. (2009). Cyclin A is redundant in fibroblasts but essential in hematopoietic and embryonic stem cells. *Cell*, *138*(2), 352–365. http://dx.doi.org/10.1016/j.cell.2009.04.062.

Kaldis, P. (1999). The cdk-activating kinase (CAK): From yeast to mammals. *Cellular and Molecular Life Sciences*, *55*(2), 284–296.

Kalitsis, P., Earle, E., Fowler, K. J., & Choo, K. H. (2000). Bub3 gene disruption in mice reveals essential mitotic spindle checkpoint function during early embryogenesis. *Genes & Development*, *14*(18), 2277–2282.

Kamjoo, M., Brison, D. R., & Kimber, S. J. (2002). Apoptosis in the preimplantation mouse embryo: Effect of strain difference and in vitro culture. *Molecular Reproduction and Development*, *61*(1), 67–77. http://dx.doi.org/10.1002/mrd.1132.

Kellogg, D. R. (2003). Wee1-dependent mechanisms required for coordination of cell growth and cell division. *Journal of Cell Science*, *116*(Pt. 24), 4883–4890. http://dx.doi.org/10.1242/jcs.00908.

Kiessling, A. A., Bletsa, R., Desmarais, B., Mara, C., Kallianidis, K., & Loutradis, D. (2009). Evidence that human blastomere cleavage is under unique cell cycle control. *Journal of Assisted Reproduction and Genetics*, *26*(4), 187–195. http://dx.doi.org/10.1007/s10815-009-9306-x.

Kiessling, A. A., Bletsa, R., Desmarais, B., Mara, C., Kallianidis, K., & Loutradis, D. (2010). Genome-wide microarray evidence that 8-cell human blastomeres over-express cell cycle drivers and under-express checkpoints. *Journal of Assisted Reproduction and Genetics*, *27*(6), 265–276. http://dx.doi.org/10.1007/s10815-010-9407-6.

Kiyokawa, H., Kineman, R. D., Manova-Todorova, K. O., Soares, V. C., Hoffman, E. S., Ono, M., … Koff, A. (1996). Enhanced growth of mice lacking the cyclin-dependent kinase inhibitor function of p27(Kip1). *Cell*, *85*(5), 721–732.

Knudsen, E. S., & Wang, J. Y. (1997). Dual mechanisms for the inhibition of E2F binding to RB by cyclin-dependent kinase-mediated RB phosphorylation. *Molecular and Cellular Biology*, *17*(10), 5771–5783.

Kohn, M. J., Bronson, R. T., Harlow, E., Dyson, N. J., & Yamasaki, L. (2003). Dp1 is required for extra-embryonic development. *Development*, *130*(7), 1295–1305.

Kohoutek, J., Li, Q., Blazek, D., Luo, Z., Jiang, H., & Peterlin, B. M. (2009). Cyclin T2 is essential for mouse embryogenesis. *Molecular and Cellular Biology*, *29*(12), 3280–3285. http://dx.doi.org/10.1128/MCB.00172-09.

Koseoglu, M. M., Graves, L. M., & Marzluff, W. F. (2008). Phosphorylation of threonine 61 by cyclin a/Cdk1 triggers degradation of stem-loop binding protein at the end of S phase. *Molecular and Cellular Biology*, *28*(14), 4469–4479. http://dx.doi.org/10.1128/MCB.01416-07.

Kozar, K., Ciemerych, M. A., Rebel, V. I., Shigematsu, H., Zagozdzon, A., Sicinska, E., … Sicinski, P. (2004). Mouse development and cell proliferation in the absence of D-cyclins. *Cell*, *118*(4), 477–491. http://dx.doi.org/10.1016/j.cell.2004.07.025.

Krishna, M., & Generoso, W. M. (1977). Timing of sperm penetration, pronuclear formation, pronuclear DNA synthesis, and first cleavage in naturally ovulated mouse eggs.

The Journal of Experimental Zoology, *202*(2), 245–252. http://dx.doi.org/10.1002/jez.1402020214.

Krzywda, S., Brzozowski, A. M., Higashitsuji, H., Fujita, J., Welchman, R., Dawson, S., … Wilkinson, A. J. (2004). The crystal structure of gankyrin, an oncoprotein found in complexes with cyclin-dependent kinase 4, a 19 S proteasomal ATPase regulator, and the tumor suppressors Rb and p53. *The Journal of Biological Chemistry*, *279*(2), 1541–1545. http://dx.doi.org/10.1074/jbc.M310265200.

Kumagai, A., & Dunphy, W. G. (1996). Purification and molecular cloning of Plx1, a Cdc25-regulatory kinase from Xenopus egg extracts. *Science*, *273*(5280), 1377–1380.

Labbe, J. C., Martinez, A. M., Fesquet, D., Capony, J. P., Darbon, J. M., Derancourt, J., … Doree, M. (1994). p40MO15 associates with a p36 subunit and requires both nuclear translocation and Thr176 phosphorylation to generate cdk-activating kinase activity in Xenopus oocytes. *The EMBO Journal*, *13*(21), 5155–5164.

Lammer, C., Wagerer, S., Saffrich, R., Mertens, D., Ansorge, W., & Hoffmann, I. (1998). The cdc25B phosphatase is essential for the G2/M phase transition in human cells. *Journal of Cell Science*, *111*(Pt. 16), 2445–2453.

Lara-Gonzalez, P., Westhorpe, F. G., & Taylor, S. S. (2012). The spindle assembly checkpoint. *Current Biology*, *22*(22), R966–R980. http://dx.doi.org/10.1016/j.cub.2012.10.006.

Lee, M. T., Bonneau, A. R., Takacs, C. M., Bazzini, A. A., DiVito, K. R., Fleming, E. S., & Giraldez, A. J. (2013). Nanog, Pou5f1 and SoxB1 activate zygotic gene expression during the maternal-to-zygotic transition. *Nature*, *503*(7476), 360–364. http://dx.doi.org/10.1038/nature12632.

Lee, E. Y., Chang, C. Y., Hu, N., Wang, Y. C., Lai, C. C., Herrup, K., … Bradley, A. (1992). Mice deficient for Rb are nonviable and show defects in neurogenesis and haematopoiesis. *Nature*, *359*(6393), 288–294. http://dx.doi.org/10.1038/359288a0.

Lee, H., Lee, D. J., Oh, S. P., Park, H. D., Nam, H. H., Kim, J. M., & Lim, D. S. (2006). Mouse emi1 has an essential function in mitotic progression during early embryogenesis. *Molecular and Cellular Biology*, *26*(14), 5373–5381. http://dx.doi.org/10.1128/MCB.00043-06.

Lee, M. G., & Nurse, P. (1987). Complementation used to clone a human homologue of the fission yeast cell cycle control gene cdc2. *Nature*, *327*(6117), 31–35. http://dx.doi.org/10.1038/327031a0.

Lee, M. H., Reynisdottir, I., & Massague, J. (1995). Cloning of p57KIP2, a cyclin-dependent kinase inhibitor with unique domain structure and tissue distribution. *Genes & Development*, *9*(6), 639–649.

Lee, G., White, L. S., Hurov, K. E., Stappenbeck, T. S., & Piwnica-Worms, H. (2009). Response of small intestinal epithelial cells to acute disruption of cell division through CDC25 deletion. *Proceedings of the National Academy of Sciences of the United States of America*, *106*(12), 4701–4706. http://dx.doi.org/10.1073/pnas.0900751106.

Lee, M. H., Williams, B. O., Mulligan, G., Mukai, S., Bronson, R. T., Dyson, N., … Jacks, T. (1996). Targeted disruption of p107: Functional overlap between p107 and Rb. *Genes & Development*, *10*(13), 1621–1632.

Leone, G., Nuckolls, F., Ishida, S., Adams, M., Sears, R., Jakoi, L., … Nevins, J. R. (2000). Identification of a novel E2F3 product suggests a mechanism for determining specificity of repression by Rb proteins. *Molecular and Cellular Biology*, *20*(10), 3626–3632.

Leone, G., Sears, R., Huang, E., Rempel, R., Nuckolls, F., Park, C. H., … Nevins, J. R. (2001). Myc requires distinct E2F activities to induce S phase and apoptosis. *Molecular Cell*, *8*(1), 105–113.

Li, N., Fassl, A., Chick, J., Inuzuka, H., Li, X., Mansour, M. R., … Sicinski, P. (2014). Cyclin C is a haploinsufficient tumour suppressor. *Nature Cell Biology*, *16*(11), 1080–1091. http://dx.doi.org/10.1038/ncb3046.

Li, J., Ran, C., Li, E., Gordon, F., Comstock, G., Siddiqui, H., … de Bruin, A. (2008). Synergistic function of E2F7 and E2F8 is essential for cell survival and embryonic development. *Developmental Cell*, *14*(1), 62–75. http://dx.doi.org/10.1016/j.devcel.2007.10.017.

Li, M., York, J. P., & Zhang, P. (2007). Loss of Cdc20 causes a securin-dependent metaphase arrest in two-cell mouse embryos. *Molecular and Cellular Biology*, *27*(9), 3481–3488. http://dx.doi.org/10.1128/MCB.02088-06.

Lincoln, A. J., Wickramasinghe, D., Stein, P., Schultz, R. M., Palko, M. E., De Miguel, M. P., … Donovan, P. J. (2002). Cdc25b phosphatase is required for resumption of meiosis during oocyte maturation. *Nature Genetics*, *30*(4), 446–449. http://dx.doi.org/10.1038/ng856.

Liu, Q., Guntuku, S., Cui, X. S., Matsuoka, S., Cortez, D., Tamai, K., … Elledge, S. J. (2000). Chk1 is an essential kinase that is regulated by Atr and required for the G(2)/M DNA damage checkpoint. *Genes & Development*, *14*(12), 1448–1459.

Liu, D., Matzuk, M. M., Sung, W. K., Guo, Q., Wang, P., & Wolgemuth, D. J. (1998). Cyclin A1 is required for meiosis in the male mouse. *Nature Genetics*, *20*(4), 377–380. http://dx.doi.org/10.1038/3855.

Lu, H., Chang, D. J., Baratte, B., Meijer, L., & Schulze-Gahmen, U. (2005). Crystal structure of a human cyclin-dependent kinase 6 complex with a flavonol inhibitor, fisetin. *Journal of Medicinal Chemistry*, *48*(3), 737–743. http://dx.doi.org/10.1021/jm049353p.

Lu, L. Y., Wood, J. L., Minter-Dykhouse, K., Ye, L., Saunders, T. L., Yu, X., & Chen, J. (2008). Polo-like kinase 1 is essential for early embryonic development and tumor suppression. *Molecular and Cellular Biology*, *28*(22), 6870–6876. http://dx.doi.org/10.1128/MCB.00392-08.

Luthardt, F. W., & Donahue, R. P. (1975). DNA synthesis in developing two-cell mouse embryos. *Developmental Biology*, *44*(1), 210–216.

Malumbres, M., Harlow, E., Hunt, T., Hunter, T., Lahti, J. M., Manning, G., … Wolgemuth, D. J. (2009). Cyclin-dependent kinases: A family portrait. *Nature Cell Biology*, *11*(11), 1275–1276. http://dx.doi.org/10.1038/ncb1109-1275.

Malumbres, M., Sotillo, R., Santamaria, D., Galan, J., Cerezo, A., Ortega, S., … Barbacid, M. (2004). Mammalian cells cycle without the D-type cyclin-dependent kinases Cdk4 and Cdk6. *Cell*, *118*(4), 493–504. http://dx.doi.org/10.1016/j.cell.2004.08.002.

Manejwala, F. M., Cragoe, E. J., Jr., & Schultz, R. M. (1989). Blastocoel expansion in the preimplantation mouse embryo: Role of extracellular sodium and chloride and possible apical routes of their entry. *Developmental Biology*, *133*(1), 210–220.

Martin, A., Odajima, J., Hunt, S. L., Dubus, P., Ortega, S., Malumbres, M., & Barbacid, M. (2005). Cdk2 is dispensable for cell cycle inhibition and tumor suppression mediated by p27(Kip1) and p21(Cip1). *Cancer Cell*, *7*(6), 591–598. http://dx.doi.org/10.1016/j.ccr.2005.05.006.

Matsuoka, S., Edwards, M. C., Bai, C., Parker, S., Zhang, P., Baldini, A., … Elledge, S. J. (1995). p57KIP2, a structurally distinct member of the p21CIP1 Cdk inhibitor family, is a candidate tumor suppressor gene. *Genes & Development*, *9*(6), 650–662.

Mayer, W., Niveleau, A., Walter, J., Fundele, R., & Haaf, T. (2000). Demethylation of the zygotic paternal genome. *Nature*, *403*(6769), 501–502. http://dx.doi.org/10.1038/35000654.

Mayer, W., Smith, A., Fundele, R., & Haaf, T. (2000). Spatial separation of parental genomes in preimplantation mouse embryos. *The Journal of Cell Biology*, *148*(4), 629–634.

Menisser-de Murcia, J., Mark, M., Wendling, O., Wynshaw-Boris, A., & de Murcia, G. (2001). Early embryonic lethality in PARP-1 Atm double-mutant mice suggests a functional synergy in cell proliferation during development. *Molecular and Cellular Biology*, *21*(5), 1828–1832. http://dx.doi.org/10.1128/MCB.21.5.1828-1832.2001.

Messerschmidt, D. M., Knowles, B. B., & Solter, D. (2014). DNA methylation dynamics during epigenetic reprogramming in the germline and preimplantation embryos. *Genes & Development*, *28*(8), 812–828. http://dx.doi.org/10.1101/gad.234294.113.

Milan, M., Campuzano, S., & Garcia-Bellido, A. (1996). Cell cycling and patterned cell proliferation in the Drosophila wing during metamorphosis. *Proceedings of the National Academy of Sciences of the United States of America*, *93*(21), 11687–11692.

Millar, J. B., Blevitt, J., Gerace, L., Sadhu, K., Featherstone, C., & Russell, P. (1991). p55CDC25 is a nuclear protein required for the initiation of mitosis in human cells. *Proceedings of the National Academy of Sciences of the United States of America*, *88*(23), 10500–10504.

Molinari, M., Mercurio, C., Dominguez, J., Goubin, F., & Draetta, G. F. (2000). Human Cdc25 A inactivation in response to S phase inhibition and its role in preventing premature mitosis. *EMBO Reports*, *1*(1), 71–79. http://dx.doi.org/10.1038/sj.embor.embor619.

Molls, M., Pon, A., Streffer, C., van Beuningen, D., & Zamboglou, N. (1983). The effects of lead and X-rays, alone or in combination, on blastocyst formation and cell kinetics of preimplantation mouse embryos in vitro. *International Journal of Radiation Biology and Related Studies in Physics, Chemistry, and Medicine*, *43*(1), 57–69.

Molls, M., Zamboglou, N., & Streffer, C. (1983). A comparison of the cell kinetics of preimplantation mouse embryos from two different mouse strains. *Cell and Tissue Kinetics*, *16*(3), 277–283.

Morgan, D. O. (1997). Cyclin-dependent kinases: Engines, clocks, and microprocessors. *Annual Review of Cell and Developmental Biology*, *13*, 261–291. http://dx.doi.org/10.1146/annurev.cellbio.13.1.261.

Morgan, D. (2007). *The cell cycle: Principles of control*. London: New Science Press Ltd.

Morrow, C. J., Tighe, A., Johnson, V. L., Scott, M. I., Ditchfield, C., & Taylor, S. S. (2005). Bub1 and aurora B cooperate to maintain BubR1-mediated inhibition of APC/CCdc20. *Journal of Cell Science*, *118*(Pt. 16), 3639–3652. http://dx.doi.org/10.1242/jcs.02487.

Muller, H., Moroni, M. C., Vigo, E., Petersen, B. O., Bartek, J., & Helin, K. (1997). Induction of S-phase entry by E2F transcription factors depends on their nuclear localization. *Molecular and Cellular Biology*, *17*(9), 5508–5520.

Murga, M., Fernandez-Capetillo, O., Field, S. J., Moreno, B., Borlado, L. R., Fujiwara, Y., … Zubiaga, A. M. (2001). Mutation of E2F2 in mice causes enhanced T lymphocyte proliferation, leading to the development of autoimmunity. *Immunity*, *15*(6), 959–970.

Murphy, M., Stinnakre, M. G., Senamaud-Beaufort, C., Winston, N. J., Sweeney, C., Kubelka, M., … Sobczak-Thepot, J. (1997). Delayed early embryonic lethality following disruption of the murine cyclin A2 gene. *Nature Genetics*, *15*(1), 83–86. http://dx.doi.org/10.1038/ng0197-83.

Musacchio, A. (2015). The molecular biology of spindle assembly checkpoint signaling dynamics. *Current Biology*, *25*(20), R1002–R1018. http://dx.doi.org/10.1016/j.cub.2015.08.051.

Myer, D. L., Robbins, S. B., Yin, M., Boivin, G. P., Liu, Y., Greis, K. D., … Stambrook, P. J. (2011). Absence of polo-like kinase 3 in mice stabilizes Cdc25A after DNA damage but is not sufficient to produce tumors. *Mutation Research*, *714*(1–2), 1–10. http://dx.doi.org/10.1016/j.mrfmmm.2011.02.006.

Myers, L. C., & Kornberg, R. D. (2000). Mediator of transcriptional regulation. *Annual Review of Biochemistry*, *69*, 729–749. http://dx.doi.org/10.1146/annurev.biochem.69.1.729.

Najm, F. J., Chenoweth, J. G., Anderson, P. D., Nadeau, J. H., Redline, R. W., McKay, R. D., & Tesar, P. J. (2011). Isolation of epiblast stem cells from preimplantation mouse embryos. *Cell Stem Cell*, *8*(3), 318–325. http://dx.doi.org/10.1016/j.stem.2011.01.016.

Nakajima, H., Toyoshima-Morimoto, F., Taniguchi, E., & Nishida, E. (2003). Identification of a consensus motif for Plk (Polo-like kinase) phosphorylation reveals Myt1 as a Plk1 substrate. *The Journal of Biological Chemistry*, *278*(28), 25277–25280. http://dx.doi.org/10.1074/jbc.C300126200.

Nakayama, K., Ishida, N., Shirane, M., Inomata, A., Inoue, T., Shishido, N., … Nakayama, K. (1996). Mice lacking p27(Kip1) display increased body size, multiple organ hyperplasia, retinal dysplasia, and pituitary tumors. *Cell*, *85*(5), 707–720.

Newport, J., & Kirschner, M. (1982). A major developmental transition in early Xenopus embryos: I. Characterization and timing of cellular changes at the midblastula stage. *Cell*, *30*(3), 675–686.

Nguyen, T. B., Manova, K., Capodieci, P., Lindon, C., Bottega, S., Wang, X. Y., … Koff, A. (2002). Characterization and expression of mammalian cyclin b3, a prepachytene meiotic cyclin. *The Journal of Biological Chemistry*, *277*(44), 41960–41969. http://dx.doi.org/10.1074/jbc.M203951200.

Nigg, E. A. (1998). Polo-like kinases: Positive regulators of cell division from start to finish. *Current Opinion in Cell Biology*, *10*(6), 776–783.

O'Connell, M. J., Raleigh, J. M., Verkade, H. M., & Nurse, P. (1997). Chk1 is a wee1 kinase in the G2 DNA damage checkpoint inhibiting cdc2 by Y15 phosphorylation. *The EMBO Journal*, *16*(3), 545–554. http://dx.doi.org/10.1093/emboj/16.3.545.

O'Farrell, P. H., Stumpff, J., & Su, T. T. (2004). Embryonic cleavage cycles: How is a mouse like a fly? *Current Biology*, *14*(1), R35–R45.

Ortega, S., Prieto, I., Odajima, J., Martin, A., Dubus, P., Sotillo, R., … Barbacid, M. (2003). Cyclin-dependent kinase 2 is essential for meiosis but not for mitotic cell division in mice. *Nature Genetics*, *35*(1), 25–31. http://dx.doi.org/10.1038/ng1232.

Oswald, J., Engemann, S., Lane, N., Mayer, W., Olek, A., Fundele, R., … Walter, J. (2000). Active demethylation of the paternal genome in the mouse zygote. *Current Biology*, *10*(8), 475–478.

Otero, G., Fellows, J., Li, Y., de Bizemont, T., Dirac, A. M., Gustafsson, C. M., … Svejstrup, J. Q. (1999). Elongator, a multisubunit component of a novel RNA polymerase II holoenzyme for transcriptional elongation. *Molecular Cell*, *3*(1), 109–118.

Padmakumar, V. C., Aleem, E., Berthet, C., Hilton, M. B., & Kaldis, P. (2009). Cdk2 and Cdk4 activities are dispensable for tumorigenesis caused by the loss of p53. *Molecular and Cellular Biology*, *29*(10), 2582–2593. http://dx.doi.org/10.1128/MCB.00952-08.

Pagano, M., Pepperkok, R., Lukas, J., Baldin, V., Ansorge, W., Bartek, J., & Draetta, G. (1993). Regulation of the cell cycle by the cdk2 protein kinase in cultured human fibroblasts. *The Journal of Cell Biology*, *121*(1), 101–111.

Pagano, M., Pepperkok, R., Verde, F., Ansorge, W., & Draetta, G. (1992). Cyclin A is required at two points in the human cell cycle. *The EMBO Journal*, *11*(3), 961–971.

Palancade, B., & Bensaude, O. (2003). Investigating RNA polymerase II carboxyl-terminal domain (CTD) phosphorylation. *European Journal of Biochemistry*, *270*(19), 3859–3870.

Parisi, T., Beck, A. R., Rougier, N., McNeil, T., Lucian, L., Werb, Z., & Amati, B. (2003). Cyclins E1 and E2 are required for endoreplication in placental trophoblast giant cells. *The EMBO Journal*, *22*(18), 4794–4803. http://dx.doi.org/10.1093/emboj/cdg482.

Parisi, E., Filosa, S., De Petrocellis, B., & Monroy, A. (1978). The pattern of cell division in the early development of the sea urchin. Paracentrotus lividus. *Developmental Biology*, *65*(1), 38–49.

Patel, S. A., & Simon, M. C. (2010). Functional analysis of the Cdk7.cyclin H.Mat1 complex in mouse embryonic stem cells and embryos. *The Journal of Biological Chemistry*, *285*(20), 15587–15598. http://dx.doi.org/10.1074/jbc.M109.081687.

Pavletich, N. P. (1999). Mechanisms of cyclin-dependent kinase regulation: Structures of Cdks, their cyclin activators, and Cip and INK4 inhibitors. *Journal of Molecular Biology*, *287*(5), 821–828. http://dx.doi.org/10.1006/jmbi.1999.2640.

Peters, J. M. (2002). The anaphase-promoting complex: Proteolysis in mitosis and beyond. *Molecular Cell, 9*(5), 931–943.

Pfleger, C. M., Lee, E., & Kirschner, M. W. (2001). Substrate recognition by the Cdc20 and Cdh1 components of the anaphase-promoting complex. *Genes & Development, 15*(18), 2396–2407. http://dx.doi.org/10.1101/gad.918201.

Pines, J. (1999). Four-dimensional control of the cell cycle. *Nature Cell Biology, 1*(3), E73–E79. http://dx.doi.org/10.1038/11041.

Plusa, B., Frankenberg, S., Chalmers, A., Hadjantonakis, A. K., Moore, C. A., Papalopulu, N., … Zernicka-Goetz, M. (2005). Downregulation of Par3 and aPKC function directs cells towards the ICM in the preimplantation mouse embryo. *Journal of Cell Science, 118*(Pt. 3), 505–515. http://dx.doi.org/10.1242/jcs.01666.

Polyak, K., Lee, M. H., Erdjument-Bromage, H., Koff, A., Roberts, J. M., Tempst, P., & Massague, J. (1994). Cloning of p27Kip1, a cyclin-dependent kinase inhibitor and a potential mediator of extracellular antimitogenic signals. *Cell, 78*(1), 59–66.

Poon, R. Y., Yamashita, K., Adamczewski, J. P., Hunt, T., & Shuttleworth, J. (1993). The cdc2-related protein p40MO15 is the catalytic subunit of a protein kinase that can activate p33cdk2 and p34cdc2. *The EMBO Journal, 12*(8), 3123–3132.

Power, M. A., & Tam, P. P. (1993). Onset of gastrulation, morphogenesis and somitogenesis in mouse embryos displaying compensatory growth. *Anatomy and Embryology, 187*(5), 493–504.

Price, D. H. (2000). P-TEFb, a cyclin-dependent kinase controlling elongation by RNA polymerase II. *Molecular and Cellular Biology, 20*(8), 2629–2634.

Raff, J. W., & Glover, D. M. (1988). Nuclear and cytoplasmic mitotic cycles continue in Drosophila embryos in which DNA synthesis is inhibited with aphidicolin. *The Journal of Cell Biology, 107*(6 Pt. 1), 2009–2019.

Rane, S. G., Dubus, P., Mettus, R. V., Galbreath, E. J., Boden, G., Reddy, E. P., & Barbacid, M. (1999). Loss of Cdk4 expression causes insulin-deficient diabetes and Cdk4 activation results in beta-islet cell hyperplasia. *Nature Genetics, 22*(1), 44–52. http://dx.doi.org/10.1038/8751.

Ray, D., Terao, Y., Nimbalkar, D., Hirai, H., Osmundson, E. C., Zou, X., … Kiyokawa, H. (2007). Hemizygous disruption of Cdc25A inhibits cellular transformation and mammary tumorigenesis in mice. *Cancer Research, 67*(14), 6605–6611. http://dx.doi.org/10.1158/0008-5472.CAN-06-4815.

Reik, W., & Walter, J. (2001). Genomic imprinting: Parental influence on the genome. *Nature Reviews. Genetics, 2*(1), 21–32. http://dx.doi.org/10.1038/35047554.

Reimann, J. D., Freed, E., Hsu, J. Y., Kramer, E. R., Peters, J. M., & Jackson, P. K. (2001). Emi1 is a mitotic regulator that interacts with Cdc20 and inhibits the anaphase promoting complex. *Cell, 105*(5), 645–655.

Reimann, J. D., Gardner, B. E., Margottin-Goguet, F., & Jackson, P. K. (2001). Emi1 regulates the anaphase-promoting complex by a different mechanism than Mad2 proteins. *Genes & Development, 15*(24), 3278–3285. http://dx.doi.org/10.1101/gad.945701.

Rempel, R. E., Saenz-Robles, M. T., Storms, R., Morham, S., Ishida, S., Engel, A., … Nevins, J. R. (2000). Loss of E2F4 activity leads to abnormal development of multiple cellular lineages. *Molecular Cell, 6*(2), 293–306.

Ren, S., & Rollins, B. J. (2004). Cyclin C/cdk3 promotes Rb-dependent G0 exit. *Cell, 117*(2), 239–251.

Richardson, H. E., Wittenberg, C., Cross, F., & Reed, S. I. (1989). An essential G1 function for cyclin-like proteins in yeast. *Cell, 59*(6), 1127–1133.

Rieder, C. L., Cole, R. W., Khodjakov, A., & Sluder, G. (1995). The checkpoint delaying anaphase in response to chromosome monoorientation is mediated by an inhibitory signal produced by unattached kinetochores. *The Journal of Cell Biology, 130*(4), 941–948.

Rieder, C. L., Khodjakov, A., Paliulis, L. V., Fortier, T. M., Cole, R. W., & Sluder, G. (1997). Mitosis in vertebrate somatic cells with two spindles: Implications for the metaphase/anaphase transition checkpoint and cleavage. *Proceedings of the National Academy of Sciences of the United States of America, 94*(10), 5107–5112.

Roshak, A. K., Capper, E. A., Imburgia, C., Fornwald, J., Scott, G., & Marshall, L. A. (2000). The human polo-like kinase, PLK, regulates cdc2/cyclin B through phosphorylation and activation of the cdc25C phosphatase. *Cellular Signalling, 12*(6), 405–411.

Rossi, D. J., Londesborough, A., Korsisaari, N., Pihlak, A., Lehtonen, E., Henkemeyer, M., & Makela, T. P. (2001). Inability to enter S phase and defective RNA polymerase II CTD phosphorylation in mice lacking Mat1. *The EMBO Journal, 20*(11), 2844–2856. http://dx.doi.org/10.1093/emboj/20.11.2844.

Russo, A. A., Jeffrey, P. D., & Pavletich, N. P. (1996). Structural basis of cyclin-dependent kinase activation by phosphorylation. *Nature Structural Biology, 3*(8), 696–700.

Russo, A. A., Tong, L., Lee, J. O., Jeffrey, P. D., & Pavletich, N. P. (1998). Structural basis for inhibition of the cyclin-dependent kinase Cdk6 by the tumour suppressor p16INK4a. *Nature, 395*(6699), 237–243. http://dx.doi.org/10.1038/26155.

Ruzankina, Y., Pinzon-Guzman, C., Asare, A., Ong, T., Pontano, L., Cotsarelis, G., … Brown, E. J. (2007). Deletion of the developmentally essential gene ATR in adult mice leads to age-related phenotypes and stem cell loss. *Cell Stem Cell, 1*(1), 113–126. http://dx.doi.org/10.1016/j.stem.2007.03.002.

Sacristan, C., & Kops, G. J. (2015). Joined at the hip: Kinetochores, microtubules, and spindle assembly checkpoint signaling. *Trends in Cell Biology, 25*(1), 21–28. http://dx.doi.org/10.1016/j.tcb.2014.08.006.

Sage, J. (2004). Cyclin C makes an entry into the cell cycle. *Developmental Cell, 6*(5), 607–608.

Santamaria, D., Barriere, C., Cerqueira, A., Hunt, S., Tardy, C., Newton, K., … Barbacid, M. (2007). Cdk1 is sufficient to drive the mammalian cell cycle. *Nature, 448*(7155), 811–815. http://dx.doi.org/10.1038/nature06046.

Santos, F., Hendrich, B., Reik, W., & Dean, W. (2002). Dynamic reprogramming of DNA methylation in the early mouse embryo. *Developmental Biology, 241*(1), 172–182. http://dx.doi.org/10.1006/dbio.2001.0501.

Sathananthan, A. H. (1997). Mitosis in the human embryo: The vital role of the sperm centrosome (centriole). *Histology and Histopathology, 12*(3), 827–856.

Satyanarayana, A., Hilton, M. B., & Kaldis, P. (2008). p21 Inhibits Cdk1 in the absence of Cdk2 to maintain the G1/S phase DNA damage checkpoint. *Molecular Biology of the Cell, 19*(1), 65–77. http://dx.doi.org/10.1091/mbc.E07-06-0525.

Satyanarayana, A., & Kaldis, P. (2009). Mammalian cell-cycle regulation: Several Cdks, numerous cyclins and diverse compensatory mechanisms. *Oncogene, 28*(33), 2925–2939. http://dx.doi.org/10.1038/onc.2009.170.

Sawicki, W., Abramczuk, J., & Blaton, O. (1978). DNA synthesis in the second and third cell cycles of mouse preimplantation development. A cytophotometric study. *Experimental Cell Research, 112*(1), 199–205.

Serizawa, H., Makela, T. P., Conaway, J. W., Conaway, R. C., Weinberg, R. A., & Young, R. A. (1995). Association of Cdk-activating kinase subunits with transcription factor TFIIH. *Nature, 374*(6519), 280–282. http://dx.doi.org/10.1038/374280a0.

Serrano, M., Hannon, G. J., & Beach, D. (1993). A new regulatory motif in cell-cycle control causing specific inhibition of cyclin D/CDK4. *Nature, 366*(6456), 704–707. http://dx.doi.org/10.1038/366704a0.

Shaltiel, I. A., Krenning, L., Bruinsma, W., & Medema, R. H. (2015). The same, only different—DNA damage checkpoints and their reversal throughout the cell cycle. *Journal of Cell Science, 128*(4), 607–620. http://dx.doi.org/10.1242/jcs.163766.

Sherr, C. J., & Roberts, J. M. (1995). Inhibitors of mammalian G1 cyclin-dependent kinases. *Genes & Development*, *9*(10), 1149–1163.

Sherr, C. J., & Roberts, J. M. (2004). Living with or without cyclins and cyclin-dependent kinases. *Genes & Development*, *18*(22), 2699–2711. http://dx.doi.org/10.1101/gad.1256504.

Shiekhattar, R., Mermelstein, F., Fisher, R. P., Drapkin, R., Dynlacht, B., Wessling, H. C., ... Reinberg, D. (1995). Cdk-activating kinase complex is a component of human transcription factor TFIIH. *Nature*, *374*(6519), 283–287. http://dx.doi.org/10.1038/374283a0.

Shim, E. Y., Walker, A. K., Shi, Y., & Blackwell, T. K. (2002). CDK-9/cyclin T (P-TEFb) is required in two postinitiation pathways for transcription in the C. elegans embryo. *Genes & Development*, *16*(16), 2135–2146. http://dx.doi.org/10.1101/gad.999002.

Sicinska, E., Aifantis, I., Le Cam, L., Swat, W., Borowski, C., Yu, Q., ... Sicinski, P. (2003). Requirement for cyclin D3 in lymphocyte development and T cell leukemias. *Cancer Cell*, *4*(6), 451–461.

Sicinski, P., Donaher, J. L., Geng, Y., Parker, S. B., Gardner, H., Park, M. Y., ... Weinberg, R. A. (1996). Cyclin D2 is an FSH-responsive gene involved in gonadal cell proliferation and oncogenesis. *Nature*, *384*(6608), 470–474. http://dx.doi.org/10.1038/384470a0.

Sicinski, P., Donaher, J. L., Parker, S. B., Li, T., Fazeli, A., Gardner, H., ... Weinberg, R. A. (1995). Cyclin D1 provides a link between development and oncogenesis in the retina and breast. *Cell*, *82*(4), 621–630.

Siracusa, G., Coletta, M., & Monesi, V. (1975). Duplication of DNA during the first cell cycle in the mouse embryo. *Journal of Reproduction and Fertility*, *42*(2), 395–398.

Slingerland, J. M., Hengst, L., Pan, C. H., Alexander, D., Stampfer, M. R., & Reed, S. I. (1994). A novel inhibitor of cyclin-Cdk activity detected in transforming growth factor beta-arrested epithelial cells. *Molecular and Cellular Biology*, *14*(6), 3683–3694.

Smith, R. K., & Johnson, M. H. (1986). Analysis of the third and fourth cell cycles of mouse early development. *Journal of Reproduction and Fertility*, *76*(1), 393–399.

Smith, J., Tho, L. M., Xu, N., & Gillespie, D. A. (2010). The ATM-Chk2 and ATR-Chk1 pathways in DNA damage signaling and cancer. *Advances in Cancer Research*, *108*, 73–112. http://dx.doi.org/10.1016/B978-0-12-380888-2.00003-0.

Snow, M. L. H. (1977). Gastrulation in the mouse: Growth and regionalization of the epiblast. *Development*, *42*, 293–303.

Solomon, M. J., Harper, J. W., & Shuttleworth, J. (1993). CAK, the p34cdc2 activating kinase, contains a protein identical or closely related to p40MO15. *The EMBO Journal*, *12*(8), 3133–3142.

Solter, D., Skreb, N., & Damjanov, I. (1971). Cell cycle analysis in the mouse EGG-cylinder. *Experimental Cell Research*, *64*(2), 331–334.

Spindle, A., Nagano, H., & Pedersen, R. A. (1985). Inhibition of DNA replication in preimplantation mouse embryos by aphidicolin. *The Journal of Experimental Zoology*, *235*(2), 289–295. http://dx.doi.org/10.1002/jez.1402350216.

Stead, E., White, J., Faast, R., Conn, S., Goldstone, S., Rathjen, J., ... Dalton, S. (2002). Pluripotent cell division cycles are driven by ectopic Cdk2, cyclin A/E and E2F activities. *Oncogene*, *21*(54), 8320–8333. http://dx.doi.org/10.1038/sj.onc.1206015.

Storre, J., Elsasser, H. P., Fuchs, M., Ullmann, D., Livingston, D. M., & Gaubatz, S. (2002). Homeotic transformations of the axial skeleton that accompany a targeted deletion of E2f6. *EMBO Reports*, *3*(7), 695–700. http://dx.doi.org/10.1093/embo-reports/kvf141.

Strausfeld, U. P., Howell, M., Descombes, P., Chevalier, S., Rempel, R. E., Adamczewski, J., ... Blow, J. J. (1996). Both cyclin A and cyclin E have S-phase promoting (SPF) activity in Xenopus egg extracts. *Journal of Cell Science*, *109*(Pt. 6), 1555–1563.

Streffer, C., van Beuningen, D., Molls, M., Zamboglou, N., & Schulz, S. (1980). Kinetics of cell proliferation in the pre-implanted mouse embryo in vivo and in vitro. *Cell and Tissue Kinetics*, *13*(2), 135–143.

Sudakin, V., Chan, G. K., & Yen, T. J. (2001). Checkpoint inhibition of the APC/C in HeLa cells is mediated by a complex of BUBR1, BUB3, CDC20, and MAD2. *The Journal of Cell Biology*, *154*(5), 925–936. http://dx.doi.org/10.1083/jcb.200102093.

Sweeney, K. J., Sarcevic, B., Sutherland, R. L., & Musgrove, E. A. (1997). Cyclin D2 activates Cdk2 in preference to Cdk4 in human breast epithelial cells. *Oncogene*, *14*(11), 1329–1340. http://dx.doi.org/10.1038/sj.onc.1200951.

Takai, H., Naka, K., Okada, Y., Watanabe, M., Harada, N., Saito, S., … Motoyama, N. (2002). Chk2-deficient mice exhibit radioresistance and defective p53-mediated transcription. *The EMBO Journal*, *21*(19), 5195–5205.

Takaki, T., Echalier, A., Brown, N. R., Hunt, T., Endicott, J. A., & Noble, M. E. (2009). The structure of CDK4/cyclin D3 has implications for models of CDK activation. *Proceedings of the National Academy of Sciences of the United States of America*, *106*(11), 4171–4176. http://dx.doi.org/10.1073/pnas.0809674106.

Tarkowski, A. K., & Wroblewska, J. (1967). Development of blastomeres of mouse eggs isolated at the 4- and 8-cell stage. *Journal of Embryology and Experimental Morphology*, *18*(1), 155–180.

Tassan, J. P., Jaquenoud, M., Fry, A. M., Frutiger, S., Hughes, G. J., & Nigg, E. A. (1995). In vitro assembly of a functional human CDK7-cyclin H complex requires MAT1, a novel 36 kDa RING finger protein. *The EMBO Journal*, *14*(22), 5608–5617.

Taylor, W. R., & Stark, G. R. (2001). Regulation of the G2/M transition by p53. *Oncogene*, *20*(15), 1803–1815. http://dx.doi.org/10.1038/sj.onc.1204252.

Tetsu, O., & McCormick, F. (2003). Proliferation of cancer cells despite CDK2 inhibition. *Cancer Cell*, *3*(3), 233–245.

Thomas, F. C., Sheth, B., Eckert, J. J., Bazzoni, G., Dejana, E., & Fleming, T. P. (2004). Contribution of JAM-1 to epithelial differentiation and tight-junction biogenesis in the mouse preimplantation embryo. *Journal of Cell Science*, *117*(Pt. 23), 5599–5608. http://dx.doi.org/10.1242/jcs.01424.

Tominaga, Y., Li, C., Wang, R. H., & Deng, C. X. (2006). Murine Wee1 plays a critical role in cell cycle regulation and pre-implantation stages of embryonic development. *International Journal of Biological Sciences*, *2*(4), 161–170.

Tong, Z. B., Gold, L., Pfeifer, K. E., Dorward, H., Lee, E., Bondy, C. A., … Nelson, L. M. (2000). Mater, a maternal effect gene required for early embryonic development in mice. *Nature Genetics*, *26*(3), 267–268. http://dx.doi.org/10.1038/81547.

Toyoshima, H., & Hunter, T. (1994). p27, a novel inhibitor of G1 cyclin-Cdk protein kinase activity, is related to p21. *Cell*, *78*(1), 67–74.

Toyoshima-Morimoto, F., Taniguchi, E., Shinya, N., Iwamatsu, A., & Nishida, E. (2001). Polo-like kinase 1 phosphorylates cyclin B1 and targets it to the nucleus during prophase. *Nature*, *410*(6825), 215–220. http://dx.doi.org/10.1038/35065617.

Trimarchi, J. M., Fairchild, B., Wen, J., & Lees, J. A. (2001). The E2F6 transcription factor is a component of the mammalian Bmi1-containing polycomb complex. *Proceedings of the National Academy of Sciences of the United States of America*, *98*(4), 1519–1524. http://dx.doi.org/10.1073/pnas.041597698.

Trimarchi, J. M., & Lees, J. A. (2002). Sibling rivalry in the E2F family. *Nature Reviews. Molecular Cell Biology*, *3*(1), 11–20. http://dx.doi.org/10.1038/nrm714.

Tsai, L. H., Lees, E., Faha, B., Harlow, E., & Riabowol, K. (1993). The cdk2 kinase is required for the G1-to-S transition in mammalian cells. *Oncogene*, *8*(6), 1593–1602.

Tsutsui, T., Hesabi, B., Moons, D. S., Pandolfi, P. P., Hansel, K. S., Koff, A., & Kiyokawa, H. (1999). Targeted disruption of CDK4 delays cell cycle entry with enhanced p27(Kip1) activity. *Molecular and Cellular Biology*, *19*(10), 7011–7019.

Usheva, A., Maldonado, E., Goldring, A., Lu, H., Houbavi, C., Reinberg, D., & Aloni, Y. (1992). Specific interaction between the nonphosphorylated form of RNA polymerase II and the TATA-binding protein. *Cell*, *69*(5), 871–881.
van de Weerdt, B. C., & Medema, R. H. (2006). Polo-like kinases: A team in control of the division. *Cell Cycle*, *5*(8), 853–864.
van den Heuvel, S., & Harlow, E. (1993). Distinct roles for cyclin-dependent kinases in cell cycle control. *Science*, *262*(5142), 2050–2054.
van der Meer, T., Chan, W. Y., Palazon, L. S., Nieduszynski, C., Murphy, M., Sobczak-Thepot, J., ... Colledge, W. H. (2004). Cyclin A1 protein shows haplo-insufficiency for normal fertility in male mice. *Reproduction*, *127*(4), 503–511. http://dx.doi.org/10.1530/rep.1.00131.
Villa-Moruzzi, E. (1993). Activation of the cdc25C phosphatase in mitotic HeLa cells. *Biochemical and Biophysical Research Communications*, *196*(3), 1248–1254.
Wang, H., & Dey, S. K. (2006). Roadmap to embryo implantation: Clues from mouse models. *Nature Reviews. Genetics*, 7(3), 185–199. http://dx.doi.org/10.1038/nrg1808.
Wang, Q., Liu, T., Fang, Y., Xie, S., Huang, X., Mahmood, R., ... Dai, W. (2004). BUBR1 deficiency results in abnormal megakaryopoiesis. *Blood*, *103*(4), 1278–1285. http://dx.doi.org/10.1182/blood-2003-06-2158.
Wang, Q. T., Piotrowska, K., Ciemerych, M. A., Milenkovic, L., Scott, M. P., Davis, R. W., & Zernicka-Goetz, M. (2004). A genome-wide study of gene activity reveals developmental signaling pathways in the preimplantation mouse embryo. *Developmental Cell*, *6*(1), 133–144.
Watanabe, N., Arai, H., Nishihara, Y., Taniguchi, M., Watanabe, N., Hunter, T., & Osada, H. (2004). M-phase kinases induce phospho-dependent ubiquitination of somatic Wee1 by SCFbeta-TrCP. *Proceedings of the National Academy of Sciences of the United States of America*, *101*(13), 4419–4424. http://dx.doi.org/10.1073/pnas.0307700101.
Westerling, T., Kuuluvainen, E., & Makela, T. P. (2007). Cdk8 is essential for preimplantation mouse development. *Molecular and Cellular Biology*, *27*(17), 6177–6182. http://dx.doi.org/10.1128/MCB.01302-06.
Winston, N., Bourgain-Guglielmetti, F., Ciemerych, M. A., Kubiak, J. Z., Senamaud-Beaufort, C., Carrington, M., ... Sobczak-Thepot, J. (2000). Early development of mouse embryos null mutant for the cyclin A2 gene occurs in the absence of maternally derived cyclin A2 gene products. *Developmental Biology*, *223*(1), 139–153. http://dx.doi.org/10.1006/dbio.2000.9721.
Wright, S. J., & Longo, F. J. (1988). Sperm nuclear enlargement in fertilized hamster eggs is related to meiotic maturation of the maternal chromatin. *The Journal of Experimental Zoology*, *247*(2), 155–165. http://dx.doi.org/10.1002/jez.1402470207.
Wu, L., Timmers, C., Maiti, B., Saavedra, H. I., Sang, L., Chong, G. T., ... Leone, G. (2001). The E2F1-3 transcription factors are essential for cellular proliferation. *Nature*, *414*(6862), 457–462. http://dx.doi.org/10.1038/35106593.
Xie, Y., Sun, T., Wang, Q. T., Wang, Y., Wang, F., Puscheck, E., & Rappolee, D. A. (2005). Acquisition of essential somatic cell cycle regulatory protein expression and implied activity occurs at the second to third cell division in mouse preimplantation embryos. *FEBS Letters*, *579*(2), 398–408. http://dx.doi.org/10.1016/j.febslet.2004.10.109.
Xiong, Y., Zhang, H., & Beach, D. (1992). D type cyclins associate with multiple protein kinases and the DNA replication and repair factor PCNA. *Cell*, *71*(3), 505–514.
Xu, Y., Ashley, T., Brainerd, E. E., Bronson, R. T., Meyn, M. S., & Baltimore, D. (1996). Targeted disruption of ATM leads to growth retardation, chromosomal fragmentation during meiosis, immune defects, and thymic lymphoma. *Genes & Development*, *10*(19), 2411–2422.

Yamamoto, K., Wang, Y., Jiang, W., Liu, X., Dubois, R. L., Lin, C. S., ... Zha, S. (2012). Kinase-dead ATM protein causes genomic instability and early embryonic lethality in mice. *The Journal of Cell Biology*, *198*(3), 305–313. http://dx.doi.org/10.1083/jcb.201204098.

Yankulov, K. Y., & Bentley, D. L. (1997). Regulation of CDK7 substrate specificity by MAT1 and TFIIH. *The EMBO Journal*, *16*(7), 1638–1646. http://dx.doi.org/10.1093/emboj/16.7.1638.

Ye, X., Zhu, C., & Harper, J. W. (2001). A premature-termination mutation in the Mus musculus cyclin-dependent kinase 3 gene. *Proceedings of the National Academy of Sciences of the United States of America*, *98*(4), 1682–1686. http://dx.doi.org/10.1073/pnas.041596198.

Zhang, Y., & Chellappan, S. P. (1995). Cloning and characterization of human DP2, a novel dimerization partner of E2F. *Oncogene*, *10*(11), 2085–2093.

Zhao, H., Watkins, J. L., & Piwnica-Worms, H. (2002). Disruption of the checkpoint kinase 1/cell division cycle 25A pathway abrogates ionizing radiation-induced S and G2 checkpoints. *Proceedings of the National Academy of Sciences of the United States of America*, *99*(23), 14795–14800. http://dx.doi.org/10.1073/pnas.182557299.

Zhou, B. B., & Elledge, S. J. (2000). The DNA damage response: Putting checkpoints in perspective. *Nature*, *408*(6811), 433–439. http://dx.doi.org/10.1038/35044005.

Zindy, F., Lamas, E., Chenivesse, X., Sobczak, J., Wang, J., Fesquet, D., ... Brechot, C. (1992). Cyclin A is required in S phase in normal epithelial cells. *Biochemical and Biophysical Research Communications*, *182*(3), 1144–1154.

CHAPTER TWO

Genome Duplication at the Beginning of Mammalian Development

M.L. DePamphilis[1]

Eunice Kennedy Shriver National Institute of Child Health and Human Development, National Institutes of Health, Bethesda, MD United States

[1]Corresponding author: e-mail address: depamphm@mail.nih.gov

Contents

Current Topics in Developmental Biology, Volume 120
ISSN 0070-2153
http://dx.doi.org/10.1016/bs.ctdb.2016.04.003

Abstract

Nothing is more fundamental to mammalian development than the ability to accurately reproduce its genome once—but only once—each time a cell divides. In fact, the basic mechanism for replicating DNA has been conserved throughout evolution, even though the magnitude of the problem became monumental. A human cell contains 670 times the DNA in an *E. coli* cell, and human development requires trillions of cell divisions that produce about 37 billion miles of DNA! But instead of increasing the speed of replication forks to compensate for increasing genome size and organism complexity, evolution simply increased the number of replication origins. This allowed mammalian development regulate initiation of DNA replication during cell proliferation without interfering with the ever-changing demands of gene expression during cell differentiation. Moreover, it allowed developing tissues to complete genome duplication before beginning mitosis and to restrict genome duplication to once per cell division. And yet, to overproduce gene products during development, some cells are allowed to differentiate into nonproliferating polyploid cells. This chapter summarizes the mechanisms that make these events possible. Ironically, aberrations in these mechanisms are linked to cancer. In fact, the pluripotent cells produced during preimplantation development not only share characteristics of cancer cells, but they can also initiate cancer.

1. BASIC CONCEPTS

1.1 Genome Duplication Is a Problem of Astronomical Proportions

A single diploid human cell contains approximately 2.09 m of DNA (6.16 billion base pairs, 6.3 picograms) that encodes 99.9995% of the genome. The remaining 0.0005% is encoded in mitochondrial DNA, which is maternally inherited, replicates independently of nuclear DNA replication, and varies widely in copy number among cell types. Nuclear DNA replication, on the other hand, is stringently regulated to ensure that genome duplication always precedes mitosis, and that it is restricted to providing each sibling cell with exactly one copy of the genome except when developmentally programmed to produce polyploid cells. This means that development of a fertilized human egg into an adult of 75 kg requires producing 29 trillion cells or about 60 trillion meters of DNA. Do the math. That is enough DNA to stretch from the Earth to the Sun and back 200 times! Include the approximately 1 trillion hematopoietic cells produced per day in adults together with the turnover in epithelial cell populations, and the number of cell divisions in a human life span is truly astronomical!

How is this possible? For a comprehensive description of the concepts and mechanisms of DNA replication in the three domains of life (bacteria, archaea, and eukarya), their evolution, and the importance of DNA replication in human disease consult the text book "Genome Duplication" (DePamphilis & Bell, 2010). For more detailed information, consult the Cold Spring Harbor series on DNA replication (Bell, Méchali, & DePamphilis, 2013; DePamphilis, 1996a, 2006). What you will discover is that all living things encode their genomes in DNA, and they all use the "replication fork" paradigm.

1.2 All Living Organisms Use the Replication Fork Paradigm

The mechanism by which DNA sequence is reproduced is conserved throughout all three domains of life (bacteria, archaea, and eukarya), in principle, if not in detail. DNA replication occurs by a semiconservative mechanism termed the "replication fork," in which each of the two strands of the double helix acts as a template on which a new complementary strand is synthesized. Thus, each of the two sibling cells receives one new and one old DNA strand. The essential events at replication forks are well established (Fig. 1). The first step in DNA replication is to separate the two DNA templates by unwinding the DNA. The second step is initiation of RNA-primed DNA synthesis on each strand. Since none of the DNA polymerases can initiate DNA synthesis de novo; they can only extend the 3′-end of either an RNA or DNA strand. Therefore, all living organisms employ a DNA primase to synthesize a short oligoribonucleotide that serves as a primer for the replicative DNA polymerase. Notably, these events occur in opposite directions because the two strands of DNA have opposite polarity (3′→5′:5′←3′) and all RNA and DNA polymerases synthesize either RNA or DNA only in one direction (5′→3′). Therefore, RNA-primed DNA synthesis initiated at the replication origin can continue without interruption only on one template (termed the leading strand template whose polarity is 3′→5′). DNA synthesis on the complementary template of each replication fork is, by necessity, a discontinuous process that occurs by the repeated initiation of short RNA-primed nascent DNA strands termed

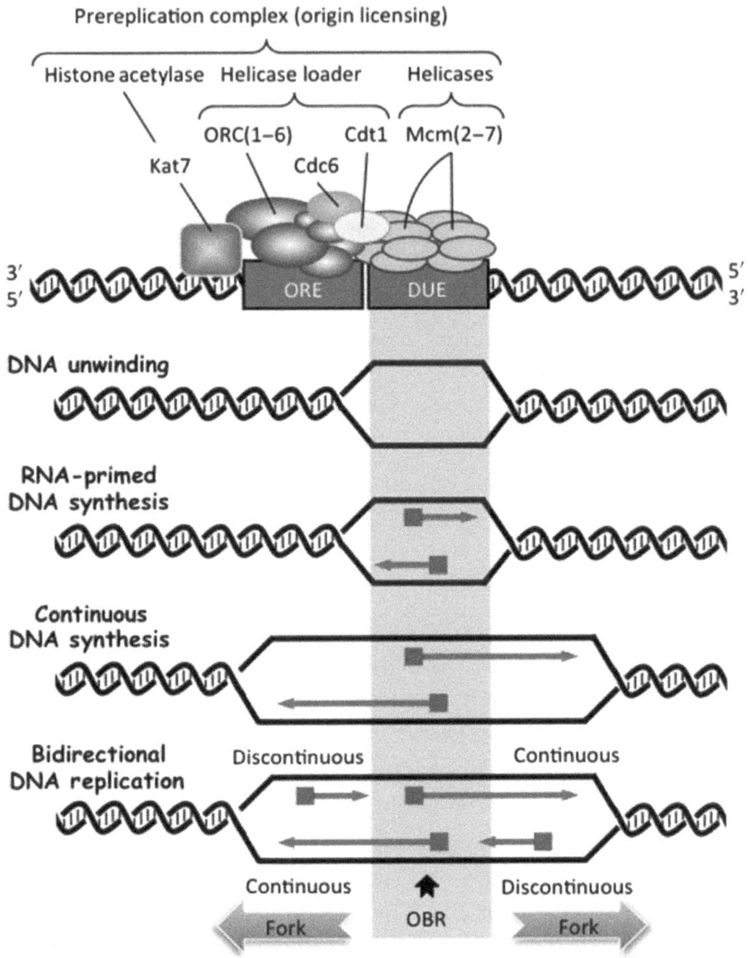

Fig. 1 Initiation of bidirectional DNA replication occurs at sites within double-stranded DNA. DNA replicators and replication origins contain at least two essential components: an origin recognition element (ORE) to which one or more initiator proteins bind, and a DNA unwinding element (DUE) consisting of an easily unwound DNA sequence. RNA primers are indicated as *green* (*gray* in the print version) *squares*, and nascent DNA chains as *red* (*gray* in the print version) *arrows*. DNA synthesis occurs continuously on the "leading strand template" of each replication fork and discontinuously on the "lagging strand template." The transition between continuous and discontinuous DNA synthesis on each template marks the origin of bidirectional replication (OBR). The recognition complex (ORC(1–6)) together with the Cdc6 protein and Cdt1 binds to and thereby "licenses" that sequence as a replication origin. This constitutes the helicase loader that recruits the eukaryotic replicative DNA helicase Mcm(2–7) (MCM helicase). Loading of two MCM helicases as head-to-head double hexamers during origin licensing is crucial for ensuring once per cell cycle DNA replication in eukaryotic cells. The histone acetyltransferase, Kat7/Hbo1, is required only when DNA is organized into chromatin.

"Okazaki fragments," after their discoverer Reiji Okazaki. When the 3′-end of the growing Okazaki fragment encounters the 5′-end of the downstream RNA-primed nascent DNA strand, the RNA primer is excised and the Okazaki fragment is ligated to the 5′-end of the long nascent DNA strand. Thus, DNA synthesis occurs continuously on the "leading strand template" of a replication fork and discontinuously on the "lagging strand template."

To be sure, other strategies for genome duplication exist, but they are confined to bacteriophage, bacterial plasmids, animal and plant viruses, and mitochondria. The essential point is that during metazoan development, the mechanism for replicating nuclear DNA is the same from zygote to adult. In fact, the sequence of events at replication forks is the same in all living cells, and the proteins that execute these events in metazoa are orthologs or analogs of the proteins used in archaea and bacteria.

1.3 Nuclear DNA Replication Begins in the Zygote

Mammalian oocytes are designed for long-term storage of nuclear DNA. They have the ability to repair DNA damage, but they cannot initiate replication either of their own nuclear DNA or of DNA plasmids injected into their nucleus (Wirak et al., 1985). Nuclear DNA replication begins asynchronously in the two pronuclei of the 1-cell embryo (Ferreira & Carmo-Fonseca, 1997). Specific chromatin regions in preimplantation mouse embryos are replicated at different times during S phase, revealing that mechanisms controlling the temporal and spatial replication of DNA are present in the haploid pronuclei of the mammalian zygote. Mitochondrial DNA replication is not governed by the mitotic cell division cycle. Variations in mouse mitochondrial DNA copy number from fertilization to birth are linked to changes in the cell's energy requirements (Aiken, Cindrova-Davies, & Johnson, 2008).

As with cells derived from adult mammalian tissues, sequence-dependent DNA replication also occurs in preimplantation mouse embryos (Wirak et al., 1985). Although the circular, double-stranded DNA molecules injected into the nuclei of mouse 1-cell or 2-cell embryos were stable, they did not replicate unless they contained a polyomavirus DNA replication origin and were provided with the origin's cognate initiation protein, the polyomavirus DNA helicase. These results reveal that initiation of DNA replication in mammalian preimplantation embryos, like cells from adults, requires particular DNA sequences, topologies, or chromatin.

1.4 The Same Replication Machinery Is Used from Zygote to Adult

Gene expression profiles for mouse oocytes and preimplantation embryos reveal that all of the genes required for nuclear DNA replication in cells derived from embryos or adults are expressed in the zygote (National Center for Biotechnology Information (NCBI), Gene Expression Omnibus (GEO) profiles, MOE430A,B). Since the DNA chips were normalized to an average expression value in order to reveal the relative abundance of RNAs, an increase between the oocyte and 1-cell stage represents a maternal mRNA recruited during maturation (Aiba, Carter, Matoba, & Ko, 2006; Pan, O'Brien, Wigglesworth, Eppig, & Schultz, 2005; Zeng & Schultz, 2005). The ability to replicate DNA is acquired during meiotic maturation, which ensures that DNA replication does not occur until after fertilization. In *Xenopus*, the only component missing for DNA replication competence is Cdc6, which is synthesized from a dormant maternal mRNA recruited during oocyte maturation (Lemaitre, Bocquet, & Mechali, 2002). In mice, both Cdc6 and Orc6 are absent in oocytes, but expressed in zygotes (Lemaitre et al., 2004; Murai, Stein, Buffone, Yamashita, & Schultz, 2010).

1.5 Not All Replication Origins Are Replicators

DNA replication is not initiated randomly throughout the genome, but at specific sites termed replication origins. Replication origins are the sites where DNA replication begins. Since most initiation events result in two replication forks that travel in opposite directions, replication origins are also termed "origins of bidirectional replication" (OBR). In bacteria, bacterial plasmids, bacteriophage, mitochondria, DNA animal viruses, and DNA plant viruses, the nucleotide sequence at the replication origin is essential for DNA unwinding to begin. Therefore, these replication origins are also termed as "replicators" because they are essential for initiation of DNA replication. However, replication origins in multicellular eukarya (the metazoa) are clearly not essential for initiation of DNA replication, because eggs and egg extracts from flies, fish, sea urchins, and amphibians can initiate DNA replication on virtually any DNA sequence. Nevertheless, as development proceeds, some DNA sequences are clearly preferred over others, because initiation events can be mapped to specific sites along the chromosome, the same sites appear in different cells of the same organism, and the same sites are used in consecutive cell division cycles. Archaea appear to have replicators similar to bacteria (Jalasvuori & Koonin, 2015), and single-cell eukarya appear to have replication origins similar to the metazoa (Leonard & Mechali, 2013).

2. REPLICATION ORIGINS ARE ESTABLISHED IN TWO STEPS

Initiation of nuclear DNA replication begins when a prereplication complex (preRC) is assembled at sites distributed throughout the nuclear genome. This event, termed "origin licensing" (Chevalier & Blow, 1996), occurs as cells transit from anaphase in mitosis to G1 phase of the mitotic cell division cycle (Fig. 2) (Baldinger & Gossen, 2009; Noguchi, Vassilev, Ghosh, Yates, & DePamphilis, 2006). Thus, preRC assembly also is referred to as "origin licensing" because it permits replication origins to initiate DNA synthesis. However, replicative DNA helicases in preRCs are not active until additional proteins convert the preRCs into preinitiation complexes (preICs), an event that occurs during the G1 to S phase transition. The purpose of preIC assembly is to activate the replicative DNA helicases, which then unwind sufficient DNA to allow loading of DNA polymerase-α·primase, the enzyme complex that initiates RNA-primed DNA synthesis on each template.

This two-step process allows mammalian cells to change the number and chromosomal locations of active replication origins each time a cell divides. Although many potential sites are licensed for DNA replication during the M to G1 phase transition, only a comparatively few are activated during S phase, as predicted by the "Jesuit Model" for origin selection (described later). Moreover, it allows mammals to regulate the number and locations of active replication origins during cell proliferation in order not to interfere with the ever-changing demands of gene expression during cell differentiation.

2.1 Prereplication Complex Assembly (Origin Licensing)

In vivo, the Kat7/Hbo1 histone acetylase activity is essential for origin licensing (Fig. 1) because assembly of prereplication complexes is facilitated by histone acetylation (Miotto & Struhl, 2010; Unnikrishnan, Gafken, & Tsukiyama, 2010). In the absence of a repressive chromatin structure, preRC assembly occurs by sequential addition of proteins onto DNA in a pathway driven by ATP hydrolysis. Initiation of nuclear DNA replication begins with assembly of the heterotypic six subunit "origin recognition complex" (ORC(1–6)) onto DNA (Fig. 3). This event occurs through

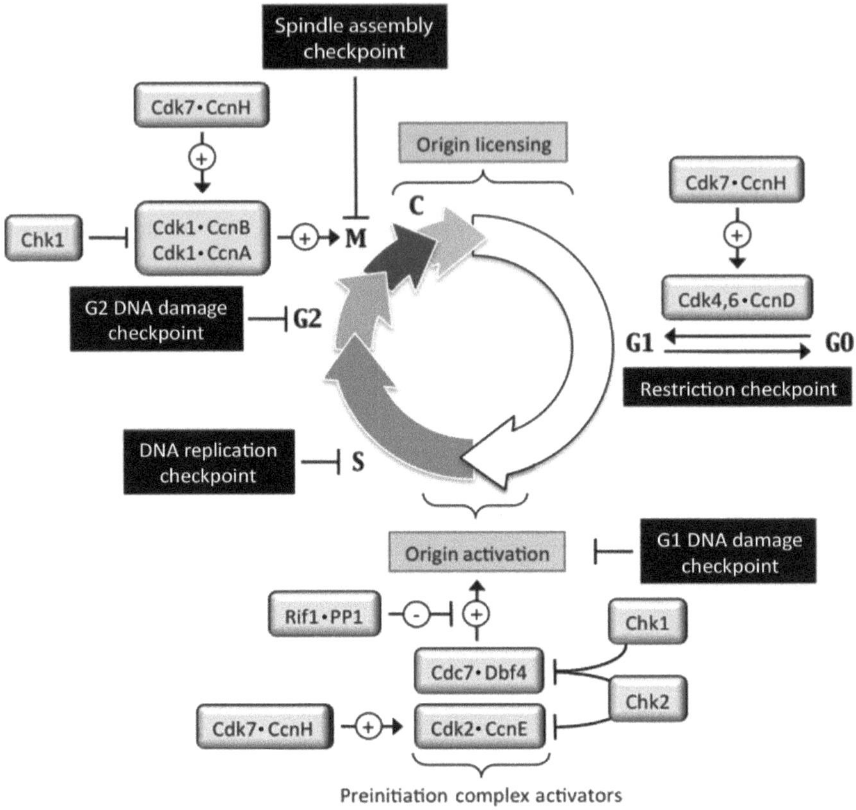

Fig. 2 The mitotic cell division cycle insures that mitosis follows nuclear DNA replication. Replication origins are licensed to initiate replication by preRC assembly (Fig. 3) during the anaphase to G1 phase transition and activated during the G1/S phase transition by converting the preRC into a preIC (Fig. 4). Replication forks (Figs. 1 and 5) duplicate nuclear DNA during S phase. These events are regulated directly by eight protein kinases (Depamphilis, de Renty, Ullah, & Lee, 2012), five of which are cyclin-dependent kinases (Cdk) that require a cyclin (Ccn) partner. Cdk7·CcnH is essential to activate Cdk1, Cdk2, Cdk4, and Cdk6. Cdk4·CcnD and Cdk6·CcnD drive cells from a quiescent state (G0) into a proliferative state. Cdk1·CcnB1 and Cdk1·CcnA2 initiate and maintain mitosis. Cdk2·CcnE and the Dbf4-dependent kinase Cdc7 are the engines that drive the assembly and activation of preinitiation complexes. Cdk2·CcnA2 prevents DNA rereplication during S phase. Cdk1·CcnA2 prevents premature initiation during G2 phase. Rif1 is a novel PP1 substrate targeting subunit that counteracts DDK-mediated phosphorylation during replication (Hiraga et al., 2014). The remaining two protein kinases are components of the cell's DNA damage response. Five cell cycle checkpoints can arrest progress if nutrient conditions are inadequate (restriction point), if DNA damage is detected, or if the metaphase spindle is not assembled correctly (DePamphilis & Bell, 2010).

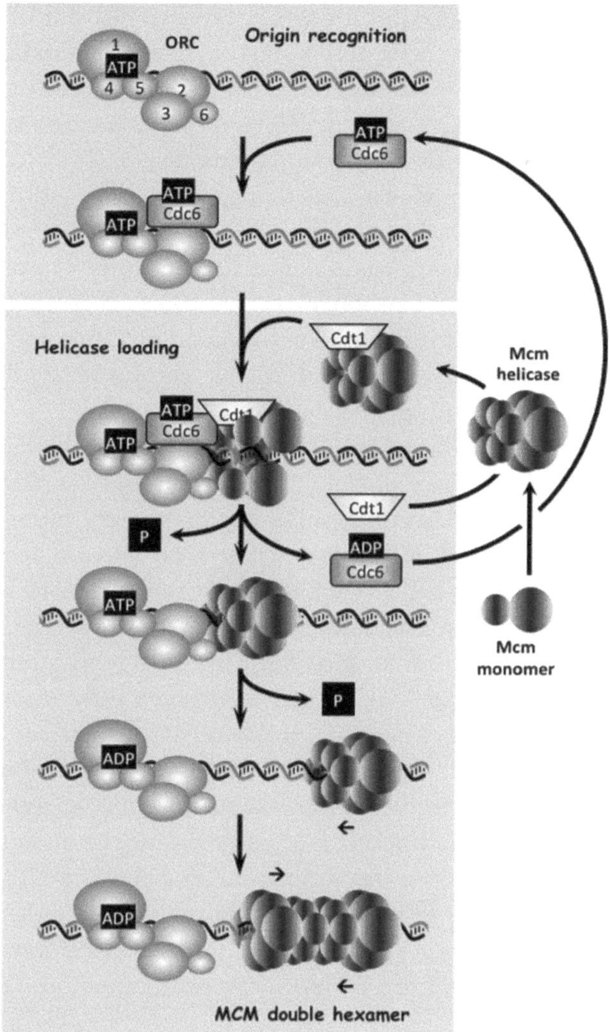

Fig. 3 Assembly of prereplication complexes (preRCs) begins during mitotic exit. PreRC assembly begins with the binding of the six subunit origin recognition complex (ORC) to DNA. ORC contains adenosine triphosphate (ATP) bound to subunits 1, 4, and 5. This event allows Cdc6·ATP to form a stable complex with the ORC·DNA site. Cdt1 can then chaperone the six subunit MCM helicase then to the DNA·ORC·Cdc6·ATP site. Cdc6·ADP is released with concomitant loading of the MCM helicase onto the DNA. It is the interaction of Cdt1·MCM with a DNA·ORC·Cdc6·ATP site that causes the MCM hexamer to open and expose its central channel. DNA then associates with the central channel within the six Mcm subunits and activates the Cdc6 ATPase. Hydrolysis of the Cdc6-bound ATP releases Cdt1 and Cdc6·ADP to be recycled into Cdc6·ATP and Cdt1·MCM. Once loaded, the MCM helicase can slide along double-strand DNA without unwinding it, because MCM helicase activity is not activated until the preRC is converted into a

(Continued)

independent nuclear localization of Orc1, ORC(2–5), and Orc6 (Ghosh, Vassilev, Zhang, Zhao, & DePamphilis, 2011). The BAH domain in human Orc1 facilitates the ability of Orc1 to activate replication origins in vivo, triggers assembly of ORC(1–6) and facilitates the interaction of Orc1 with nucleosomes (Kara, Hossain, Prasanth, & Stillman, 2015; Noguchi et al., 2006; Zhang, Sankaran, Gozani, & Song, 2015). Mutations in Orc1, Orc4, Orc6, Cdt1, or Cdc6 are detected in 67–78% of patients with Meier–Gorlin syndrome, a genetic defect leading to primordial dwarfism (de Munnik et al., 2015). The symptoms are most severe with Orc1 and Orc4 mutations. Mutations in the BAH domain of Orc1 link histone H4 methylation to origin licensing and to Meier–Gorlin syndrome (Kuo et al., 2012).

Cdc6 stabilizes the binding of ORC(1–6) to DNA and allows recruitment of Cdt1 to form a "helicase loader." The six subunit "minichromosome maintenance complex" (Mcm2–7) is the eukaryotic replicative DNA helicase (MCM helicase) (Bell & Kaguni, 2013). Cdt1 protein chaperones the MCM helicase onto the helicase loader. Each replication fork requires at least one MCM helicase. Therefore, at least two MCM helicase must be loaded in order to produce an origin of bidirectional replication. However, just like DNA polymerases, MCM helicases travels only in one direction along the template strand. Therefore, the two MCM helicases are oriented in opposite directions (termed a double hexamer). Upon binding, each MCM helicase separates the two complementary strands, keeping one strand within the central cavity of the helicase and extruding the other strand to the outside of the helicase. Given the antiparallel arrangement of the two complementary DNA strands ($5'\rightarrow3':3'\rightarrow5'$) and the unique polarity of DNA helicases (they travel only $3'\rightarrow5'$ along the leading strand template), these two enzymes will unwind the DNA in opposite directions (bidirectional replication).

Fig. 3—Cont'd preinitiation complex. Loading of the MCM helicase stimulates hydrolysis of ORC-bound ATP. This event allows the MCM helicase to migrate along the DNA. Exchange of ADP for ATP resets ORC for a new round of preRC assembly. In multicellular eukarya, this event may also trigger release of the Orc1 subunit from ORC·chromatin sites and its subsequent ubiquitin-dependent degradation. A second MCM helicase must be loaded in opposite orientation to the first in order to achieve bidirectional DNA replication. *See Bell, S. P., & Kaguni, J. M. (2013). Helicase loading at chromosomal origins of replication.* Cold Spring Harbor Perspectives in Biology *5; Coster, G., Frigola, J., Beuron, F., Morris, E. P., & Diffley, J. F. (2014). Origin licensing requires ATP binding and hydrolysis by the MCM replicative helicase.* Molecular Cell *55, 666–677.* (See the color plate.)

2.2 Preinitiation Complex Assembly (Origin Activation)

PreIC assembly involves two critical transitions. The first is changing an inactive MCM with the 3′→5′ template strand passing through it center into an active MCM helicase that proceeds to unwind the DNA duplex (Fig. 4). The second transition is loading the replicative DNA polymerases onto the exposed DNA templates (Fig. 5). These events involve another 24 proteins in addition to those that comprise the preRC (Figs. 1 and 3) and the activity of two protein kinases, Cdc7·Dbf4 and Cdk2·CcnE that drive cells from G1 into S phase (Fig. 2).

In the first step, the "Dbf4-dependent protein kinase" (DDK) Cdc7·Dbf4 phosphorylates the MCM helicase. This event enhances the association between Ticrr, Cdc45, and DNA replication origins. In the second step, cyclin-dependent kinase-2 activated by cyclin E (Cdk2·CcnE), phosphorylates RecQ4 and Ticrr. This event allows the four subunit GINS complex, Cdc45 and Ticrr to bind to replication origins. The resulting protein·DNA complex now contains an active MCM helicase. The Mcm (2–7) complex alone is not an efficient DNA helicase. It is the Cdc45-MCM-GINS complex that unwinds duplex DNA at replication forks. GINS also forms links with other proteins at the replication fork that stabilize the replication machinery.

As the MCM helicases unwind the DNA, the single-strand DNA that is generated binds RP-A, the primary, if not exclusive eukaryotic single-strand DNA binding protein. The ssDNA·RP-A complex is the template for DNA polymerases. The first DNA polymerase to mount the template is DNA polymerase-epsilon (Pol-ε), the enzyme that will eventually synthesize nascent DNA continuously on the leading strand template. However, Pol-ε, like all DNA polymerases, requires either a DNA or RNA primer annealed to the template. Therefore, the enzyme that synthesizes the first nascent DNA chain at replication origins is DNA polymerase-α·primase (Pol-α·primase).

As the Cdc45–MCM–GINS complex unwinds the duplex DNA, it travels along one of the DNA templates in the 3′→5′ direction. When sufficient ssDNA·RP-A becomes available, Pol-α·primase will initiate RNA-primed DNA synthesis on the same template to which the helicase is traveling (Fig. 5). This is the first "Okazaki fragment" on what becomes the leading strand DNA template; the template on which DNA synthesis is continuous (Fig. 1). After 50–300 hundred nucleotides of synthesis by Pol-α·primase, Pol-ε takes over and continues to synthesize DNA until it

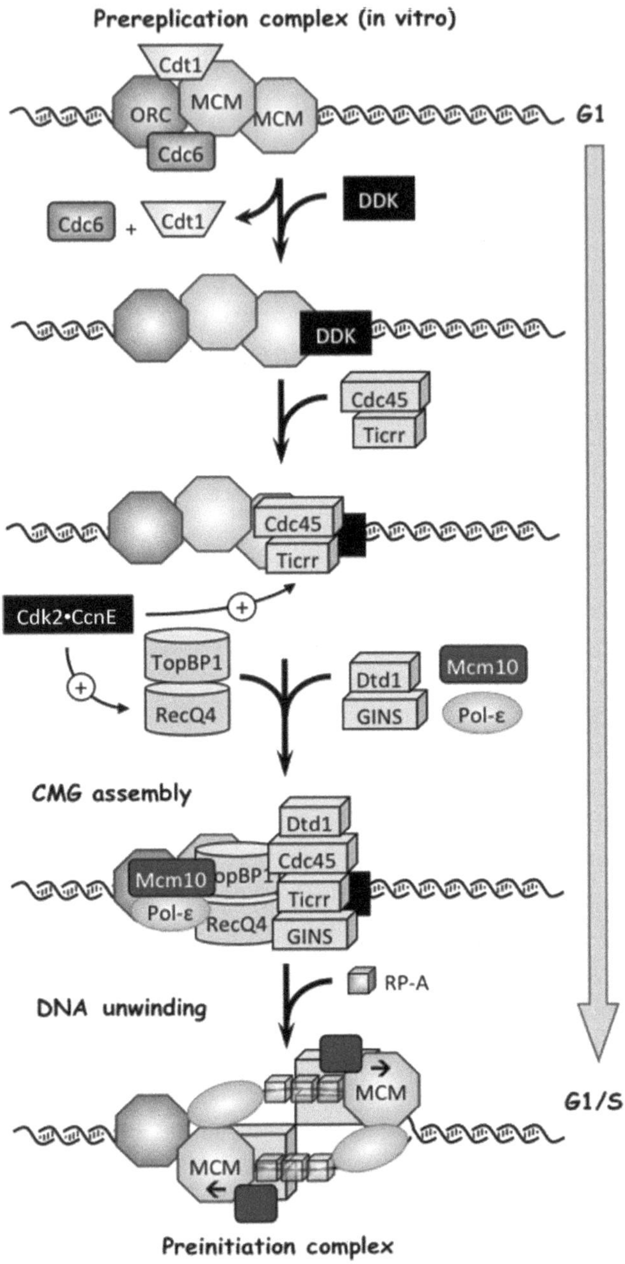

Fig. 4 Conversion of prereplication complexes (preRCs) into preinitiation complexes (preICs) occurs at the end of G1 phase. PreICs are assembled at the G1 to S phase transition. Except for the two MCM helicases, the stoichiometry of other proteins is not indicated, only their order of addition. Ticrr is also known as Treslin or Sld3, and Dtd1 as DNA unwinding element-binding protein B or DUE-B. TopBP1 is a topoisomerase II binding protein. RecQ4 is a DNA helicase. GINS is a heterotypic four subunit complex that, together with Cdc45, is essential for efficient MCM helicase activity. CDK·cyclin is the only direct participant not bound to chromatin. *See Tanaka, S., & Araki, H. (2013). Helicase activation and establishment of replication forks at chromosomal origins of replication.* Cold Spring Harbor Perspectives in Biology *5, a010371.* (See the color plate.)

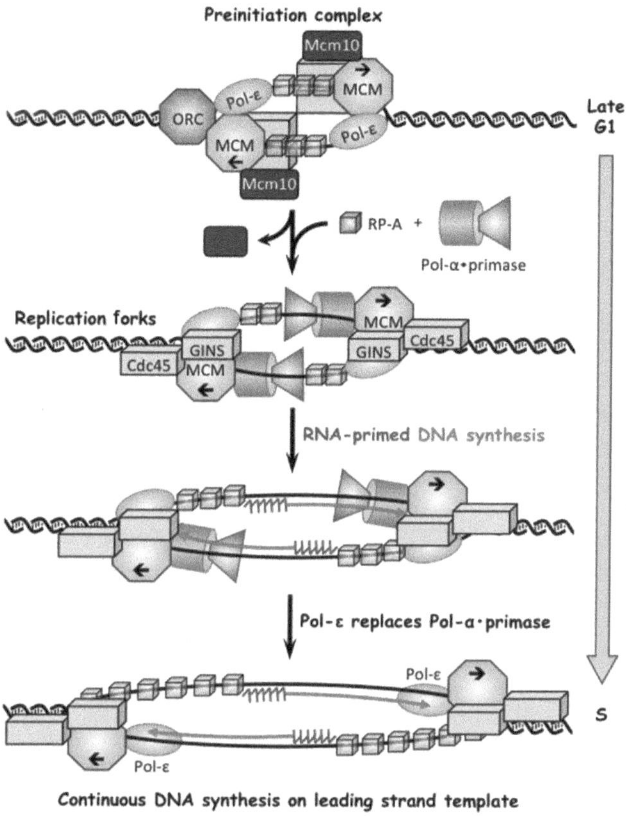

Fig. 5 Conversion of preinitiation complexes into replication forks occurs at the beginning of S phase. RNA-primed DNA synthesis begins with the loading of the DNA polymerase-α·primase. DNA synthesis is linked to DNA unwinding by a physical interaction between the MCM–GINS–Cdc45 complex and DNA polymerase-α·primase. After synthesis of 100–200 nucleotides, DNA polymerase-ε replaces DNA polymerase-α·primase and DNA synthesis continues without interruption on the leading strand template. DNA polymerase-α·primase initiates RNA-primed DNA synthesis on the lagging strand template to produce an Okazaki fragment (Fig. 1). (See the color plate.)

encounters a replication fork coming in the opposite direction or the end of the chromosome. Pol-α·primase relocates to the rear where it initiates a new Okazaki fragment, and the replication fork advances.

DNA primase synthesizes the first RNA primer on each of the two DNA templates so that DNA polymerase-α can initiate synthesis of the complementary strand by extending these primers using deoxyribonucleotides. Pol-α is quickly displaced by DNA polymerase-ε on one template and DNA polymerase-δ on the other. These enzymes make fewer errors in copying the DNA template sequence than does Pol-α. Since all DNA polymerases synthesize DNA in only one direction (5′→3′), DNA synthesis occurs continuously on one template strand and discontinuously on the other through repeated initiation, synthesis and joining of short RNA-primed nascent DNA chains. These transient replication intermediates are termed "Okazaki fragments" after the person who discovered them in bacteria, Reiji Okazaki. The RNA primers are excised, the 3′-end of the nascent DNA chain is extended to the 5′-end of the downstream fragment, and a DNA ligase joins the newly synthesized DNA fragments together. Other proteins facilitate these events.

Once nuclear DNA replication begins (S phase), the sequence of events and the proteins that execute them are the same in all eukarya and in each cell of a developing organism from the fertilized egg to the adult.

3. THE FIRST MITOTIC CELL DIVISION IS UNIVERSAL

The mitotic cell division cycle is a sequence of events in all eukarya that ensures that nuclear DNA replication (S phase) is always followed by mitosis (M phase) and then cytokinesis, and that the nuclear genome is duplicated once, but only once, during each cell division (Fig. 2). It applies to single-cell organisms as well as metazoa. It applies to zygotes, embryos and adults. Progress through the mitotic cell cycle in mammals is regulated by at least 34 different genes (Ciemerych & Sicinski, 2005), but only eight protein kinases directly regulate DNA replication, of which only cyclin-dependent kinases Cdk1 and Cdk7, the Dbf4-dependent kinase Cdc7, and the DNA damage checkpoint kinase Chek1/Chk1 are essential for mammalian development (Depamphilis et al., 2012). The other four kinases have roles that can be fulfilled by other proteins.

Among the metazoa, the first mitotic cell division is the same (Fig. 6). Oocytes are terminally differentiated cells that are arrested after completing the first stage of meiosis. They support RNA and protein synthesis but not DNA replication. When oocytes complete meiosis the resulting eggs are not only competent for fertilization; they are competent for DNA replication. Eggs are transcriptionally silent, but translationally active. Cdc6 and Orc6 are the only DNA replication protein missing in the oocytes of flies, frogs, and mice whose translation during meiotic maturation is necessary and sufficient to confer DNA replication competence before fertilization (Lemaitre et al., 2002, 2004; Murai et al., 2010; Whitmire, Khan, & Coue, 2002). In addition to their role in origin licensing, Cdc6 also behaves as a Cdk1

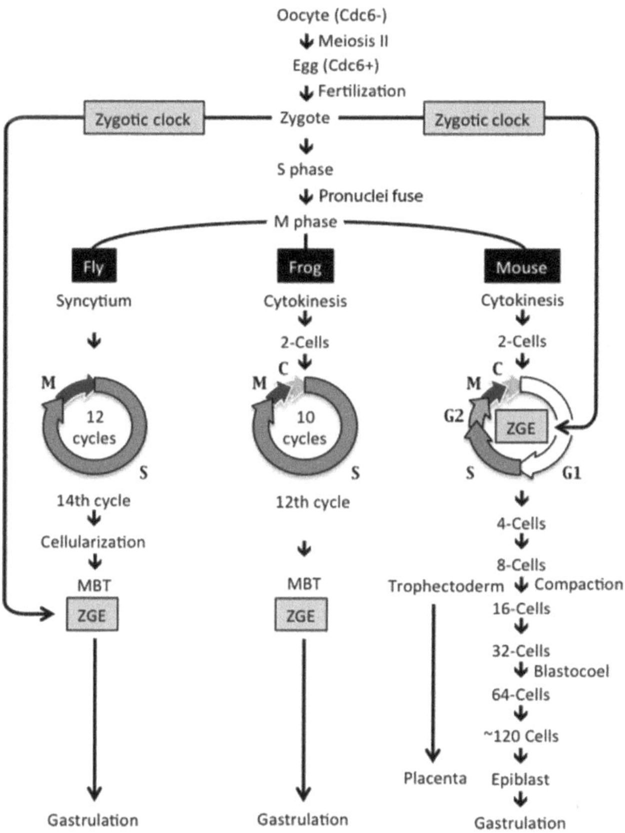

Fig. 6 Developmentally programmed changes in the mitotic cell cycle distinguish mammals from other species. The first mitotic cell division is universal, as exemplified in *Drosophila melanogaster* (fly), *Xenopus laevis* (frog), and *Mus musculus* (mouse). Oocyte maturation into eggs includes expression of the Cdc6 protein that is essential for initiation of nuclear DNA replication. The second mitotic cell division distinguishes flies from frogs and mammals from nonmammals. Flies and other insects continue to rapidly undergo cycles of DNA replication (S phase) followed by mitosis (M phase) without undergoing cell cleavage. Frogs, fish, and echinoderms continue to rapidly undergo cycles of DNA replication followed by mitosis followed by cell cleavage (cell division in the absence of cell growth). Mice and other mammals continue to undergo slow cycles of DNA replication followed by mitosis followed by cell cleavage. The most significant change is that zygotic gene expression (ZGE), the onset of transcription and the degradation of maternally inherited RNA, occurs in 2-cell embryos of mice and the 4-cell to 16-cell embryos of rabbits, cows, pigs, and humans. Activation of the bulk of ZGE in both flies and mice is determined by the time elapsed after fertilization (a "zygotic clock"), rather than by DNA replication or mitosis.

inhibitor that regulates entry into and progression through mitosis by limiting the level of Cdk1 activity (El Dika et al., 2014). Fertilization triggers formation of two pronuclei followed by nuclear DNA replication and then mitosis, all of which happens in the absence of DNA transcription. Here is where the first cell cycles diverge. Whereas the zygotes of frogs and mice undergo cytokinesis to form a 2-cell embryo, the zygotes of flies simply continue nuclear DNA replication and mitosis within a syncytium.

4. THE SECOND CELL DIVISION DISTINGUISHES MAMMALS FROM NONMAMMALS

4.1 Cell Cleavage Gives Way to Cell Growth

The embryos of flies and frogs, like representatives of all the major phyla that deposit their eggs externally, undergo a series of rapid cleavage cell cycles in the absence of cell growth, so that cells become progressively smaller (Lee, Bonneau, & Giraldez, 2014; O'Farrell, Stumpff, & Su, 2004). Rapid cell cleavages lack G1 and G2 phases. When DNA replication in either *Xenopus* or *Drosophila* embryos is blocked, cells progress to mitosis with catastrophic consequences (Glover et al., 1989; Newport & Dasso, 1989). Mammalian zygotes also undergo a series of cleavage cell cycles, but the first cell cleavage event in mouse zygotes is about 123 times longer than in the zygotes of flies and 37 times longer than those of frogs (Table 1; Fig. 6). Similarly, fly genomes (represented by *Drosophila melanogaster*) can be replicated in as few as 3.4 min, frog genomes (represented by *Xenopus laevis*) in 20 min, and rodent genomes in 120 min, but the fastest replication rates in flies and frogs occur prior to gastrulation whereas in mammals they occur during gastrulation. As a rule, external fertilization leads to rapid cell cleavage cycles, whereas internal fertilization leads to slow cell cleavage cycles.

Rapid cell cleavage cycles occur in the absence of DNA transcription. Therefore, DNA replication and mitosis depend entirely on maternally inherited proteins and translation of maternally inherited mRNA. Rapid cleavage cycles terminate with the "midblastula transition" (MBT), which is recognized by the onset of transcription and translation of zygotic genes (zygotic gene expression, ZGE), degradation of maternally inherited RNAs, and an extended cell division cycle. Frog embryos undergo 12 cleavage events before the MBT. Fly embryos cycle through a series of 14 synchronous S and M phases in the absence of cytokinesis before cellularization occurs, followed quickly by the MBT.

Table 1 Changes in the Time Required for Cell Division During Animal Development

	Time		Velocity	
Organism	**Cell Cycle**	**S Phase**	**Replication (Mb/min)**	**Fork (Kb/min)**
Flies (180 Mbp DNA)				
Drosophila syncytial embryo	8.3 → 23 m	3.4 m	53	2.6
Kc cells	8 h	6 h		0.8
Frogs (3100 Mbp DNA)				
Xenopus cleavage embryos	30 m	20 m	155	0.3 → 1.1
Rodents (2600 Mbp DNA)				
Mouse preimplantation embryo				
Cycles 1 and 2	17 → 20 h	6 h	7.2	
Cycles 3 and 4	10 → 14 h	6 h		
Mouse gastrula				
Proliferative zone	2.2 h	~1.9 h	~23	
Rat gastrula				
Anterior primitive streak	3 → 3.5 h	2 → 2.75 h	16 → 22	
Mouse fibroblasts				1.8
Human fibroblasts	18 → 25 h	6 → 8 h	6.4 → 8.6	0.6 → 3.6

Abbreviations: *Mb*, Mega (10^6) bases; *Kb*, kilo (10^3) bases. The 3.4 min S phase in flies is for the most rapid early cell cycles. Duration of metaphase/anaphase in mice is ~20 min. Note that DNA replication rates depend strongly on temperature in both animals and cell cultures. Cultured *Xenopus* cells appear similar to cultured mammalian cells, although specific data were not available.

Embryo data are from Blumenthal, A. B., Kriegstein, H. J., & Hogness, D. S. (1974). The units of DNA replication in Drosophila melanogaster chromosomes. *Cold Spring Harbor Symposia on Quantitative Biology*, 38, 205–223; Bolton, V. N., Oades, P. J., & Johnson, M. H. (1984). The relationship between cleavage, DNA replication, and gene expression in the mouse 2-cell embryo. *Journal of Embryology and Experimental Morphology*, 79, 139–163; Howlett, S. K. (1986). A set of proteins showing cell cycle dependent modification in the early mouse embryo. *Cell*, 45, 387–396; Howlett, S. K., & Bolton, V. N. (1985). Sequence and regulation of morphological and molecular events during the first cell cycle of mouse embryogenesis. *Journal of Embryology and Experimental Morphology*, 87, 175–206; Mac Auley, A., Werb, Z., & Mirkes, P. E. (1993). Characterization of the unusually rapid cell cycles during rat gastrulation. *Development*, 117, 873–883; O'Farrell, P. H., Stumpff, J., & Su, T. T. (2004). Embryonic cleavage cycles: How is a mouse like a fly? *Current Biology*, 14, R35–R45; Snow, M. H. L. (1977). Gastrulation in the mouse: Growth and regionalization of the epiblast. *Journal of Embryology and Experimental Morphology*, 42, 293–303, and references therein. Replication fork data are from Cayrou, C., Coulombe, P., Vigneron, A., Stanojcic, S., Ganier, O., Peiffer, I., et al. (2011). Genome-scale analysis of metazoan replication origins reveals their organization in specific but flexible sites defined by conserved features. *Genome Research*, 21, 1438–1449; Conti, C., Sacca, B., Herrick, J., Lalou, C., Pommier, Y., & Bensimon, A. (2007). Replication fork velocities at adjacent replication origins are coordinately modified during DNA replication in human cells. *Molecular Biology of the Cell*, 18, 3059–3067; Conti, C., Seiler, J. A., & Pommier, Y. (2007). The mammalian DNA replication elongation checkpoint: Implication of Chk1 and relationship with origin firing as determined by single DNA molecule and single cell analyses. *Cell Cycle*, 6, 2760–2767.

Slow cell cleavages occur in mammals. Fertilization of mouse eggs triggers pronuclei formation, followed by DNA replication, mitosis, and cytokinesis. These events require 17–20 h with subsequent cleavage cell cycles occurring every 10–14 h. These slow cleavage cell cycles exhibit G1, S, G2, and M phases. When DNA replication is blocked in mouse preimplantation embryos, cells do not undergo mitosis, cytokinesis or apoptosis (Bolton, Oades, & Johnson, 1984). ZGE occurs in the 2-cell embryos of mice and in the 4- to 16-cell embryos of other mammals. Despite this rapid transition from maternal to zygotic control, cell cleavage continues after activation of ZGE to produce smaller cells with an increased nuclear to cytoplasmic ratio until implantation (Aiken, Swoboda, Skepper, & Johnson, 2004). This change in strategy allows mammals to begin cell differentiation prior to implantation, which insures that placenta and fetus will develop concurrently (O'Farrell et al., 2004).

4.2 Cell Cycle Checkpoints Are Acquired at Different Stages in Development

Five cell cycle checkpoints ensure the accuracy of mitotic cell division in metazoan cells (Fig. 2), but cleavage stage metazoan embryos do not acquire any of them until the midblastula transition (Zhang, Kothari, & Lampson, 2015). In contrast, at least two of them are essential to mammalian preimplantation development.

The "spindle assembly checkpoint" (SAC) prevents the onset of anaphase until the kinetochores are attached correctly to the mitotic spindle (Lara-Gonzalez, Westhorpe, & Taylor, 2012). SAC prevents production of abnormal mature oocytes, thereby protecting preimplantation development against both aneuploidy and DNA damage (Marangos et al., 2015). SAC is also essential during preimplantation development, although it appears to utilize different members of the same gene family than it does during postimplantation development (Fernandez-Miranda et al., 2011; Wei et al., 2011).

Nuclear DNA damage is assessed before beginning S phase by the G1 DNA damage checkpoint, during S phase by the DNA replication checkpoint and before beginning M phase by the G2 DNA damage checkpoint. All three DNA damage pathways respond to an excess of single-stranded DNA complexed with replication protein-A produced by DNA damage or stalled replication forks. The ATR kinase senses damaged DNA and stalled replication forks, and then activates the Chk1 kinase, which then phosphorylates and inhibits Cdc7 (a kinase required for origin activation) and Cdc25 (a phosphatase required for activation of Cdk1 and Cdk2). Chk1 is essential for preimplantation development (Liu et al., 2000; Takai et al., 2000).

Double-stranded DNA breaks are detected by the ATM kinase, which then activates the Chek2/Chk2 kinase to phosphorylate Cdc7, Cdc25, and tumor suppressor Tp53, a transcription factor that triggers expression of genes involved in either the DNA damage response or apoptosis. Remarkably, *Chk2−/−* mice are viable and fertile (Takai et al., 2002).

In the absence of sufficient nutrients, growth factors or optimal temperature, cells in G1 phase enter a quiescent state (G0). This transition is regulated by the "restriction checkpoint," which consists of two signal detection pathways (Foster, Yellen, Xu, & Saqcena, 2010). One senses growth factors, and the other senses nutrients. The G1 to G0 transition is regulated by action of Cdk4·CcnD and Cdk6·CcnD. When cells are arrested in G0, their DNA replication machines are dismantled, and they eventually enter a senescent state from which they cannot reenter the mitotic cell cycle (Campisi, 2013). Conversely, once a cell passes the restriction checkpoint, it is committed to division; it must divide, or terminally differentiate into a polyploid cell, or die. Mice lacking either Cdk4 and Cdk6, or all three D-cyclins traverse preimplantation development, but die during late embryonic development due to severe anemia (Kozar et al., 2004; Malumbres et al., 2004).

4.3 DNA Methylation Patterns Inherited from Gametes Are Erased

DNA methylation suppresses gene transcription. Therefore, changes in the DNA methylation pattern during development or aging will affect which genes are expressed and which are not. In mice, the paternal genome in 1-cell embryos is rapidly demethylated before DNA replication begins, indicative of active enzymatic demethylation, whereas the maternal genome undergoes DNA replication-dependent demethylation during subsequent cleavage divisions, indicative of passive demethylation (Chen & Riggs, 2011). Since methylated nucleotides are not substrates for mammalian DNA polymerases, sites of DNA methylation are lost in the nascent strand, but not in the template strand, as a consequence of DNA replication. New DNA methylation patterns are established de novo beginning with the blastocyst stage by the DNA (cytosine-5)methyltransferases DNMT3A and DNMT3B, and they are maintained throughout somatic cell divisions by DNMT1 (Messerschmidt, Knowles, & Solter, 2014). DNMT1 maintains the methylation pattern by acting preferentially on the hemimethylated CpG sites generated by DNA replication.

4.4 Activation of ZGE Does Not Depend on DNA Replication

Since ZGE in mammals (as well as in flies and frogs) begins only after DNA replication is completed in the 1-cell embryo, it would not be surprising if activation of ZGE depended on prior DNA replication. In fact, this is not the case. A small group of zygotic genes are expressed in the embryos of flies and frogs when the nuclear to cytoplasmic ratio reaches a critical stage, presumably because the exponential increase in DNA content during cell cleavages results in the titration and inactivation of a hypothetical cytoplasmic factor that inhibits the MBT (Lee et al., 2014). However, most zygotic transcription, as well as the accompanying maternal RNA degradation, does not rely on the nuclear to cytoplasmic ratio in fly embryos, but instead occurs at a strict time interval following fertilization, indicating the presence of a timing mechanism (Lu, Li, Elemento, Tavazoie, & Wieschaus, 2009). The presence of this "zygotic clock" was demonstrated earlier in mice (Aoki, Worrad, & Schultz, 1997; Bolton et al., 1984; Wiekowski, Miranda, & DePamphilis, 1991). Activation of ZGE occurs in mouse embryos at the same time postfertilization (the time required to form a 2-cell embryo) regardless of whether or not DNA replication is inhibited. Zygotes remained morphologically as 1-cell embryos, but they activate ZGE as if they had reached the 2-cell stage. This clock applies to genes on plasmid expression vectors as well as endogenous genes, to promoters driven by any one of the three RNA polymerases, and to both transcription and translation (Nothias, Miranda, & DePamphilis, 1996).

The zygotic clock delays transcription until both paternal and maternal genomes are replicated and remodeled from a postmeiotic state to one in which transcription is repressed by chromatin structure in a manner that can be relieved by enhancers at appropriate times during development (Nothias, Majumder, Kaneko, & DePamphilis, 1995). Injection of DNA into the nuclei of mouse oocytes and embryos revealed that transcription promoters and origins of DNA replication are repressed in the maternal pronucleus of oocytes and zygotes, and in the nuclei of 2-cell embryos, regardless of their parental origin or ploidy. This repression is linked to changes in chromatin, and the ability to relieve this repression with embryo-responsive transcription enhancers requires a unique coactivator that first appears with formation of a 2-cell embryo (Majumder, Miranda, & DePamphilis, 1993;

Majumder, Zhao, Kaneko, & DePamphilis, 1997). A TATA box was required for enhancer stimulation of promoter activity only after cells began to differentiate (Majumder & DePamphilis, 1994). These enhancers respond to TEAD transcription factors present in preimplantation embryos and embryonic stem cells (Melin, Miranda, Montreau, DePamphilis, & Blangy, 1993). Tead2 is expressed at the beginning of preimplantation development (Kaneko, Cullinan, Latham, & DePamphilis, 1997), and Tead4 is expressed at the 8-cell stage where it is essential for trophectoderm development (Kaneko & DePamphilis, 2013; Yagi et al., 2007). All four TEAD transcription factors bind the same DNA sequence, utilize the Yap1/Yap65 transcriptional coactivator to activate transcription, and are regulated through cytoplasmic localization of Yap1 by 14–3–3 protein (Vassilev, Kaneko, Shu, Zhao, & DePamphilis, 2001). In this manner, Yap1 and its cousin Taz regulate TEAD-dependent transcription in response to mitogenic signals via the HIPPO signal transduction pathway, a regulator of tissue regeneration and stem cell biology (Hansen, Moroishi, & Guan, 2015).

5. REPLICATION ORIGINS ARE DEVELOPMENTALLY REGULATED

As genomes grew larger with the development of complex multicellular organisms (eg, humans), replication fork velocity did not increase to compensate for increasing length of DNA in chromosomes; in fact, replication forks are about 60 times slower in humans than in bacteria! Instead, the mechanisms that determine where and when DNA replication begins evolved amazing flexibility in order to accommodate the ever-increasing demands of large genomes encoding thousands of different genes that must be turned on and off at different times during cell proliferation and differentiation. In other words, the requirements for DNA replication evolved so that they did not compete with the requirements for DNA transcription.

In 1974, David Hogness and Harold Callan discovered that the density of replication origins throughout the chromosomes of flies, newts, and frogs decreased significantly as rapidly cleaving nuclei in the fertilized eggs of these organisms developed into adults. At the beginning, replication bubbles were closely spaced, but as cells differentiated, the number of active replication origins in the genome decreased dramatically. With the development of origin mapping techniques, it eventually became clear that selection of initiation sites for DNA replication could change during metazoan development, from virtually random initiation throughout the nuclear DNA to site-specific replication origins that can be activated either in clusters or individually. Replication bubbles in *fly* embryos or *frog* eggs or egg extracts are distributed throughout the DNA with periodic spacing of 5–15 kbp that depend on the amount of initiator protein (ORC(1–6)) available (Blow, Gillespie, Francis, & Jackson, 2001; Blumenthal, Kriegstein, & Hogness, 1974; Hyrien & Mechali, 1993). This allows the frog genome to be duplicated at a rate of ~155 Mbp/min (Table 1). As development proceeds, however, some initiation sites disappear while others are enhanced, resulting in a clear pattern of preferred initiation sites at specific genomic loci (Hyrien, Maric, & Mechali, 1995; Sasaki, Sawado, Yamaguchi, & Shinomiya, 1999). Developmental acquisition of origin specification also occurs during mammalian development, although the changes are less dramatic (Borowiec & Schildkraut, 2011).

5.1 The "Jesuit Model" for Origin Selection

The simplest explanation for the observations described earlier is that metazoan genomes contain many potential replication origins, but that as multicellular organisms develop, epigenetic changes mask some origins and activate others. In 1990, two independent efforts to locate specific DNA loci that acted as replication origins in hamster chromosomes resulted in paradoxical conclusions: one study concluded that initiation events were essentially "randomly" distributed throughout the 55 kbp intergenic region adjacent to the DHFR gene, termed an "initiation zone" (Vaughn, Dijkwel, & Hamlin, 1990), whereas the other study concluded that most initiation events began at a few specific sites located within this region, termed OBR (Burhans, Vassilev, Caddle, Heintz, & DePamphilis, 1990). How might such disparate results be reconciled?

The simplest explanation was that replication complexes assembled at many sites along the chromosomes, but that only a subset of these sites were activated at the onset of S phase (DePamphilis, 1993a, 1993b). This hypothesis was termed the "Jesuit Model" because it was the Jesuits who remind us that "many are called, but few are chosen" (Matthew 22:14). If different cells within a population select the same initiation site (a strongly preferred replication origin), then a narrowly defined initiation site will be observed (Fig. 7A) termed an "origin of bidirectional replication." On the other hand, if different cells within a population select different initiation sites (weakly

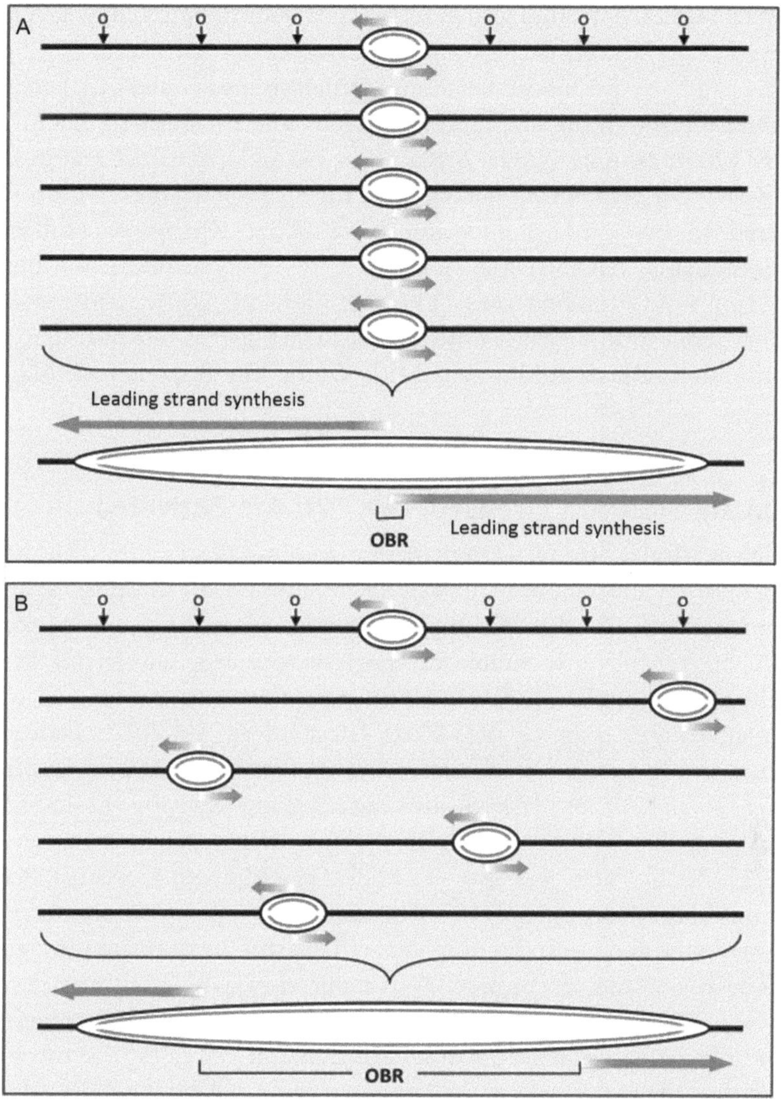

Fig. 7 Initiation sites, initiation zones, and origins of bidirectional replication are different views of the same event. Replication initiation sites can appear as well-defined replication origins or broadly defined initiation zones, depending on the relative efficiency of individual origins and the resolution of the origin mapping technology employed. (A) If different cells within a population select the same replication origin (o), then the initiation site will map as a discreet well-defined origin of bidirectional replication (OBR). (B) If different cells within a population select different replication origins within a particular DNA locus, then the initiation site will appear as a broad initiation zone with a poorly defined OBR.

preferred replication origins) then a broadly defined initiation site will be observed (Fig. 7B) termed an "initiation zone."

The ability to distinguish between "initiation zones" and OBR depends on the resolution of the mapping technology and its ability to quantify the relative activities of individual origins. For example, the ura4 region in fission yeast contains three genetically distinct replicators within what appeared to be a 5 kbp initiation zone (Kim, Zhang, & Huberman, 2001). Similarly, ARS101 and ARS310 in budding yeast contain multiple redundant ORC binding sites (Theis & Newlon, 2001), and the 8 kbp β-globin replication origin consists of at least two genetically distinguishable initiation sites (Okada & Shimizu, 2013; Wang, Lin, Lopreiato, & Aladjem, 2006).

5.2 Many Sites Are Licensed, but Few Are Activated

The Jesuit Model provides a remarkably accurate description of how cells regulate when and where to initiate DNA replication throughout animal development. Cells derived from human and mouse tissues contain 30,000–50,000 fully active replication origins, but deep sequencing of short nascent DNA strands can detect 10 times as many replication sites with a mean interorigin distance of 11 kbp (Besnard et al., 2012; Leonard & Mechali, 2013). Similarly, cells derived from flies contain 62 initiation sites with a mean size of 32 kbp distributed along 22 Mbp of chromatin (MacAlpine, Rodriguez, & Bell, 2004), but the median distance between ORC binding sites is also 11 kbp (MacAlpine, Gordan, Powell, Hartemink, & MacAlpine, 2010). The spacing of replication origins in cells from mammals and insects is comparable to the spacing during rapid cleavage cycles in frogs. Thus, the bulk of DNA replication is driven by a small subset (~10%) of the total number of available replication origins. Moreover, most of the fully active origins exist in several different cell lines, and more than 80% of the fully active origins used in one mitotic cell cycle are reused in the following cell cycle (Li, Chen, Solessio, & Gilbert, 2003). Thus, replication origins in metazoan chromosomes appear to consist of primary initiation sites that are used at a high frequency and secondary initiation sites that are used at a low frequency.

The existence of a plethora of minor replication origins provides cells with the ability to respond to genotoxic stress. Replication forks frequently stall, for example, when encountering tightly bound protein–DNA complexes, transcription machinery, repetitive sequences, or DNA lesions. Even changes in deoxyribonucleotide pools can change the frequency of initiation at minor replication origins (Anglana, Apiou, Bensimon, & Debatisse, 2003). Minor replication origins remain dormant until activated in response to replication stress (Ge et al., 2015; McIntosh & Blow, 2013).

Dormant replication origins are licensed origins that become fully active only under conditions that inhibit ongoing DNA replication forks. Thus, dormant origins ensure genome integrity by replicating regions between stalled replication forks. Dormant origins confer two important advantages to the development of complex organisms. First, they allow flexibility in determining where DNA replication begins, thereby avoiding conflicts between the need to replicate the entire genome, and the need to express specific genes, at specific sites, at a specific time. Second, they allow flexibility in determining how many replication origins are necessary to complete the job.

5.3 Both Genetic and Epigenetic Parameters Define Replication Origins

Of the three domain of life, only the replication origins in the chromosomes of bacteria and archaea, as well as plasmids and viruses that replicate in these cells, constitute replicators: specific DNA sequences that are essential for initiation of DNA replication (DePamphilis & Bell, 2010; Leonard & Mechali, 2013). In contrast, DNA replication origins in the metazoa are conditional: they facilitate—but they are not essential—for initiation of DNA replication. Artificial assembly of a transcription complex or binding of an ORC or Cdc6 protein to DNA is sufficient to create a site-specific replication origin in frog eggs and mammalian cells (Crevel & Cotterill, 2012; Danis et al., 2004; Takeda, Shibata, Parvin, & Dutta, 2005). Otherwise DNA replication in either rapid cleavage embryos or egg extracts from either frogs or flies begins at virtually any sequence, foreign or domestic, introduced into these cells (Hyrien & Mechali, 1993; Shinomiya & Ina, 1991). In contrast, DNA injected into the nucleus of mouse zygotes does not replicate unless it contains a known replicator (polyomavirus) and the cognate origin recognition protein (polyomavirus T-antigen) is provided (Martinez-Salas, Cupo, & DePamphilis, 1988). The simplest explanation is that the ratio of replication proteins to DNA in rapid cleavage zygotes is much greater than in mouse zygotes, thereby negating any advantage of specific DNA sequences. The volume of a frog egg (1.2 mm diameter), for example, is about 2800 times greater than a mouse egg (0.085 mm diameter). As cell cleavage reduces the ratio of maternally inherited replication proteins to DNA, some sites will be more effective as replication origins than others.

Metazoan replication origins are simply chromosomal loci that permit assembly of prereplication complexes (Fig. 1) that can later be converted into preinitiation complexes and activated by Cdk2·CcnE and Cdc7·Dbf4 (Fig. 2). Such sites are determined by both genetic and epigenetic parameters (Aladjem, Falaschi, & Kowalski, 2006; DePamphilis, 1996b; Gilbert, 2001; Leonard & Mechali, 2013). Genetic parameters include easily unwound DNA sequences (Aladjem et al., 2006) such as asymmetric A:T-rich sequences (Stanojcic, Lemaitre, Brodolin, Danis, & Mechali, 2008; You et al., 2003) and consensus G-quadruplex forming motifs (Besnard et al., 2012; Cayrou et al., 2011; Foulk, Urban, Casella, & Gerbi, 2015; Picard et al., 2014). Epigenetic parameters include nuclear structure (Gilbert, Miyazawa, & DePamphilis, 1995), histone modifications (Dorn & Cook, 2011), chromosome structure, and transcription.

The effect of epigenetic parameters on regulatory DNA sequences begins in mammals during the onset of ZGE when changes in chromatin structure result in the need for enhancers to activate DNA replication replicators as well as transcription promoters. The polyomavirus replicator is active in the paternal pronucleus of a mouse zygote as long as the polyomavirus origin recognition protein (the T-antigen DNA helicase) is provided (Martinez-Salas, Linney, Hassell, & DePamphilis, 1989). However, beginning with the 2-cell embryo, the polyomavirus replicator is inactive unless linked to a cell-specific enhancer and provided with T-antigen. Enhancers are required to relieve chromatin-mediated repression of both promoters and replicators (Majumder & DePamphilis, 1995; Majumder et al., 1997; Wiekowski, Miranda, Nothias, & DePamphilis, 1997). Thus, as changes in chromatin structure prevent access to some DNA sites but not to others, the ability to license and then activate replication origins becomes more restricted, as postulated by the "Jesuit Model."

6. GENOME DUPLICATION IS RESTRICTED TO ONCE PER CELL DIVISION

From zygote to adult, proliferating cells must produce one, but only one, copy of their nuclear DNA prior to cell division. To do otherwise, results either in cell death or in genomic instability and aneuploidy, two hallmarks of cancer cells (Abbas, Keaton, & Dutta, 2013). To prevent these catastrophes, cells must navigate successfully through eight specific cell cycle

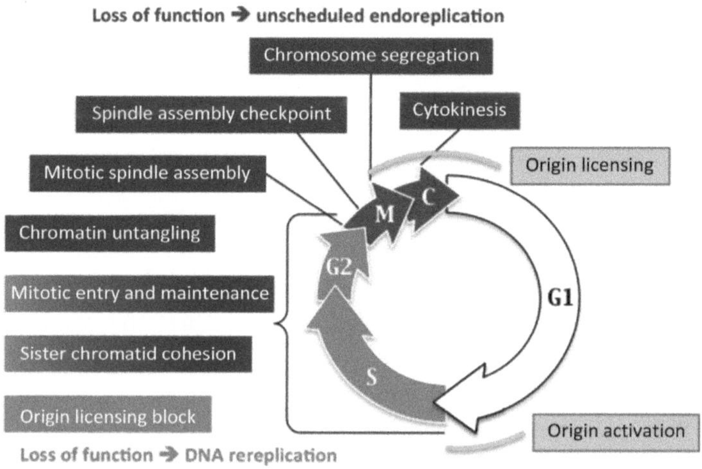

Fig. 8 Specific cell cycle events that prevent either DNA rereplication or unscheduled endoreplication. Eight cell cycle events restrict genome duplication to once per cell division. Events shaded in *red* (*gray* in the print version) prevent DNA rereplication, whereas events shaded in *blue* (*dark gray* in the print version) prevent unscheduled endoreplication. Events shaded in both *red* (*gray* in the print version) and *blue* (*dark gray* in the print version) (sister chromatid cohesion and mitotic entrance and maintenance) prevent either event, depending on whether the cells are in S phase or in mitosis when that event is interrupted. Origin licensing refers to preRC assembly. Origin activation refers to preIC assembly.

events (Fig. 8). Forty-two genes have been identified whose activity is essential for one or more of these events (Vassilev et al., 2016). If any one of these genes is suppressed, the result is either unscheduled endoreplication or DNA rereplication.

6.1 Preventing Unscheduled Endoreplication

A repeated S phase without an intervening mitosis or cytokinesis is termed endoreplication. Although developmentally programmed endoreplication occurs frequently in ferns, flowering plants, mollusks, arthropods, amphibians, and fish, it is rare in mammals (Zielke, Edgar, & DePamphilis, 2013). However, "unscheduled endoreplication" occurs when cells have completed S phase, but are prevented from entering or completing either mitosis or cytokinesis by the introduction of chemotherapeutic drugs or siRNA. Such cells eventually "slip through" mitosis and cytokinesis and into G1 phase to exist as single cells with a tetraploid nucleus. These cells can then enter S phase to produce octoploid cells, particularly if they lack DNA damage checkpoints. Unscheduled endoreplication occurs whenever the complete separation of sister chromatids is prevented after DNA replication is completed. The critical steps are maintaining cohesion between sister chromatids throughout S phase and up to metaphase, untangling sister chromatids following termination of DNA replication, correct assembly of the mitotic spindle, and separation of sister chromatids during anaphase.

6.2 Preventing DNA Rereplication

When a region of DNA that has already been replicated once during S phase is replicated a second time during the same S phase, the result is "DNA rereplication." This situation would occur if either the same replication origin was relicensed and activated, or a new replication origin was licensed within DNA that had already been replicated and then activated. DNA rereplication results in an accumulation of stalled replication forks that are easily damaged to produce double-strand DNA breaks that trigger apoptosis. Therefore, once S phase begins, origin licensing must prevented blocked until mitosis is completed.

At least seven concerted pathways exist that prevent DNA rereplication in mammalian cells (Fig. 9) by inactivating the helicase loader, thereby preventing both the reloading of MCM helicases at activated replication origins and the licensing of new replication origins. These pathways, which fall into two biological cycles, exist in flies, frogs, nematodes, and mammals (Blow & Dutta, 2005; DePamphilis et al., 2006; Siddiqui, On, & Diffley, 2013; Sonneville, Querenet, Craig, Gartner, & Blow, 2012).

6.3 The ORC Cycle

The "ORC Cycle" begins when the Orc1 subunit of the origin recognition complex [ORC(1–6)] is selectively targeted during S phase for inactivation by posttranslational CDK-dependent phosphorylation and then ubiquitin-dependent degradation (DePamphilis, 2003; DePamphilis et al., 2006). Since the Orc1 subunit is essential for ORC binding to DNA, loss of the Orc1 subunit results in destabilization of the remaining ORC subunits (Lee et al., 2012; Siddiqui & Stillman, 2007). Since Cdc6 binding to DNA is dependent on ORC(1–6) (Fig. 3), destabilization of the ORC·DNA interaction will destabilize the Cdc6·DNA interaction. Cdc6 then becomes a target for phosphorylation by Cdk2·CcnA, which results in its nuclear exclusion. These events should prevent premature licensing of replication origins during S phase. Reassembly of the helicase loader occurs by reassembly of Orc1, ORC(2–5), Orc6, and Cdc6 onto chromatin

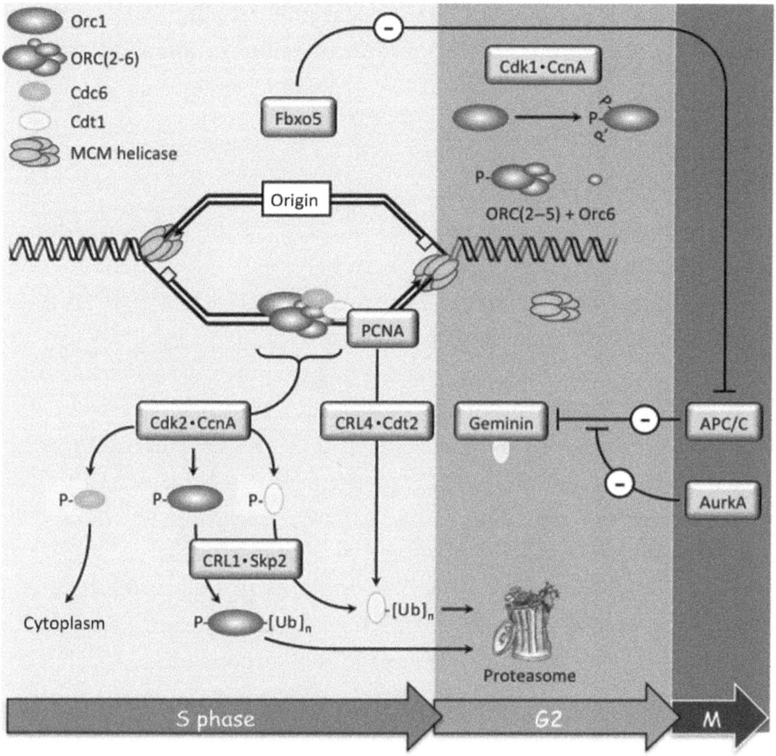

Fig. 9 Multiple concerted pathways prevent DNA rereplication. When cells enter S phase, seven concerted pathways exist can inactivate the helicase loader, thereby preventing rereplication of DNA prior to mitosis. Cdk2·CcnA phosphorylates Orc1, Cdc6, and Cdt1, thereby inhibiting their ability to form helicase loaders. In addition, Cdc6-P is localized to the cytoplasm where it cannot facilitate origin licensing, and Orc1-P is ubiquitinated and targeted for destruction. Inactivation of Orc1 results in dissociation of ORC(2–5) and Orc6 from chromosomes. Cdt1-P is ubiquitinated by the CRL1·Skp2 ubiquitin ligase. Cdt1 that is not phosphorylated is ubiquitinated by the CRL4·Cdt2 ubiquitin ligase bound to PCNA and DNA. PCNA is the eukaryotic sliding clamp protein that facilitates DNA synthesis by DNA polymerases-δ and -ε. The ubiquitinated proteins are then degraded by the 26S proteasome. Geminin is a protein that binds specifically to Cdt1 and inhibits its activity. These activities are available from S through early M phase. As cells exit mitosis, geminin, and cyclin A are ubiquitinated by the anaphase-promoting complex (originally termed cyclosome, APC/C), an activity that is inhibited during S to early M phase by the APC/C specific inhibitor Fbxo5/Emi1. Reassembly of the helicase loader occurs by reassembly of Orc1, ORC(2–5), Orc6, and Cdc6 onto chromatin, an event that is triggered by the Orc1 subunit during the anaphase to G1 phase transition. (See the color plate.)

(Ghosh et al., 2011; Kara et al., 2015; Sonneville et al., 2012), an event that appears to be triggered by the Orc1 subunit during the anaphase to G1 phase transition (Noguchi et al., 2006).

6.4 The Cdt1 Cycle

The "Cdt1 Cycle" begins when Cdt1 is targeted for ubiquitin-dependent degradation during S phase. Two pathways exist: CDK-dependent phosphorylation followed by ubiquitination of Cdt1-P and PCNA-DNA-dependent ubiquitination of Cdt1. Cdt1 activity is also inhibited by binding to geminin protein, but the importance of geminin is confined primarily to late S, G2, and early mitosis (Klotz-Noack, McIntosh, Schurch, Pratt, & Blow, 2012). Geminin binding to Cdt1 and CDK-dependent phosphorylation of Cdc6 prevent Cdt1 and Cdc6 degradation by the APC/C during mitosis, thereby allowing them to participate in origin licensing as cells exit mitosis (Ballabeni et al., 2004; Mailand & Diffley, 2005).

The Cdt1 cycle is critical at the beginning of animal development. During early development of frogs and mammals only geminin and the CRL4 ubiquitin ligase are essential to prevent DNA rereplication. Cdt1 is a stable abundant protein in *Xenopus* egg extracts. It is not degraded substantially until cycle 10 (Kisielewska & Blow, 2012). Geminin is only fully degraded by the APC/C after the MBT (cell cycle 12) (Hodgson, Li, Tada, & Blow, 2002; Li & Blow, 2005). Therefore, two concerted pathways prevent DNA rereplication in cleavage stage embryos by regulating both Cdt1 activity and Cdt1 proteolysis. In the beginning, when the nuclear to cytoplasmic ratio is low, Cdt1 activity is blocked by geminin and regulation of geminin activity appears to involve both the APC/C and nuclear import of geminin. As the ratio of nuclei to cytoplasm increases, Cdt1 binds to CRL4·PCNA·DNA complexes that assemble at newly minted replication forks and at sites of DNA repair where Cdt1 is then targeted for ubiquitin-dependent degradation (Arias & Walter, 2005, 2007). Thus, suppressing geminin RNA levels in *Xenopus* egg extracts or 2-cell embryos cannot immediately deplete the supply of geminin protein. Consequently, DNA rereplication and apoptosis are not manifested until the MBT and beyond (Kerns, Schultz, Barry, Thorne, & McGarry, 2012; Kerns, Torke, Benjamin, & McGarry, 2007).

Once the developmental program activates ZGE, totipotent and pluripotent cells depend primarily, if not exclusively, on Geminin to prevent DNA rereplication. Geminin is essential for gastrulation in frogs (Kerns et al., 2012), and probably in flies as well (Takada, Kwak, Koppetsch, & Theurkauf, 2007). Most notably, geminin is essential to prevent DNA rereplication-dependent apoptosis in mouse preimplantation embryos (Hara, Nakayama, & Nakayama, 2006; Huang, Kaneko, Pan, & DePamphilis, 2015), in the epiblast (Huang et al., 2015; Patterson, Waller, & Kroll, 2014), and in embryonic stem cells (Huang et al., 2015), which are derived from the epiblast. However, as pluripotent cells differentiate, geminin is no longer essential for viability, although it continues to contribute to preventing DNA rereplication (Dorn, Chastain, Hall, & Cook, 2009; Huang et al., 2015). In effect, other pathways, such as Cdk2·CcnA, contribute significantly (Zhu & Depamphilis, 2009). Thus, whether or not a particular cell, such as a cancer cell, requires geminin to prevent DNA rereplication depends on the relative levels of relevant gene products. For example, suppression of CRL4 in HeLa cells induces DNA rereplication (Jin, Arias, Chen, Harper, & Walter, 2006), whereas suppression of geminin does not, unless Cdt1 is overexpressed (Nishitani et al., 2006). Thus, geminin is essential for viability when the cellular levels of Cdt1 are high, as in totipotent blastomeres, pluripotent stem cells, some cancer cells, and perhaps some rapidly proliferating cells in adult animals, but the role of geminin in most cells of the adult organism facilitates proliferation without being essential for viability. Similarly, the primary role of CRL4 in mammals and presumably other animals is to facilitate DNA repair by inhibiting origin licensing (Havens & Walter, 2011).

7. DEVELOPMENTALLY PROGRAMMED ENDOREPLICATION

Given the extraordinary problem of preventing excess genome duplication during metazoan development, it is remarkable that some cells are developmentally programmed to exit their mitotic cell cycle in response to environmental signals, injury or stress, and to differentiate into nonproliferating, viable, polyploid cells. Although this occurs frequently in ferns, flowering plants, mollusks, arthropods, amphibians, and fish, it is rare among mammals (Fox & Duronio, 2013; Lacroix & Maddox, 2012; Ullah, Lee, & Depamphilis, 2009). Developmentally programmed polyploidy first occurs in mammals during peri-implantation development with the formation of trophoblast giant cells.

7.1 Cell Cycle Arrest in Either G2 or Mitosis Triggers Endoreplication

The earliest cell differentiation event in mammalian development occurs when the outer blastomeres of the preimplantation embryo form a monolayer of epithelial cells (trophectoderm) that envelops the remaining blastomeres (inner cell mass). In mice, TE specification occurs in the 8-cell to 16-cell compacted morula (Fig. 6; see book Preface, Fig. 1). The inner cell mass provides fibroblast growth factor-4 (FGF4) to the polar trophectoderm, but when the mural trophectoderm in late stage blastocysts becomes deprived of FGF4 the trophoblast cells terminally differentiate into the mononuclear, polyploid, nonproliferating trophoblast giant cells (TGCs) that are essential for implantation and placentation (Roberts & Fisher, 2011) (see book Preface, Fig. 2).

Trophoblast stem cells (TSCs) are derived from the trophectoderm of the blastocyst and give rise exclusively to all of the trophoblast lineages in the placenta (Oda, Shiota, & Tanaka, 2010; Rielland, Brochard, Lacroix, Renard, & Jouneau, 2009; Tanaka, Kunath, Hadjantonakis, Nagy, & Rossant, 1998). TSCs proliferate as tightly packed colonies when cultured in medium conditioned by primary embryonic fibroblasts and supplemented with FGF4. When cultured in the absence of FGF4 and conditioned medium, TSCs differentiate into TGCs. Thus, FGF4 deprivation of TSCs in vitro mimics the induction of trophoblast differentiation into TGCs in vivo (reviewed in Hu & Cross, 2010). FGF4 governs TSC differentiation by regulating mitotic entry (Fig. 10).

Differentiation of TSCs into TGCs can be triggered in the presence of FGF4 by selective inhibition of Cdk1 with RO3306, suggesting that FGF4 deprivation inhibits the G2 to M phase transition by selectively inhibiting Cdk1, the protein kinase essential for mitotic entry (Ullah, Kohn, Yagi, Vassilev, & DePamphilis, 2008). In fact, FGF4 deprivation of TSCs induces expression of p57, a protein that inhibits Cdk1 activity, and a gene that is essential for TSC differentiation (Ullah et al., 2008). Since p57 also inhibits Cdk2, p57 levels in giant cells oscillate during multiple rounds of endoreplication (high during G phase, low during S phase) so that Cdk2·CcnE can initiate S phase (Fig. 2) and Cdk2·CcnA can prevent DNA rereplication (Fig. 9).

FGF4 regulates expression of p57 protein in TSCs by regulating the level of Chk1 kinase (Ullah, de Renty, & Depamphilis, 2011). When TSCs proliferate without differentiating (termed self-renewal), they transcribe and translate the p57 gene, but the basal level of Chk1 kinase activity phosphorylates p57 protein, thereby targeting it for ubiquitin-dependent degradation. When TSCs are deprived of FGF4, Chk1 protein disappears with the concomitant appearance of p57 protein and formation of TGCs. Thus,

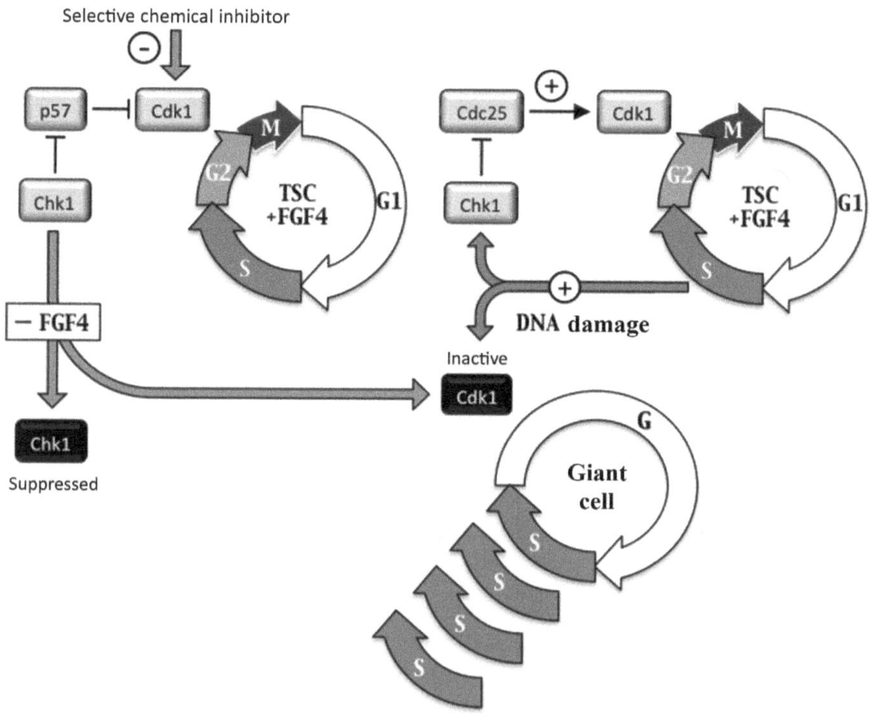

Fig. 10 Inhibition of Cdk1 activity triggers endoreplication in trophoblast cells. Selective inhibition of Cdk1 activity in trophoblast stem cells by RO3306, FGF4 deprivation, or induction of DNA damage triggers multiple S phases without an intervening mitosis or cytokinesis to produce giant cells with a single enlarged nucleus containing as many as several hundred copies of each chromosome.

inhibition of Cdk1 activity either by RO3306 or by the p57 protein induces TSC differentiation into TGCs.

When megakaryoblasts differentiate into megakaryocytes the nuclear envelope breaks down and parts of anaphase occur without nuclear division, thereby producing mononucleated cells with separated chromatids. This phenomenon is termed endomitosis, but the end result is equivalent to endoreplication—mononucleated polyploid cells. The mechanism responsible for endomitosis remains speculative, but selective inactivation of Cdk1 readily converts endomitosis into endoreplication (Trakala et al., 2015), just as inactivation of Cdk1·CcnB induces unscheduled endoreplication in mouse embryonic fibroblasts, chicken DT40 cells, or human cancer cells (Diril et al., 2012; Hochegger et al., 2007; Vassilev et al., 2016).

7.2 DNA Damage Can Trigger Endoreplication, but Not Apoptosis

DNA damage will also induce differentiation of TSCs into TGCs, because DNA damage induces high levels of Chk1 kinase activity, which inhibit Cdc25, a protein phosphatase that is essential to maintain Cdk1 in its active form (Fig. 2). Thus, DNA damage arrests cells in G2 phase by inhibiting Cdk1 activity, which induces TSC differentiation into TGCs (Fig. 10). Terminally differentiated cells must remain viable, but since they cannot proliferate, they cannot pass on genetic mutations, so repairing DNA damage is not a concern. Thus, induction of DNA rereplication-induced DNA damage in *Drosophila* does not induce apoptosis in follicle cells undergoing endoreplication, whereas it does in cells undergoing proliferation (Mehrotra, Maqbool, Kolpakas, Murnen, & Calvi, 2008).

Suppressing Chk1 activity in TSCs not only triggers endoreplication, but also it facilitates initiation of DNA replication, because activation of a DNA damage response prevents origin licensing. This would account for the fact that developmentally programmed endoreplication in trophoblast cells and megakaryoblasts efficiently duplicates the entire genome (Sher et al., 2013). In addition, the tumor suppressor protein Tp53 is also downregulated in TGCs (Soloveva & Linzer, 2004). Although Tp53 arrests cell proliferation in response to DNA damage, it will induce apoptosis if the damage is extensive and repair efforts fail. Thus, the combined actions of CHK1 and Tp53 maintain genomic integrity in proliferating mammalian cells by preventing them from exiting the mitotic cell cycle under conditions that trigger endoreplication. Conversely, cells that are developmentally programmed to terminally differentiate into polyploid cells must suppress both CHK1 and Tp53.

7.3 DNA Damage Is Suppressed During Developmentally Programmed Endoreplication

Ablation or suppression of the geminin gene in TSCs proliferating in the presence of FGF4 mimics the events triggered by FGF4 deprivation (Fig. 11): arrest of cell proliferation, formation of giant cells, excess DNA replication in the absence of DNA damage and apoptosis, and changes in gene expression that include loss of Chk1 with concomitant upregulation of p57 and p21 (de Renty, Kaneko, et al., 2014). The p57 protein localizes to the nucleus where it inhibits Cdk1, and together with p27 regulates endocycles, but the p21 protein is phosphorylated by the Akt1 kinase and

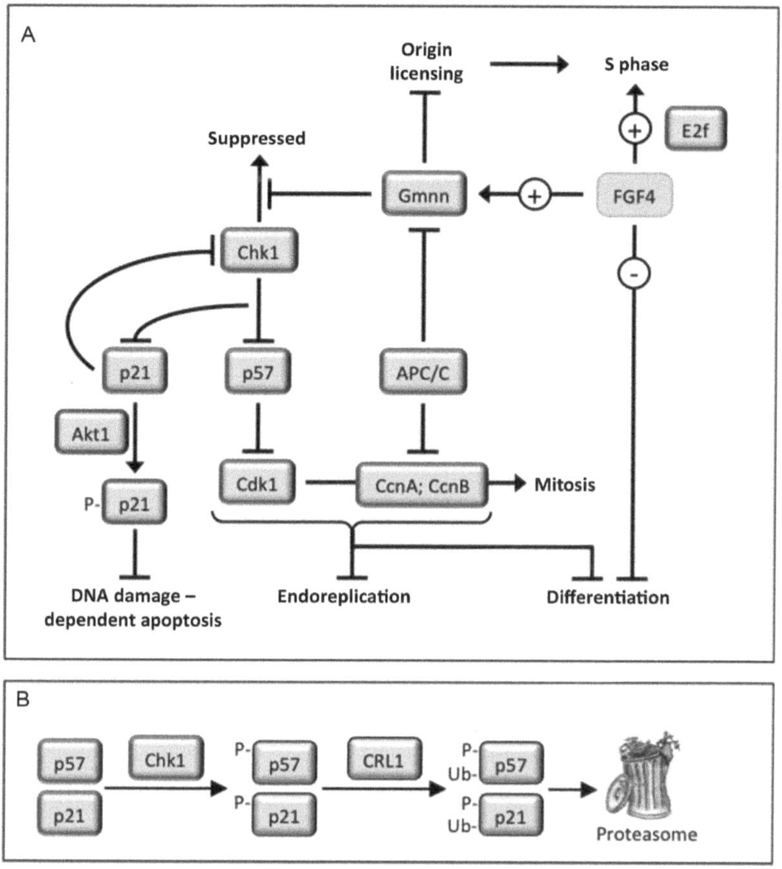

Fig. 11 The FGF4 signal transduction pathway governs trophoblast proliferation and differentiation. FGF4 (and probably other mitogenic proteins as well) is essential for trophoblast proliferation. This mitogenic activity is likely mediated by E2F-dependent gene expression (Chen et al., 2010), and possibly directed at regulating the activity of the anaphase-promoting complex (APC) (Yang et al., 2012). FGF4 deprivation results in downregulation of geminin activity to a level that maintains endocycles (de Renty, Kaneko, & DePamphilis, 2014), but that does not prevent downregulation of Chk1 protein. The loss of Chk1 kinase activity results in expression of two CDK-specific inhibitors, p57 and p21 (Ullah et al., 2011). The p57 protein prevents the onset of mitosis by selectively inhibiting Cdk1 activity, thereby triggering the first round of endoreplication (Hattori, Davies, Anson-Cartwright, & Cross, 2000; Ullah et al., 2008). This event activates the G1 phase APC·Cdh1 ubiquitin ligase, which targets geminin, cyclin B, and cyclin A for degradation, thereby allowing licensing of replication origins and the onset of S phase without passing through mitosis (Ullah, Lee, Lilly, & DePamphilis, 2009). Inhibition of Cdk1 triggers both endoreplication and TSC differentiation. In the absence of p57, FGF4 deprivation produces multinucleated TGCs, revealing the existence of alternative mechanisms to trigger TSC differentiation (Ullah et al., 2008). Endocycles also require p57, which is expressed during G phase and then suppressed during S phase to allow sequential assembly and activation of prereplication complexes (Ullah et al., 2008). Geminin maintains endocycles by preventing DNA rereplication. The p21/Cdkn1a protein localizes to the cytoplasm in TGCs where it prevents DNA damage-induced apoptosis (de Renty, DePamphilis, & Ullah, 2014). It might also maintain suppression of Chk1 by reducing *Chk1* RNA levels (Gottifredi, Karni-Schmidt, Shieh, & Prives, 2001), as observed during FGF4 deprivation (Fig. 9).

localized to the cytoplasm where it suppresses DNA damage-dependent apoptosis (de Renty, DePamphilis, et al., 2014).

7.4 P57 Is Essential for Mammalian Development

In the absence of p57/CDKN1C, FGF4 deprivation of TSCs does not trigger terminal differentiation into TGCs, but instead results in the accumulation of multinucleated giant cells, because mitosis, which is driven by Cdk1·CcnB and Cdk1·CcnA, continues in the absence of cytokinesis (Ullah et al., 2011, 2008). This phenotype could account for the clear association of p57 deficiency with preeclampsia, placental changes that result in the onset of hypertension with proteinuria during human pregnancy after 20 weeks of gestation (Romanelli et al., 2009). In fact, p57 deficiency in a mouse fetus induces the symptoms of preeclampsia (Kanayama et al., 2002; Knox & Baker, 2007). Furthermore, p57 deficiency is closely associated with Beckwith–Wiedemann syndrome, tissue overgrowth characterized by macroglossia, macrosomia, and abdominal wall defects (Romanelli et al., 2010), and an imprinted p57 gene is mutated in patients with this syndrome (Hatada et al., 1996).

8. CANCER AND HUMAN DEVELOPMENT

8.1 Gene Amplification in Mammals Is Unique to Cancer Cells

In some organisms, such as protozoa, flies, and frogs, developmentally programmed endoreplication is followed by selectively over replicating genes in specific cells in order to produce large quantities of specific proteins (Calvi, 2006; Foulk et al., 2013; Tower, 2004). It appears that the replication origins used in these organisms to amplify specific genes utilize most, if not all, of the same proteins that initiate DNA replication during normal cell proliferation. However, in contrast to the replication origins that initiate DNA replication once per cell division, amplification origins continue to initiate DNA replication without an intervening mitosis or cytokinesis.

Developmentally programmed gene amplification has not been reported in mammals. However, mammalian cells with multiple copies of a selected DNA locus can be isolated by passaging them in vitro in the presence of a chemical inhibitor of a protein required for DNA replication. These loci invariably contain one or more genes whose cognate protein counters the effect of the chemical inhibitor. For example, cells with multiple copies of the dihydrofolate reductase (DHFR) gene can be selected in the presence of methotrexate, a competitive inhibitor of DHFR. Since DHFR is essential for cell viability, only those cells that produce high levels of DHFR will proliferate in the presence of low levels of methotrexate. Gene amplification is a typical genetic alteration in cancer, and many oncogenes have been identified in the amplified regions. This type of gene amplification depends on nonspecific DNA damage and recombination rather than initiation of DNA replication at specific DNA loci (Matsui, Ihara, Suda, Mikami, & Semba, 2013).

8.2 Genome Instability Is an Integral Part of Cancer Development

Cells isolated from human cancers typically vary widely in chromosomal content with both structural and numerical alterations, and they exhibit high rates of genome instability, as demonstrated by the rapid restoration of genomic heterogeneity following clonal selection (Masramon et al., 2006) and the accumulation of polyploid and aneuploid cells (Geigl, Obenauf, Schwarzbraun, & Speicher, 2008). Genome sequencing reveals that up to 37% of all tumors have transitioned through a polyploid state during their development (Zack et al., 2013), suggesting that tumorigenesis is accelerated by transition through the inherently unstable polyploid state. This transition from a normal cell to a cancer cell must involve changes in the mechanisms that restrict genome duplication to once per cell division. In other words, fluctuations in the activity of a protein that prevents accumulation of excess DNA can produce aneuploid/polyploid cells. In fact, of the 42 genes that are required in 1 or more of the 8 cell cycle events that prevent either DNA rereplication or unscheduled endoreplication in vitro (Fig. 8), 14 have been shown to prevent aneuploidy or polyploidy in mice, and 8 have been shown to prevent tumorigenesis (Vassilev et al., 2016).

8.3 Pluripotent Stem Cells Can Give Rise to Cancer

Stem cells are defined as undifferentiated cells that retain the potential to proliferate indefinitely (self-renewal) without loosing the ability to give rise to other types of cells through differentiation. Embryonic stem cells (ESCs) are derived from the epiblast (Boroviak, Loos, Bertone, Smith, & Nichols, 2014; Tang et al., 2010), the tissue formed during blastocyst development (see Preface) that gives rise to the three embryonic germ layers (ectoderm, mesoderm, and endoderm) and to germ cells. ESCs are termed pluripotent, because they retain the ability to form these cell types both in embryos and in culture.

Cancer arises either by multiple mutations in normal cells, or by activation of cancer stem cells (CSCs). CSCs are simply stem cells that drive tumorigenesis instead of tissue formation (Valent et al., 2012). Other cells within a tumor can proliferate under conditions where normal cells cannot, but only CSCs can initiate a tumor de novo. CSCs are analogous to ESCs in that both CSCs and ESCs undergo "self-renewal," ESC gene expression signatures also are found in many human cancers and in mouse cancer models (Kim & Orkin, 2011; Palmer, Schmid, Berger, & Kohane, 2012), and both CSCs and ESCs can produce malignant tumors (Ben-David & Benvenisty, 2011).

Remarkably, when either mouse or human ESCs are placed at ectopic sites in immuno-compromised adult mice, they behave like CSCs by producing either benign (teratomas) or malignant (teratocarcinomas) tumors consisting of various types of normal cells. Teratoma and teratocarcinoma are generic names for a variety of germ cell tumors. Although commonly associated with ovarian or testicular tissues, they also appear in lungs, peritoneum, cervix, and cranium. Testicular cancer is the most common cancer in middle-aged males, and ovarian cancer causes more deaths than any other cancer of the female reproductive system. Sacrococcygeal teratomas are the most common tumors in newborns. Tumors experimentally derived from normal or induced human pluripotent stem cells, as well as mouse ESCs, closely resemble spontaneous teratomas/teratocarcinomas that occur in both mice and humans. Thus, ESCs—or some cell produced during ESC differentiation—behaves as the CSC for these tumors. Moreover, the primary role of geminin in totipotent blastomeres and pluripotent cells of mammalian preimplantation development, as well as in germ cells, is to prevent DNA rereplication-dependent apoptosis (Barry, Schultz, Payne, & McGarry, 2012; Huang et al., 2015), a role previously believed to be essential only in cancer cells (Zhu & Depamphilis, 2009). The requirement for geminin to prevent DNA rereplication in mammals is diminished or lost during cell differentiation (Schultz et al., 2011; Shinnick, Eklund, & McGarry, 2010).

Consider the possibility that the seeds for cancer might well be planted during embryonic development. Pluripotent stem cells are designed to proliferate, differentiate, or die. The alternative is to become quiescent. Those that either continue self-renewal or become quiescent would be passively distributed among tissues during animal development. If they were subsequently activated at ectopic sites in response to environmental signals, they would produce tumors rather than tissues, unless they were eliminated by the host's immune system.

9. EPILOG

Genome duplication marks both the beginning and the end of mammalian existence. It begins in the pronuclei of the zygote. The rules that govern it are established during preimplantation development. It ends when cells undergo senescence, and when it is no longer regulated, cells either undergo apoptosis or produce a cancer. Understanding how genome duplication is regulated is essential to human health and longevity.

REFERENCES

Abbas, T., Keaton, M. A., & Dutta, A. (2013). Genomic instability in cancer. *Cold Spring Harbor Perspectives in Biology*, *5*, a012914.

Aiba, K., Carter, M. G., Matoba, R., & Ko, M. S. (2006). Genomic approaches to early embryogenesis and stem cell biology. *Seminars in Reproductive Medicine*, *24*, 330–339.

Aiken, C. E., Cindrova-Davies, T., & Johnson, M. H. (2008). Variations in mouse mitochondrial DNA copy number from fertilization to birth are associated with oxidative stress. *Reproductive Biomedicine Online*, *17*, 806–813.

Aiken, C. E., Swoboda, P. P., Skepper, J. N., & Johnson, M. H. (2004). The direct measurement of embryogenic volume and nucleo-cytoplasmic ratio during mouse preimplantation development. *Reproduction*, *128*, 527–535.

Aladjem, M. I., Falaschi, A., & Kowalski, D. (2006). Eukaryotic DNA replication origins. In M. L. DePamphilis (Ed.), *DNA replication and human disease* (pp. 31–62). Cold Spring Harbor, NY: Cold Spring Harbor Laboratory Press.

Anglana, M., Apiou, F., Bensimon, A., & Debatisse, M. (2003). Dynamics of DNA replication in mammalian somatic cells: Nucleotide pool modulates origin choice and interorigin spacing. *Cell*, *114*, 385–394.

Aoki, F., Worrad, D. M., & Schultz, R. M. (1997). Regulation of transcriptional activity during the first and second cell cycles in the preimplantation mouse embryo. *Developmental Biology*, *181*, 296–307.

Arias, E. E., & Walter, J. C. (2005). Replication-dependent destruction of Cdt1 limits DNA replication to a single round per cell cycle in Xenopus egg extracts. *Genes and Development*, *19*, 114–126.

Arias, E. E., & Walter, J. C. (2007). Strength in numbers: Preventing rereplication via multiple mechanisms in eukaryotic cells. *Genes and Development*, *21*, 497–518.

Baldinger, T., & Gossen, M. (2009). Binding of Drosophila ORC proteins to anaphase chromosomes requires cessation of mitotic cyclin-dependent kinase activity. *Molecular and Cellular Biology*, *29*, 140–149.

Ballabeni, A., Melixetian, M., Zamponi, R., Masiero, L., Marinoni, F., & Helin, K. (2004). Human geminin promotes pre-RC formation and DNA replication by stabilizing CDT1 in mitosis. *EMBO Journal*, *23*, 3122–3132.

Barry, K. A., Schultz, K. M., Payne, C. J., & McGarry, T. J. (2012). Geminin is required for mitotic proliferation of spermatogonia. *Developmental Biology*, *371*, 35–46.

Bell, S. P., & Kaguni, J. M. (2013). Helicase loading at chromosomal origins of replication. In S. D. Bell, M. Mechali, & M. L. DePamphilis (Eds.), *DNA replication*. Cold Spring Harbor Perspectives in Biology (Vol. 5, pp. 61–80).

Bell, S. D., Méchali, M., & DePamphilis, M. L. (2013). *DNA replication*. Cold Spring Harbor, NY: Cold Spring Harbor Laboratory Press. pp. 576.

Ben-David, U., & Benvenisty, N. (2011). The tumorigenicity of human embryonic and induced pluripotent stem cells. *Nature Reviews Cancer*, *11*, 268–277.

Besnard, E., Babled, A., Lapasset, L., Milhavet, O., Parrinello, H., Dantec, C., et al. (2012). Unraveling cell type-specific and reprogrammable human replication origin signatures associated with G-quadruplex consensus motifs. *Nature Structural & Molecular Biology, 19*, 837–844.

Blow, J. J., & Dutta, A. (2005). Preventing re-replication of chromosomal DNA. *Nature Reviews Molecular Cell Biology, 6*, 476–486.

Blow, J. J., Gillespie, P. J., Francis, D., & Jackson, D. A. (2001). Replication origins in Xenopus egg extract are 5-15 kilobases apart and are activated in clusters that fire at different times. *Journal of Cell Biology, 152*, 15–25.

Blumenthal, A. B., Kriegstein, H. J., & Hogness, D. S. (1974). The units of DNA replication in Drosophila melanogaster chromosomes. *Cold Spring Harbor Symposia on Quantitative Biology, 38*, 205–223.

Bolton, V. N., Oades, P. J., & Johnson, M. H. (1984). The relationship between cleavage, DNA replication, and gene expression in the mouse 2-cell embryo. *Journal of Embryology and Experimental Morphology, 79*, 139–163.

Boroviak, T., Loos, R., Bertone, P., Smith, A., & Nichols, J. (2014). The ability of inner-cell-mass cells to self-renew as embryonic stem cells is acquired following epiblast specification. *Nature Cell Biology, 16*, 516–528.

Borowiec, J. A., & Schildkraut, C. L. (2011). Open sesame: Activating dormant replication origins in the mouse immunoglobulin heavy chain (Igh) locus. *Current Opinion in Cell Biology, 23*, 284–292.

Burhans, W. C., Vassilev, L. T., Caddle, M. S., Heintz, N. H., & DePamphilis, M. L. (1990). Identification of an origin of bidirectional DNA replication in mammalian chromosomes. *Cell, 62*, 955–965.

Calvi, B. R. (2006). Developmental gene amplification. In M. L. DePamphilis (Ed.), *DNA replication and human disease* (pp. 233–255). Cold Spring Harbor, NY: Cold Spring Harbor Laboratory Press.

Campisi, J. (2013). Aging, cellular senescence, and cancer. *Annual Review of Physiology, 75*, 685–705.

Cayrou, C., Coulombe, P., Vigneron, A., Stanojcic, S., Ganier, O., Peiffer, I., et al. (2011). Genome-scale analysis of metazoan replication origins reveals their organization in specific but flexible sites defined by conserved features. *Genome Research, 21*, 1438–1449.

Chen, L., Chan, S. W., Zhang, X., Walsh, M., Lim, C. J., Hong, W., et al. (2010). Structural basis of YAP recognition by TEAD4 in the hippo pathway. *Genes and Development, 24*, 290–300.

Chen, Z. X., & Riggs, A. D. (2011). DNA methylation and demethylation in mammals. *The Journal of Biological Chemistry, 286*, 18347–18353.

Chevalier, S., & Blow, J. J. (1996). Cell cycle control of replication initiation in eukaryotes. *Current Opinion in Cell Biology, 8*, 815–821.

Ciemerych, M. A., & Sicinski, P. (2005). Cell cycle in mouse development. *Oncogene, 24*, 2877–2898.

Crevel, G., & Cotterill, S. (2012). Forced binding of the origin of replication complex to chromosomal sites in Drosophila S2 cells creates an origin of replication. *Journal of Cell Science, 125*, 965–972.

Danis, E., Brodolin, K., Menut, S., Maiorano, D., Girard-Reydet, C., & Mechali, M. (2004). Specification of a DNA replication origin by a transcription complex. *Nature Cell Biology, 6*, 721–730.

de Munnik, S. A., Hoefsloot, E. H., Roukema, J., Schoots, J., Knoers, N. V., Brunner, H. G., et al. (2015). Meier-Gorlin syndrome. *Orphanet Journal of Rare Diseases, 10*, 114.

de Renty, C., DePamphilis, M. L., & Ullah, Z. (2014). Cytoplasmic localization of p21 protects trophoblast giant cells from DNA damage induced apoptosis. *PloS One, 9*, e97434.

de Renty, C., Kaneko, K. J., & DePamphilis, M. L. (2014). The dual roles of geminin during trophoblast proliferation and differentiation. *Developmental Biology, 387*, 49–63.

DePamphilis, M. L. (1993a). Eukaryotic DNA replication: Anatomy of an origin. *Annual Review of Biochemistry, 62*, 29–63.

DePamphilis, M. L. (1993b). Origins of DNA replication in metazoan chromosomes. *The Journal of Biological Chemistry, 268*, 1–4.

DePamphilis, M. L. (1996a). *DNA replication in eukaryotic cells*. Plainview, NY: Cold Spring Harbor Laboratory Press.

DePamphilis, M. L. (1996b). Origins of DNA replication. In M. L. DePamphilis (Ed.), *DNA replication in eukaryotic cells* (pp. 45–86). New York: Cold Spring Harbor Laboratory Press.

DePamphilis, M. L. (2003). The 'ORC cycle': A novel pathway for regulating eukaryotic DNA replication. *Gene, 310*, 1–15.

DePamphilis, M. L. (2006). *DNA replication and human disease*. Cold Spring Harbor, NY: Cold Spring Harbor Laboratory Press. pp. 814.

DePamphilis, M. L., & Bell, S. D. (2010). *Genome duplication (concepts, mechanisms, evolution and disease)*. London/New York: Garland Science.

DePamphilis, M. L., Blow, J. J., Ghosh, S., Saha, T., Noguchi, K., & Vassilev, A. (2006). Regulating the licensing of DNA replication origins in metazoa. *Current Opinion in Cell Biology, 18*, 231–239.

Depamphilis, M. L., de Renty, C. M., Ullah, Z., & Lee, C. Y. (2012). "The Octet": Eight protein kinases that control mammalian DNA replication. *Frontiers in Physiology, 3*, 368.

Diril, M. K., Ratnacaram, C. K., Padmakumar, V. C., Du, T., Wasser, M., Coppola, V., et al. (2012). Cyclin-dependent kinase 1 (Cdk1) is essential for cell division and suppression of DNA re-replication but not for liver regeneration. *Proceedings of the National Academy of Sciences of the United States of America, 109*, 3826–3831.

Dorn, E. S., Chastain, P. D., 2nd., Hall, J. R., & Cook, J. G. (2009). Analysis of re-replication from deregulated origin licensing by DNA fiber spreading. *Nucleic Acids Research, 37*, 60–69.

Dorn, E. S., & Cook, J. G. (2011). Nucleosomes in the neighborhood: New roles for chromatin modifications in replication origin control. *Epigenetics, 6*, 552–559.

El Dika, M., Laskowska-Kaszub, K., Koryto, M., Dudka, D., Prigent, C., Tassan, J. P., et al. (2014). CDC6 controls dynamics of the first embryonic M-phase entry and progression via CDK1 inhibition. *Developmental Biology, 396*, 67–80.

Fernandez-Miranda, G., Trakala, M., Martin, J., Escobar, B., Gonzalez, A., Ghyselinck, N. B., et al. (2011). Genetic disruption of aurora B uncovers an essential role for aurora C during early mammalian development. *Development, 138*, 2661–2672.

Ferreira, J., & Carmo-Fonseca, M. (1997). Genome replication in early mouse embryos follows a defined temporal and spatial order. *Journal of Cell Science, 110*(Pt. 7), 889–897.

Foster, D. A., Yellen, P., Xu, L., & Saqcena, M. (2010). Regulation of G1 cell cycle progression: Distinguishing the restriction point from a nutrient-sensing cell growth checkpoint(s). *Genes & Cancer, 1*, 1124–1131.

Foulk, M. S., Urban, J. M., Casella, C., & Gerbi, S. A. (2015). Characterizing and controlling intrinsic biases of lambda exonuclease in nascent strand sequencing reveals phasing between nucleosomes and G-quadruplex motifs around a subset of human replication origins. *Genome Research, 25*, 725–735.

Foulk, M. S., Waggener, J. M., Johnson, J. M., Yamamoto, Y., Liew, G. M., Urnov, F. D., et al. (2013). Isolation and characterization of the ecdysone receptor and its heterodimeric partner ultraspiracle through development in Sciara coprophila. *Chromosoma, 122*, 103–119.

Fox, D. T., & Duronio, R. J. (2013). Endoreplication and polyploidy: Insights into development and disease. *Development, 140*, 3–12.

Ge, X. Q., Han, J., Cheng, E. C., Yamaguchi, S., Shima, N., Thomas, J. L., et al. (2015). Embryonic stem cells license a high level of dormant origins to protect the genome against replication stress. *Stem Cell Reports*, *5*(2), 185–194.

Geigl, J. B., Obenauf, A. C., Schwarzbraun, T., & Speicher, M. R. (2008). Defining 'chromosomal instability'. *Trends in Genetics*, *24*, 64–69.

Ghosh, S., Vassilev, A. P., Zhang, J., Zhao, Y., & DePamphilis, M. L. (2011). Assembly of the human origin recognition complex occurs through independent nuclear localization of its components. *The Journal of Biological Chemistry*, *286*, 23831–23841.

Gilbert, D. M. (2001). Making sense of eukaryotic DNA replication origins. *Science*, *294*, 96–100.

Gilbert, D. M., Miyazawa, H., & DePamphilis, M. L. (1995). Site-specific initiation of DNA replication in Xenopus egg extract requires nuclear structure. *Molecular and Cellular Biology*, *15*, 2942–2954.

Glover, D. M., Alphey, L., Axton, J. M., Cheshire, A., Dalby, B., Freeman, M., et al. (1989). Mitosis in Drosophila development. *Journal of Cell Science. Supplement*, *12*, 277–291.

Gottifredi, V., Karni-Schmidt, O., Shieh, S. S., & Prives, C. (2001). p53 down-regulates CHK1 through p21 and the retinoblastoma protein. *Molecular and Cellular Biology*, *21*, 1066–1076.

Hansen, C. G., Moroishi, T., & Guan, K. L. (2015). YAP and TAZ: A nexus for Hippo signaling and beyond. *Trends Cell Biology*, *25*(9), 499–513.

Hara, K., Nakayama, K. I., & Nakayama, K. (2006). Geminin is essential for the development of preimplantation mouse embryos. *Genes to Cells*, *11*, 1281–1293.

Hatada, I., Ohashi, H., Fukushima, Y., Kaneko, Y., Inoue, M., Komoto, Y., et al. (1996). An imprinted gene p57KIP2 is mutated in Beckwith-Wiedemann syndrome. *Nature Genetics*, *14*, 171–173.

Hattori, N., Davies, T. C., Anson-Cartwright, L., & Cross, J. C. (2000). Periodic expression of the cyclin-dependent kinase inhibitor p57(Kip2) in trophoblast giant cells defines a G2-like gap phase of the endocycle. *Molecular Biology of the Cell*, *11*, 1037–1045.

Havens, C. G., & Walter, J. C. (2011). Mechanism of CRL4(Cdt2), a PCNA-dependent E3 ubiquitin ligase. *Genes and Development*, *25*, 1568–1582.

Hiraga, S., Alvino, G. M., Chang, F., Lian, H. Y., Sridhar, A., Kubota, T., et al. (2014). Rif1 controls DNA replication by directing Protein Phosphatase 1 to reverse Cdc7-mediated phosphorylation of the MCM complex. *Genes and Development*, *28*, 372–383.

Hochegger, H., Dejsuphong, D., Sonoda, E., Saberi, A., Rajendra, E., Kirk, J., et al. (2007). An essential role for Cdk1 in S phase control is revealed via chemical genetics in vertebrate cells. *Journal of Cell Biology*, *178*, 257–268.

Hodgson, B., Li, A., Tada, S., & Blow, J. J. (2002). Geminin becomes activated as an inhibitor of Cdt1/RLF-B following nuclear import. *Current Biology*, *12*, 678–683.

Hu, D., & Cross, J. C. (2010). Development and function of trophoblast giant cells in the rodent placenta. *The International Journal of Developmental Biology*, *54*, 341–354.

Huang, Y. Y., Kaneko, K. J., Pan, H., & DePamphilis, M. L. (2015). Geminin is essential to prevent DNA re-replication-dependent apoptosis in pluripotent cells, but not in differentiated cells. *Stem Cells*, *33*(11), 3239–3253.

Hyrien, O., Maric, C., & Mechali, M. (1995). Transition in specification of embryonic metazoan DNA replication origins. *Science*, *270*, 994–997.

Hyrien, O., & Mechali, M. (1993). Chromosomal replication initiates and terminates at random sequences but at regular intervals in the ribosomal DNA of Xenopus early embryos. *EMBO Journal*, *12*, 4511–4520.

Jalasvuori, M., & Koonin, E. V. (2015). Classification of prokaryotic genetic replicators: Between selfishness and altruism. *Annals of the New York Academy of Sciences*, *1341*, 96–105.

Jin, J., Arias, E. E., Chen, J., Harper, J. W., & Walter, J. C. (2006). A family of diverse Cul4-Ddb1-interacting proteins includes Cdt2, which is required for S phase destruction of the replication factor Cdt1. *Molecular Cell*, *23*, 709–721.

Kanayama, N., Takahashi, K., Matsuura, T., Sugimura, M., Kobayashi, T., Moniwa, N., et al. (2002). Deficiency in p57Kip2 expression induces preeclampsia-like symptoms in mice. *Molecular Human Reproduction*, *8*, 1129–1135.

Kaneko, K. J., Cullinan, E. B., Latham, K. E., & DePamphilis, M. L. (1997). Transcription factor mTEAD-2 is selectively expressed at the beginning of zygotic gene expression in the mouse. *Development*, *124*, 1963–1973.

Kaneko, K. J., & DePamphilis, M. L. (2013). TEAD4 establishes the energy homeostasis essential for blastocoel formation. *Development*, *140*, 3680–3690.

Kara, N., Hossain, M., Prasanth, S. G., & Stillman, B. (2015). Orc1 binding to mitotic chromosomes precedes spatial patterning during G1 phase and assembly of the origin recognition complex in human cells. *The Journal of Biological Chemistry*, *290*, 12355–12369.

Kerns, S. L., Schultz, K. M., Barry, K. A., Thorne, T. M., & McGarry, T. J. (2012). Geminin is required for zygotic gene expression at the Xenopus mid-blastula transition. *PloS One*, *7*, e38009.

Kerns, S. L., Torke, S. J., Benjamin, J. M., & McGarry, T. J. (2007). Geminin prevents rereplication during xenopus development. *The Journal of Biological Chemistry*, *282*, 5514–5521.

Kim, J., & Orkin, S. H. (2011). Embryonic stem cell-specific signatures in cancer: Insights into genomic regulatory networks and implications for medicine. *Genome Medicine*, *3*, 75.

Kim, S. M., Zhang, D. Y., & Huberman, J. A. (2001). Multiple redundant sequence elements within the fission yeast ura4 replication origin enhancer. *BMC Molecular Biology*, *2*, 1.

Kisielewska, J., & Blow, J. J. (2012). Dynamic interactions of high Cdt1 and geminin levels regulate S phase in early Xenopus embryos. *Development*, *139*, 63–74.

Klotz-Noack, K., McIntosh, D., Schurch, N., Pratt, N., & Blow, J. J. (2012). Re-replication induced by geminin depletion occurs from G2 and is enhanced by checkpoint activation. *Journal of Cell Science*, *125*, 2436–2445.

Knox, K. S., & Baker, J. C. (2007). Genome-wide expression profiling of placentas in the p57Kip2 model of pre-eclampsia. *Molecular Human Reproduction*, *13*, 251–263.

Kozar, K., Ciemerych, M. A., Rebel, V. I., Shigematsu, H., Zagozdzon, A., Sicinska, E., et al. (2004). Mouse development and cell proliferation in the absence of D-cyclins. *Cell*, *118*, 477–491.

Kuo, A. J., Song, J., Cheung, P., Ishibe-Murakami, S., Yamazoe, S., Chen, J. K., et al. (2012). The BAH domain of ORC1 links H4K20me2 to DNA replication licensing and Meier-Gorlin syndrome. *Nature*, *484*, 115–119.

Lacroix, B., & Maddox, A. S. (2012). Cytokinesis, ploidy and aneuploidy. *The Journal of Pathology*, *226*, 338–351.

Lara-Gonzalez, P., Westhorpe, F. G., & Taylor, S. S. (2012). The spindle assembly checkpoint. *Current Biology*, *22*, R966–R980.

Lee, K. Y., Bang, S. W., Yoon, S. W., Lee, S. H., Yoon, J. B., & Hwang, D. S. (2012). Phosphorylation of ORC2 protein dissociates origin recognition complex from chromatin and replication origins. *The Journal of Biological Chemistry*, *287*, 11891–11898.

Lee, M. T., Bonneau, A. R., & Giraldez, A. J. (2014). Zygotic genome activation during the maternal-to-zygotic transition. *Annual Review of Cell and Developmental Biology*, *30*, 581–613.

Lemaitre, J. M., Bocquet, S., & Mechali, M. (2002). Competence to replicate in the unfertilized egg is conferred by Cdc6 during meiotic maturation. *Nature*, *419*, 718–722.

Lemaitre, J. M., Bocquet, S., Terret, M. E., Namdar, M., Ait-Ahmed, O., Kearsey, S., et al. (2004). The regulation of competence to replicate in meiosis by Cdc6 is conserved during evolution. *Molecular Reproduction & Development*, *69*, 94–100.
Leonard, A. C., & Mechali, M. (2013). DNA replication origins. *Cold Spring Harbor Perspectives in Biology*, *5*, a010116.
Li, A., & Blow, J. J. (2005). Cdt1 downregulation by proteolysis and geminin inhibition prevents DNA re-replication in Xenopus. *The EMBO Journal*, *24*, 395–404.
Li, F., Chen, J., Solessio, E., & Gilbert, D. M. (2003). Spatial distribution and specification of mammalian replication origins during G1 phase. *The Journal of Cell Biology*, *161*, 257–266.
Liu, Q., Guntuku, S., Cui, X. S., Matsuoka, S., Cortez, D., Tamai, K., et al. (2000). Chk1 is an essential kinase that is regulated by Atr and required for the G(2)/M DNA damage checkpoint. *Genes & Development*, *14*, 1448–1459.
Lu, X., Li, J. M., Elemento, O., Tavazoie, S., & Wieschaus, E. F. (2009). Coupling of zygotic transcription to mitotic control at the Drosophila mid-blastula transition. *Development*, *136*, 2101–2110.
MacAlpine, H. K., Gordan, R., Powell, S. K., Hartemink, A. J., & MacAlpine, D. M. (2010). Drosophila ORC localizes to open chromatin and marks sites of cohesin complex loading. *Genome Research*, *20*, 201–211.
MacAlpine, D. M., Rodriguez, H. K., & Bell, S. P. (2004). Coordination of replication and transcription along a Drosophila chromosome. *Genes & Development*, *18*, 3094–3105.
Mailand, N., & Diffley, J. F. (2005). CDKs promote DNA replication origin licensing in human cells by protecting Cdc6 from APC/C-dependent proteolysis. *Cell*, *122*, 915–926.
Majumder, S., & DePamphilis, M. L. (1994). TATA-dependent enhancer stimulation of promoter activity in mice is developmentally acquired. *Molecular and Cellular Biology*, *14*, 4258–4268.
Majumder, S., & DePamphilis, M. L. (1995). A unique role for enhancers is revealed during early mouse development. *Bioessays*, *17*, 879–889.
Majumder, S., Miranda, M., & DePamphilis, M. L. (1993). Analysis of gene expression in mouse preimplantation embryos demonstrates that the primary role of enhancers is to relieve repression of promoters. *EMBO Journal*, *12*, 1131–1140.
Majumder, S., Zhao, Z., Kaneko, K., & DePamphilis, M. L. (1997). Developmental acquisition of enhancer function requires a unique coactivator activity. *EMBO Journal*, *16*, 1721–1731.
Malumbres, M., Sotillo, R., Santamaria, D., Galan, J., Cerezo, A., Ortega, S., et al. (2004). Mammalian cells cycle without the D-type cyclin-dependent kinases Cdk4 and Cdk6. *Cell*, *118*, 493–504.
Marangos, P., Stevense, M., Niaka, K., Lagoudaki, M., Nabti, I., Jessberger, R., et al. (2015). DNA damage-induced metaphase I arrest is mediated by the spindle assembly checkpoint and maternal age. *Nature Communications*, *6*, 8706.
Martinez-Salas, E., Cupo, D. Y., & DePamphilis, M. L. (1988). The need for enhancers is acquired upon formation of a diploid nucleus during early mouse development. *Genes & Development*, *2*, 1115–1126.
Martinez-Salas, E., Linney, E., Hassell, J., & DePamphilis, M. L. (1989). The need for enhancers in gene expression first appears during mouse development with formation of the zygotic nucleus. *Genes & Development*, *3*, 1493–1506.
Masramon, L., Vendrell, E., Tarafa, G., Capella, G., Miro, R., Ribas, M., et al. (2006). Genetic instability and divergence of clonal populations in colon cancer cells in vitro. *Journal of Cell Science*, *119*, 1477–1482.
Matsui, A., Ihara, T., Suda, H., Mikami, H., & Semba, K. (2013). Gene amplification: Mechanisms and involvement in cancer. *Biomolecular Concepts*, *4*, 567–582.

McIntosh, D., & Blow, J. J. (2013). Dormant origins, the licensing checkpoint, and the response to replicative stresses. In S. D. Bell, M. Méchali, & M. L. DePamphilis (Eds.), *Dormant replication origins, DNA replication*. Cold Spring Harbor Perspectives in Biology (Vol. 4(10)).

Mehrotra, S., Maqbool, S. B., Kolpakas, A., Murnen, K., & Calvi, B. R. (2008). Endocycling cells do not apoptose in response to DNA rereplication genotoxic stress. *Genes & Development, 22*, 3158–3171.

Melin, F., Miranda, M., Montreau, N., DePamphilis, M. L., & Blangy, D. (1993). Transcription enhancer factor-1 (TEF-1) DNA binding sites can specifically enhance gene expression at the beginning of mouse development. *EMBO Journal, 12*, 4657–4666.

Messerschmidt, D. M., Knowles, B. B., & Solter, D. (2014). DNA methylation dynamics during epigenetic reprogramming in the germline and preimplantation embryos. *Genes & Development, 28*, 812–828.

Miotto, B., & Struhl, K. (2010). HBO1 histone acetylase activity is essential for DNA replication licensing and inhibited by Geminin. *Molecular Cell, 37*, 57–66.

Murai, S., Stein, P., Buffone, M. G., Yamashita, S., & Schultz, R. M. (2010). Recruitment of Orc6l, a dormant maternal mRNA in mouse oocytes, is essential for DNA replication in 1-cell embryos. *Developmental Biology, 341*, 205–212.

Newport, J., & Dasso, M. (1989). On the coupling between DNA replication and mitosis. *Journal of Cell Science. Supplement, 12*, 149–160.

Nishitani, H., Sugimoto, N., Roukos, V., Nakanishi, Y., Saijo, M., Obuse, C., et al. (2006). Two E3 ubiquitin ligases, SCF-Skp2 and DDB1-Cul4, target human Cdt1 for proteolysis. *The EMBO Journal, 25*, 1126–1136.

Noguchi, K., Vassilev, A., Ghosh, S., Yates, J. L., & DePamphilis, M. L. (2006). The BAH domain facilitates the ability of human Orc1 protein to activate replication origins in vivo. *The EMBO Journal, 25*, 5372–5382.

Nothias, J. Y., Majumder, S., Kaneko, K. J., & DePamphilis, M. L. (1995). Regulation of gene expression at the beginning of mammalian development. *The Journal of Biological Chemistry, 270*, 22077–22080.

Nothias, J. Y., Miranda, M., & DePamphilis, M. L. (1996). Uncoupling of transcription and translation during zygotic gene activation in the mouse. *The EMBO Journal, 15*, 5715–5725.

Oda, M., Shiota, K., & Tanaka, S. (2010). Trophoblast cell lineage in cloned mouse embryos. *Development, Growth & Differentiation, 52*, 285–291.

O'Farrell, P. H., Stumpff, J., & Su, T. T. (2004). Embryonic cleavage cycles: How is a mouse like a fly? *Current Biology, 14*, R35–R45.

Okada, N., & Shimizu, N. (2013). Dissection of the beta-globin replication-initiation region reveals specific requirements for replicator elements during gene amplification. *PloS One, 8*, e77350.

Palmer, N. P., Schmid, P. R., Berger, B., & Kohane, I. S. (2012). A gene expression profile of stem cell pluripotentiality and differentiation is conserved across diverse solid and hematopoietic cancers. *Genome Biology, 13*, R71.

Pan, H., O'Brien, M. J., Wigglesworth, K., Eppig, J. J., & Schultz, R. M. (2005). Transcript profiling during mouse oocyte development and the effect of gonadotropin priming and development in vitro. *Developmental Biology, 286*, 493–506.

Patterson, E. S., Waller, L. E., & Kroll, K. L. (2014). Geminin loss causes neural tube defects through disrupted progenitor specification and neuronal differentiation. *Developmental Biology, 393*(1), 44–56.

Picard, F., Cadoret, J. C., Audit, B., Arneodo, A., Alberti, A., Battail, C., et al. (2014). The spatiotemporal program of DNA replication is associated with specific combinations of chromatin marks in human cells. *PLoS Genetics, 10*, e1004282.

Rielland, M., Brochard, V., Lacroix, M. C., Renard, J. P., & Jouneau, A. (2009). Early alteration of the self-renewal/differentiation threshold in trophoblast stem cells derived from mouse embryos after nuclear transfer. *Developmental Biology, 334*, 325–334.

Roberts, R. M., & Fisher, S. J. (2011). Trophoblast stem cells. *Biology of Reproduction, 84*, 412–421.

Romanelli, V., Belinchon, A., Benito-Sanz, S., Martinez-Glez, V., Gracia-Bouthelier, R., Heath, K. E., et al. (2010). CDKN1C (p57(Kip2)) analysis in Beckwith-Wiedemann syndrome (BWS) patients: Genotype-phenotype correlations, novel mutations, and polymorphisms. *American Journal of Medical Genetics. Part A, 152A*, 1390–1397.

Romanelli, V., Belinchon, A., Campos-Barros, A., Heath, K. E., Garcia-Minaur, S., Martinez-Glez, V., et al. (2009). CDKN1C mutations in HELLP/preeclamptic mothers of Beckwith-Wiedemann Syndrome (BWS) patients. *Placenta, 30*, 551–554.

Sasaki, T., Sawado, T., Yamaguchi, M., & Shinomiya, T. (1999). Specification of regions of DNA replication initiation during embryogenesis in the 65-kilobase DNApolalpha-dE2F locus of Drosophila melanogaster. *Molecular and Cellular Biology, 19*, 547–555.

Schultz, K. M., Banisadr, G., Lastra, R. O., McGuire, T., Kessler, J. A., Miller, R. J., et al. (2011). Geminin-deficient neural stem cells exhibit normal cell division and normal neurogenesis. *PloS One, 6*, e17736.

Sher, N., Von Stetina, J. R., Bell, G. W., Matsuura, S., Ravid, K., & Orr-Weaver, T. L. (2013). Fundamental differences in endoreplication in mammals and Drosophila revealed by analysis of endocycling and endomitotic cells. *Proceedings of the National Academy of Sciences of the United States of America, 110*, 9368–9373.

Shinnick, K. M., Eklund, E. A., & McGarry, T. J. (2010). Geminin deletion from hematopoietic cells causes anemia and thrombocytosis in mice. *The Journal of Clinical Investigation, 120*, 4303–4315.

Shinomiya, T., & Ina, S. (1991). Analysis of chromosomal replicons in early embryos of Drosophila melanogaster by two-dimensional gel electrophoresis. *Nucleic Acids Research, 19*, 3935–3941.

Siddiqui, K., On, K. F., & Diffley, J. F. (2013). Regulating DNA replication in eukarya. In S. D. Bell, M. Méchali, & M. L. DePamphilis (Eds.), *DNA replication*. Cold Spring Harbor Perspectives in Biology (Vol. 5, pp. 361–380).

Siddiqui, K., & Stillman, B. (2007). ATP-dependent assembly of the human origin recognition complex. *The Journal of Biological Chemistry, 282*, 32370–32383.

Soloveva, V., & Linzer, D. I. (2004). Differentiation of placental trophoblast giant cells requires downregulation of p53 and Rb. *Placenta, 25*, 29–36.

Sonneville, R., Querenet, M., Craig, A., Gartner, A., & Blow, J. J. (2012). The dynamics of replication licensing in live Caenorhabditis elegans embryos. *The Journal of Cell Biology, 196*, 233–246.

Stanojcic, S., Lemaitre, J. M., Brodolin, K., Danis, E., & Mechali, M. (2008). In Xenopus egg extracts, DNA replication initiates preferentially at or near asymmetric AT sequences. *Molecular and Cellular Biology, 28*, 5265–5274.

Takada, S., Kwak, S., Koppetsch, B. S., & Theurkauf, W. E. (2007). grp (chk1) replication-checkpoint mutations and DNA damage trigger a Chk2-dependent block at the Drosophila midblastula transition. *Development, 134*, 1737–1744.

Takai, H., Naka, K., Okada, Y., Watanabe, M., Harada, N., Saito, S., et al. (2002). Chk2-deficient mice exhibit radioresistance and defective p53-mediated transcription. *The EMBO Journal, 21*, 5195–5205.

Takai, H., Tominaga, K., Motoyama, N., Minamishima, Y. A., Nagahama, H., Tsukiyama, T., et al. (2000). Aberrant cell cycle checkpoint function and early embryonic death in Chk1(-/-) mice. *Genes & Development, 14*, 1439–1447.

Takeda, D. Y., Shibata, Y., Parvin, J. D., & Dutta, A. (2005). Recruitment of ORC or CDC6 to DNA is sufficient to create an artificial origin of replication in mammalian cells. *Genes & Development, 19*, 2827–2836.

Tanaka, S., Kunath, T., Hadjantonakis, A. K., Nagy, A., & Rossant, J. (1998). Promotion of trophoblast stem cell proliferation by FGF4. *Science*, *282*, 2072–2075.

Tang, F., Barbacioru, C., Bao, S., Lee, C., Nordman, E., Wang, X., et al. (2010). Tracing the derivation of embryonic stem cells from the inner cell mass by single-cell RNA-Seq analysis. *Cell Stem Cell*, *6*, 468–478.

Theis, J. F., & Newlon, C. S. (2001). Two compound replication origins in Saccharomyces cerevisiae contain redundant origin recognition complex binding sites. *Molecular and Cellular Biology*, *21*, 2790–2801.

Tower, J. (2004). Developmental gene amplification and origin regulation. *Annual Review of Genetics*, *38*, 273–304.

Trakala, M., Rodriguez-Acebes, S., Maroto, M., Symonds, C. E., Santamaria, D., Ortega, S., et al. (2015). Functional reprogramming of polyploidization in megakaryocytes. *Developmental Cell*, *32*, 155–167.

Ullah, Z., de Renty, C., & Depamphilis, M. L. (2011). Checkpoint kinase 1 prevents cell cycle exit linked to terminal cell differentiation. *Molecular and Cellular Biology*, *31*, 4129–4143.

Ullah, Z., Kohn, M. J., Yagi, R., Vassilev, L. T., & DePamphilis, M. L. (2008). Differentiation of trophoblast stem cells into giant cells is triggered by p57/Kip2 inhibition of CDK1 activity. *Genes & Development*, *22*, 3024–3036.

Ullah, Z., Lee, C. Y., & Depamphilis, M. L. (2009a). Cip/Kip cyclin-dependent protein kinase inhibitors and the road to polyploidy. *Cell Div*, *4*, 10.

Ullah, Z., Lee, C. Y., Lilly, M. A., & DePamphilis, M. L. (2009b). Developmentally programmed endoreduplication in animals. *Cell Cycle*, *8*, 1501–1509.

Unnikrishnan, A., Gafken, P. R., & Tsukiyama, T. (2010). Dynamic changes in histone acetylation regulate origins of DNA replication. *Nature Structural and Molecular Biology*, *17*, 430–437.

Valent, P., Bonnet, D., De Maria, R., Lapidot, T., Copland, M., Melo, J. V., et al. (2012). Cancer stem cell definitions and terminology: The devil is in the details. *Nature Reviews Cancer*, *12*, 767–775.

Vassilev, A., Kaneko, K. J., Shu, H., Zhao, Y., & DePamphilis, M. L. (2001). TEAD/TEF transcription factors utilize the activation domain of YAP65, a Src/Yes-associated protein localized in the cytoplasm. *Genes & Development*, *15*, 1229–1241.

Vassilev, A., Lee, C. Y., Vassilev, B., Zhu, W., Ormanoglu, P., Martin, S. E., et al. (2016). Identification of genes that are essential to restrict genome duplication to once per cell division. *Oncotarget*. http://dx.doi.org/10.18632/oncotarget.9008. [Epub ahead of print].

Vaughn, J. P., Dijkwel, P. A., & Hamlin, J. L. (1990). Replication initiates in a broad zone in the amplified CHO dihydrofolate reductase domain. *Cell*, *61*, 1075–1087.

Wang, L., Lin, C. M., Lopreiato, J. O., & Aladjem, M. I. (2006). Cooperative sequence modules determine replication initiation sites at the human beta-globin locus. *Human Molecular Genetics*, *15*, 2613–2622.

Wei, Y., Multi, S., Yang, C. R., Ma, J., Zhang, Q. H., Wang, Z. B., et al. (2011). Spindle assembly checkpoint regulates mitotic cell cycle progression during preimplantation embryo development. *PloS One*, *6*, e21557.

Whitmire, E., Khan, B., & Coue, M. (2002). Cdc6 synthesis regulates replication competence in Xenopus oocytes. *Nature*, *419*, 722–725.

Wiekowski, M., Miranda, M., & DePamphilis, M. L. (1991). Regulation of gene expression in preimplantation mouse embryos: Effects of the zygotic clock and the first mitosis on promoter and enhancer activities. *Developmental Biology*, *147*, 403–414.

Wiekowski, M., Miranda, M., Nothias, J. Y., & DePamphilis, M. L. (1997). Changes in histone synthesis and modification at the beginning of mouse development correlate with the establishment of chromatin mediated repression of transcription. *Journal of Cell Science*, *110*(Pt. 10), 1147–1158.

Wirak, D. O., Chalifour, L. E., Wassarman, P. M., Muller, W. J., Hassell, J. A., & DePamphilis, M. L. (1985). Sequence-dependent DNA replication in preimplantation mouse embryos. *Molecular and Cellular Biology*, *5*, 2924–2935.

Yagi, R., Kohn, M. J., Karavanova, I., Kaneko, K. J., Vullhorst, D., DePamphilis, M. L., et al. (2007). Transcription factor TEAD4 specifies the trophectoderm lineage at the beginning of mammalian development. *Development*, *134*, 3827–3836.

Yang, V. S., Carter, S. A., Ng, Y., Hyland, S. J., Tachibana-Konwalski, K., Fisher, R. A., et al. (2012). Distinct activities of the anaphase-promoting complex/cyclosome (APC/C) in mouse embryonic cells. *Cell Cycle*, *11*, 846–855.

You, Z., Ishimi, Y., Mizuno, T., Sugasawa, K., Hanaoka, F., & Masai, H. (2003). Thymine-rich single-stranded DNA activates Mcm4/6/7 helicase on Y-fork and bubble-like substrates. *The EMBO Journal*, *22*, 6148–6160.

Zack, T. I., Schumacher, S. E., Carter, S. L., Cherniack, A. D., Saksena, G., Tabak, B., et al. (2013). Pan-cancer patterns of somatic copy number alteration. *Nature Genetics*, *45*, 1134–1140.

Zeng, F., & Schultz, R. M. (2005). RNA transcript profiling during zygotic gene activation in the preimplantation mouse embryo. *Developmental Biology*, *283*, 40–57.

Zhang, M., Kothari, P., & Lampson, M. A. (2015). Spindle assembly checkpoint acquisition at the mid-blastula transition. *PloS One*, *10*, e0119285.

Zhang, W., Sankaran, S., Gozani, O., & Song, J. (2015). A Meier-Gorlin syndrome mutation impairs the ORC1-nucleosome association. *ACS Chemical Biology*, *10*, 1176–1180.

Zhu, W., & Depamphilis, M. L. (2009). Selective killing of cancer cells by suppression of geminin activity. *Cancer Research*, *69*, 4870–4877.

Zielke, N., Edgar, B. A., & DePamphilis, M. L. (2013). Endoreplication. *Cold Spring Harbor Perspectives in Biology*, *5*, a012948.

CHAPTER THREE

Zygotic Genome Activation Revisited: Looking Through the Expression and Function of Zscan4

M.S.H. Ko[1]
Keio University School of Medicine, Tokyo, Japan
[1]Corresponding author: e-mail address: ko.minoru@keio.jp

Contents

Abstract

Zygotic genome activation (ZGA, a.k.a. zygotic gene activation) is a critical event in development, when the paternally derived genome and maternally derived genome begin to be activated and transcribed after fertilization. Major ZGA occurs at the two-cell stage in mice and the four- to eight-cell stage in human preimplantation embryos. It has been thought that ZGA exists to provide RNAs and proteins supporting embryonic development after supplies stored in oocytes are used up; however, this paradigm does not seem to explain recent findings. For example, many ZGA genes—once activated—are quickly turned off, and thus ZGA forms a transient wave of transcriptional activation. In addition, ZGA genes are not evolutionarily conserved. In this review, we address these issues by focusing on *Zscan4* (zinc finger and SCAN domain-containing 4),

Current Topics in Developmental Biology, Volume 120
ISSN 0070-2153
http://dx.doi.org/10.1016/bs.ctdb.2016.04.004

which was identified for its specific expression in preimplantation embryos during ZGA. Detailed molecular analyses of *Zscan4* expression and function have revealed common features of Zscan4-associated events (Z4 events) in mouse embryonic stem cells and ZGA in preimplantation embryos. One feature is a rapid derepression and rerepression of constitutive heterochromatin, which includes pericentromeric major satellites and telomeres, and facultative heterochromatin, which includes retrotransposons and Z4 event-associated genes. We propose that the Z4 event superimposed on ZGA plays a critical role in the maintenance of genome and chromosome integrity in preimplantation embryos by promoting correction of DNA damage and chromosome abnormalities.

1. INTRODUCTION

ZGA stands for zygotic genome activation or zygotic gene activation, in which the paternally derived genome and maternally derived genome are transcribed for the first time after fertilization. In mice, ZGA begins during the one-cell stage, as demonstrated by expression assays of a plasmid-borne reporter gene (Nothias, Majumder, Kaneko, & DePamphilis, 1995; Ram & Schultz, 1993) and BrUTP incorporation (Aoki, Worrad, & Schultz, 1997). This is called minor ZGA, which has been shown to be insensitive to inhibition by RNA polymerase II-blocker alpha-amanitin (Hamatani, Carter, Sharov, & Ko, 2004; Zeng & Schultz, 2005). Recent work has also shown that minor ZGA is a promiscuous expression of the entire genome (Abe et al., 2015). In contrast to the passive nature of minor ZGA, major ZGA or simply ZGA that occurs during the two-cell embryo stage in mice seems to be the product of a more active process because it takes place after forming the chromatin-based repression of the entire genome (Cho, Sakai, Nagara, & Aoki, 2002; DePamphilis, Kaneko, & Vassilev, 2002; Latham & Schultz, 2001). Unlike minor ZGA, major ZGA shows alpha-amanitin sensitivity, which identifies 1727 genes in mouse late two-cell embryos (G2 phase) in one report (Hamatani et al., 2004) and ~3000 genes in another report (Zeng & Schultz, 2005). The timing of ZGA is varied even among mammalian species; for example, major ZGA occurs in the four- to eight-cell embryo stage in humans (Braude, Bolton, & Moore, 1988; Niakan, Han, Pedersen, Simon, & Pera, 2012; Vassena et al., 2011).

As one of the major and most critical developmental events, ZGA has been studied in a variety of species and has been discussed in a number of extensive reviews (Biechele, Lin, Rinaudo, & Ramalho-Santos, 2015; Lee, Bonneau, & Giraldez, 2014; Li, Lu, & Dean, 2013; Svoboda,

Franke, & Schultz, 2015). The mainstream thinking about ZGA has not changed for long time: the major role of ZGA is to provide RNAs and proteins that are required to go through the preimplantation period and to sustain the basic function of embryonic cells such as the expression of housekeeping genes. Maternally stored RNAs and proteins are not sufficient to sustain embryonic development and new synthesis of RNAs and proteins is required. In this review, we will provide a new angle to the conventional view of ZGA by looking at ZGA through the perspectives of the expression and function of *Zscan4*, a gene that was originally identified for its specific expression during ZGA in late two-cell embryos. We discuss the possibility that ZGA occurs not only for making RNAs and proteins to sustain the basic function of embryos but also for carrying out or at least being involved in more unique embryonic functions.

2. WAVES OF GENE ACTIVATION: WHY THEY ARE TRANSIENT?

Although early molecular studies of preimplantation embryos identified a few genes that begin to be transcribed during ZGA (Davis, De Sousa, & Schultz, 1996), the advent of global gene expression profiling has dramatically increased the number of genes identified for their expression during preimplantation embryo development (Ko, 2006). Large-scale sequencing of cDNA clones isolated from preimplantation embryos revealed many genes that were identified for the first time (Evsikov et al., 2004; Ko et al., 2000; Okazaki et al., 2002). Subsequent DNA microarray studies revealed the expression changes of essentially all genes and provided a global genome-wide view of gene activities during preimplantation development (Hamatani et al., 2004; Wang, Cowan, Chipperfield, & Powers, 2005; Wang et al., 2004; Zeng, Baldwin, & Schultz, 2004).

One of the most important concepts that has been established by such studies is the waves of activation during preimplantation development (Fig. 1). Approximately one-half of all genes show dynamic expression changes during preimplantation development (Hamatani et al., 2004). Specifically, RNAs (maternal RNAs) transcribed from a few thousand genes during oocyte maturation are stored in oocytes but are degraded rapidly during preimplantation development (first wave). Next, ZGA forms a transient peak of a few thousand genes in two- to four-cell embryos (second wave). Finally, the expression of other sets of genes also form a transient peak

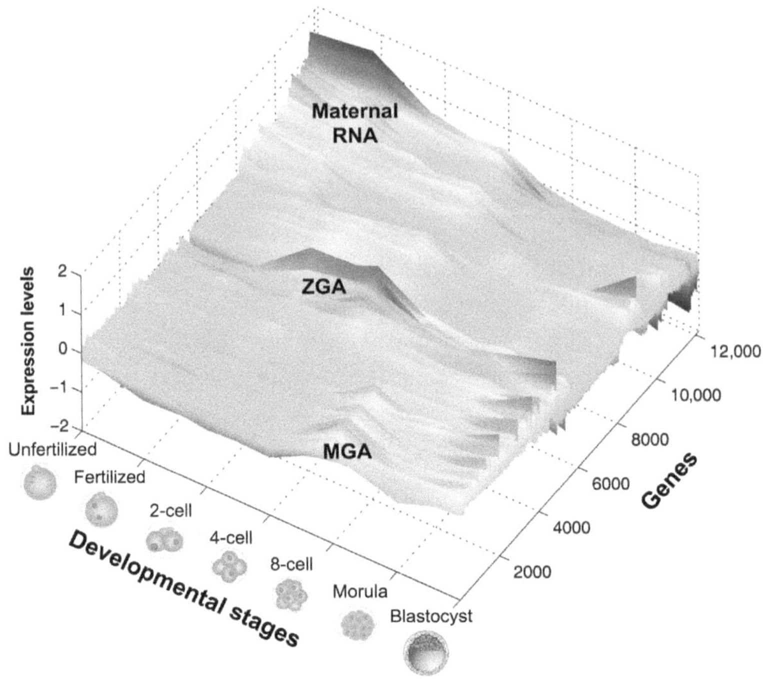

Fig. 1 A bird's eye view of gene expression patterns during mouse preimplantation development. The expression changes of 12,179 genes are shown. A *curve* represents the expression of each gene over the preimplantation period and is color coded by its expression levels (log-scale). Maternal RNA, RNAs stored in oocytes; MGA, mid-preimplantation gene activation; ZGA, zygotic genome activation. *Adapted from Ko, M. S. (2005). Molecular biology of preimplantation embryos: Primer for philosophical discussions.* Reproductive Biomedicine Online, 10*(Suppl. 1), 80–87.* (See the color plate.)

in 8-cell to morula embryos (third wave, named "mid-preimplantation gene activation (MGA)" (Hamatani et al., 2004)). We have identified a *Trim43* gene as an MGA-specific gene (Stanghellini, Falco, Lee, Monti, & Ko, 2009), but this gene will not be discussed in this review.

Though improvements in our understanding of molecular profiles of preimplantation embryos continue, we still seem to be far away from understanding the implications of these waves of activation. For example, some speculate that the cascade-like activation mechanism causes the following multiple waves: gene products from the first wave cause the activation of the second wave (ZGA), gene products of the second wave cause the third wave, and so on (Hamatani et al., 2004; Ko et al., 2000). However, no clear-cut example of such a mechanism has been found yet.

One outstanding question is why these waves such as ZGA are transient. Why does not the expression of many ZGA genes continue? The fact that ZGA forms a transient wave and does not continue its high expression is counterintuitive, particularly considering the well-accepted view of ZGA that the ZGA gene products are required for sustaining embryo development including the expression of housekeeping genes by supplementing the maternal RNAs and their protein products, which are eventually depleted during preimplantation development. If the main reason for the expression of these ZGA genes is to provide basic RNAs and proteins that are required for the maintenance of embryonic cells, why are many of them transient and why do they have to be turned off? Furthermore, it has also been shown that two- and four-cell embryos show gene expression profiles very distinct from oocytes/one-cell embryos and eight-cell/morula/ blastocyst embryos by principal component analysis of their gene expression profiles (Hamatani et al., 2004). Accordingly, it is reasonable to consider that ZGA is required not only for continuous embryonic development but also for some specific function unique to these early embryos. What are these unique functions?

3. WHY ARE ZGA GENES EVOLUTIONARILY SO DIVERSE?

The second outstanding question regards why many genes activated during ZGA are so diverse even among mammalian species. Recent bioinformatics analysis of published data has revealed that sets of genes activated during ZGA are very divergent: only 40% of human ZGA genes are shared in mice and only 8.5% of human ZGA genes are shared in both mice and bovines (Xie et al., 2010). By contrast, sets of genes whose transcripts are maternally stored are much more conserved: 69% of human genes are shared with mice and 49% of human genes are shared with both mice and bovines (Xie et al., 2010). Similar observations have been made by comparing transcripts among mice, flies, and fish; there is significant conservation of orthologous transcripts in maternally stored RNAs, but almost no conservation in ZGA transcripts (Heyn et al., 2014). These results are quite surprising, as one would expect that developmentally critical events such as ZGA to be evolutionarily conserved among species. One explanation could be the slight difference in the timing of ZGA during development. However, one could also argue that the type of genes activated may be less important than the phenomenon of massive gene activation itself. In other words, the reason for the presence of ZGA is not only to make RNAs and proteins to be used for embryo development but also to carry out some other functions. What are these functions?

4. Zscan4: A GENE IDENTIFIED FOR ITS SPECIFIC EXPRESSION DURING ZGA

To address the aforementioned questions, it is important to find genes specifically expressed during these particular events. Using a genome-wide view of gene expression changes, we first aimed to identify genes that are uniquely expressed during ZGA. Among these candidates, we finally identified such a gene: *Zscan4* (Falco et al., 2007) (Fig. 2). During preimplantation development, RNA and protein products of *Zscan4* cannot be detected in oocytes (Metaphase II) or one-cell embryos (fertilized eggs), but are abundantly present in late two-cell embryos (Fig. 2A and B). However, after one more cell division, the *Zscan4* gene product is dramatically reduced and cannot be detected anymore. *Zscan4* is not expressed in blastocysts, but when blastocysts are cultured as an outgrowth, a small number of cells begin to express *Zscan4* (Falco et al., 2007). These cells may have the potential to become mouse embryonic stem (ES) cells.

We originally named the gene "Zygotic genome activation 1 (*Zga1*)," but later started to use a structure-based gene name according to the gene nomenclature, "Zinc finger and SCAN domain-containing 4 (*Zscan4*)" (Falco et al., 2007). Detailed genome sequence analysis has revealed that six genes (*Zscan4a*, *Zscan4b*, *Zscan4c*, *Zscan4d*, *Zscan4e*, and *Zscan4f*) and three pseudogenes are clustered in a 0.85-Mb region on mouse chromosome 7 (Falco et al., 2007). *Zscan4* genes are present exclusively in mammals, and there is only one copy of the *ZSCAN4* gene in the human genome (Falco et al., 2007). We thus collectively call it a *Zscan4* gene, even in mice. (According to standard gene nomenclature, we use *Zscan4* for mouse genes and *ZSCAN4* for the human gene.) Although *Zscan4* is often called a two-cell-specific gene, it should be called a ZGA-specific gene; in human embryos, *ZSCAN4* is specifically expressed in six- to eight-cell embryos, which coincides with major ZGA in human embryos (Vassena et al., 2011).

4.1 *Zscan4* Is also Expressed in ES Cells and Tissue Stem Cells

In mouse ES cells, *Zscan4* is expressed only in a small fraction (1–5%) of ES cells at a given time (Fig. 2C; Carter et al., 2008; Falco et al., 2007). This is in sharp contrast to *Pou5f1* (a.k.a. *Oct4*, *Oct3/4*), which is ubiquitously expressed in undifferentiated mouse ES cells (Fig. 2D; Carter et al., 2008;

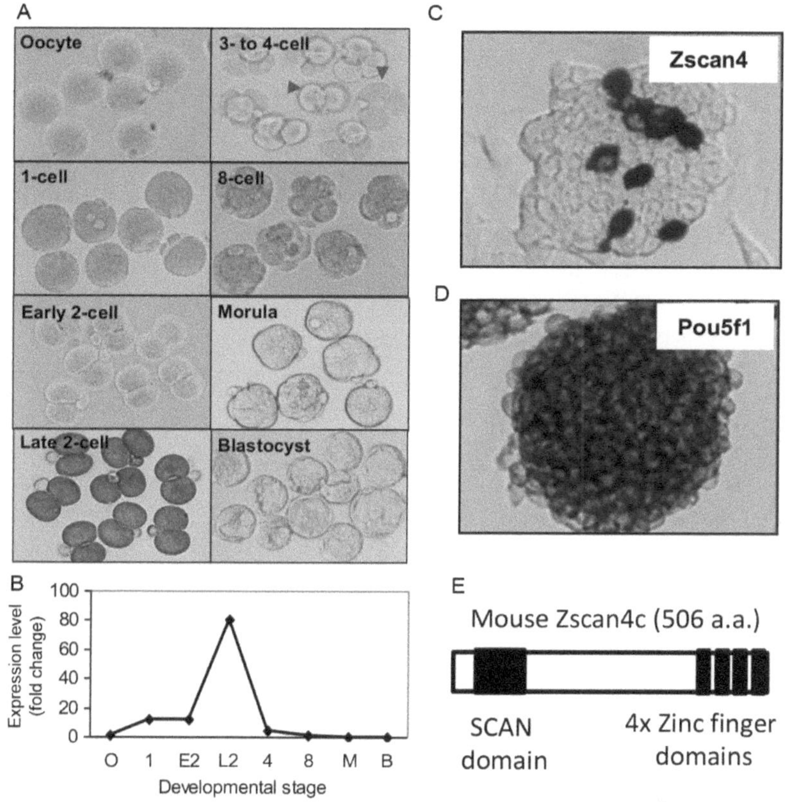

Fig. 2 Expression profiles of *Zscan4* in mouse preimplantation embryos and mouse ES cells. (A) Whole-mount RNA in situ hybridization of mouse preimplantation embryos with a *Zscan4* probe. (B) qRT-PCR analyses of *Zscan4* expression during preimplantation development. (C) Whole-mount RNA in situ hybridization of mouse ES cell colonies with a *Zscan4* probe. (D) As a control, whole-mount RNA in situ hybridization with *Pou5fa* (a.k.a. *Oct3/4* or *Oct4*). (E) Structure of mouse Zscan4 protein (Zscan4c). *Adapted from Falco, G., Lee, S. L., Stanghellini, I., Bassey, U. C., Hamatani, T., & Ko, M. S. (2007). Zscan4: A novel gene expressed exclusively in late 2-cell embryos and embryonic stem cells.* Developmental Biology, 307, *539–550.* (See the color plate.)

Falco et al., 2007). Further analysis shows that *Zscan4* is expressed transiently, and thus, undifferentiated mouse ES cells oscillate between Zscan4-negative cells (>95%) and Zscan4-positive cells (1–5%) (Zalzman et al., 2010). Mouse ES cells stay as Zscan4-positive [Zscan4(+)] cells for several hours to a half day. So mouse ES cells are mostly Zscan4-negative [Zscan4(−)] cells and occasionally become Zscan4(+) cells and then return to Zscan4(−) cells rapidly (Zalzman et al., 2010). It has been shown that essentially all cells undergo Z4 event (Zalzman et al., 2010).

Although Zscan4(+) cells were initially called being in a "Zscan4(+) state (Amano et al., 2013; Zalzman et al., 2010)," we now say that they are experiencing a "Zscan4-associated event" or "Z4 event" for short (Akiyama et al., 2015). We make a distinction between "states" and "events" due to the nomenclature similarities with mouse ES cell "states." The most well-established ES cell states are a "naïve state" (also called a "ground state") and "primed state" (Nichols & Smith, 2009). It has been shown that mouse ES cells oscillate between a naïve state and a primed state and, thus, are the mixture of these two states. The naïve state, but not primed state, expresses a *Zfp42* (a.k.a. *Rex1*) gene, so the ES cells in a naïve state can be distinguished from ES cells in a primed state. Superficially, this situation appears similar to Zscan4(+) cells and Zscan4(−) cells. However, the striking difference is that ES cells can be cultured and propagated stably in either a naïve state or primed state by using a specific cell culture media defined for each state (Nichols & Smith, 2009), whereas Zscan4(+) cells cannot be maintained and propagated stably in any culture condition tested thus far (Amano et al., 2013). As we discuss below, there are mechanistic reasons why ES cells cannot stay as Zscan4(+) cells for long time. Because it comes and goes, we believe that it is more appropriate to call it an event.

Zscan4 is also expressed in a small fraction of tissue stem cells, eg, in the adult mouse pancreas (Ko et al., 2013). In the human adult pancreas, *ZSCAN4* is also expressed in a small fraction of cells, which are mostly overlapped with other stem cell markers such as LGR5 (Ko et al., 2013). Interestingly, the number of ZSCAN4(+) cells dramatically increases when the pancreas is inflamed, eg, acute pancreatitis. When the inflammation subsides and a part of the pancreatic tissues is restored, the number of ZSCAN4(+) cells decreases and goes back to the normal level; ie, only a small number of cells are ZSCAN4(+).

4.2 *Zscan4* Functions in Mouse Preimplantation Embryos

The transient expression of *Zscan4* seems to be critical for proper development of embryos. Knocking down *Zscan4* in two-cell embryos delays embryo development for a day and resultant blastocysts do not implant (Falco et al., 2007). Sustained expression of *Zscan4* beyond the two-cell stage by overexpressing *Zscan4* also shows essentially the same phenotype (Falco et al., 2007). Thus, the timed and transient expression of *Zscan4* is critical for the proper development of preimplantation embryos.

4.3 *Zscan4* Functions in Mouse ES Cells

In mouse ES cells, preventing transient *Zscan4* expression by the continuous expression of shRNA against *Zscan4* does not cause any immediate phenotype, which is reasonable as Zscan4 is not expressed in most ES cells (Zalzman et al., 2010). However, after six passages or so, the proliferation of ES cells slows down, eventually resulting in cell culture crisis. The culture crisis is accompanied with massive karyotype abnormality, which is partially rescued by brief overexpression of an exogenous *Zscan4* gene (Zalzman et al., 2010). Detailed molecular analyses revealed that knocking down *Zscan4* causes the shortening of telomeres. Interestingly, transient overexpression of *Zscan4* causes a rapid, telomerase-independent, telomere extension in mouse ES cells. Further analyses have revealed that transient overexpression of *Zscan4* promotes telomere recombination or telomere-sister chromatid exchange (T-SCE), but suppresses spontaneous SCE in the nontelomeric genomic region (Zalzman et al., 2010). This homologous recombination is associated with meiosis-specific recombination machinery that is upregulated in Zscan4(+) cells (Zalzman et al., 2010). It has also been shown that the forced transient expression of mouse *Zscan4* improves karyotypes and developmental potential, which otherwise characteristically deteriorate during the long-term culture of mouse ES cells (Amano et al., 2013). *Zscan4* also functions as one of the reprogramming factors for mouse-induced pluripotent stem (iPS) cells and improves their quality (Hirata et al., 2012; Jiang et al., 2013). Adding *Zscan4* as a reprogramming factor reactivates preimplantation genes during iPS cell formation (Hirata et al., 2012).

4.4 Mechanism of *Zscan4* Action

The detailed molecular comparison between the rare Zscan4(+) ES cells and abundant Zscan4(−) ES cells has recently revealed that dynamic epigenetic changes occur in Zscan4(+) ES cells (Akiyama et al., 2015; Fig. 3). When ES cells transition from Zscan4(−) cells to Zscan4(+) cells, derepression of heterochromatin occurs, which is accompanied by acetylation of histones, eg, H3K27ac, demethylation of DNAs, a burst of transcription, and clustering of heterochromatin around the nucleolus (Akiyama et al., 2015). These heterochromatin regions include the constitutive heterochromatin, ie, major satellites—pericentromeric region and telomeres—as well as facultative heterochromatin that are scattered throughout the genome, ie, retrotransposons and a few hundred genes that are upregulated in

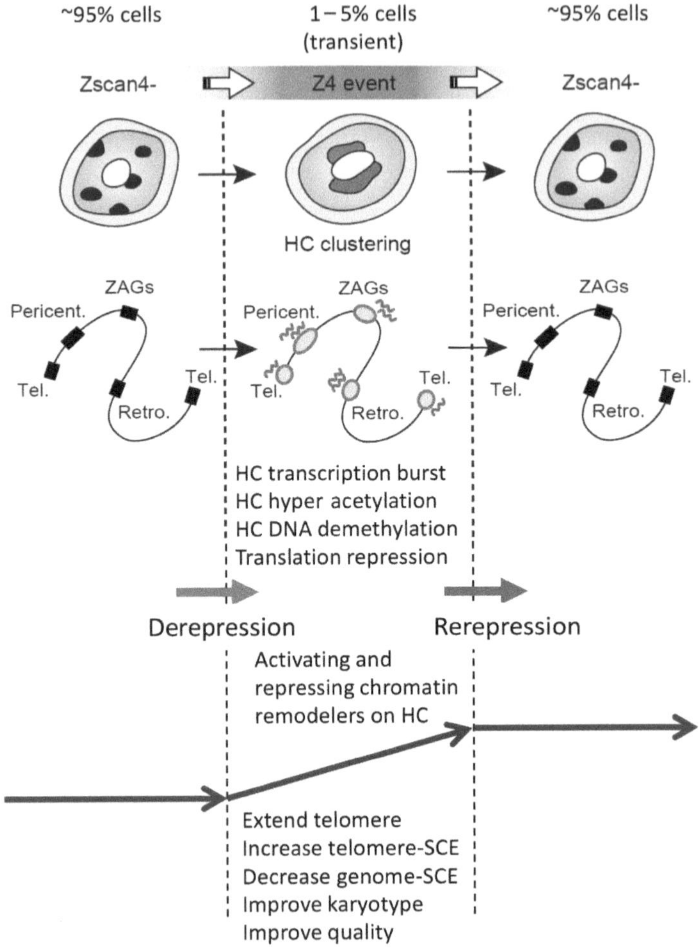

Fig. 3 A schematic summary of the Z4 event. Pericent., pericentomeres; Retro., retrotransposons; Tel., telomeres; ZAGs, Z4 event-associated genes. *Adapted from Akiyama, T., Xin, L., Oda, M., Sharov, A. A., Amano, M., Piao, Y., et al. (2015). Transient bursts of Zscan4 expression are accompanied by the rapid derepression of heterochromatin in mouse embryonic stem cells.* DNA Research, 22, *307–318.* (See the color plate.)

Zscan4(+) cells, named Z4 event-associated genes (ZAGs) (Akiyama et al., 2015). A list of ZAGs can be found in supplementary tables of previous reports (Akiyama et al., 2015; Amano et al., 2013). As mentioned earlier, ES cells stay as Zscan4(+) cells for a short time and return to Zscan4(−) cells. When ES cells transition from Zscan4(+) cells to Zscan4(−) cells, rerepression of heterochromatin occurs and the aforementioned activities in the cells return to their previous condition. These dramatic changes in heterochromatin can explain why ES cells cannot stay as Zscan4(+) cells for long, as these sweeping epigenetic changes are usually considered detrimental to cells.

5. OTHER ZGA-SPECIFIC GENES SUCH AS EIF1A-LIKE GENES AND TCSTV1/3

Are there any other ZGA-specific genes like Zscan4? Although the global gene expression analyses of preimplantation embryos identified many genes that show a peak of expression in two-cell embryos, ie, during ZGA, most of these genes are also expressed in other tissues and, thus, cannot be called two-cell-specific or ZGA-specific genes. Interestingly, global gene expression analyses of mouse ES cells provided a clue: as expected, genes coexpressed in Zscan4(+) ES cells include ZGA-specific genes (Akiyama et al., 2015; Amano et al., 2013; Macfarlan et al., 2012). Previously, some of these genes have been considered to be two-cell-specific genes and, thus, some have started to refer to Zscan4(+) cells as a two-cell-like state (2C-like state) (Macfarlan et al., 2012). However, actual qRT-PCR analyses of these genes showed that the expression of these genes is not exactly timed at the mouse two-cell embryo stage (Amano et al., 2013). For example, compared to the expression of Zscan4 that rather exclusively peaks at the late two-cell stage, the expression of *Tcstv1/3*, *Gm428*, and *BC061212* peaks at the early two-cell stage. *Arg2*, *EG627488*, and *AF06703* show a peak at the late two-cell stage, but their peak is rather broad, spanning over other preimplantation stages. Because ZGA expression has a broader peak and is a broader category, it is more appropriate to refer to these genes as ZGA-specific genes. Therefore, referring to Zscan4(+) cells as being in a "2C-like state" may be a misnomer. Whole-mount RNA in situ hybridization has revealed that these genes, eg, *Eif1a*, *Tcstv1/3*, *Pif1*, *PR23-149D11.5*, *BC061212*, *AF067063*, and *EG668777*, show a Zscan4-like expression pattern in mouse ES cells: these genes are expressed in only 1–5% of mouse ES cells (Fig. 4; Amano et al., 2013). Double-fluorescence RNA in situ hybridization has confirmed that the expression of these genes, eg, *Eif1a*, *Tcstv1/3*, *AF067063*, *EG668777*, *LOC332923*, *BC061212*, and *EG627488*, is overlapped with the expression of *Zscan4* (Amano et al., 2013).

One of these ZGA-specific genes, eukaryotic translation initiation factor 1A (*Eif1a*, previously also called *Eif4c*), has also been known for its specific expression in mouse two-cell embryos (Davis et al., 1996). Originally identified as the mouse eukaryotic elongation factor Tu in erythroleukemic cells

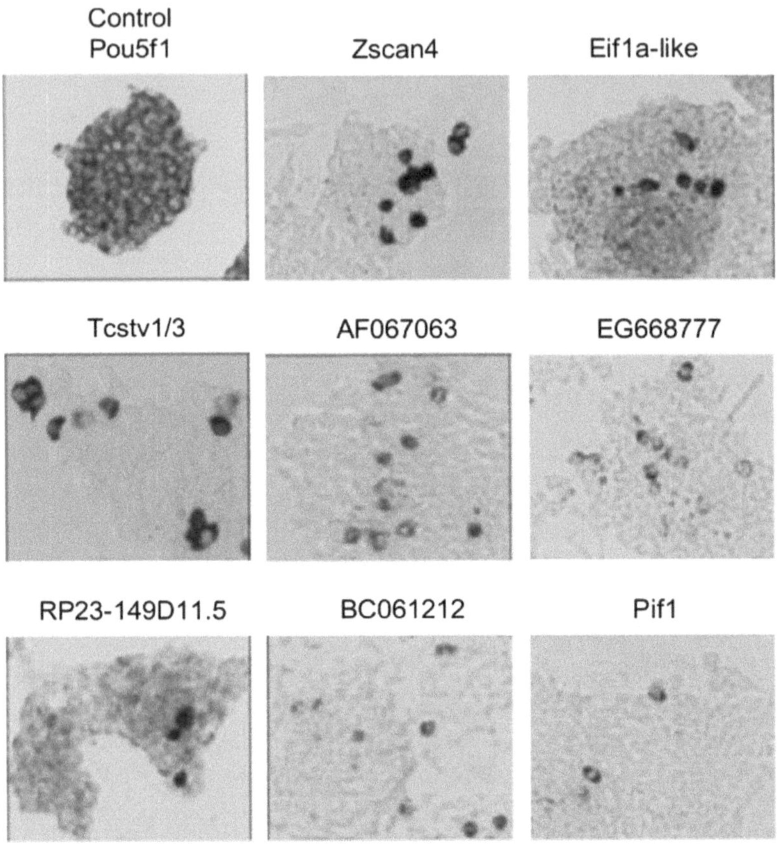

Fig. 4 Whole-mount RNA in situ hybridization of ZAGs in mouse ES cells. *Adapted from Amano, T., Hirata, T., Falco, G., Monti, M., Sharova, L. V., Amano, M., et al. (2013). Zscan4 restores the developmental potency of embryonic stem cells.* Nature Communications, 4, *1966*. (See the color plate.)

(Roth et al., 1987), it was discovered as a gene transiently expressed in mouse two-cell embryos (Davis et al., 1996) and has been used as a two-cell embryo-specific marker (Sonehara, Nagata, & Aoki, 2008). However, recent findings indicate that the genes specifically expressed in Zscan4(+) cells and also highly expressed in mouse two-cell embryos are not *Eif1a* itself, but *Eif1a-like* genes (Hung et al., 2013). *Eif1a-like* genes clustered on Chromosome 12 show high sequence similarity to *Eif1a* and consist of 10 genes (*Eif1al1–Eifa1l10*) and nine pseudogenes (*Eifa1l-ps1–Eif1al-ps9*) (Hung et al., 2013). *Eif1a-like* genes may function as a dominant negative factor and compete with endogenous *Eif1a*, as their overexpression represses global protein synthesis (Hung et al., 2013). In the mouse ES cells going through a

Z4 event, the predominant presence of *Eif1a-like* genes is consistent with the absence of nascent protein synthesis (Hung et al., 2013). Whether the nascent protein synthesis is blocked in two-cell embryos is not clear at this point.

Tcstv1 (two-cell-stage, variable group, member 1) and *Tcstv3* (two-cell-stage, variable group, member 3) were originally reported as two-cell-specific transcripts and named so by Struwe and Solter by depositing sequences in GenBank (GenBank/EMBL/DDBJ accession no. AF067057.1). As they are named two-cell-stage genes, the appearance of *Tcstv1/3* genes in gene expression analyses has been taken as an indicator of two-cell stage gene expression program (Macfarlan et al., 2012; Sharova et al., 2007; Zhang et al., 2006). A recent study has revealed that the overexpression of *Tcstv1/3* induces *Zscan4* and elongates telomeres in mouse ES cells (Zhang et al., 2016).

One striking feature of ZGA-specific genes discovered in mouse embryos is that most of them are specific to mice and their orthologues are not present in human and other mammalian species. *Zscan4* is a unique exception, as it is present widely in mammalian species (but not outside of mammals), including a single copy *ZSCAN4* gene in humans. Other ZGA-specific genes, such as *Eif1a-like* genes, *Tcstv1/3*, *GM428*, and *BC061212*, are all present only in mice, but not in humans. The species specificity of ZGA-specific genes in mice relates to one of the two questions raised in this review: why ZGA genes, which include ZGA-specific genes, are evolutionarily so diverse. As we discussed earlier, the lack of evolutional conservation of ZGA genes is not a mouse-specific issue, but rather inherent in ZGA genes across the species (Heyn et al., 2014; Xie et al., 2010).

6. HETEROCHROMATIN IS OPENED AND TRANSCRIBED DURING ZGA AND Z4 EVENTS

Why are ZGA-specific genes not evolutionarily conserved? A hint comes from the analyses of Z4 events in mouse ES cells. As we mentioned earlier, ZGA-specific genes in mouse embryos and genes upregulated in Zscan4(+) mouse ES cells, including ZAGs, overlap significantly. It has been shown that ZAGs are usually wrapped in heterochromatin and transcriptionally silent, but activated and transcribed during a Z4 event (Akiyama et al., 2015). Similarly, retrotransposons are usually wrapped in heterochromatin, but activated and transcribed during a Z4 event (Akiyama et al., 2015) and ZGA (Macfarlan et al., 2012; Svoboda et al., 2004). Because both ZAGs and retrotransposons are scattered throughout the genome and are not located in typical constitutive heterochromatin, they form islands of facultative heterochromatin buried in euchromatin. During a Z4 event, constitutive heterochromatin is also activated and transcribed (Akiyama et al., 2015). It has been well established that in two-cell mouse embryos, major satellites are actively transcribed and play a critical role for the establishment of heterochromatin (Casanova et al., 2013; Probst & Almouzni, 2011). These results indicate that activation and transcription of both constitutive and facultative heterochromatin are a common feature of ZGA in mouse preimplantation embryos and Z4 events in mouse ES cells.

Heterochromatin is rapidly evolving and DNA sequences located in heterochromatin regions are not evolutionarily well conserved (Hughes & Hawley, 2009). If a significant fraction of ZGA-specific genes and transcripts is encoded in the heterochromatin region, it would be reasonable that ZGA-specific genes are not evolutionarily conserved. Indeed, that is the case for constitutive heterochromatin, such as telomeres and pericentromeric regions, and facultative heterochromatin, such as retrotransposons. These sequences are usually wrapped in the heterochromatin and are activated and transcribed during ZGA and Z4 events.

How about newly identified ZAGs buried in heterochromatin regions? One possibility is that most of them are in the vicinity of repetitive sequences such as retrotransposons that are wrapped in heterochromatin and thus they are activated as bystanders. In this sense, it is interesting to note that there has been a report that some of these genes are transcribed as read-through products from the neighboring retrotransposons (Macfarlan et al., 2012; Peaston et al., 2004). Although it was reported that *Zscan4* transcripts are also the products of read-through from neighboring retrotransposons (Macfarlan et al., 2012), our own RNA-sequencing analysis clearly shows that *Zscan4* transcripts start at their own canonical transcription start sites (Akiyama et al., 2015). Zscan4 could be an exception in this case.

7. ZGA IS TRANSIENT: WHAT HAS BEEN OPENED MUST BE CLOSED

The opening of heterochromatin and the expression of these heterochromatin genes are generally known to compromise the genome stability and are deleterious to cells (Carone & Lawrence, 2013; Egger, Liang, Aparicio, & Jones, 2004; Ting et al., 2011). Because both events are associated with the maximum openness of the chromatin and massive transcription from the whole genome, including heterochromatin, prolonged ZGA and prolonged Z4 events are detrimental to preimplantation embryos and ES cells. That is, the opening of heterochromatin must be followed by the closing of heterochromatin. This notion seems to provide an answer to one of the questions raised in this review: why the waves of activation, more specifically ZGA, are transient. The notion also explains why Z4 events (ie, Zscan4(+) mouse ES cells) exist in a transient manner and should be distinguished from the concept of the ES cell states.

Considering the short duration for ZGA and the Z4 event, there must be an active mechanism for rapid derepression, followed by rerepression of chromatin, including heterochromatin. What kinds of mechanism could cause genome-wide rapid epigenetic changes? A hint comes again from Z4 events in mouse ES cells. As mentioned earlier, when mouse ES cells experience a Z4 event, chromatin, including heterochromatin, is rapidly opened and then closed, which is similar to what is happening during the transition from a one-cell embryo to a four-cell embryo. In the case of a Z4 event, detailed molecular analyses revealed that Zscan4 protein forms complexes with both activating chromatin-remodeling complexes (eg, SWI/SNF) and repressing chromatin-remodeling complexes (eg, HDAC1, HDAC2, LSD1/KDM1A, NuRD, and Sin3A) (Akiyama et al., 2015). These findings are consistent with the notion that chromatin is rapidly derepressed and rerepressed during Z4 events in mouse ES cells. These findings also suggest that a similar mechanism may also take place in preimplantation embryos; when present during ZGA, Zscan4 recruits chromatin-remodeling complexes, first activating and then repressing.

8. ZGA TRIGGERS DNA DAMAGES, WHICH MUST BE REPAIRED

The aforementioned notion that preimplantation embryos cannot endure ZGA for a long time and ES cells cannot endure a Z4 event for a long time raises another question: why does the chromatin, especially heterochromatin, have to be opened in the first place? Regarding this question, recent studies in *Drosophila* (Blythe & Wieschaus, 2015) and *Caenorhabditis elegans* (Butuci, Williams, Wong, Kramer, & Michael, 2015) have revealed an astonishing feature of ZGA; the authors of these papers have shown that massive transcriptional activation during ZGA causes intrinsic DNA damage, which must be corrected for proper development. These results are counterintuitive, as they indicate that ZGA itself is an extra burden for embryos; this is especially critical since it occurs at a time when embryos need to protect their genome integrity.

If this damage to the DNAs is not repaired, the embryos will not develop normally or the mutation load will be transmitted to the next generation. So it is reasonable to consider that cells at this specific point have a high capacity to repair these mutation loads. It is also reasonable to think that the repair mechanism must be something similar to the mechanism occurring in germ cells—homologous recombination-mediated DNA repairs—because all other mutation repair mechanisms that normally occur in somatic cells leave traces of undesirable DNA sequence alterations.

9. CORRECTION OF KARYOTYPE ABNORMALITY BY *ZSCAN4*

There has been anecdotal evidence that either selection or self-correction occurs during preimplantation development in human embryos (Bazrgar, Gourabi, Valojerdi, Yazdi, & Baharvand, 2013). Although the mechanism is unknown, studies using preimplantation genetic diagnosis of in vitro fertilization clinics have shown that the observed incidence of chromosome abnormality is reduced from embryos on Day 3 (before ZGA) to embryos on Day 5 (after ZGA). Therefore, it is conceivable that ZGA-specific genes are involved in the active repair of DNAs and/or the active correction of karyotype abnormality.

One of the candidates for ZGA-specific genes involved in the correction of karyotype abnormality is *Zscan4*. In mouse ES cells, transient over-expression of *Zscan4* can increase the number of mouse ES cells with normal karyotypes (Amano et al., 2013). In our earlier work, this has been hinted by the rescue of karyotype abnormality by overexpressing *Zscan4* in mouse ES cells that are deteriorated by the knockdown of *Zscan4* (Zalzman et al., 2010). Although these experiments were done using doxycycline-regulatable *Zscan4* as a transgene stably integrated into the mouse genome, recently we have demonstrated that karyotype correction in mouse ES cells can be achieved by either synthetic mRNAs or Sendai virus vector encoding mouse or human *ZSCAN4* genes (Amano et al., 2015).

A more surprising finding is that these *ZSCAN4* mRNAs or Sendai virus vectors encoding *ZSCAN4* can correct karyotype abnormalities in human primary fibroblast cells (Amano et al., 2015). We have shown this phenomenon by treating human primary fibroblast cells derived from individuals with Down syndrome (Trisomy 21) and Edwards syndrome (Trisomy 18).

Considering the fact that fibroblast cells do not express *ZSCAN4* naturally, it is surprising to see that *ZSCAN4* can function without its usual context, eg, other genes associated with ZGA.

10. COMMONALITY BETWEEN ZGA-SPECIFIC PHENOMENA AND THE Z4 EVENT

What we have presented and discussed thus far strongly indicates the commonality between ZGA-specific phenomena in preimplantation embryos and Z4 events in mouse ES cells (Table 1). The transition from one-cell embryos to two-cell embryos is accompanied by a large number of ZGA genes that begin to be transcribed for the first time after fertilization. These ZGA genes play multiple roles in preimplantation development. However, if we extract ZGA-specific phenomena or ZGA-specific genes from ZGA in general, they seem to overlap significantly with the Z4 event or ZAGs. Therefore, it is perhaps more appropriate to say that a Z4 event occurs, which is defined by mouse ES cells, also in mouse preimplantation embryos during ZGA.

What are the features of the Z4 event that takes place in mouse ES cells and in mouse preimplantation embryos during ZGA? Based on the studies in mouse ES cells, the Z4 event is involved in the integrity of the genome and

Table 1 Commonality Between ZGA and the Z4 Event

	ZGA	**Z4 Event**
	Preimplantation embryos	Mouse ES cells
	Mainly two-cell embryo in mice (four- to eight-cell embryo in humans)	Zscan4(+) cells
Heterochromatin transcription	Yes (mouse) Unknown (human)	Yes
Zscan4 transcription	Yes (late two-cell stage in mice) Yes (six- to eight-cell stage in human)	Yes
Eif1a-like genes transcription	Yes (mouse) Unknown (human)	Yes
Global translation suppression	Unknown	Yes
Telomere elongation by homologous recombination	Yes	Yes

chromosomes. When ES cells go through the Z4 event, the telomeres are extended, the number of euploid cells increased, and thus, karyotypes are corrected (Amano et al., 2013, 2015; Zalzman et al., 2010). Patently, *Zscan4* is not merely a marker of the Z4 event, but plays a critical role in the Z4 event; application of Zscan4 causes these effects on mouse ES cells. Extending the notion to mouse preimplantation embryos, the Z4 event is most likely involved in the maintenance of genome and chromosome integrity in preimplantation embryos. Furthermore, *Zscan4* must play a central and active role in the Z4 event in preimplantation development.

Mechanistically, the Z4 event is involved in the derepression and subsequent rerepression of constitutive and facultative heterochromatin. During the Z4 event, usually transcriptionally inactive heterochromatin is transiently activated and transcribed. Such activations occur in both constitutive heterochromatin, which harbors major satellites in pericentromeric region and telomeres, and facultative heterochromatin, which harbors ZAGs (also ZGA-specific genes) and retrotransposons.

The relaxation of heterochromatin may be possibly related to the extension of telomeres. Telomere extension is one of the functions of *Zscan4* in mouse ES cells (Zalzman et al., 2010). It has been shown that when DNA methylation levels of the telomeric or subtelomeric region are reduced, telomeres undergo telomere recombination and telomere extension (Gonzalo et al., 2006). Therefore, one of the possible mechanisms for telomere extension by *Zscan4* is mediated by the relaxation of telomeric heterochromatin. Interestingly, telomerase-independent telomere extension by T-SCE has also been observed in mouse preimplantation embryos (Liu et al., 2007). However, it remains to clarify whether the telomere elongation occurs during ZGA or involves *Zscan4*. The relaxation of pericentromeric heterochromatin may also explain an important phenomenon we found: correction of karyotype abnormality by *Zscan4*. The pericentromeric regions are place where the kinetochore binds, and the regulation of kinetochores is critical for segregation of chromosomes during mitosis.

11. PERSPECTIVES

Let us address the two questions posed in the beginning of this review: why is ZGA transient and why are ZGA genes so evolutionarily diverse? Given all of the things considered thus far, we would argue that these two features of ZGA are more of the features of ZAGs (also ZGA-specific genes) than ZGA genes in general. The Z4 event is intrinsically transient, as cells cannot endure an extended Z4 event and, thus, cells have a mechanism to prevent the continuous expression of ZAGs. It may be conceivable that the majority of ZGA genes are by-products, which are not required for the proper function of embryo development. They are transcribed when the zygotic genomes go through the widespread, sweeping checking and correction of the zygotic genome. This may be the reason why ZGA genes are not evolutionarily conserved. In other words, the main reason for the presence of the Z4 event during ZGA is not to make RNAs and proteins to be used for embryo development, but merely the results in the reorganization or reprogramming of the whole genome, particularly heterochromatin. When the genomes and chromatin are reprogrammed, transcriptions from various genomic regions are accompanied and form the majority of transcripts in ZGA. This may also explain why so many genes are transcribed during ZGA, as they are more a reflection of the reorganization of the whole genome. This may also explain why they are transient, as these transcripts per se are not critical and thus their transcription does not have to continue.

ACKNOWLEDGMENTS

The author would like to thank past and present members of the lab for contributing the data and concept presented in this review. The authors thank the editorial assistance of Vanessa B. Harrington and Akihiro C. Ko.

Conflict of Interest: The author is a co-founder, a member, and a chief scientific officer for Elixirgen, LLC, which promotes clinical application of ZSCAN4.

REFERENCES

Abe, K., Yamamoto, R., Franke, V., Cao, M., Suzuki, Y., Suzuki, M. G., et al. (2015). The first murine zygotic transcription is promiscuous and uncoupled from splicing and 3′ processing. *The EMBO Journal*, *34*, 1523–1537.

Akiyama, T., Xin, L., Oda, M., Sharov, A. A., Amano, M., Piao, Y., et al. (2015). Transient bursts of Zscan4 expression are accompanied by the rapid derepression of heterochromatin in mouse embryonic stem cells. *DNA Research*, *22*, 307–318.

Amano, T., Hirata, T., Falco, G., Monti, M., Sharova, L. V., Amano, M., et al. (2013). Zscan4 restores the developmental potency of embryonic stem cells. *Nature Communications*, *4*, 1966.

Amano, T., Jeffries, E., Amano, M., Ko, A. C., Yu, H., & Ko, M. S. (2015). Correction of Down syndrome and Edwards syndrome aneuploidies in human cell cultures. *DNA Research*, *22*, 331–342.

Aoki, F., Worrad, D. M., & Schultz, R. M. (1997). Regulation of transcriptional activity during the first and second cell cycles in the preimplantation mouse embryo. *Developmental Biology*, *181*, 296–307.

Bazrgar, M., Gourabi, H., Valojerdi, M. R., Yazdi, P. E., & Baharvand, H. (2013). Self-correction of chromosomal abnormalities in human preimplantation embryos and embryonic stem cells. *Stem Cells and Development*, *22*, 2449–2456.

Biechele, S., Lin, C.-J., Rinaudo, P. F., & Ramalho-Santos, M. (2015). Unwind and transcribe: Chromatin reprogramming in the early mammalian embryo. *Current Opinion in Genetics & Development*, *34*, 17–23.

Blythe, S. A., & Wieschaus, E. F. (2015). Zygotic genome activation triggers the DNA replication checkpoint at the midblastula transition. *Cell*, *160*, 1169–1181.

Braude, P., Bolton, V., & Moore, S. (1988). Human gene expression first occurs between the four- and eight-cell stages of preimplantation development. *Nature*, *332*, 459–461.

Butuci, M., Williams, A. B., Wong, M. M., Kramer, B., & Michael, W. M. (2015). Zygotic genome activation triggers chromosome damage and checkpoint signaling in *C. elegans* primordial germ cells. *Developmental Cell*, *34*, 85–95.

Carone, D. M., & Lawrence, J. B. (2013). Heterochromatin instability in cancer: From the Barr body to satellites and the nuclear periphery. *Seminars in Cancer Biology*, *23*, 99–108.

Carter, M. G., Stagg, C. A., Falco, G., Yoshikawa, T., Bassey, U. C., Aiba, K., et al. (2008). An in situ hybridization-based screen for heterogeneously expressed genes in mouse ES cells. *Gene Expression Patterns*, *8*, 181–198.

Casanova, M., Pasternak, M., El Marjou, F., Le Baccon, P., Probst, A. V., & Almouzni, G. (2013). Heterochromatin reorganization during early mouse development requires a single-stranded noncoding transcript. *Cell Reports*, *4*, 1156–1167.

Cho, T., Sakai, S., Nagara, M., & Aoki, F. (2002). Involvement of chromatin structure in the regulation of mouse zygotic gene activation. *Animal Science Journal*, *73*, 113–122.

Davis, W., Jr., De Sousa, P. A., & Schultz, R. M. (1996). Transient expression of translation initiation factor eIF-4C during the 2-cell stage of the preimplantation mouse embryo: Identification by mRNA differential display and the role of DNA replication in zygotic gene activation. *Developmental Biology*, *174*, 190–201.

DePamphilis, M. L., Kaneko, K. J., & Vassilev, A. (2002). Activation of zygotic gene expression in mammals. In M. L. DePamphilis (Ed.), *Gene expression at the beginning of animal development* (pp. 55–84). The Netherlands: Elsevier Science.

Egger, G., Liang, G., Aparicio, A., & Jones, P. A. (2004). Epigenetics in human disease and prospects for epigenetic therapy. *Nature*, *429*, 457–463.

Evsikov, A. V., de Vries, W. N., Peaston, A. E., Radford, E. E., Fancher, K. S., Chen, F. H., et al. (2004). Systems biology of the 2-cell mouse embryo. *Cytogenetic and Genome Research*, *105*, 240–250.

Falco, G., Lee, S. L., Stanghellini, I., Bassey, U. C., Hamatani, T., & Ko, M. S. (2007). Zscan4: A novel gene expressed exclusively in late 2-cell embryos and embryonic stem cells. *Developmental Biology*, *307*, 539–550.

Gonzalo, S., Jaco, I., Fraga, M. F., Chen, T., Li, E., Esteller, M., et al. (2006). DNA methyltransferases control telomere length and telomere recombination in mammalian cells. *Nature Cell Biology*, *8*, 416–424.

Hamatani, T., Carter, M. G., Sharov, A. A., & Ko, M. S. (2004). Dynamics of global gene expression changes during mouse preimplantation development. *Developmental Cell*, *6*, 117–131.

Heyn, P., Kircher, M., Dahl, A., Kelso, J., Tomancak, P., Kalinka, A. T., et al. (2014). The earliest transcribed zygotic genes are short, newly evolved, and different across species. *Cell Reports*, *6*, 285–292.

Hirata, T., Amano, T., Nakatake, Y., Amano, M., Piao, Y., Hoang, H. G., et al. (2012). Zscan4 transiently reactivates early embryonic genes during the generation of induced pluripotent stem cells. *Scientific Reports*, *2*, 208.

Hughes, S. E., & Hawley, R. S. (2009). Heterochromatin: A rapidly evolving species barrier. *PLoS Biology*, 7, e1000233.

Hung, S. S., Wong, R. C., Sharov, A. A., Nakatake, Y., Yu, H., & Ko, M. S. (2013). Repression of global protein synthesis by eif1a-like genes that are expressed specifically in the two-cell embryos and the transient zscan4-positive state of embryonic stem cells. *DNA Research*, *20*, 391–402.

Jiang, J., Lv, W., Ye, X., Wang, L., Zhang, M., Yang, H., et al. (2013). Zscan4 promotes genomic stability during reprogramming and dramatically improves the quality of iPS cells as demonstrated by tetraploid complementation. *Cell Research*, *23*, 92–106.

Ko, M. S. (2006). Expression profiling of the mouse early embryo: Reflections and perspectives. *Developmental Dynamics*, *235*, 2437–2448.

Ko, S. B., Azuma, S., Yokoyama, Y., Yamamoto, A., Kyokane, K., Niida, S., et al. (2013). Inflammation increases cells expressing ZSCAN4 and progenitor cell markers in the adult pancreas. *American Journal of Physiology. Gastrointestinal and Liver Physiology*, *304*, G1103–G1116.

Ko, M. S., Kitchen, J. R., Wang, X., Threat, T. A., Wang, X., Hasegawa, A., et al. (2000). Large-scale cDNA analysis reveals phased gene expression patterns during preimplantation mouse development. *Development*, *127*, 1737–1749.

Latham, K. E., & Schultz, R. M. (2001). Embryonic genome activation. *Frontiers in Bioscience: A Journal and Virtual Library*, *6*, D748–D759.

Lee, M. T., Bonneau, A. R., & Giraldez, A. J. (2014). Zygotic genome activation during the maternal-to-zygotic transition. *Annual Review of Cell and Developmental Biology*, *30*, 581–613.

Li, L., Lu, X., & Dean, J. (2013). The maternal to zygotic transition in mammals. *Molecular Aspects of Medicine*, *34*, 919–938.

Liu, L., Bailey, S. M., Okuka, M., Munoz, P., Li, C., Zhou, L., et al. (2007). Telomere lengthening early in development. *Nature Cell Biology*, *9*, 1436–1441.

Macfarlan, T. S., Gifford, W. D., Driscoll, S., Lettieri, K., Rowe, H. M., Bonanomi, D., et al. (2012). Embryonic stem cell potency fluctuates with endogenous retrovirus activity. *Nature*, *487*, 57–63.

Niakan, K. K., Han, J., Pedersen, R. A., Simon, C., & Pera, R. A. (2012). Human pre-implantation embryo development. *Development*, *139*, 829–841.

Nichols, J., & Smith, A. (2009). Naive and primed pluripotent states. *Cell Stem Cell*, *4*, 487–492.

Nothias, J. Y., Majumder, S., Kaneko, K. J., & DePamphilis, M. L. (1995). Regulation of gene expression at the beginning of mammalian development. *The Journal of Biological Chemistry*, *270*, 22077–22080.

Okazaki, Y., Furuno, M., Kasukawa, T., Adachi, J., Bono, H., Kondo, S., et al. (2002). Analysis of the mouse transcriptome based on functional annotation of 60,770 full-length cDNAs. *Nature*, *420*, 563–573.

Peaston, A. E., Evsikov, A. V., Graber, J. H., de Vries, W. N., Holbrook, A. E., Solter, D., et al. (2004). Retrotransposons regulate host genes in mouse oocytes and preimplantation embryos. *Developmental Cell*, *7*, 597–606.

Probst, A. V., & Almouzni, G. (2011). Heterochromatin establishment in the context of genome-wide epigenetic reprogramming. *Trends in Genetics*, *27*, 177–185.

Ram, P. T., & Schultz, R. M. (1993). Reporter gene expression in G2 of the 1-cell mouse embryo. *Developmental Biology*, *156*, 552–556.

Roth, W. W., Bragg, P. W., Corrias, M. V., Reddy, N. S., Dholakia, J. N., & Wahba, A. J. (1987). Expression of a gene for mouse eucaryotic elongation factor Tu during murine erythroleukemic cell differentiation. *Molecular and Cellular Biology*, 7, 3929–3936.

Sharova, L. V., Sharov, A. A., Piao, Y., Shaik, N., Sullivan, T., Stewart, C. L., et al. (2007). Global gene expression profiling reveals similarities and differences among mouse pluripotent stem cells of different origins and strains. *Developmental Biology*, *307*, 446–459.

Sonehara, H., Nagata, M., & Aoki, F. (2008). Roles of the first and second round of DNA replication in the regulation of zygotic gene activation in mice. *The Journal of Reproduction and Development*, *54*, 381–384.

Stanghellini, I., Falco, G., Lee, S. L., Monti, M., & Ko, M. S. (2009). Trim43a, Trim43b, and Trim43c: Novel mouse genes expressed specifically in mouse preimplantation embryos. *Gene Expression Patterns*, *9*, 595–602.

Svoboda, P., Franke, V., & Schultz, R. M. (2015). Sculpting the transcriptome during the oocyte-to-embryo transition in mouse. *Current Topics in Developmental Biology*, *113*, 305–349.

Svoboda, P., Stein, P., Anger, M., Bernstein, E., Hannon, G. J., & Schultz, R. M. (2004). RNAi and expression of retrotransposons MuERV-L and IAP in preimplantation mouse embryos. *Developmental Biology*, *269*, 276–285.

Ting, D. T., Lipson, D., Paul, S., Brannigan, B. W., Akhavanfard, S., Coffman, E. J., et al. (2011). Aberrant overexpression of satellite repeats in pancreatic and other epithelial cancers. *Science*, *331*, 593–596.

Vassena, R., Boue, S., Gonzalez-Roca, E., Aran, B., Auer, H., Veiga, A., et al. (2011). Waves of early transcriptional activation and pluripotency program initiation during human preimplantation development. *Development*, *138*, 3699–3709.

Wang, S., Cowan, C. A., Chipperfield, H., & Powers, R. D. (2005). Gene expression in the preimplantation embryo: In-vitro developmental changes. *Reproductive Biomedicine Online*, *10*, 607–616.

Wang, Q. T., Piotrowska, K., Ciemerych, M. A., Milenkovic, L., Scott, M. P., Davis, R. W., et al. (2004). A genome-wide study of gene activity reveals developmental signaling pathways in the preimplantation mouse embryo. *Developmental Cell*, *6*, 133–144.

Xie, D., Chen, C. C., Ptaszek, L. M., Xiao, S., Cao, X., Fang, F., et al. (2010). Rewirable gene regulatory networks in the preimplantation embryonic development of three mammalian species. *Genome Research*, *20*, 804–815.

Zalzman, M., Falco, G., Sharova, L. V., Nishiyama, A., Thomas, M., Lee, S. L., et al. (2010). Zscan4 regulates telomere elongation and genomic stability in ES cells. *Nature*, *464*, 858–863.

Zeng, F., Baldwin, D. A., & Schultz, R. M. (2004). Transcript profiling during preimplantation mouse development. *Developmental Biology*, *272*, 483–496.

Zeng, F., & Schultz, R. M. (2005). RNA transcript profiling during zygotic gene activation in the preimplantation mouse embryo. *Developmental Biology*, *283*, 40–57.

Zhang, Q., Dan, J., Wang, H., Guo, R., Mao, J., Fu, H., et al. (2016). Tcstv1 and Tcstv3 elongate telomeres of mouse ES cells. *Scientific Reports*, *6*, 19852.

Zhang, W., Walker, E., Tamplin, O. J., Rossant, J., Stanford, W. L., & Hughes, T. R. (2006). Zfp206 regulates ES cell gene expression and differentiation. *Nucleic Acids Research*, *34*, 4780–4790.

CHAPTER FOUR

From Meiosis to Mitosis: The Astonishing Flexibility of Cell Division Mechanisms in Early Mammalian Development

L. Bury[1,2], P.A. Coelho, D.M. Glover
University of Cambridge, Cambridge, United Kingdom
[1]Corresponding author: e-mail address: dmg25@hermes.cam.ac.uk

Contents

Abstract

The execution of female meiosis and the establishment of the zygote is arguably the most critical stage of mammalian development. The egg can be arrested in the prophase of meiosis I for decades, and when it is activated, the spindle is assembled de novo. This spindle must function with the highest of fidelity and yet its assembly is unusually achieved in the absence of conventional centrosomes and with minimal

[2] Present address: Whitehead Institute and Department of Biology, MIT, Nine Cambridge Center, Cambridge, Massachusetts 02142, USA.

Current Topics in Developmental Biology, Volume 120
ISSN 0070-2153
http://dx.doi.org/10.1016/bs.ctdb.2016.04.011

influence of chromatin. Moreover, its dramatic asymmetric positioning is achieved through remarkable properties of the actin cytoskeleton to ensure elimination of the polar bodies. The second meiotic arrest marks a uniquely prolonged metaphase eventually interrupted by egg activation at fertilization to complete meiosis and mark a period of preparation of the male and female pronuclear genomes not only for their entry into the mitotic cleavage divisions but also for the imminent prospect of their zygotic expression.

1. INTRODUCTION

The meiotic divisions of female mammals are extraordinary cell cycles. In contrast to male meiosis, which takes place continuously from the time of puberty, meiosis in females begins in fetal development. The duplicated maternal and paternal chromosomes pair and initiate recombination with the homologues being held together at chiasmata, the points of chromosome crossovers, to form a structure known as bivalent. The oocyte then arrests in meiosis until sexual maturity. This prophase arrest permits oocytes to grow as is necessary to support meiotic maturation, fertilization, and early embryonic development (Moore, 1975; Moore & Lintern-Moore, 1978; Wassarman, Josefowicz, & Letourneau, 1976). Meiotic maturation of the oocyte is then triggered by hormonal stimulation. It is characterized by breakdown of the nuclear envelope (also known as the germinal vesicle, GV) and formation of the first meiotic spindle, that in the absence of centrosomes, is triggered around the chromosomes. In contrast to the alternating phases of DNA replication and chromosome segregation of mitotic cell cycles, meiosis requires two successive rounds of division without an intervening S-phase.

Before undertaking the first division, the spindle and its attached chromosomes migrate toward to the cortex of the oocyte. Sister kinetochores are cooriented during the first division allowing homologous chromosomes to segregate. This asymmetric division produces one small, nonfunctional daughter cell, the first polar body, while the oocyte retains the great bulk of cytoplasm containing maternal stores accumulated during oogenesis. Polar body extrusion is immediately followed by assembly of the meiosis II spindle, at which stage the oocyte remains arrested for a second time, now awaiting fertilization. This division is completed when the two haploid gametes fuse to form the zygote, and a second polar body containing sister chromatids is extruded.

The zygote has two characteristic pronuclei, male and female, around which a spindle begins to assemble for the first zygotic mitosis, that will now cleave the fertilized egg into two equal parts. This will be the first of a series of subsequent cleavage divisions, that give rise to progressively smaller blastomeres. The 8-cell stage marks an important transition in the early embryo's life, at which cells compact and polarize. In the fourth and fifth rounds of cleavage, a proportion of cells then undergo asymmetric divisions, in regard to their cell fate, to give rise to distinct inside and outside lineages. Polar cells residing on the outside of the embryo will develop into the extraembryonic trophectoderm tissue, while the embryo proper, the epiblast, and a second extraembryonic tissue, the primitive endoderm, develop from apolar inside cells. Around the 16–32 cell stage, a cavity begins to form inside the embryo, which is now called a blastocyst.

In this chapter, we will examine how the major cellular changes that accompany these events are orchestrated to occur in concert with developmental progression. This requires the functions of signaling mechanisms that regulate differing modes of cell cycle progression, and control the behavior of the microtubule and actin cytoskeletons. Here, we explore how the functions of component cell biological systems are coordinated during progression through these complex transitions.

2. COORDINATING CELL CYCLE REGULATION WITH DEVELOPMENT

2.1 Prophase Arrest in Meiosis

The mammalian oocyte enters its first meiotic prophase during fetal development, which in mice occurs between embryonic days 12.5 and 13.5 (Kocer, Reichmann, Best, & Adams, 2009; McLaren, 2003; Sánchez & Smitz, 2012). Prophase I is subdivided into the leptotene, zygotene, and pachytene stages, that characterize pairing of homologous chromosomes, synapsis, and recombination, and which are followed by dipolotene and diakinesis. Upon reaching the dipolotene stage, oocytes enter a resting phase also known as dictyate (Borum, 1961). Prophase I arrested oocytes are surrounded by somatic cells within the ovary to form functional units termed primordial follicles. Starting around the time of birth, primordial follicles become activated to enter folliculogenesis. During this process, follicles either undergo degeneration by atresia or grow to become preovulatory follicles with full developmental competence, able to reinitiate meiosis and to support embryonic development. In the course of this growth phase, oocytes increase their volume by over 100-fold, reaching a final size of approximately 80 μm in diameter at which stage they have accumulated transcripts required for subsequent meiotic progression (Mehlmann, 2005). Finally, reinitiation of meiosis is activated at puberty by periodic pre-ovulatory surge of luteinizing hormone (LH) released from the pituitary gland (Fig. 1).

In all species, progression through meiotic and mitotic cell cycles is controlled by maturation-promoting factor (MPF), initially identified in frog and starfish oocytes (Gautier, Matsukawa, Nurse, & Maller, 1989; Gerhart, 1984; Labbe et al., 1989). MPF is a serine/threonine kinase composed of the catalytic subunit CDK1 and the regulatory subunit cyclin B (Fig. 2). The activation of MPF requires dephosphorylation of CDK1 at residues Thr14 and Tyr15 and its association with cyclin B (Clarke & Karsenti, 1991). The phosphorylation state of these two residues on CDK1 is determined by a balance of specific isoforms of the WEE1 kinases and CDC25 phosphatases

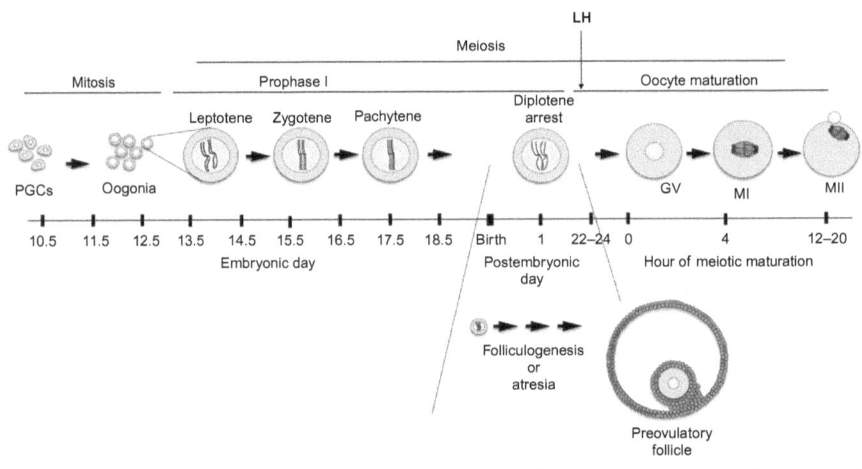

Fig. 1 Stages of oogenesis and folliculogenesis in the mouse. Primordial germ cells (PGCs) divide mitotically giving rise to oogonia that populate the genital ridge. Around embryonic day 12.5–13.5, oocytes enter prophase I of meiosis, that includes the stages zyogotene, pachytene, and diplotene upon which they enter a resting state, dictyate. This arrest begins before birth and is characterized by oocyte growth and follicle formation. Within the large, preovulatory follicle, the oocyte and surrounding granulosa cells reside in the fluid filed antrum. After the onset of puberty, matured follicles can be activated by a periodic surge of luteinizing hormone (LH), which promotes maturation of the oocyte. As oocytes reenter meiosis I, they undergo breakdown of their germinal vesicle, congression of homologous chromosomes at metaphase I, followed by their segregation at anaphase I. Upon polar body extrusion, oocytes become arrested in meiosis II until fertilization. (See the color plate.)

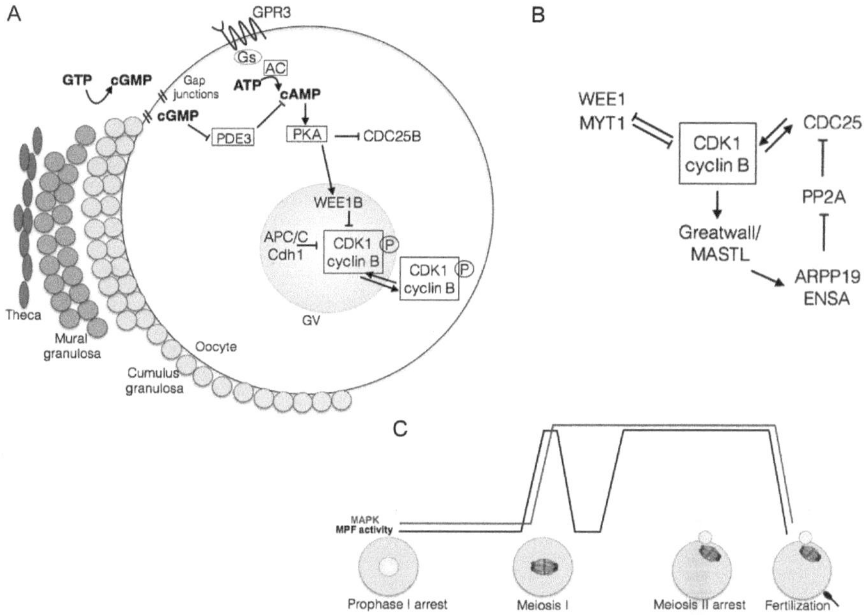

Fig. 2 Pathways regulating meiotic and mitotic entry. (A) Regulatory events underlying entry into meiosis I. A G-protein-coupled membrane-bound receptor (GPR3) stimulates adenylate cyclase (AC) to produce cAMP. cAMP hydrolysis is prevented by surrounding cumulus cells (indicated in *gray*) via production of cGMP by guanylyl cyclase NPR2 (not shown). NPR2 itself is stimulated in response to the hormone Natriuretic peptide precursor C (NPPC), produced by granulosa cells. cAMP activates protein kinase A (PKA), which in turn phosphorylates and activates the CDK1 inhibitory kinase WEE1B/WEE2 and also inhibits the CDK1-activating phosphatase CDC25B. The APC/C with its subunit Cdh1 mediates degradation of cyclin B1. Both WEE1B and APC/C localize mainly to the nucleus during prophase I arrest, while cyclin B1 and CDC25B are sequestered within the cytoplasm. (B) Signaling pathways mediating mitotic entry. A complex regulatory network is responsible for M-phase initiation by activation of CDK1. CDK1 activity depends on association with its regulatory subunit cyclin B, whose levels increase during G2. Cyclin B/CDK1 complexes remain initially inactive, due to two inhibitory phosphorylations (T14/Y15) mediated by WEE1/MYT1 kinases. Full CDK1 activation requires removal of these inhibitory phosphates by phosphatase CDC25. Positive feedback loops to reinforce CDK1 activation involve phosphorylation and activation of Greatwall kinase that in turn phosphorylates two inhibitors (ARPP19 and ENSA) of the Cdk1 antagonizing phosphatase PP2A. (C) Changes in MPF and CSF activity during mouse oocyte meiotic maturation. MPF (cyclin B-CDK1) is activated at germinal vesicle breakdown (GVBD) until it reaches a plateau at the end of the first meiotic division. An APC/C^{CDC20}-dependent drop in MFP activity takes place between meiosis I and II. MPF is rapidly reactivated to enter meiosis II and remains at high levels during the metaphase II arrest. This arrest is maintained by CSF (MAPK and EMI2) until fertilization.

(Lew & Kornbluth, 1996). Although mouse oocytes contain the three CDC25 isoforms, only CDC25A and B appear to be important to trigger meiotic resumption (Lincoln et al., 2002; Solc et al., 2008). Opposing these phosphatases, the WEE1B (WEE2) and MYT1 kinases are implicated in meiotic arrest (Han, Chen, Paronetto, & Conti, 2005; Oh, Han, & Conti, 2010). Once activated, the CDK1 kinase phosphorylates a wide range of target proteins to mediate nuclear envelope breakdown (NEBD), chromosome condensation, and remodeling of the actin and microtubule cytoskeleton.

Prophase arrest in meiosis I is regulated by the second messenger cAMP (cyclic adenosine 3′5′-monophosphate) (Eppig & Downs, 1984; Eppig, Freter, Ward-Bailey, & Schultz, 1983; Schultz, Montgomery, & Belanoff, 1983). High levels of cAMP within the oocyte are crucial for maintaining arrest and a drop in cAMP is associated with meiotic entry. cAMP is generated in the oocyte through the constitutive action of a membrane-associated G-protein-coupled receptor (GPR3) that stimulates G-proteins to activate adenylyl cyclase (Mehlmann et al., 2004). Surrounding follicular cells also contribute to maintaining high oocyte cAMP levels by producing cGMP (cyclic guanosine 3′5′-monophosphate) (Norris et al., 2009; Sela-Abramovich, Galiani, Nevo, & Dekel, 2008). cGMP diffuses into the oocyte through gap junctions and inhibits the oocyte-specific cAMP phosphodiesterase-3A (PDE3A), thereby preventing cAMP hydrolysis (Conti et al., 2002; Hambleton et al., 2005; Norris et al., 2009; Vaccari, Weeks, Hsieh, Menniti, & Conti, 2009) (see Fig. 2A). This molecular crosstalk between follicular cells and the oocytes is crucial, as oocytes spontaneously resume meiosis when released from preovulatory antral follicles, a process that can be prevented by addition of membrane permeable cAMP analogs or inhibitors of the phosphodiesesterase PDE3A (Fig. 2) (Conti et al., 2002; Dekel & Beers, 1978; Eppig, Schroeder, & O'Brien, 1992; Eppig, Wigglesowrth, & O'Brien, 1992; Vivarelli, Conti, De Felici, & Siracusa, 1983).

cAMP exerts its function via protein kinase A (PKA) signaling (Conti et al., 2002; Kovo et al., 2006). PKA directly activates the WEE1B kinase and inhibits CDC25B dual specificity phosphatase (Han et al., 2005; Lincoln et al., 2002; Oh et al., 2010; Pirino, Wescott, & Donovan, 2009; Zhang et al., 2008) thus suppressing activation of CDK1 (Fig. 2B). This regulatory mechanism appears to be meiosis specific because PKA does not phosphorylate these substrates in somatic cells.

In response to hormonal LH stimulation, gap junctions throughout the follicle are disrupted. This prevents entry of cGMP into the oocyte, causing cAMP degradation and inactivation of PKA. Concomitant changes in localization of CDK1/cyclin B as well their regulatory kinases and phosphatases are responsible for an accumulation of active nuclear MPF complexes able to drive events associated with meiosis I entry (Holt, Weaver, & Jones, 2010; Oh et al., 2010).

As an additional control mechanism to maintain low MPF activity throughout the prophase arrest stage, cyclin B1 is continuously targeted for ubiquitin-mediated degradation by the anaphase-promoting complex/cyclasome (APC/C) E3 ubiquitin-protein ligase associated with Cdh1. This mechanism is essential as its inhibition is able to cause meiotic resumption (Reis, Chang, Levasseur, & Jones, 2006). Antagonizing activity is also provided by the Cdc14b phosphatase that acts to prevent meiotic resumption by countering Cdk1 activity and promoting cyclin B degradation, functioning in an APC/C-dependent manner (Schindler & Schultz, 2009a, 2009b).

During meiotic arrest, cyclin B is excluded from the nucleus (Marangos & Carroll, 2004; Mitra & Schultz, 1996), as is the phosphatase CDC25B whose activity is counteracted by cytoplasmic kinase MYT1. CDK1 activity is maintained at low levels by nuclear WEE2 (Oh et al., 2010) and nuclear APC/C^{Cdh1} (Holt et al., 2010). Oocyte cAMP levels fall prior to NEBD and both cyclin B and CDC25B are translocated to the nucleus (Marangos & Carroll, 2004; Mitra & Schultz, 1996; Pirino et al., 2009; Solc et al., 2008; Zhang et al., 2008), causing activation of a fraction of MPF. MPF in turn promotes export of WEE2 to the cytoplasm, further reenforcing the active state of CDK1 (see Fig. 2A).

2.2 Meiotic Maturation

Unlike mitosis, where the G2/M-transition is associated with a rapid, switch-like activation of CDK1, meiosis I is characterized by a slow rise in CDK1-cyclin B activity resulting in a protracted prometaphase I that can last between 6 and 7 h in mouse oocytes (Davydenko, Schultz, & Lampson, 2013; Jones, 2008; Ledan, Polanski, Terret, & Maro, 2001; Polanski et al., 1998). This gradual increase in CDK1 activity is accompanied by progressive accumulation of cyclin B protein. This is achieved partially through de novo translation of cyclin B mRNA (Hampl & Eppig, 1995; Winston, 1997). The rate of cyclin B synthesis directly determines the duration of prometaphase (Polanski et al., 1998). However, control of cyclin B degradation is equally important. This requires that CDK1 acts in concert with MAPK to suppress APC/C activity allowing accumulation of the APC/C's M-phase substrates. CDK1/MAPK appears to act to stabilize the spindle assembly checkpoint (SAC), which when inhibited leads to increased APC/C activity (Fig. 2) (Nabti, Marangos, Bormann, Kudo, & Carroll, 2014).

Although CDK1/cyclin B complexes have long been considered the sole components of MPF activity sufficient for driving mitosis and meiosis, studies in various systems have shown that mitotic progression also requires the simultaneous suppression of the antagonizing B55 subunit associated form of protein phosphatase 2A (PP2A). This is mediated by the Greatwall (Gwl) kinase also known as MASTL (microtubule-associated serine/threonine kinase-like) that is also required for mitotic entry and progression (Alvarez-Fernández et al., 2013; Castilho, Williams, Mochida, Zhao, & Goldberg, 2009; Gharbi-Ayachi et al., 2010; Mochida, Maslen, Skehel, & Hunt, 2010; Rangone et al., 2011). Gwl phosphorylates and thereby stimulates the activity of cAMP-regulated phosphoprotein 19 (ARPP19) and α-endosulfine (ENSA), two inhibitors of the PP2A regulatory subunit B55. However, in mouse oocytes it appears that MASTL activity is not necessary for entry into meiosis I and progression through prometaphase I occurs without a need to suppress PP2A (Adhikari, Liu, & Kaldis, 2015). This mechanism, unique to the first meiotic division, might be required to allow the gradual rise in CDK1 activity during the protracted prometaphase (Davydenko et al., 2013; Jones, 2008; Ledan et al., 2001; Polanski et al., 1998; Verlhac, Kubiak, Clarke, & Maro, 1994; Verlhac et al., 1996).

CDK1 activity finally peaks at metaphase, functioning as important regulator of the timing of kinetochore–microtubule (KT-MT) attachment (see later). Once bivalents have become properly aligned on the metaphase plate and attached to spindle MTs, anaphase onset is initiated by loss of SAC-mediated repression of the APC/C. The APC/C, together with its coactivator CDC20, promotes ubiquitin-mediated degradation of securin and cyclin B1 (Jin et al., 2010; Reis et al., 2007), thus promoting cleavage of cohesion subunits by separase and lowering of CDK1 activity, respectively, allowing chromosome segregation and polar body extrusion (Herbert et al., 2003; Jones, 2008; Kudo et al., 2006; Verlhac, Terret, & Pintard, 2010).

2.3 Metaphase II Arrest: Cytostatic Factor

Following extrusion of the first polar body, CDK1 activity is then rapidly upregulated to promote entry into MII. After rapidly progressing through prometaphase of MII, oocytes become arrested at metaphase II. The molecular basis for the MII arrest was first determined in *Xenopus* oocytes in the classical cytoplasmic transfer experiments of Masui and Markert (1971). This work demonstrated that MII oocytes contain an activity termed cytostatic factor (CSF). CSF functions as an inhibitor of the APC/C^{CDC20} to stabilize cyclin B and hence MPF activity. Two active components of CSF are the proteins Emi2 and Mos/MAPK (Inoue, Ohe, Kanemori, Nobui, & Sagata, 2007; Jones, 2005; Nishiyama, Ohsumi, & Kishimoto, 2007; Shoji et al., 2006; Tung et al., 2005; Wu et al., 2007). Accordingly, depletion of EMI2 from mouse oocytes results in MII release (Amanai, Shoji, Yoshida, Brahmajosyula, & Perry, 2006; Shoji et al., 2006) and contributions of MAPK to CSF activity are indicated by the sterility of MAPK knockout female mice. Oocytes from maternal Mos knockout females (Colledge, Carlton, Udy, & Evans, 1994; Hashimoto et al., 1994; Verlhac et al., 1994) or from which Mos have been depleted by RNAi (Wianny & Zernicka-Goetz, 2000) fail to arrest at meiosis II and parthogenetically activate. However, inhibition of the APC/C is not the only mechanism by which MII arrest is achieved. Other pathways act directly upon CDK1. As with entry into meiosis I, the CDK1-activating phosphatase CDC25A, is necessary to stabilize meiosis II arrest, and down-regulation of its activity causes CDK1 phosphorylation, MPF inactivation, and MII exit without cyclin B degradation (Oh, Susor, Schindler, Schultz, & Conti, 2013). In contrast to MI, activity of the CDK1 antagonizing phosphatase PP2A/B55 also needs to be suppressed through activity of the MASTL kinase (Adhikari et al., 2014).

2.4 Fertilization: Egg Activation

Sperm entry induces the resumption and completion of meiosis II, release of cortical granules, and translation of maternal mRNAs. These phenomena are collectively known as "egg activation" (Ducibella et al., 2002; Schultz & Kopf, 1995; Stricker, 1999). An increase in levels of intracellular calcium ($[Ca^{2+}]_i$) provides a universal trigger of egg activation in all species studied so far (Stricker, 1999). In mammals this occurs as set of brief but periodical increases in $[Ca^{2+}]_i$ that last for several hours after sperm entry (Miyazaki et al., 1986).

Already in the 19th century it was proposed that intracellular ionic increases contributed to the release of oocytes from meiotic arrest. By studying sea urchin eggs, Loeb and colleagues observed that initiation of development in these animals was possible simply by varying the concentration and composition of the fertilizing medium (Loeb, 1907).

Increase in $[Ca^{2+}]_i$ is thought to be triggered by phospholipase Cζ (PLCζ), an enzyme delivered by the sperm (Saunders et al., 2002). PLCζ is required for production of the second messenger inositol triphosphate (IP_3), which then binds its receptor, IP_3R, the main intracellular Ca^{2+}-release channel in almost all mammalian cell types (reviewed in Berridge, Lipp, & Bootman, 2000; Bootman, Lipp, & Berridge, 2001).

One result of Ca^{2+} release is cortical granule exocytosis (Kline & Kline, 1992; Sardet, Prodon, Dumollard, Chang, & Chenevert, 2002), also known as the cortical reaction. This secretory process is dependent on SNARE proteins and is responsible for blocking polyspermy by modifying the extracellular matrix of the oocyte, the zona pellucida. Cortical granules are secretory organelles stored in the cortex of unfertilized oocytes and are derived from Golgi complexes during oocyte growth (Austin, 1956; Gulyas, Stouffer, & Hodgen, 1979). How is the change in Ca^{2+} translated into an exocytotic event?

One likely mediator is the calcium sensor synaptotagmin, important for stimulated secretion in many cell types and shown to regulate cortical granule exocytosis in sea urchin oocytes (Leguia, Conner, Berg, & Wessel, 2006). Ca^{2+} ions also complex with calmodulin (CaM), which can in turn bind to and activate other protein kinases such as the Ca^{2+}/calmodulin-dependent protein kinase II (CAMKII) and myosin light chain kinase (MLCK). MLCK is a potentially important regulator of myosin II in non-muscle cells (Bresnick, 1999), and a requirement for MLCK activity in cytokinesis has been reported in both sea urchin and mouse eggs (Fig. 3) (Deng, Williams, & Schultz, 2005; Groigno & Whitaker, 1998; Matson, Markoulaki, & Ducibella, 2006).

Another consequence of calcium-dependent signaling is the exit from meiosis II. This requires CAMKII, which phosphorylates the CSF component Emi2 (Hansen, Tung, & Jackson, 2006, see earlier). EMI2 is further phosphorylated by P1 (Hansen et al., 2006; Liu & Maller, 2005; Rauh, Schmidt, Bormann, Nigg, & Mayer, 2005; Schmidt, Rauh, Nigg, & Mayer, 2006), and subsequently targeted for SCF-mediated degradation (Schmidt et al., 2006). This leads to the subsequent activation of the APC/C, destruction of cyclin B and securin, and completion of meiosis II. Accordingly, blocking cyclin B degradation by D-box mutation prevents exit from metaphase II despite a Ca^{2+} signal (Madgwick et al., 2004). However, activation of APC/C through the overexpression of CDC20 is not sufficient to overcome the MII arrest (Tsurumi, Hoffmann, Geley, Graeser, & Polanski, 2004), suggesting that the inactivation of MPF might require mechanisms in addition to cyclin B degradation. Indeed it appears that

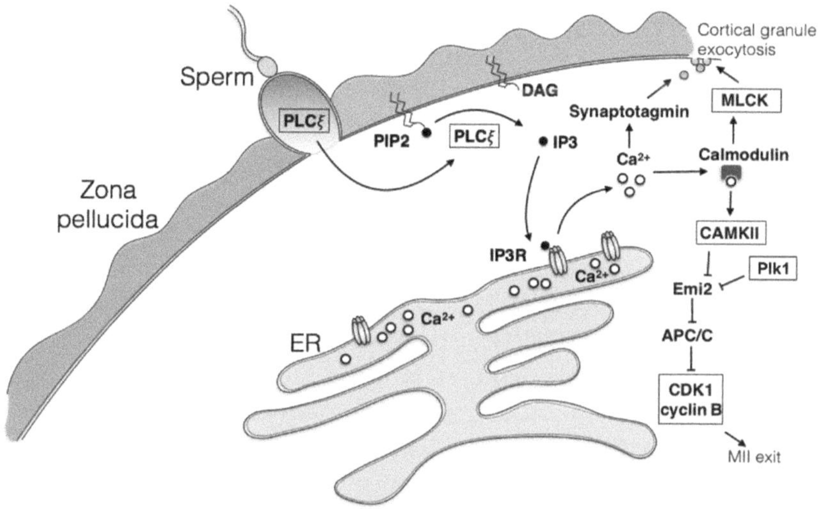

Fig. 3 Events triggered by fertilization. Following sperm and oocyte fusion at fertilization, phospholipase Cζ (PLCζ) is released into the cytoplasm of the oocyte, where it facilitates the hydrolysis of membrane-bound phosphatidylinositol 4,5-bisphosphate (PIP2) to diacylglycerol (DAG) and the second messenger inositol-1,4,5 trisphosphate (IP3). IP3 triggers Ca^{2+} release from intracellular Ca^{2+} stores in the endoplasmic reticulum (ER) by binding to its inositol-trisphosphate receptor (IP3R) localized in the ER membrane. Ca^{2+} oscillations have several downstream effects. Via calmodulin, Ca^{2+} activates the Ca^{2+}/calmodulin-dependent kinase II (CAMKII), causing an Emi2-dependent decrease in MPF activity (CDK1/cyclin B) that promotes cell cycle resumption. Ca^{2+} also triggers cortical granule exocytosis by stimulating the activity of synaptotagmin and myosin light chain kinase (MLCK).

reactivation of WEE2, and thus the inhibitory phosphorylation of Cdk1, is also necessary to initiate MPF inactivation during exit from meiosis II in mouse oocytes (Oh, Susor, & Conti, 2011).

The important function of CAMKII as calcium sensor upon fertilization is reflected by the fact that expression of constitutively active forms of CAMKII into mouse eggs is able to initiate all events of downstream egg activation and to promote development to the blastocyst stage (Knott et al., 2006; Madgwick, Levasseur, & Jones, 2005).

3. FIDELITY OF CHROMOSOME TRANSMISSION

3.1 The SAC

It is particularly important that chromosomes are transmitted with the highest possible fidelity in the germ line and this requires effective checkpoint controls. The regulation of chromosome segregation has been studied predominantly in mitosis but as similar regulatory proteins are used in meiosis it is helpful first to review our knowledge of the mitotic surveillance mechanisms that ensure accurate chromosome transmission (Fig. 4).

Attachments of MTs onto bioriented kinetochore pairs are selectively stabilized and erroneous attachments destabilized, through a mechanism requiring the catalytic subunit of the chromosomal passenger complex (CPC), Aurora B kinase. If kinetochores are unattached, they provide a primary signal to activate the SAC. The SAC comprises the proteins MPS1, Mitotic-Arrest-Deficient 1 (MAD1), MAD2, Budding-Uninhibited-by-Benzimidazole 1 (BUB1), BUBR1, and BUB3 (Musacchio & Hardwick, 2002; Taylor, Scott, & Holland, 2004). It functions to arrest the cell cycle by sequestering CDC20, activator of the APC/C E3 ubiquitin-protein ligase, into the mitotic checkpoint complex (MCC), thereby delaying exit from mitosis. In the absence of SAC signaling, cohesion between the sister chromatids is lost and cells exit from mitosis regardless of chromosome–spindle attachments, resulting in an increased frequency of chromosome missegregation.

The SAC monitors proper attachments between MTs and the kinetochore to which it is intimately connected. The outer kinetochore comprises a supercomplex, the KMN network, consisting of the subcomplexes KNL1, MIS12, and NDC80. If MTs are incorrectly attached to kinetochores this brings the NDC80 complex into the vicinity of centromeric Aurora B activity which destabilizes kinetochore–microtubule interactions by phosphorylating specific kinetochore substrates (Santaguida, Vernieri, Villa, Ciliberto, & Musacchio, 2011). Tension on attached sister kinetochores pulls the kinetochore substrates out of reach of Aurora B, preventing their phosphorylation and consequently allowing stabilization of kinetochore fibers (Liu, Vader, Vromans, Lampson, & Lens, 2009) and silencing of the SAC.

3.1.1 The SAC in Meiosis

In contrast to the rapid kinetochore–microtubule capture in mitosis, bivalents become more gradually attached to spindle MTs during prometaphase. MT nucleation initially occurs around chromosomes at NEBD leading to formation of an apolar spindle intermediate (see later). Subsequently, bivalents congress in a belt-like configuration on the spindle as it is becoming bipolar (Kitajima, Ohsugi, & Ellenberg, 2011). This phase is followed by homologous chromosomes adopting a stretched configuration along the long spindle axis that is maintained until anaphase onset. During this process,

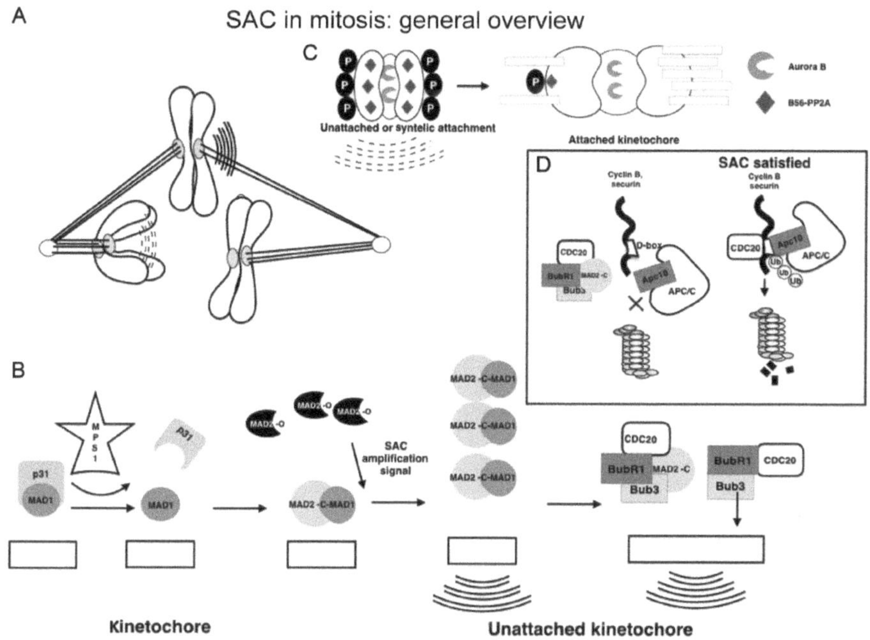

Fig. 4 Spindle assembly checkpoint in mammalian mitotic cells SAC. (A) The spindle assembly checkpoint is a surveillance mechanism that is active during the entire mitosis and monitors the presence of unattached kinetochores, providing a primary signal that delays mitotic exit. (B) MPS1 phosphorylates $p31^{comet}$ allowing the recruitment of O-MAD2 and amplification of a SAC signal. MAD1 at kinetochores is required to recruit MAD2 to the kinetochores during the early steps of mitosis. MAD2 is known to adopt two distinct conformations; when unbound it has an open conformation (O-MAD2) but upon binding to MAD1 or CDC20, MAD2 has a closed conformation (C-MAD2). The MAD1–C-MAD2 core complex is recruited to kinetochores after mitotic entry; and cytosolic O-MAD2 can then be recruited to kinetochore-bound MAD1 and MAD1–C-MAD2. O-MAD2, once bound to MAD1–C-MAD2, captures the APC/C cofactor CDC20 to form a C-MAD2–CDC20 complex. MAD2-CDC20 also promotes the interaction between BUBR1 and CDC20. SAC controls APC/C activity and it does so by sequestering CDC20 into a complex called the mitotic checkpoint complex (MCC). (C) Correct attachments of MTs onto bioriented kinetochore must be selectively stabilized while erroneous attachments are destabilized through a mechanism requiring the CPC's catalytic subunit, Aurora B kinase. A gradient of Aurora B kinase activity from the inner centromere destabilizes kinetochore–microtubule interactions on tensionless kinetochores by phosphorylating specific substrates at the kinetochore. Tension on amphitelically attached sister kinetochores pulls the kinetochore substrates out of reach of Aurora B, preventing their phosphorylation and consequently allowing stabilization of kinetochore fibers. KT-MT attachments are stabilized through kinetochore protein dephosphorylation by PP2A-B56 phosphatase that counteracts Aurora B kinase activity. (D) When the SAC is satisfied, CDC20 is able to bind the APC/C and cyclin B as well as securin can be proteolytically cleaved by the proteasome. Cohesion between the sister chromatids is lost and cells exit mitosis.

close to 90% of all chromosomes undergo an average of three rounds of error correction of their kinetochore–microtubule attachments before achieving correct orientation.

In mouse oocytes, the CPC may have one of two related catalytic subunits, Aurora kinases B or C, that participate in destabilizing incorrect KT-MT attachments (Balboula & Schindler, 2014; Davydenko et al., 2013; Li et al., 2004; Rattani et al., 2013; Sasai et al., 2004; Sharif et al., 2010; Yang et al., 2010). Importantly, even after bivalent stretching has occurred (approximately 4 h after NEBD), Aurora B/C activity is still able to destabilize attachments, and MT fibers do not form stable end-on attachments to kinetochores until 6–8 h after NEBD (Brunet et al., 1999; Gui & Homer, 2012; Kitajima et al., 2011). PP2A-B56 phosphatase is progressively recruited to kinetochores and this correlates with decreased phosphorylation of Aurora B/C substrates (Yoshida, Kaido, & Kitajima, 2015). Recruitment of PP2A-B56 appears to be independent of kinetochore stretching, but downstream of CDK1-dependent phosphorylation of the BUBR1 kinase. Thus the ability to form KT-MT fibers only after the spindle is bipolar (Breuer et al., 2010; Lane, Yun, & Jones, 2012) reflects the slow ramping-up of CDK1 activity during meiosis I as discussed earlier (Choi et al., 1991; Gavin, Cavadore, & Schorderet-Slatkine, 1994; Polanski et al., 1998). These studies demonstrate that in contrast to somatic cells and meiosis II (Yoshida et al., 2015), oocytes lack a mechanism for coordinating stretching and kinetochore phosphoregulation during MI, and this may explain the high frequency of KT-MT attachment errors (Fig. 5) (Yoshida et al., 2015).

Despite the fact that the meiosis I division remains APC/C-dependent and has a full complement of SAC proteins (Hached et al., 2011; Homer, McDougall, Levasseur, Murdoch, & Herbert, 2005; Lane, Chang, Jennings, & Jones, 2010; McGuinness et al., 2009; Tsurumi et al., 2004), oocytes are able to silence the SAC and undergo anaphase I in the presence of misaligned bivalents (Gui & Homer, 2012; Kolano, Brunet, Silk, Cleveland, & Verlhac, 2012; Lane et al., 2012). Thus, preventing and correcting attachment errors is particularly important, and if not performed sufficiently can contribute to early embryonic aneuploidy, birth defects, and spontaneous abortions (discussed later).

Meiosis I has the additional complexity that the sister chromatids of one chromosome are oriented toward the same pole to enable homologue segregation (coorientation). This contrasts to meiosis II or mitosis where sister chromatids biorient. Cohesion between sister chromatids is generated by the

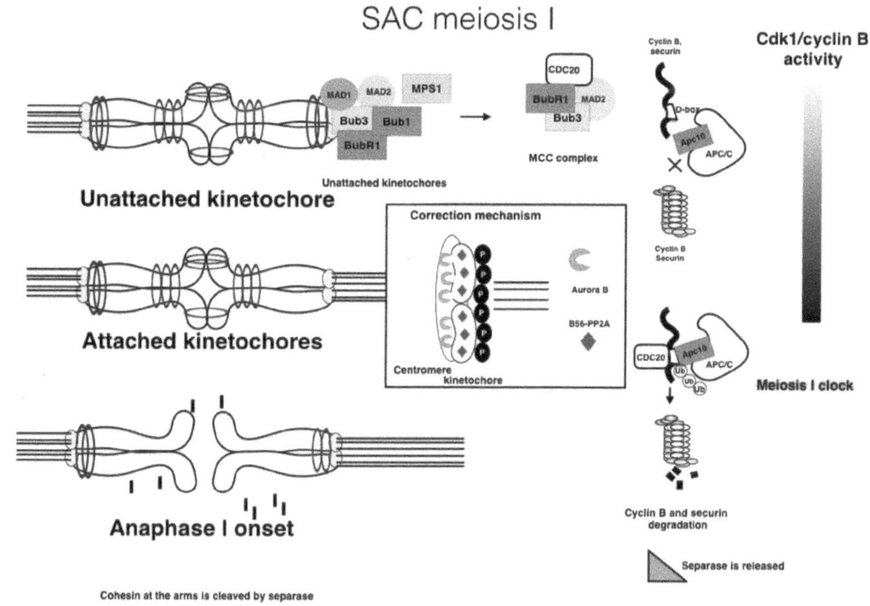

Fig. 5 Spindle assembly checkpoint in mouse meiosis I. SAC proteins localize to unattached kinetochores and prevent metaphase I-to-anaphase I transition in mouse oocytes. In metaphase I, correct attachments lead to removal of SAC proteins from kinetochores. APC/C binds CDC20, and cyclin B and securin are targeted for degradation by the proteasome, Proteolytic cleavage of securin allows Separase release that then cleaves the cohesin subunit REC8 on chromosome arms, leading to cohesin removal and metaphase I- to-anaphase I transition. Sister chromatids are still connected by cohesin. Aurora B/C activity destabilizes correct attachments even after bivalent stretching. KT-MT attachments are stabilized through kinetochore protein dephosphorylation by PP2A-B56 phosphatase that counteracts Aurora B kinase activity. PP2A-B56 phosphatase is progressively recruited to kinetochores independent of bivalent stretching. Thus it appears that the SAC is blind to stretching as a result of proper KT-MT attachment at the metaphase plate. Instead, the time given for establishment and correction of MT–kinetochore attachments is dictated by increasing levels of cyclin B. Once cyclin B degradation starts, that is before metaphase I, MT attachments are stabilized.

four-subunit complex cohesin (reviewed in consisting of the proteins SMC1, SMC3, SCC3, and α-kleisin (SCC1)) (reviewed in Nasmyth & Haering, 2005). Once correct KT-MT attachments are achieved and the SAC is satisfied, proteasomal degradation of securin frees separase to cleave the SCC1 cohesin subunit, triggering anaphase motion of sister chromatids to opposite poles. In meiosis I cohesion is lost at chromosome arms but it must be retained at centromeres to ensure that sisters remain together until

MII. This is achieved through a unique and conserved feature of meiotic cohesion; the replacement of SCC1 by its meiosis-specific counterpart REC8. REC8, which localizes along the length of chromosomes, is lost from chromosome arms during MI but remains at centromeres until meiosis II (Eijpe, Offenberg, Jessberger, Revenkova, & Heyting, 2003; Klein et al., 1999; Lee, Iwai, Yokota, & Yamashita, 2003; Lee et al., 2006; Pasierbek et al., 2001; Watanabe & Nurse, 1999). Considering Separase is active in both meiotic anaphases, mechanisms must exist to protect cohesins at the centromere region in MI. This function is carried out by the protein MEI-S332 in *Drosophila* and its homologues, collectively known as shugoshins (Clarke & Orr-Weaver, 2006; Kitajima, Kawashima, & Watanabe, 2004; Rabitsch et al., 2004). Mammals and fission yeast, unlike budding yeast and *Drosophila*, contain two shugoshin family members, Sgo1 and Sgo2, the latter being required for centromeric cohesion protection in mouse oocytes. SGO2 functions by recruiting the phosphatase PP2A to kinetochores (Kitajima et al., 2006), allowing PP2A to antagonize PLK1-dependent phosphorylation of cohesin and shugoshin, ensuring their centromeric retention.

It is key for the separation of homologues in MI that monoorientation of the unified kinetochores is established. The proteins Spo13 and Mam1 are crucial for this in budding yeast; as is Moa1 in fission yeast (Katis et al., 2004; Kitajima et al., 2011; Lee, Kiburz, & Amon, 2004; Tóth et al., 2000; Yokobayashi & Watanabe, 2005). In the mouse, MEIKIN (meiosis-specific kinetochore protein), their functional homologue, plays a crucial role in monoorientation and protection of centromeric cohesion. MEIKIN achieves this by cooperating with PLK1 kinase to stabilize the localization of the cohesin protector shugoshin (Kim et al., 2015).

In humans, oocytes may remain arrested for decades in prophase I before entering the meiotic divisions and chances of chromosomally abnormal pregnancy increase dramatically with advancing maternal age (Nagaoka, Hassold, & Hunt, 2012). As the first meiotic division is particularly error prone (see earlier; Hassold & Hunt, 2001), the majority of early embryonic aneuploidies result from errors in meiosis I. Oocytes of aged mice also display an increased percentage of chromosomes in which sister kinetochores have separated forming two distinct foci instead of a single unit. This is likely to reflect loss of centromeric cohesin (Chiang, Duncan, Schindler, Schultz, & Lampson, 2010), which is not renewed during oocyte growth but deteriorates with time (Chiang et al., 2010; Jessberger, 2012; Liu & Keefe, 2008; Revenkova, Herrmann, Adelfalk, & Jessberger, 2010; Tachibana-Konwalski et al., 2010). However, it seems that unlike younger mice where kinetochores are held together tightly during MI, separated sister kinetochores are an intrinsic feature of human MI oocytes even in young women (Sakakibara et al., 2015; Zielinska, Holubcová, Blayney, Elder, & Schuh, 2015). As oocytes age, this becomes even more apparent, with an increased fraction of bivalents disintegrating precociously into univalents, and displaying an increased interkinetochore distance (Sakakibara et al., 2015; Zielinska et al., 2015).

3.2 The SAC in Meiosis II and in the Early Embryonic Cleavage Divisions

Studies of the SAC in meiosis II are difficult because oocytes are only released from metaphase II at fertilization by CSF (see earlier and reviewed in Wu & Kornbluth, 2008). Therefore, SAC functionality in MII has to be analyzed under conditions of CSF release. However, it has been shown using several BUBR1 and MPS1 mutants, together with expression of dominant negative BUB1 (Tsurumi et al., 2004), that CSF arrest and release per se do not require the SAC (Touati et al., 2015).

Little is known about the SAC in the embryonic cleavage divisions, other than the fact that it is functional (Maciejewska, Polanski, Kisiel, Kubiak, & Ciemerych, 2009; Wei et al., 2011). In contrast to the human embryo where aneuploidy leads to a high frequency of spontaneous abortions, cleavage proceeds with high fidelity in the mouse embryo. Aneuploidy has been modeled in mouse embryos deficient for the BUBR1 checkpoint protein (Schmid et al., 2014), which causes growth retardation, apoptosis, disruption of bilateral symmetry, and embryonic lethality between embryonic days 7.5 and 13.5. Another study has used the small molecule reversine to inhibit the checkpoint kinase MPS1 and so model the effects of inducing high degrees of aneuploidy in the mouse embryo (Bolton et al., 2016). Interestingly, this latter study has shown that aneuploidy can be tolerated in the extraembryonic trophectoderm that is destined to become placental tissues but results in apoptosis in the epiblast, thus protecting the pool of pluripotent cells required to build the body.

4. ACENTRIOLAR SPINDLES

4.1 Meiotic Spindle Assembly

The equitable segregation of chromosomes depends upon the establishment of a functional spindle that begins in meiosis I when MPF is not fully activated and also has to take place in the absence of centrosomes. This contrasts with most mitotically dividing cells that have two centrosomes that become the spindle poles and which nucleate two asters of MTs that "search and capture" kinetochores (Kirschner & Mitchison, 1986; Mitchison & Kirschner, 1985). In the majority of metazoans centrioles are eliminated during oogenesis and so conventional centrosomes are unable to form (Delattre, 2004; Manandhar, Schatten, & Sutovsky, 2005). Mammals are no exception; electron microscopy studies have been able to detect centrioles in fetal mouse oocytes at the pachytene stage, but not at later stages, indicating the existence of an active and rapid mechanism for centriole elimination during oogenesis (Calarco, Donahue, & Szollosi, 1972). The mechanism of centriole elimination is essentially unknown. In most mammals, the incoming sperm contributes a centriole in the form of the basal body of the sperm axoneme. However, the mouse and other rodents are exceptions in that the basal body does not contribute to the zygote. Thus not only meiosis but also the early mitotic divisions proceed in the absence of centrioles.

Instead of using centrosomes to organize the meiotic spindle, mouse oocytes nucleate their MTs from multiple acentriolar microtubule-organizing centers (MTOCs) (Maro, Howlett, & Webb, 1985; Schuh & Ellenberg, 2007). These MTOCs contain components such as pericentrin and γ-tubulin and thus thought to molecularly resemble the PCM of centrosomes (Calarco-Gillam, Siebert, Hubble, Mitchison, & Kirschner, 1983; Carabatsos, Sellitto, Goodenough, & Albertini, 2000). The functional MTOCs present in prophase-arrested oocytes closely surround the nucleus at the time of NEBD (Luksza, Queguigner, Verlhac, & Brunet, 2013; Schuh & Ellenberg, 2007), whereupon they undergo a process of dramatic reorganization. Initially they elongate within large invaginations of the nuclear envelope (Luksza et al., 2013) and then fragment into smaller MTOC structures before finally merging at the poles of the nascent spindle. This reorganization is dependent on the kinase PLK1, dynein, and MTs (Luksza et al., 2013) and may act to facilitate the process of acentrosomal bipolar spindle formation. Initial stochastic MTOC distribution causes formation of a ball of MTs upon initiation of MT nucleation (Schuh & Ellenberg, 2007) and this is eventually resolved into a barrel-shaped bipolar array.

What are the pathways that regulate the nucleation of MTs in the absence of centrosomes? Acentrosomal spindle assembly has been particularly well studied in *Xenopus* egg extracts, in which MTs are nucleated in the vicinity of mitotic chromatin (Heald et al., 1996). In this system, chromatin-mediated microtubule nucleation relies on the small GTPase

Ran (Figure 1.15) (Carazo-Salas et al., 1999; Ohba, Nakamura, Nishitani, & Nishimoto, 1999), recognized to have essential functions during various stages of the cell cycle. Like other GTPases, Ran exists in two forms, an active GTP-bound and an inactive GDP-bound form. Chromatin localization of the Ran nucleotide exchange factor RCC1 and cytoplasmic localization of its GTPase-activating protein Ran-GAP, causes an enrichment of Ran-GTP in the interphase nucleus and surrounding mitotic chromosomes (Kalab, Weis, & Heald, 2002). During interphase, the Ran-GTP gradient provides a directional bias to nuclear transport, while in mitosis it promotes spindle assembly and allows nuclear envelope and nuclear pore reassembly (Hetzer, Gruss, & Mattaj, 2002). The functions of Ran are mediated via its interaction with transport receptors of the importin β superfamily. Nuclear cargoes and a number of spindle assembly factors containing nuclear localization domains are sequestered by importin β, and released upon Ran-GTP binding. As the Ran-GTP pathway involves both direct and indirect effectors, Ran-GTP is thought to induce a positive feedback loop to promote rapid spindle assembly (O'Connell & Khodjakov, 2007). Multiple proteins have been identified as direct or indirect targets of Ran-GTP (Meunier & Vernos, 2012). Among these are the MT-associated protein TPX2 (Gruss et al., 2001), which acts as cofactor for Aurora A. TPX2 targets Aurora A to the mitotic spindle where it contributes to the activation of the kinase (Bayliss, Sardon, Vernos, & Conti, 2003; Eyers & Maller, 2004; Kufer et al., 2002). Aurora A in turn triggers phosphorylation of TACC3 that functions with the MAP XMAP215/chTOG to stabilize nucleated MTs thus promoting spindle assembly.

Direct observations using a FRET-based sensor in mouse oocytes have demonstrated the presence Ran-gradient around chromatin after NEBD (Dumont, Petri, et al., 2007). However, although inhibiting Ran-GTP reduces microtubule density around chromosomes and delays spindle assembly, it does not prevent formation of a bipolar spindle (Dumont, Petri, et al., 2007; Schuh & Ellenberg, 2007). In mouse oocytes, levels of the Ran-effector TPX2 are low at NEBD, a result of APC/C^{Cdh1}-mediated degradation, and gradually increase in the course of prometaphase and metaphase I (Brunet et al., 2008). High levels of TPX2 at late prometaphase act via TACC3 and Aurora A to maintain the integrity of the spindle poles. However, the role of Ran-GTP gradient at earlier stages of spindle assembly in oocytes is currently unknown. Despite the limiting levels of TPX2, Aurora A appears to be activated even before NEBD, as indicated by the presence of its phosphorylated form on MTOCs prior to NEBD (Saskova et al., 2008).

Overexpression of Aurora A causes an increase in the number of MTOCs in prophase-arrested oocytes (Saskova et al., 2008), suggesting a function of this kinase that might be independent of its activating partner TPX2. Aurora A has also been shown to phosphorylate Nedd1 (Pinyol, Scrofani, & Vernos, 2013) and thereby regulate recruitment of γ-tubulin, and to promote MT nucleation by inhibition of the MT depolymerizing activity of MCAK (Katayama, Sasai, Kloc, Brinkley, & Sen, 2008), mechanisms that could also apply in the mouse oocyte.

One pathway that is able to act at least partially independent of Ran-GTP in MT nucleation lies downstream of the CPC. The CPC promotes MT nucleation at kinetochores independently of the spindle poles, a mechanism that has been characterized using *Xenopus* egg extracts where beads coated with reconstituted CPC could promote MT assembly and spindle formation (Kelly et al., 2007; Sampath et al., 2004). It is unclear whether the contribution of the CPC to acentrosomal spindle assembly in this system reflects an ability to promote MT nucleation or stabilization. Two substrates of the CPC kinase Aurora B have been identified that point toward a MT stabilizing role. These are the MT depolymerase, MCAK, and the MT-associated protein, OP18/Stathmin, that are both negatively regulated by Aurora B phosphorylation (Gadea & Ruderman, 2006; Ohi, Sapra, Howard, & Mitchison, 2004). The female sterility of *Drosophila* INCENP mutants (Colombié et al., 2008) has led support to the idea that the CPC pathway might also contribute to acentrosomal spindle assembly in oocytes in vivo. However, in the mouse oocyte, bipolar spindles are still able to form upon inactivation of either of the CPC catalytic subunits Aurora B or C, and instead there are defects in chromosome alignment and cytokinesis in meiosis I (Sharif et al., 2010). This can be attributed to a requirement for Auroras B and C in KT-MT error correction as discussed earlier (Liu et al., 2010; Welburn et al., 2010) and in cytokinesis.

PLK1 is present in nuclei of prophase-arrested mouse oocytes. It is also present at MTOCs and kinetochores from NEBD until metaphase I and relocalizes onto the spindle midzone in anaphase I (Clift & Schuh, 2015; Pahlavan et al., 2000; Solc et al., 2015; Tong et al., 2002; Wianny, Tavares, Evans, Glover, & Zernicka-Goetz, 1998; Yuan et al., 2007). PLK1 promotes nuclear envelope permeabilization in the oocyte, and recruitment of centrosomal proteins to MTOCs (Solc et al., 2015). It is also essential for reorganization of MTOCs at NEBD (Clift & Schuh, 2015), and in promoting efficient attachment of microtubules to the kinetochores to satisfy the SAC (Solc et al., 2015). However, although PLK1 contributes to proper spindle formation in the mouse oocyte, it does not appear to have a direct role in MT stabilization, as it does in *Xenopus* egg extracts.

This contrasts to PLK4 that surprisingly, as the master regulator of centriole formation, is present in the acentriolar mouse oocyte and early embryo. In the oocyte, PLK4 localizes around chromatin and at acentriolar MTOCs. Inhibition of PLK4 delays initiation of MT growth at the time of NEBD. As Plk4 is able to phosphorylate and potentiate Aurora A activity in vitro, this might constitute a potential downstream mechanism (our unpublished observations). A role of PLK4 in regulating spindle assembly in the oocyte is consistent with our findings in the early mouse embryo where PLK4 stimulates MT nucleation (see later; Coelho et al., 2013). Interestingly, the dependence of spindle assembly upon PLK4 function is far greater in the early embryo, suggesting availability of substrates or presence of compensatory pathways might differ in these divisions. Indeed, the oocyte might have developed redundant mechanisms to ensure spindle formation during its uniquely protracted metaphase of meiosis I that also exhibits different requirements for both Ran-GTP (Dumont, Petri, et al., 2007) and its effector TPX2 (Fig. 6) (Brunet et al., 2008) compared to meiosis II.

A third acentrosomal MT nucleation pathway has been described for meiotic spindle formation that, unlike Ran-GTP or CPC pathways, is not chromatin-mediated. Instead it involves MT-dependent MT nucleation within the spindle. Evidence for this pathway arose from observations reporting localization of γ-tubulin to spindle MTs, and the distribution of minus ends of MTs throughout the entire length of the spindle (Burbank, Groen, Perlman, Fisher, & Mitchison, 2006). This pool of spindle-associated γ-tubulin was demonstrated to be an important contributor to spindle formation (Khodjakov & Rieder, 1999; Lüders, Patel, & Stearns, 2006). A regulatory function in this pathway has been attributed to a protein complex termed Augmin in *Drosophila* (Goshima, Mayer, Zhang, Stuurman, & Vale, 2008) or HAUS (homologous to Augmin subunits) in human cells (Lawo et al., 2009). By promoting γ-tubulin-dependent branching nucleation of MTs, the Augmin complex is able to amplify the spindle MT network (Petry, Groen, Ishihara, Mitchison, & Vale, 2013; Zhu, Coppinger, Jang, Yates, & Fang, 2008; Zhu, Fang, & Fang, 2009). The effect of Augmin inhibition varies in different systems, and its importance becomes

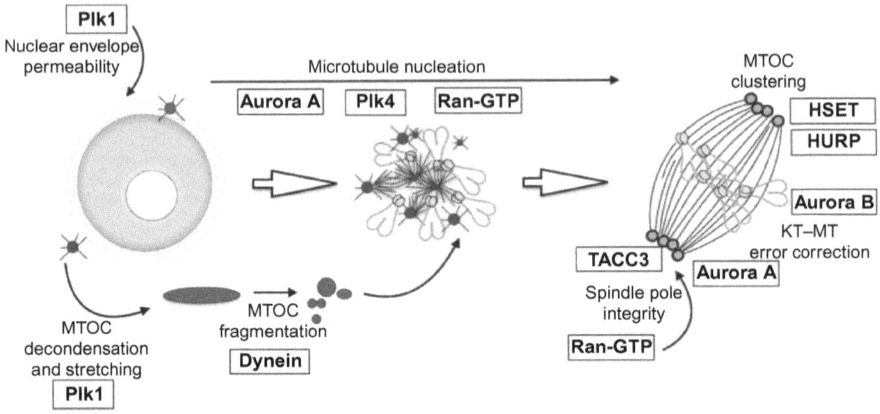

Fig. 6 Acentrosomal spindle assembly in the mouse oocyte is a multistep process. Upon nuclear envelope breakdown in the mouse oocyte, microtubule-organizing centers (MTOCs) closely surround the nuclear envelope. The kinase activity of PLK1 on the one hand promotes nuclear envelope permeability (Solc et al., 2015), and on the other hand c-Nap (not shown)-dependent MTOC decondensation and stretching (Clift & Schuh, 2015). Fragmented MTOCs are reorganized around the forming spindle, in a dynein and microtubule-dependent manner (Luksza et al., 2013). At the same time, microtubule nucleation is initiated. This partly relies on an active Ran-gradient around the chromosomes (Dumont, Petri, et al., 2007; Schuh & Ellenberg, 2007) and rapid MT growth is triggered by cooperative activities of the kinases Aurora A and PLK4 that localize to acentriolar MTOCs. During later stages of spindle assembly, Ran-GTP becomes crucial to establish the integrity of spindle poles and this relies on its downstream effector transforming acidic coiled-coil-containing protein 3 (TACC3). Gradients of Aurora A at the spindle poles and Aurora B at kinetochores are essential for proper kinetochore–microtubule (KT-MT) error correction. The kinesin-14 motor protein HSET and the microtubule-associated protein HURP are both required to efficiently cluster the multiple MTOCs at the spindle poles and enable formation of the typical barrel-shaped meiosis I spindle (Breuer et al., 2010; Mountain et al., 1999).

particularly relevant in the absence of centrosomes (Petry, Pugieux, Nédélec, & Vale, 2011; Uehara et al., 2009; Wainman et al., 2009). However, its role in acentrosomal MT nucleation in the mouse oocyte remains to be determined.

Once MT nucleation is initiated upon NEBD, the establishment of spindle bipolarity is only gradually achieved by clustering MTOCs at the two opposing poles (Breuer et al., 2010; Clift & Schuh, 2013). In somatic cells, two centrosomes usually determine spindle bipolarity and supernumerary centrosomes can cause formation of a multipolar spindle (Ganem, Godinho, & Pellman, 2009) as is frequently observed in cancer cells (Zyss & Gergely, 2009). However, even when multiple centrosomes are present, they usually become clustered at the poles of a bipolar spindle—a process that permits division and ensures cell viability (Ganem et al., 2009). Interestingly, the mechanisms required for centrosome clustering and focusing of acentriolar MTOCs appear to be dependent on shared mechanisms. In particular, HURP, a MAP and Ran-effector, is essential for the formation of a robust central MT domain, acting as scaffold for the sorting of MTOCs at the poles. By influencing the activity of kinesin-5, HURP enables the establishment and maintenance of meiotic spindle bipolarity While HURP is essential for meiosis, it is dispensable in somatic cell mitosis (Breuer et al., 2010). However, HURP becomes crucial in cancer cells harboring multiple centrosomes. Similar observations have been reported for the minus-end directed motor protein HSET. Originally identified as Ncd in *Drosophila* and shown to be required for meiotic spindle formation in oocytes, this kinesin is also important for acentrosomal female meiosis in the mouse (Mountain et al., 1999). Although nonessential for cell division in normal cells, Ncd/HSET is able to suppress multipolar spindle formation in cancer cells by promoting the clustering of multiple centrosomes at the poles (Godinho, Kwon, & Pellman, 2009; Kwon et al., 2008). Clustering of multiple centrosomes also requires other factors, such as members of the Augmin complex, or factors involved in KT-MT attachment and sister chromatid cohesion (Leber et al., 2010), suggesting that a balance of forces generated by kinetochore and spindle components is necessary to maintain centrosome clustering at the poles. Finally, the actin cytoskeleton together with cell adhesion and polarity proteins have also been found to facilitate centrosome clustering but it remains to be tested, if any of these mechanisms contribute to spindle bipolarization in the mouse oocyte.

4.2 Acentriolar Mitoses of the Early Embryo

The very first rounds of mitosis in the embryo occur in the absence of centrioles (Delattre, 2004; Manandhar, Sutovsky, Joshi, Stearns, & Schatten, 1998; Schatten, 1994; Woolley & Fawcett, 1973) and the de novo formation of centrioles does not begin until the 64-cell stage (Calarco-Gillam et al., 1983; Gueth-Hallonet et al., 1993; Howe & FitzHarris, 2013, and our observations figure N).

It is well known that centrioles are not essential for mitotic spindle assembly (Azimzadeh, Wong, Downhour, Sánchez Alvarado, & Marshall, 2012; Berns & Richardson, 1977; Debec, Détraves, Montmory, Géraud, &

Wright, 1995; Khodjakov, Cole, Oakley, & Rieder, 2000; Mahoney, Goshima, Douglass, & Vale, 2006; Megraw, Kao, & Kaufman, 2001), however their presence increases mitotic fidelity (Khodjakov & Rieder, 2001; Zamora & Marshall, 2005). Even in cells containing centrosomes, both centrosomal and acentrosomal spindle assembly pathways contribute to spindle assembly (Meunier & Vernos, 2012). Thus the first five rounds of mitotic acentriolar divisions, previously regarded as exhibiting exceptional mechanisms for spindle assembly, can be properly regarded as a natural system in which to study the acentrosomal spindle assembly pathway in mitosis in mammals. However, it appears that the process from meiotic to mitotic acentrosomal divisions is a gradual one.

During the first mitotic division of the embryo, the spindle assembles from multiple acentriolar MTOCs, in a similar manner to the oocyte (Calarco-Gillam et al., 1983; Courtois, Schuh, Ellenberg, & Hiiragi, 2012; Gueth-Hallonet et al., 1993; Hiraoka, Golden, & Magnuson, 1989; Maro et al., 1985). Initially the zygote contains as many as 80 MTOCs distributed between the cytoplasm and the pronuclear envelopes (Courtois et al., 2012). MTs nucleated from these MTOCs grow considerably during the first interphase (Houliston, Pickering, & Maro, 1987; Schatten, Simerly, & Schatten, 1985; Szollosi, Calarco, & Donahue, 1972) and at NEBD, the MTOCs surrounding the pronuclear envelope associate with chromatin. This results in the formation of a multipolar spindle in the center of the cell (contrasting to the apolar MT ball that develops as the acentriolar spindle forms in the oocyte) and is rapidly sorted into a bipolar array by minus-end-directed dynein and plus-end-directed kinesin (Courtois et al., 2012).

At each successive cleavage division until the 64-cell stage, the number of acentriolar MTOCs participating in spindle assembly gradually decreases as spindles become more focused at the poles (Courtois et al., 2012; Schatten et al., 1985). Starting from the fourth division, the length of the spindle decreases, as it scales with the size of the blastomeres (Courtois et al., 2012). At the 32–64 cell division, cells begin to display centrin-positive staining at the poles, consistent with the appearance of centrioles (Gueth-Hallonet et al., 1993), and spindle poles attain an increasingly focused morphology. Finally, in the blastocyst, all cells assemble a spindle from two centrin-positive poles. These changes correlate with a shift in the requirement for kinesin-5 for spindle polarity. Whereas oocytes and early mitotic divisions depend on kinesin-5, later cleavage divisions do not (FitzHarris, 2009). This is in accordance with the observation that centrosomal spindles do not rely on kinesin-5 to maintain bipolarity (Kollu, Bakhoum, & Compton, 2009). This appears to represent part of a more general switch in motor protein dependence, as expression of kinesin-12, that is involved in formation and maintenance of mitotic spindles (Tanenbaum et al., 2009), begins to be detected only after the 32- to 64-cell blastocyst stage (Courtois et al., 2012).

Unlike meiosis, where pathways regulating MT nucleation are beginning to be elucidated (see earlier), much remains to be learned about the mechanisms required for spindle assembly in this exceptional acentriolar mitotic system. Similar to the oocyte, inhibition of the Ran-pathway in the mouse zygote effects but does not prevent formation of a bipolar spindle (Courtois et al., 2012).

However, recent studies have introduced one unlikely player in this process—the centrosomal kinase PLK4. Despite the absence of centrioles and in accordance with our observations in the mouse oocyte (see earlier), PLK4 is present at acentriolar MTOCs in the early mouse embryo from the zygote stage onward, where it is absolutely required to promote microtubule nucleation and acentrosomal spindle assembly (Coelho et al., 2013). Depletion of maternal Plk4 prevents nucleation and growth of microtubules, leading to monopolar spindle formation, cytokinesis failure and consequently, developmental arrest. To mediate centriole formation, PLK4 requires its partner protein CEP152/Asterless in all systems tested (Bonaccorsi, Giansanti, & Gatti, 1998; Cizmecioglu et al., 2010; Dzhindzhev et al., 2010; Hatch, Kulukian, Holland, Cleveland, & Stearns, 2010; Varmark et al., 2007). Intriguingly, bipolar spindle formation also requires Plk4's association with CEP152 in the mouse embryo. Although CEP152 is nuclear during interphase, it is present at the MTOCs when MT nucleation begins. Artificially tethering CEP152 to cellular membranes sequesters PLK4 and is sufficient to trigger spindle assembly from these ectopic membranous sites. These results provide new players in the process of acentrosomal spindle assembly, suggesting an unexpected function for centriolar proteins during mitosis (Fig. 7).

In *Drosophila*, the overexpression of PLK4 leads to supernumerary centrioles in both fertilized and unfertilized eggs (Rodrigues-Martins, Riparbelli, Callaini, Glover, & Bettencourt-Dias, 2007). As unfertilized eggs contain centrioles from neither mother nor father, this must represent de novo centriole formation. Why then does elevated PLK4 not lead to centriole formation in the mouse zygote? The answer seems to lie in the fact that

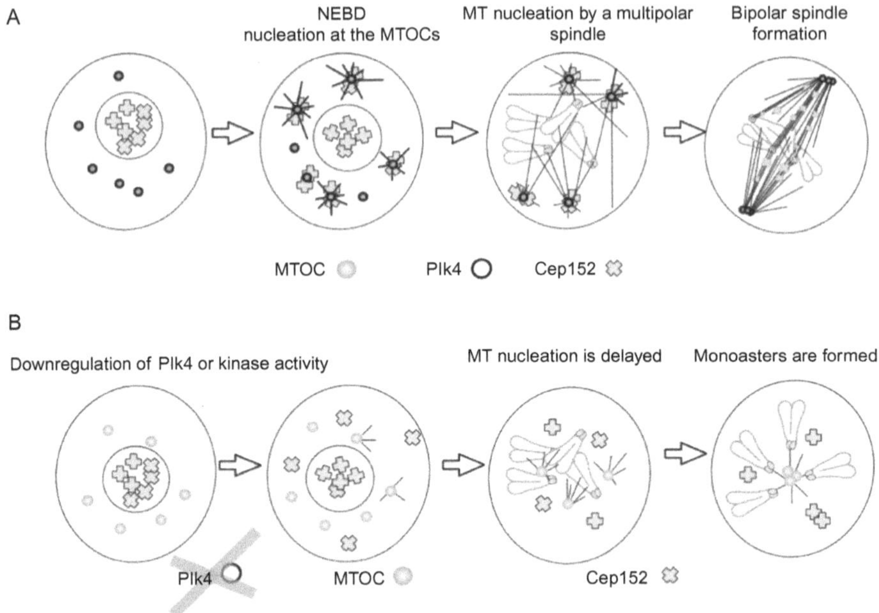

Fig. 7 Acentrosomal spindle assembly in early embryo. (A) PLK4 localizes at MTOCs from interphase onward; in contrast to CEP152, which is maintained in the nucleus until the onset of mitosis. Once CEP152 is released from the nucleus at nuclear envelope breakdown, it interacts with PLK4 at the MTOCs. CEP152 could provide a platform for PLK4 to either interact with its substrates, and/or to be activated itself, gaining the ability to trigger MT nucleation. MT-associated proteins are likely to be activated during this process, either as direct or indirect consequence of PLK4 phosphorylation. (B) Lack of PLK4 or PLK4 kinase activity delays microtubule nucleation and compromises bipolar spindle assembly in the early mouse embryo.

several centriole components necessary for procentriole formation are missing at this developmental stage, or are inappropriately localized (our unpublished observations). Thus it seems likely that the first appearance of centrioles in the blastocyst reflects the time at which the genes for its component proteins are first expressed. The presence of some centrosome components such as centrin2, CP110, and ODF2 can be observed in spindle poles after 16–64-cell stage embryos (Courtois et al., 2012). Moreover, centrin2 the cartwheel component SAS6, and the PCM component γ-tubulin also colocalize at spindle poles after this stages (Coelho et al., 2013). This accords well with the timing of the first appearance of centrioles at blastocyst stage detected by electron microscopy (Fig. 8) (Gueth-Hallonet et al., 1993; Howe & FitzHarris, 2013).

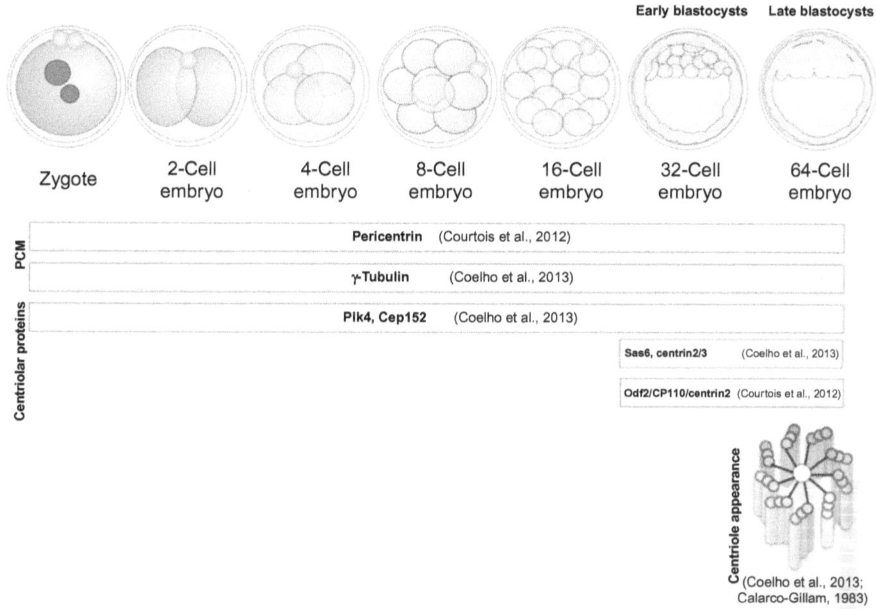

Fig. 8 Acentrosomal spindle assembly in the early mitotic divisions. Schematic representation of the embryonic stages that exhibit acentrosomal spindle assembly before de novo centriole formation at the late blastocyst stage. A summary of PCM and centriolar proteins expressed in these divisions is shown. Essential centriolar components such as SAS6 only start to be expressed at the 32-cell stage embryo, one division before centriole formation.

5. POSITIONING THE SPINDLE

5.1 Roles of the Actin Cytoskeleton

The asymmetric divisions that are the hallmark of meiosis are essential to achieve the elimination of one set of homologues in meiosis I and a set of chromatids in meiosis II and so retain maternal components in the oocyte to support successful fertilization and early embryonic development.

In the absence of astral microtubules and centrosomes, symmetry-breaking is achieved through in intrinsic properties of the cytoplasm in response to the actin cytoskeleton (Fig. 9) (Azoury, Lee, Georget, Hikal, & Verlhac, 2011; Verlhac, Lefebvre, Guillaud, Rassinier, & Maro, 2000). The main driving force for intracellular movements of the meiosis I spindle is dynamic actin reorganization in a process that involves two different pathways of actin nucleation. Unlike somatic cells where actin-rich

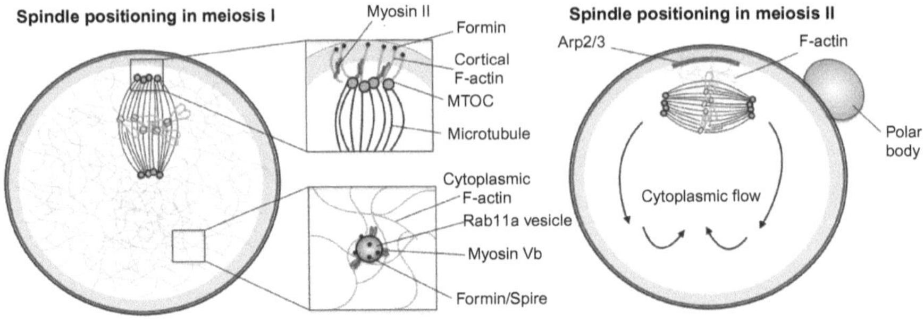

Fig. 9 Spindle positioning in meiosis I and II in mouse oocytes. In meiosis I, a cytoplasmic meshwork is nucleated by formin-2 and spire 1 and 2 proteins from RAB11A coated cytoplasmic vesicles around the spindle. Cortical F-actin is nucleated by actin-related protein 2/3 (ARP2/3). Myosin II at the spindle poles locally pulls on the actin meshwork, moving the spindle toward the cortex. At metaphase II, cortical ARP2/3 nucleates an inwardly flowing actin network, which drives cytoplasmic streaming that pushes the spindle toward the cortex.

retraction-fibers exert forces on astral microtubules to influence spindle positioning (Mitchison, 1992; Piel & Joanny, 2011; Thery, Jiménez-Dalmaroni, Racine, Bornens, & JUlicher, 2007; Thery et al., 2005), mouse oocytes are isolated from external geometric cues and show the first signs of polarization during prophase I (Dalton & Carroll, 2013; FitzHarris, Marangos, & Carroll, 2007; Halet & Carroll, 2007).

The asymmetric positioning of the meiotic spindle relies on different F-actin pools. The first includes a cytoplasmic actin meshwork nucleated by formin-2, spire 1, and spire 2 proteins (Azoury et al., 2008; Dumont, Million, et al., 2007; Leader et al., 2002; Pfender, Kuznetsov, Pleiser, Kerkhoff, & Schuh, 2011; Schuh & Ellenberg, 2008). These actin nucleators are clustered on RAB11A-positive cytoplasmic vesicles associated with the plus-end-directed motor myosin Vb and surround the meiotic spindle. Following nucleation of the cytoplasmic meshwork, the spindle becomes embedded in a so-called actin cage that is connected to the cortex. Absence or inhibition of formin or spire proteins impairs formation of the actin network (Azoury et al., 2008; Pfender et al., 2011; Schuh & Ellenberg, 2008) and prevents spindle migration (Dumont, Million, et al., 2007; Pfender et al., 2011). The direction of spindle migration along its long axis is determined by the fact that one of the spindle poles is closer to the cortex than the other (Verlhac et al., 2000). Phosphomyosin II, present at both poles of the spindle, pulls on cytoplasmic F-actin. As the leading pole of the spindle is pulled with greater strength than the opposite pole, this brings the leading pole closer to the cortex increasing the asymmetry. Consistent with the fact that myosin II at the spindle poles is required for pulling forces, treatment with a myosin II inhibitor impairs spindle migration (Fig. 9) (Chaigne et al., 2013; Schuh & Ellenberg, 2008).

A second actin pool is cortical F-actin, nucleated by the ARP2/3 complex. This cortical pool appears only around 3 h after meiotic resumption, a consequence of Mos kinase activity that initiates a downstream MAP kinase cascade (Verlhac et al., 1996, 2000). Activation of the Mos-MAPK pathway, via Wave 2-dependent ARP2/3 activation, promotes thickening of cortical F-actin. At the same time, the Mos-MAPK pathway also triggers exclusion of myosin II from the cortex (Chaigne et al., 2013). These two events cause softening of the oocyte cortex, a characteristic absent from *mos−/−* oocytes. Cortex softening appears essential for asymmetric positioning of the first meiotic spindle (Chaigne et al., 2013)—consistent with the finding that artificial stiffening of the cortex prevents spindle migration. Because a soft cortex is more responsive to pulling forces exerted by myosin II, spindle migration is enhanced, further amplifying the asymmetry generated at spindle poles by myosin II (Schuh & Ellenberg, 2008).

Upon reaching the oocyte cortex, proximity of the chromosomes and their associated Ran-GTP gradient signal the assembly of a microvilli-free membrane domain also known as an "actin cap" (Halet & Carroll, 2007; Maro, Johnson, Webb, & Flach, 1986). Ran has been shown to promote actin nucleation by causing CDC42-dependent polarized accumulation of the N-WASP-ARP2/3 complex to the cortex (Dehapiot, Carrière, Carroll, & Halet, 2013; Deng, Suraneni, Schultz, & Li, 2007; Dumont, Million, et al., 2007). Live imaging studies have revealed that actin filaments that are continuously nucleated from the cortical cap are instrumental to produce cytoplasmic streaming. Cytoplasmic streaming in turn promotes a directional pushing force onto the spindle, to maintain its off-center positioning at the cortex (Fig. 9) (Yi & Li, 2012; Yi et al., 2011). Following extrusion of the first polar body, the asymmetrically positioned MII spindle and associated chromosomes via Ran-GTP induce and maintain the actomyosin cap, required for the second polar body extrusion (Kunda & Baum, 2009).

5.2 Repositioning the Spindle in the Zygote

Mouse zygotes are not polarized, and the positioning of the first cleavage plane depends only on the position of the two pronuclei. It is important for the first zygotic mitosis to take place symmetrically that the two pronuclei become centrally positioned. There are no astral microtubules to be utilized for spindle positioning as has been documented in cells with centrosomes. Instead, the zygote has to rely upon the actin cytoskeleton (Chaigne et al., 2016), similar to spindle positioning in the oocyte (see earlier). The first step involves the coarse positioning of nuclei and this is dependent on the dynamics of an F-actin/myosin Vb meshwork but not on cortical tension. By contrast, the clustering of condensed chromosomes require microtubules and the fine positioning of the spindle is regulated by F-actin and a myosin II-dependent increase in cortical tension. The final clustering of chromosomes per se relies on microtubules, but the spindle position depends on an increase of cortical tension and the presence of F-actin. Once the spindle is positioned, it seems that actin is less important and the viscosity of the cytoplasm provides a passive mechanism for holding the spindle in a central position. Exactly how the viscosity of the cytoplasm is controlled and how this might differ between the oocyte and the zygote are questions that demand more attention.

5.3 Positioning the Spindle for Asymmetric Fate-Determining Divisions

Until the fourth round of cleavage, all cells divide symmetrically. At the 8-cell stage, the embryo compacts and cells polarize. Asymmetric divisions of a subset of cells in the 4- to 8-cell and the 8- to 16-cell divisions become key for assigning fate as they direct cells into the interior of the embryo to become the pluripotent epiblast and the primitive endoderm (reviewed by Lueng et al. in chapter "Polarity in Cell Fate Acquisition in the Early Mouse Embryo" this volume). The mechanism that orients these key divisions is poorly characterized. A recent study found that as the 8-cell embryo progresses through the cell cycle, many nuclei move from apical to more basal positions, in a microtubule- and kinesin-dependent manner. Asymmetric divisions take place when nuclei are located basally whereas cells with apical nuclei divide symmetrically. Nuclear positioning responds to both expression of the trophectoderm determining transcription factor Cdx2 and cell polarity, but the mechanism that underlies this is unclear. In most polarized somatic cells, the centrosome plays a crucial role in the reorganization of the cytoplasm in response to polarization cues and in directing division symmetry. It is therefore of great interest to determine exactly how organized microtubules respond to polarity cues in these acentriolar cells.

6. CONCLUDING REMARKS

Our current knowledge of cell division during meiosis and in the cleavage divisions of the embryo provides a solid foundation upon which we can build. It is important that our knowledge of these events continues to grow as these divisions are arguably the most important ones to establish our lives. It is crucial that we understand the mechanisms regulating chromosome transmission throughout these stages as aneuploidy presents a growing problem with the demographic tendency for an increased age of motherhood. We have much more to find out about the pathways that regulate the equitable partitioning of chromosomes throughout these stages in relation to the unusual mechanisms whereby the actin and microtubule cytoskeletons manipulate the positioning of chromatin. This continues to present a challenge since, because of its very nature, this is cell biology that can only be carried out in oocytes and early embryos of model systems where material is often limiting. Nevertheless, there is optimism for the future through approaches such as those of Pfender and colleagues (2015) in their recent screen to downregulate almost 800 genes in oocytes that identified genes required for chromosome segregation and spindle function (Pfender et al., 2015). An exciting path lies ahead.

REFERENCES

Adhikari, D., Diril, M. K., Busayavalasa, K., Risal, S., Nakagawa, S., Lindkvist, R., et al. (2014). Mastl is required for timely activation of APC/C in meiosis I and Cdk1 reactivation in meiosis II. *The Journal of Cell Biology*, *206*(7), 843–853. http://doi.org/10.1083/jcb.201406033.

Adhikari, D., Liu, K., & Kaldis, P. (2015). Mastl/PP2A regulate Cdk1 in oocyte maturation. *Oncotarget*, *6*(22), 18734–18735. http://doi.org/10.18632/oncotarget.4916.

Alvarez-Fernández, M., Sánchez-Martínez, R., Sanz-Castillo, B., Gan, P. P., Sanz-Flores, M., Trakala, M., et al. (2013). Greatwall is essential to prevent mitotic collapse after nuclear envelope breakdown in mammals. *Proceedings of the National Academy of Sciences*, *110*(43), 17374–17379. http://doi.org/10.1073/pnas.1310745110.

Amanai, M., Shoji, S., Yoshida, N., Brahmajosyula, M., & Perry, A. C. F. (2006). Injection of mammalian metaphase II oocytes with short interfering RNAs to dissect meiotic and early mitotic events. *Biology of Reproduction*, *75*(6), 891–898. http://doi.org/10.1095/biolreprod.106.054213.

Austin, C. R. (1956). Cortical granules in hamster eggs. *Experimental Cell Research*, *10*(2), 533–540.

Azimzadeh, J., Wong, M. L., Downhour, D. M., Sánchez Alvarado, A., & Marshall, W. F. (2012). Centrosome loss in the evolution of planarians. *Science*, *335*(6067), 461–463. http://doi.org/10.1126/science.1214457.

Azoury, J., Lee, K. W., Georget, V., Hikal, P., & Verlhac, M.-H. (2011). Symmetry breaking in mouse oocytes requires transient F-actin meshwork destabilization. *Development, 138*(14), 2903–2908. http://doi.org/10.1242/dev.060269.

Azoury, J., Lee, K. W., Georget, V., Rassinier, P., Leader, B., & Verlhac, M.-H. (2008). Spindle positioning in mouse oocytes relies on a dynamic meshwork of actin filaments. *Current Biology: CB, 18*(19), 1514–1519. http://doi.org/10.1016/j.cub.2008.08.044.

Balboula, A. Z., & Schindler, K. (2014). Selective disruption of Aurora C kinase reveals distinct functions from Aurora B kinase during meiosis in mouse oocytes. *PLoS Genetics, 10*(2), e1004194. http://doi.org/10.1371/journal.pgen.1004194.

Bayliss, R., Sardon, T., Vernos, I., & Conti, E. (2003). Structural basis of Aurora-A activation by TPX2 at the mitotic spindle. *Molecular Cell, 12*(4), 851–862.

Berns, M. W., & Richardson, S. M. (1977). Continuation of mitosis after selective laser microbeam destruction of the centriolar region. *The Journal of Cell Biology, 75*(3), 977–982.

Berridge, M. J., Lipp, P., & Bootman, M. D. (2000). The versatility and universality of calcium signalling. *Nature Reviews. Molecular Cell Biology, 1*(1), 11–21. http://doi.org/10.1038/35036035.

Bolton, H., Graham, S. J. L., Van der Aa, N., Kumar, P., Theunis, K., Fernandez Gallardo, E., et al. (2016). Mouse model of chromosome mosaicism reveals lineage-specific depletion of aneuploid cells and normal developmental potential. *Nature Communications,* 7, 11165. http://doi.org/10.1038/ncomms11165.

Bonaccorsi, S., Giansanti, M. G., & Gatti, M. (1998). Spindle self-organization and cytokinesis during male meiosis in asterless mutants of Drosophila melanogaster. *The Journal of Cell Biology, 142*(3), 751–761.

Bootman, M. D., Lipp, P., & Berridge, M. J. (2001). The organisation and functions of local Ca(2+) signals. *Quarterly Journal of Microscopical Science, 114*(Pt. 12), 2213–2222.

Borum, K. (1961). Oogenesis in the mouse. A study of the meiotic prophase. *Experimental Cell Research, 24*, 495–507.

Bresnick, A. R. (1999). Molecular mechanisms of nonmuscle myosin-II regulation. *Current Opinion in Cell Biology, 11*(1), 26–33.

Breuer, M., Kolano, A., Kwon, M., Li, C.-C., Tsai, T.-F., Pellman, D., et al. (2010). HURP permits MTOC sorting for robust meiotic spindle bipolarity, similar to extra centrosome clustering in cancer cells. *The Journal of Cell Biology, 191*(7), 1251–1260. http://doi.org/10.1083/jcb.201005065.

Brunet, S., Dumont, J., Lee, K. W., Kinoshita, K., Hikal, P., Gruss, O. J., et al. (2008). Meiotic regulation of TPX2 protein levels governs cell cycle progression in mouse oocytes. *PloS One, 3*(10), e3338. http://doi.org/10.1371/journal.pone.0003338.

Brunet, S., Maria, A. S., Guillaud, P., Dujardin, D., Kubiak, J. Z., & Maro, B. (1999). Kinetochore fibers are not involved in the formation of the first meiotic spindle in mouse oocytes, but control the exit from the first meiotic M phase. *The Journal of Cell Biology, 146*(1), 1–12.

Burbank, K. S., Groen, A. C., Perlman, Z. E., Fisher, D. S., & Mitchison, T. J. (2006). A new method reveals microtubule minus ends throughout the meiotic spindle. *The Journal of Cell Biology, 175*(3), 369–375. http://doi.org/10.1083/jcb.200511112.

Calarco, P. G., Donahue, R. P., & Szollosi, D. (1972). Germinal vesicle breakdown in the mouse oocyte. *Quarterly Journal of Microscopical Science, 10*(2), 369–385.

Calarco-Gillam, P. D., Siebert, M. C., Hubble, R., Mitchison, T., & Kirschner, M. W. (1983). Centrosome development in early mouse embryos as defined by an autoantibody against pericentriolar material. *Cell, 35*(3 Pt. 2), 621–629.

Carabatsos, M. J., Sellitto, C., Goodenough, D. A., & Albertini, D. F. (2000). Oocyte-granulosa cell heterologous gap junctions are required for the coordination of nuclear and cytoplasmic meiotic competence. *Developmental Biology, 226*(2), 167–179. http://doi.org/10.1006/dbio.2000.9863.

Carazo-Salas, R. E., Guarguaglini, G., Gruss, O. J., Segref, A., Karsenti, E., & Mattaj, I. W. (1999). Generation of GTP-bound Ran by RCC1 is required for chromatin-induced-mitotic spindle formation. *Nature, 400*(6740), 178–181. http://doi.org/10.1038/22133.

Castilho, P. V., Williams, B. C., Mochida, S., Zhao, Y., & Goldberg, M. L. (2009). The M phase kinase Greatwall (Gwl) promotes inactivation of PP2A/B55delta, a phosphatase directed against CDK phosphosites. *Molecular Biology of the Cell, 20*(22), 4777–4789. http://doi.org/10.1091/mbc.E09-07-0643.

Chaigne, A., Campillo, C., Gov, N. S., Voituriez, R., Azoury, J., Umaña-Diaz, C., et al. (2013). A soft cortex is essential for asymmetric spindle positioning in mouse oocytes. *Nature Cell Biology, 15*(8), 958–966. http://doi.org/10.1038/ncb2799.

Chaigne, A., Campillo, C., Voituriez, R., Gov, N. S., Sykes, C., Verlhac, M.-H., et al. (2016). F-actin mechanics control spindle centring in the mouse zygote. *Nature Communications, 7*, 10253. http://doi.org/10.1038/ncomms10253.

Chiang, T., Duncan, F. E., Schindler, K., Schultz, R. M., & Lampson, M. A. (2010). Evidence that weakened centromere cohesion is a leading cause of age-related aneuploidy in oocytes. *Current Biology: CB, 20*(17), 1522–1528. http://doi.org/10.1016/j.cub.2010.06.069.

Choi, T., Aoki, F., Mori, M., Yamashita, M., Nagahama, Y., & Kohmoto, K. (1991). Activation of p34cdc2 protein kinase activity in meiotic and mitotic cell cycles in mouse oocytes and embryos. *Development, 113*(3), 789–795.

Cizmecioglu, O., Arnold, M., Bahtz, R., Settele, F., Ehret, L., Haselmann-Weiss, U., et al. (2010). Cep152 acts as a scaffold for recruitment of Plk4 and CPAP to the centrosome. *The Journal of Cell Biology, 191*(4), 731–739. http://doi.org/10.1083/jcb.201007107.

Clarke, P. R., & Karsenti, E. (1991). Regulation of p34cdc2 protein kinase: New insights into protein phosphorylation and the cell cycle. *Quarterly Journal of Microscopical Science, 100*(Pt 3), 409–414.

Clarke, A., & Orr-Weaver, T. L. (2006). Sister chromatid cohesion at the centromere: Confrontation between kinases and phosphatases? *Developmental Cell, 10*(5), 544–547. http://doi.org/10.1016/j.devcel.2006.04.017.

Clift, D., & Schuh, M. (2013). Restarting life: Fertilization and the transition from meiosis to mitosis. *Nature Reviews Molecular Cell Biology, 14*(9), 549–562. http://doi.org/10.1038/nrm3643.

Clift, D., & Schuh, M. (2015). A three-step MTOC fragmentation mechanism facilitates bipolar spindle assembly in mouse oocytes. *Nature Communications, 6*, 7217. http://doi.org/10.1038/ncomms8217.

Coelho, P. A., Bury, L., Sharif, B., Riparbelli, M. G., Fu, J., Callaini, G., et al. (2013). Spindle formation in the mouse embryo requires Plk4 in the absence of centrioles. *Developmental Cell, 27*(5), 586–597. http://doi.org/10.1016/j.devcel.2013.09.029.

Colledge, W. H., Carlton, M. B., Udy, G. B., & Evans, M. J. (1994). Disruption of c-mos causes parthenogenetic development of unfertilized mouse eggs. *Nature, 370*(6484), 65–68. http://doi.org/10.1038/370065a0.

Colombié, N., Cullen, C. F., Brittle, A. L., Jang, J. K., Earnshaw, W. C., Carmena, M., et al. (2008). Dual roles of Incenp crucial to the assembly of the acentrosomal metaphase spindle in female meiosis. *Development, 135*(19), 3239–3246. http://doi.org/10.1242/dev.022624.

Conti, M., Andersen, C. B., Richard, F., Mehats, C., Chun, S. Y., Horner, K., et al. (2002). Role of cyclic nucleotide signaling in oocyte maturation. *Molecular and Cellular Endocrinology, 187*(1-2), 153–159.

Courtois, A., Schuh, M., Ellenberg, J., & Hiiragi, T. (2012). The transition from meiotic to mitotic spindle assembly is gradual during early mammalian development. *The Journal of Cell Biology, 198*(3), 357–370. http://doi.org/10.1083/jcb.201202135.

Dalton, C. M., & Carroll, J. (2013). Biased inheritance of mitochondria during asymmetric cell division in the mouse oocyte. *Journal of Cell Science*, *126*(Pt. 13), 2955–2964. http://doi.org/10.1242/jcs.128744.

Davydenko, O., Schultz, R. M., & Lampson, M. A. (2013). Increased CDK1 activity determines the timing of kinetochore-microtubule attachments in meiosis I. *The Journal of Cell Biology*, *202*(2), 221–229. http://doi.org/10.1083/jcb.201303019.

Debec, A., Détraves, C., Montmory, C., Géraud, G., & Wright, M. (1995). Polar organization of gamma-tubulin in acentriolar mitotic spindles of Drosophila melanogaster cells. *Quarterly Journal of Microscopical Science*, *108*(Pt 7), 2645–2653.

Dehapiot, B., Carrière, V., Carroll, J., & Halet, G. (2013). Polarized Cdc42 activation promotes polar body protrusion and asymmetric division in mouse oocytes. *Developmental Biology*, *377*(1), 202–212. http://doi.org/10.1016/j.ydbio.2013.01.029.

Dekel, N., & Beers, W. H. (1978). Rat oocyte maturation in vitro: Relief of cyclic AMP inhibition by gonadotropins. *Proceedings of the National Academy of Sciences of the United States of America*, *75*(9), 4369–4373.

Delattre, M. (2004). The arithmetic of centrosome biogenesis. *Quarterly Journal of Microscopical Science*, *117*(9), 1619–1630. http://doi.org/10.1242/jcs.01128.

Deng, M., Suraneni, P., Schultz, R. M., & Li, R. (2007). The ran GTPase mediates chromatin signaling to control cortical polarity during polar body extrusion in mouse oocytes. *Developmental Cell*, *12*(2), 301–308. http://doi.org/10.1016/j.devcel.2006.11.008.

Deng, M., Williams, C. J., & Schultz, R. M. (2005). Role of MAP kinase and myosin light chain kinase in chromosome-induced development of mouse egg polarity. *Developmental Biology*, *278*(2), 358–366. http://doi.org/10.1016/j.ydbio.2004.11.013.

Ducibella, T., Huneau, D., Angelichio, E., Xu, Z., Schultz, R. M., Kopf, G. S., et al. (2002). Egg-to-embryo transition is driven by differential responses to Ca(2+) oscillation number. *Developmental Biology*, *250*(2), 280–291.

Dumont, J., Million, K., Sunderland, K., Rassinier, P., Lim, H., Leader, B., et al. (2007). Formin-2 is required for spindle migration and for the late steps of cytokinesis in mouse oocytes. *Developmental Biology*, *301*(1), 254–265. http://doi.org/10.1016/j.ydbio.2006.08.044.

Dumont, J., Petri, S., Pellegrin, F., Terret, M.-E., Bohnsack, M. T., Rassinier, P., et al. (2007). A centriole- and RanGTP-independent spindle assembly pathway in meiosis I of vertebrate oocytes. *The Journal of Cell Biology*, *176*(3), 295–305. http://doi.org/10.1083/jcb.200605199.

Dzhindzhev, N. S., Yu, Q. D., Weiskopf, K., Tzolovsky, G., Cunha-Ferreira, I., Riparbelli, M., et al. (2010). Asterless is a scaffold for the onset of centriole assembly. *Nature*, *467*(7316), 714–718. http://doi.org/10.1038/nature09445.

Eijpe, M., Offenberg, H., Jessberger, R., Revenkova, E., & Heyting, C. (2003). Meiotic cohesin REC8 marks the axial elements of rat synaptonemal complexes before cohesins SMC1beta and SMC3. *The Journal of Cell Biology*, *160*(5), 657–670. http://doi.org/10.1083/jcb.200212080.

Eppig, J. J., & Downs, S. M. (1984). Chemical signals that regulate mammalian oocyte maturation. *Biology of Reproduction*, *30*(1), 1–11.

Eppig, J. J., Freter, R. R., Ward-Bailey, P. F., & Schultz, R. M. (1983). Inhibition of oocyte maturation in the mouse: Participation of cAMP, steroid hormones, and a putative maturation-inhibitory factor. *Developmental Biology*, *100*(1), 39–49.

Eppig, J. J., Schroeder, A. C., & O'Brien, M. J. (1992). Developmental capacity of mouse oocytes matured in vitro: Effects of gonadotrophic stimulation, follicular origin and oocyte size. *Journal of Reproduction and Fertility*, *95*(1), 119–127.

Eppig, J. J., Wigglesowrth, K., & O'Brien, M. J. (1992). Comparison of embryonic developmental competence of mouse oocytes grown with and without serum. *Molecular Reproduction and Development*, *32*(1), 33–40. http://doi.org/10.1002/mrd.1080320107.

Eyers, P. A., & Maller, J. L. (2004). Regulation of Xenopus Aurora A activation by TPX2. *Journal of Biological Chemistry*, *279*(10), 9008–9015. http://doi.org/10.1074/jbc.M312424200.

FitzHarris, G. (2009). A shift from kinesin 5-dependent metaphase spindle function during preimplantation development in mouse. *Development*, *136*(12), 2111–2119. http://doi.org/10.1242/dev.035089.

FitzHarris, G., Marangos, P., & Carroll, J. (2007). Changes in endoplasmic reticulum structure during mouse oocyte maturation are controlled by the cytoskeleton and cytoplasmic dynein. *Developmental Biology*, *305*(1), 133–144. http://doi.org/10.1016/j.ydbio.2007.02.006.

Gadea, B. B., & Ruderman, J. V. (2006). Aurora B is required for mitotic chromatin-induced phosphorylation of Op18/Stathmin. *Proceedings of the National Academy of Sciences of the United States of America*, *103*(12), 4493–4498. http://doi.org/10.1073/pnas.0600702103.

Ganem, N. J., Godinho, S. A., & Pellman, D. (2009). A mechanism linking extra centrosomes to chromosomal instability. *Nature*, *460*(7252), 278–282. http://doi.org/10.1038/nature08136.

Gautier, J., Matsukawa, T., Nurse, P., & Maller, J. (1989). Dephosphorylation and activation of Xenopus p34cdc2 protein kinase during the cell cycle. *Nature*, *339*(6226), 626–629. http://doi.org/10.1038/339626a0.

Gavin, A. C., Cavadore, J. C., & Schorderet-Slatkine, S. (1994). Histone H1 kinase activity, germinal vesicle breakdown and M phase entry in mouse oocytes. *Quarterly Journal of Microscopical Science*, *107*(Pt 1), 275–283.

Gerhart, J. (1984). Cell cycle dynamics of an M-phase-specific cytoplasmic factor in Xenopus laevis oocytes and eggs. *The Journal of Cell Biology*, *98*(4), 1247–1255. http://doi.org/10.1083/jcb.98.4.1247.

Gharbi-Ayachi, A., Labbé, J.-C., Burgess, A., Vigneron, S., Strub, J.-M., Brioudes, E., et al. (2010). The substrate of Greatwall kinase, Arpp19, controls mitosis by inhibiting protein phosphatase 2A. *Science*, *330*(6011), 1673–1677. http://doi.org/10.1126/science.1197048.

Godinho, S. A., Kwon, M., & Pellman, D. (2009). Centrosomes and cancer: How cancer cells divide with too many centrosomes. *Cancer Metastasis Reviews*, *28*(1-2), 85–98. http://doi.org/10.1007/s10555-008-9163-6.

Goshima, G., Mayer, M., Zhang, N., Stuurman, N., & Vale, R. D. (2008). Augmin: A protein complex required for centrosome-independent microtubule generation within the spindle. *The Journal of Cell Biology*, *181*(3), 421–429. http://doi.org/10.1083/jcb.200711053.

Groigno, L., & Whitaker, M. (1998). An anaphase calcium signal controls chromosome disjunction in early sea urchin embryos. *Cell*, *92*(2), 193–204.

Gruss, O. J., Carazo-Salas, R. E., Schatz, C. A., Guarguaglini, G., Kast, J., Wilm, M., et al. (2001). Ran induces spindle assembly by reversing the inhibitory effect of importin alpha on TPX2 activity. *Cell*, *104*(1), 83–93.

Gueth-Hallonet, C., Antony, C., Aghion, J., Santa-Maria, A., Lajoie-Mazenc, I., Wright, M., et al. (1993). gamma-Tubulin is present in acentriolar MTOCs during early mouse development. *Quarterly Journal of Microscopical Science*, *105*(Pt 1), 157–166. http://doi.org/10.1083/jcb.72.2.368.

Gui, L., & Homer, H. (2012). Spindle assembly checkpoint signalling is uncoupled from chromosomal position in mouse oocytes. *Development*, *139*(11), 1941–1946. http://doi.org/10.1242/dev.078352.

Gulyas, B. J., Stouffer, R. L., & Hodgen, G. D. (1979). Progesterone synthesis and fine structure of dissociated monkey (Macaca mulatta) luteal cells maintained in culture. *Biology of Reproduction*, *20*(4), 779–792.

Hached, K., Xie, S. Z., Buffin, E., Cladière, D., Rachez, C., Sacras, M., et al. (2011). Mps1 at kinetochores is essential for female mouse meiosis I. *Development, 138*(11), 2261–2271. http://doi.org/10.1242/dev.061317.

Halet, G., & Carroll, J. (2007). Rac activity is polarized and regulates meiotic spindle stability and anchoring in mammalian oocytes. *Developmental Cell, 12*(2), 309–317. http://doi.org/10.1016/j.devcel.2006.12.010.

Hambleton, R., Krall, J., Tikishvili, E., Honeggar, M., Ahmad, F., Manganiello, V. C., et al. (2005). Isoforms of cyclic nucleotide phosphodiesterase PDE3 and their contribution to cAMP hydrolytic activity in subcellular fractions of human myocardium. *Journal of Biological Chemistry, 280*(47), 39168–39174. http://doi.org/10.1074/jbc.M506760200.

Hampl, A., & Eppig, J. J. (1995). Analysis of the mechanism(s) of metaphase I arrest in maturing mouse oocytes. *Development, 121*(4), 925–933.

Han, S. J., Chen, R., Paronetto, M. P., & Conti, M. (2005). Wee1B is an oocyte-specific kinase involved in the control of meiotic arrest in the mouse. *Current Biology: CB, 15*(18), 1670–1676. http://doi.org/10.1016/j.cub.2005.07.056.

Hansen, D. V., Tung, J. J., & Jackson, P. K. (2006). CaMKII and polo-like kinase 1 sequentially phosphorylate the cytostatic factor Emi2/XErp1 to trigger its destruction and meiotic exit. *Proceedings of the National Academy of Sciences of the United States of America, 103*(3), 608–613. http://doi.org/10.1073/pnas.0509549102.

Hashimoto, N., Watanabe, N., Furuta, Y., Tamemoto, H., Sagata, N., Yokoyama, M., et al. (1994). Parthenogenetic activation of oocytes in c-mos-deficient mice. *Nature, 370*(6484), 68–71. http://doi.org/10.1038/370068a0.

Hassold, T., & Hunt, P. (2001). To err (meiotically) is human: The genesis of human aneuploidy. *Nature Reviews. Genetics, 2*(4), 280–291. http://doi.org/10.1038/35066065.

Hatch, E. M., Kulukian, A., Holland, A. J., Cleveland, D. W., & Stearns, T. (2010). Cep152 interacts with Plk4 and is required for centriole duplication. *The Journal of Cell Biology, 191*(4), 721–729. http://doi.org/10.1083/jcb.201006049.

Heald, R., Tournebize, R., Blank, T., Sandaltzopoulos, R., Becker, P., Hyman, A., et al. (1996). Self-organization of microtubules into bipolar spindles around artificial chromosomes in Xenopus egg extracts. *Nature, 382*(6590), 420–425. http://doi.org/10.1038/382420a0.

Herbert, M., Levasseur, M., Homer, H., Yallop, K., Murdoch, A., & McDougall, A. (2003). Homologue disjunction in mouse oocytes requires proteolysis of securin and cyclin B1. *Nature Cell Biology, 5*(11), 1023–1025. http://doi.org/10.1038/ncb1062.

Hetzer, M., Gruss, O. J., & Mattaj, I. W. (2002). The Ran GTPase as a marker of chromosome position in spindle formation and nuclear envelope assembly. *Nature Cell Biology, 4*(7), E177–E184. http://doi.org/10.1038/ncb0702-e177.

Hiraoka, L., Golden, W., & Magnuson, T. (1989). Spindle-pole organization during early mouse development. *Developmental Biology, 133*(1), 24–36. http://doi.org/10.1016/0012-1606(89)90293-5.

Holt, J. E., Weaver, J., & Jones, K. T. (2010). Spatial regulation of APCCdh1-induced cyclin B1 degradation maintains G2 arrest in mouse oocytes. *Development, 137*(8), 1297–1304. http://doi.org/10.1242/dev.047555.

Homer, H. A., McDougall, A., Levasseur, M., Murdoch, A. P., & Herbert, M. (2005). Mad2 is required for inhibiting securin and cyclin B degradation following spindle depolymerisation in meiosis I mouse oocytes. *Reproduction, 130*(6), 829–843. http://doi.org/10.1530/rep.1.00856.

Houliston, E., Pickering, S. J., & Maro, B. (1987). Redistribution of microtubules and pericentriolar material during the development of polarity in mouse blastomeres. *The Journal of Cell Biology, 104*(5), 1299–1308. http://doi.org/10.1083/jcb.104.5.1299.

Howe, K., & FitzHarris, G. (2013). A non-canonical mode of microtubule organization operates throughout pre-implantation development in mouse. *Cell Cycle (Georgetown, Tex.)*, *12*(10), 1616–1624. http://doi.org/10.4161/cc.24755.

Inoue, D., Ohe, M., Kanemori, Y., Nobui, T., & Sagata, N. (2007). A direct link of the Mos-MAPK pathway to Erp1/Emi2 in meiotic arrest of Xenopus laevis eggs. *Nature*, *446*(7139), 1100–1104. http://doi.org/10.1038/nature05688.

Jessberger, R. (2012). Age-related aneuploidy through cohesion exhaustion. *EMBO Reports*, *13*(6), 539–546. http://doi.org/10.1038/embor.2012.54.

Jin, F., Hamada, M., Malureanu, L., Jeganathan, K. B., Zhou, W., Morbeck, D. E., et al. (2010). Cdc20 is critical for meiosis I and fertility of female mice. *PLoS Genetics*, *6*(9), e1001147. http://doi.org/10.1371/journal.pgen.1001147.

Jones, K. T. (2005). Mammalian egg activation: From Ca^{2+} spiking to cell cycle progression. *Reproduction*, *130*(6), 813–823. http://doi.org/10.1530/rep.1.00710.

Jones, K. T. (2008). Meiosis in oocytes: Predisposition to aneuploidy and its increased incidence with age. *Human Reproduction Update*, *14*(2), 143–158. http://doi.org/10.1093/humupd/dmm043.

Kalab, P., Weis, K., & Heald, R. (2002). Visualization of a Ran-GTP gradient in interphase and mitotic Xenopus egg extracts. *Science*, *295*(5564), 2452–2456. http://doi.org/10.1126/science.1068798.

Katayama, H., Sasai, K., Kloc, M., Brinkley, B. R., & Sen, S. (2008). Aurora kinase-A regulates kinetochore/chromatin associated microtubule assembly in human cells. *Cell Cycle (Georgetown, Tex.)*, 7(17), 2691–2704.

Katis, V. L., Matos, J., Mori, S., Shirahige, K., Zachariae, W., & Nasmyth, K. (2004). Spo13 facilitates monopolin recruitment to kinetochores and regulates maintenance of centromeric cohesion during yeast meiosis. *Current Biology: CB*, *14*(24), 2183–2196. http://doi.org/10.1016/j.cub.2004.12.020.

Kelly, A. E., Sampath, S. C., Maniar, T. A., Woo, E. M., Chait, B. T., & Funabiki, H. (2007). Chromosomal enrichment and activation of the aurora B pathway are coupled to spatially regulate spindle assembly. *Developmental Cell*, *12*(1), 31–43. http://doi.org/10.1016/j.devcel.2006.11.001.

Khodjakov, A. L., Cole, R. W., Oakley, B. R., & Rieder, C. L. (2000). Centrosome-independent mitotic spindle formation in vertebrates. *Current Biology: CB*, *10*(2), 59–67.

Khodjakov, A. L., & Rieder, C. L. (1999). The sudden recruitment of gamma-tubulin to the centrosome at the onset of mitosis and its dynamic exchange throughout the cell cycle, do not require microtubules. *The Journal of Cell Biology*, *146*(3), 585–596.

Khodjakov, A. L., & Rieder, C. L. (2001). Centrosomes enhance the fidelity of cytokinesis in vertebrates and are required for cell cycle progression. *The Journal of Cell Biology*, *153*(1), 237–242.

Kim, J., Ishiguro, K.-I., Nambu, A., Akiyoshi, B., Yokobayashi, S., Kagami, A., et al. (2015). Meikin is a conserved regulator of meiosis-I-specific kinetochore function. *Nature*, *517*(7535), 466–471. http://doi.org/10.1038/nature14097.

Kirschner, M. W., & Mitchison, T. (1986). Microtubule dynamics. *Nature*, *324*(6098), 621. http://doi.org/10.1038/324621a0.

Kitajima, T. S., Kawashima, S. A., & Watanabe, Y. (2004). The conserved kinetochore protein shugoshin protects centromeric cohesion during meiosis. *Nature*, *427*(6974), 510–517. http://doi.org/10.1038/nature02312.

Kitajima, T. S., Ohsugi, M., & Ellenberg, J. (2011). Complete kinetochore tracking reveals error-prone homologous chromosome biorientation in mammalian oocytes. *Cell*, *146*(4), 568–581. http://doi.org/10.1016/j.cell.2011.07.031.

Kitajima, T. S., Sakuno, T., Ishiguro, K.-I., Iemura, S.-I., Natsume, T., Kawashima, S. A., et al. (2006). Shugoshin collaborates with protein phosphatase 2A to protect cohesin. *Nature*, *441*(7089), 46–52. http://doi.org/10.1038/nature04663.

Klein, F., Mahr, P., Gálová, M., Buonomo, S. B., Michaelis, C., Nairz, K., et al. (1999). A central role for cohesins in sister chromatid cohesion, formation of axial elements, and recombination during yeast meiosis. *Cell*, *98*(1), 91–103. http://doi.org/10.1016/S0092-8674(00)80609-1.

Kline, D., & Kline, J. T. (1992). Repetitive calcium transients and the role of calcium in exocytosis and cell cycle activation in the mouse egg. *Developmental Biology*, *149*(1), 80–89.

Knott, J. G., Gardner, A. J., Madgwick, S., Jones, K. T., Williams, C. J., & Schultz, R. M. (2006). Calmodulin-dependent protein kinase II triggers mouse egg activation and embryo development in the absence of Ca^{2+} oscillations. *Developmental Biology*, *296*(2), 388–395. http://doi.org/10.1016/j.ydbio.2006.06.004.

Kocer, A., Reichmann, J., Best, D., & Adams, I. R. (2009). Germ cell sex determination in mammals. *Molecular Human Reproduction*, *15*(4), 205–213. http://doi.org/10.1093/molehr/gap008.

Kolano, A., Brunet, S., Silk, A. D., Cleveland, D. W., & Verlhac, M.-H. (2012). Error-prone mammalian female meiosis from silencing the spindle assembly checkpoint without normal interkinetochore tension. *Proceedings of the National Academy of Sciences*, *109*(27), E1858–E1867. http://doi.org/10.1073/pnas.1204686109.

Kollu, S., Bakhoum, S. F., & Compton, D. A. (2009). Interplay of microtubule dynamics and sliding during bipolar spindle formation in mammalian cells. *Current Biology: CB*, *19*(24), 2108–2113. http://doi.org/10.1016/j.cub.2009.10.056.

Kovo, M., Kandli-Cohen, M., Ben-Haim, M., Galiani, D., Carr, D. W., & Dekel, N. (2006). An active protein kinase A (PKA) is involved in meiotic arrest of rat growing oocytes. *Reproduction*, *132*(1), 33–43. http://doi.org/10.1530/rep.1.00824.

Kudo, N. R., Wassmann, K., Anger, M., Schuh, M., Wirth, K. G., Xu, H., et al. (2006). Resolution of chiasmata in oocytes requires separase-mediated proteolysis. *Cell*, *126*(1), 135–146. http://doi.org/10.1016/j.cell.2006.05.033.

Kufer, T. A., Sillјé, H. H. W., Körner, R., Gruss, O. J., Meraldi, P., & Nigg, E. A. (2002). Human TPX2 is required for targeting Aurora-A kinase to the spindle. *The Journal of Cell Biology*, *158*(4), 617–623. http://doi.org/10.1083/jcb.200204155.

Kunda, P., & Baum, B. (2009). The actin cytoskeleton in spindle assembly and positioning. *Trends in Cell Biology*, *19*(4), 174–179. http://doi.org/10.1016/j.tcb.2009.01.006.

Kwon, M., Godinho, S. A., Chandhok, N. S., Ganem, N. J., Azioune, A., Thery, M., et al. (2008). Mechanisms to suppress multipolar divisions in cancer cells with extra centrosomes. *Genes & Development*, *22*(16), 2189–2203. http://doi.org/10.1101/gad.1700908.

Labbe, J. C., Picard, A., Peaucellier, G., Cavadore, J. C., Nurse, P., & Dorée, M. (1989). Purification of MPF from starfish: Identification as the H1 histone kinase p34cdc2 and a possible mechanism for its periodic activation. *Cell*, *57*(2), 253–263. http://doi.org/10.1016/0092-8674(89)90963-X.

Lane, S. I. R., Chang, H. Y., Jennings, P. C., & Jones, K. T. (2010). The Aurora kinase inhibitor ZM447439 accelerates first meiosis in mouse oocytes by overriding the spindle assembly checkpoint. *Reproduction*, *140*(4), 521–530. http://doi.org/10.1530/REP-10-0223.

Lane, S. I. R., Yun, Y., & Jones, K. T. (2012). Timing of anaphase-promoting complex activation in mouse oocytes is predicted by microtubule-kinetochore attachment but not by bivalent alignment or tension. *Development*, *139*(11), 1947–1955. http://doi.org/10.1242/dev.077040.

Lawo, S., Bashkurov, M., Mullin, M., Ferreria, M. G., Kittler, R., Habermann, B., et al. (2009). HAUS, the 8-subunit human augmin complex, regulates centrosome and spindle integrity. *Current Biology: CB*, *19*(10), 816–826. http://doi.org/10.1016/j.cub.2009.04.033.

Leader, B., Lim, H., Carabatsos, M. J., Harrington, A., Ecsedy, J., Pellman, D., et al. (2002). Formin-2, polyploidy, hypofertility and positioning of the meiotic spindle in mouse oocytes. *Nature Cell Biology*, *4*(12), 921–928. http://doi.org/10.1038/ncb880.

Leber, B., Maier, B., Fuchs, F., Chi, J., Riffel, P., Anderhub, S., et al. (2010). Proteins required for centrosome clustering in cancer cells. *Science Translational Medicine*, *2*(33), 33ra38. http://doi.org/10.1126/scitranslmed.3000915.

Ledan, E., Polanski, Z., Terret, M. E., & Maro, B. (2001). Meiotic maturation of the mouse oocyte requires an equilibrium between cyclin B synthesis and degradation. *Developmental Biology*, *232*(2), 400–413. http://doi.org/10.1006/dbio.2001.0188.

Lee, J., Iwai, T., Yokota, T., & Yamashita, M. (2003). Temporally and spatially selective loss of Rec8 protein from meiotic chromosomes during mammalian meiosis. *Quarterly Journal of Microscopical Science*, *116*(Pt 13), 2781–2790. http://doi.org/10.1242/jcs.00495.

Lee, B. H., Kiburz, B. M., & Amon, A. (2004). Spo13 maintains centromeric cohesion and kinetochore coorientation during meiosis I. *Current Biology: CB*, *14*(24), 2168–2182. http://doi.org/10.1016/j.cub.2004.12.033.

Lee, J., Okada, K., Ogushi, S., Miyano, T., Miyake, M., & Yamashita, M. (2006). Loss of Rec8 from chromosome arm and centromere region is required for homologous chromosome separation and sister chromatid separation, respectively, in mammalian meiosis. *Cell Cycle (Georgetown, Tex.)*, *5*(13), 1448–1455.

Leguia, M., Conner, S., Berg, L., & Wessel, G. M. (2006). Synaptotagmin I is involved in the regulation of cortical granule exocytosis in the sea urchin. *Molecular Reproduction and Development*, *73*(7), 895–905. http://doi.org/10.1002/mrd.20454.

Lew, D. J., & Kornbluth, S. (1996). Regulatory roles of cyclin dependent kinase phosphorylation in cell cycle control. *Current Opinion in Cell Biology*, *8*(6), 795–804.

Li, X., Sakashita, G., Matsuzaki, H., Sugimoto, K., Kimura, K., Hanaoka, F., et al. (2004). Direct association with inner centromere protein (INCENP) activates the novel chromosomal passenger protein, Aurora-C. *Journal of Biological Chemistry*, *279*(45), 47201–47211. http://doi.org/10.1074/jbc.M403029200.

Lincoln, A. J., Wickramasinghe, D., Stein, P., Schultz, R. M., Palko, M. E., De Miguel, M. P., et al. (2002). Cdc25b phosphatase is required for resumption of meiosis during oocyte maturation. *Nature Genetics*, *30*(4), 446–449. http://doi.org/10.1038/ng856.

Liu, L., & Keefe, D. L. (2008). Defective cohesin is associated with age-dependent misaligned chromosomes in oocytes. *Reproductive BioMedicine Online*, *16*(1), 103–112.

Liu, J., & Maller, J. L. (2005). Calcium elevation at fertilization coordinates phosphorylation of XErp1/Emi2 by Plx1 and CaMK II to release metaphase arrest by cytostatic factor. *Current Biology: CB*, *15*(16), 1458–1468. http://doi.org/10.1016/j.cub.2005.07.030.

Liu, D., Vader, G., Vromans, M. J. M., Lampson, M. A., & Lens, S. M. A. (2009). Sensing chromosome bi-orientation by spatial separation of aurora B kinase from kinetochore substrates. *Science*, *323*(5919), 1350–1353. http://doi.org/10.1126/science.1167000.

Liu, D., Vleugel, M., Backer, C. B., Hori, T., Fukagawa, T., Cheeseman, I. M., et al. (2010). Regulated targeting of protein phosphatase 1 to the outer kinetochore by KNL1 opposes Aurora B kinase. *The Journal of Cell Biology*, *188*(6), 809–820. http://doi.org/10.1083/jcb.201001006.

Loeb, J. (1907). On the chemical character of the process of fertilization and its bearing upon the theory of life phenomena. *Science*, *26*(666), 425–437. http://doi.org/10.1126/science.26.666.425.

Lüders, J., Patel, U. K., & Stearns, T. (2006). GCP-WD is a gamma-tubulin targeting factor required for centrosomal and chromatin-mediated microtubule nucleation. *Nature Cell Biology*, *8*(2), 137–147. http://doi.org/10.1038/ncb1349.

Luksza, M., Queguigner, I., Verlhac, M.-H., & Brunet, S. (2013). Rebuilding MTOCs upon centriole loss during mouse oogenesis. *Developmental Biology*, *382*(1), 48–56. http://doi.org/10.1016/j.ydbio.2013.07.029.
Maciejewska, Z., Polanski, Z., Kisiel, K., Kubiak, J. Z., & Ciemerych, M. A. (2009). Spindle assembly checkpoint-related failure perturbs early embryonic divisions and reduces reproductive performance of LT/Sv mice. *Reproduction*, *137*(6), 931–942. http://doi.org/10.1530/REP-09-0011.
Madgwick, S., Levasseur, M., & Jones, K. T. (2005). Calmodulin-dependent protein kinase II, and not protein kinase C, is sufficient for triggering cell-cycle resumption in mammalian eggs. *Quarterly Journal of Microscopical Science*, *118*(Pt 17), 3849–3859. http://doi.org/10.1242/jcs.02506.
Madgwick, S., Nixon, V. L., Chang, H.-Y., Herbert, M., Levasseur, M., & Jones, K. T. (2004). Maintenance of sister chromatid attachment in mouse eggs through maturation-promoting factor activity. *Developmental Biology*, *275*(1), 68–81. http://doi.org/10.1016/j.ydbio.2004.07.024.
Mahoney, N. M., Goshima, G., Douglass, A. D., & Vale, R. D. (2006). Making microtubules and mitotic spindles in cells without functional centrosomes. *Current Biology: CB*, *16*(6), 564–569. http://doi.org/10.1016/j.cub.2006.01.053.
Manandhar, G., Schatten, H., & Sutovsky, P. (2005). Centrosome reduction during gametogenesis and its significance. *Biology of Reproduction*, *72*(1), 2–13. http://doi.org/10.1095/biolreprod.104.031245.
Manandhar, G., Sutovsky, P., Joshi, H. C., Stearns, T., & Schatten, G. (1998). Centrosome reduction during mouse spermiogenesis. *Developmental Biology*, *203*(2), 424–434. http://doi.org/10.1006/dbio.1998.8947.
Marangos, P., & Carroll, J. (2004). The dynamics of cyclin B1 distribution during meiosis I in mouse oocytes. *Reproduction*, *128*(2), 153–162. http://doi.org/10.1530/rep.1.00192.
Maro, B., Howlett, S. K., & Webb, M. (1985). Non-spindle microtubule organizing centers in metaphase II-arrested mouse oocytes. *The Journal of Cell Biology*, *101*(5 Pt 1), 1665–1672. http://doi.org/10.1083/jcb.101.5.1665.
Maro, B., Johnson, M., Webb, M., & Flach, G. (1986). Mechanism of polar body formation in the mouse oocyte—An interaction between the chromosomes, the cytoskeleton and the plasma-membrane. *Journal of Embryology and Experimental Morphology*, *92*, 11–32.
Masui, Y., & Markert, C. L. (1971). Cytoplasmic control of nuclear behavior during meiotic maturation of frog oocytes. *The Journal of Experimental Zoology*, *177*(2), 129–145. http://doi.org/10.1002/jez.1401770202.
Matson, S., Markoulaki, S., & Ducibella, T. (2006). Antagonists of myosin light chain kinase and of myosin II inhibit specific events of egg activation in fertilized mouse eggs. *Biology of Reproduction*, *74*(1), 169–176. http://doi.org/10.1095/biolreprod.105.046409.
McGuinness, B. E., Anger, M., Kouznetsova, A., Gil-Bernabé, A. M., Helmhart, W., Kudo, N. R., et al. (2009). Regulation of APC/C activity in oocytes by a Bub1-dependent spindle assembly checkpoint. *Current Biology: CB*, *19*(5), 369–380. http://doi.org/10.1016/j.cub.2009.01.064.
McLaren, A. (2003). Primordial germ cells in the mouse. *Developmental Biology*, *262*(1), 1–15.
Megraw, T. L., Kao, L. R., & Kaufman, T. C. (2001). Zygotic development without functional mitotic centrosomes. *Current Biology: CB*, *11*(2), 116–120.
Mehlmann, L. M. (2005). Stops and starts in mammalian oocytes: Recent advances in understanding the regulation of meiotic arrest and oocyte maturation. *Reproduction*, *130*(6), 791–799. http://doi.org/10.1530/rep.1.00793.
Mehlmann, L. M., Saeki, Y., Tanaka, S., Brennan, T. J., Evsikov, A. V., Pendola, F. L., et al. (2004). The Gs-linked receptor GPR3 maintains meiotic arrest in mammalian oocytes. *Science*, *306*(5703), 1947–1950. http://doi.org/10.1126/science.1103974.

Meunier, S., & Vernos, I. (2012). Microtubule assembly during mitosis—From distinct origins to distinct functions? *Quarterly Journal of Microscopical Science*, *125*(12), 2805–2814. http://doi.org/10.1242/jcs.092429.
Mitchison, T. J. (1992). Compare and contrast actin filaments and microtubules. *Molecular Biology of the Cell*, *3*(12), 1309–1315.
Mitchison, T. J., & Kirschner, M. W. (1985). Properties of the kinetochore in vitro. II. Microtubule capture and ATP-dependent translocation. *The Journal of Cell Biology*, *101*(3), 766–777.
Mitra, J., & Schultz, R. M. (1996). Regulation of the acquisition of meiotic competence in the mouse: Changes in the subcellular localization of cdc2, cyclin B1, cdc25C and wee1, and in the concentration of these proteins and their transcripts. *Quarterly Journal of Microscopical Science*, *109*(Pt. 9), 2407–2415.
Miyazaki, S., Hashimoto, N., Yoshimoto, Y., Kishimoto, T., Igusa, Y., & Hiramoto, Y. (1986). Temporal and spatial dynamics of the periodic increase in intracellular free calcium at fertilization of golden hamster eggs. *Developmental Biology*, *118*(1), 259–267.
Mochida, S., Maslen, S. L., Skehel, M., & Hunt, T. (2010). Greatwall phosphorylates an inhibitor of protein phosphatase 2A that is essential for mitosis. *Science*, *330*(6011), 1670–1673. http://doi.org/10.1126/science.1195689.
Moore, G. P. (1975). The RNA polymerase activity of the preimplantation mouse embryo. *Journal of Embryology and Experimental Morphology*, *34*(2), 291–298.
Moore, G. P., & Lintern-Moore, S. (1978). Transcription of the mouse oocyte genome. *Biology of Reproduction*, *18*(5), 865–870.
Mountain, V., Simerly, C., Howard, L., Ando, A., Schatten, G., & Compton, D. A. (1999). The kinesin-related protein, HSET, opposes the activity of Eg5 and cross-links microtubules in the mammalian mitotic spindle. *The Journal of Cell Biology*, *147*(2), 351–366.
Musacchio, A., & Hardwick, K. G. (2002). The spindle checkpoint: Structural insights into dynamic signalling. *Nature Reviews. Molecular Cell Biology*, *3*(10), 731–741. http://doi.org/10.1038/nrm929.
Nabti, I., Marangos, P., Bormann, J., Kudo, N. R., & Carroll, J. (2014). Dual-mode regulation of the APC/C by CDK1 and MAPK controls meiosis I progression and fidelity. *The Journal of Cell Biology*, *204*(6), 891–900. http://doi.org/10.1083/jcb.201305049.
Nagaoka, S. I., Hassold, T. J., & Hunt, P. A. (2012). Human aneuploidy: Mechanisms and new insights into an age-old problem. *Nature Reviews Genetics*, *13*(7), 493–504. http://doi.org/10.1038/nrg3245.
Nasmyth, K., & Haering, C. H. (2005). The structure and function of SMC and kleisin complexes. *Annual Review of Biochemistry*, *74*(1), 595–648. http://doi.org/10.1146/annurev.biochem.74.082803.133219.
Nishiyama, T., Ohsumi, K., & Kishimoto, T. (2007). Phosphorylation of Erp1 by p90rsk is required for cytostatic factor arrest in Xenopus laevis eggs. *Nature*, *446*(7139), 1096–1099. http://doi.org/10.1038/nature05696.
Norris, R. P., Ratzan, W. J., Freudzon, M., Mehlmann, L. M., Krall, J., Movsesian, M. A., et al. (2009). Cyclic GMP from the surrounding somatic cells regulates cyclic AMP and meiosis in the mouse oocyte. *Development*, *136*(11), 1869–1878. http://doi.org/10.1242/dev.035238.
O'Connell, C. B., & Khodjakov, A. L. (2007). Cooperative mechanisms of mitotic spindle formation. *Journal of Cell Science*, *120*(10), 1717–1722. http://doi.org/10.1242/jcs.03442.
Oh, J. S., Han, S. J., & Conti, M. (2010). Wee1B, Myt1, and Cdc25 function in distinct compartments of the mouse oocyte to control meiotic resumption. *The Journal of Cell Biology*, *188*(2), 199–207. http://doi.org/10.1083/jcb.200907161.

Oh, J. S., Susor, A., & Conti, M. (2011). Protein tyrosine kinase Wee1B is essential for metaphase II exit in mouse oocytes. *Science*, *332*(6028), 462–465. http://doi.org/10.1126/science.1199211.

Oh, J. S., Susor, A., Schindler, K., Schultz, R. M., & Conti, M. (2013). Cdc25A activity is required for the metaphase II arrest in mouse oocytes. *Journal of Cell Science*, *126*(Pt. 5), 1081–1085. http://doi.org/10.1242/jcs.115592.

Ohba, T., Nakamura, M., Nishitani, H., & Nishimoto, T. (1999). Self-organization of microtubule asters induced in Xenopus egg extracts by GTP-bound Ran. *Science*, *284*(5418), 1356–1358. http://doi.org/10.1126/science.284.5418.1356.

Ohi, R., Sapra, T., Howard, J., & Mitchison, T. J. (2004). Differentiation of cytoplasmic and meiotic spindle assembly MCAK functions by Aurora B-dependent phosphorylation. *Molecular Biology of the Cell*, *15*(6), 2895–2906. http://doi.org/10.1091/mbc.E04-02-0082.

Pahlavan, G., Polanski, Z., Kalab, P., Golsteyn, R., Nigg, E. A., & Maro, B. (2000). Characterization of polo-like kinase 1 during meiotic maturation of the mouse oocyte. *Developmental Biology*, *220*(2), 392–400. http://doi.org/10.1006/dbio.2000.9656.

Pasierbek, P., Jantsch, M., Melcher, M., Schleiffer, A., Schweizer, D., & Loidl, J. (2001). A Caenorhabditis elegans cohesion protein with functions in meiotic chromosome pairing and disjunction. *Genes & Development*, *15*(11), 1349–1360. http://doi.org/10.1101/gad.192701.

Petry, S., Groen, A. C., Ishihara, K., Mitchison, T. J., & Vale, R. D. (2013). Branching microtubule nucleation in Xenopus egg extracts mediated by augmin and TPX2. *Cell*, *152*(4), 768–777. http://doi.org/10.1016/j.cell.2012.12.044.

Petry, S., Pugieux, C., Nédélec, F. J., & Vale, R. D. (2011). Augmin promotes meiotic spindle formation and bipolarity in Xenopus egg extracts. *Proceedings of the National Academy of Sciences*, *108*(35), 14473–14478. http://doi.org/10.1073/pnas.1110412108.

Pfender, S., Kuznetsov, V., Pasternak, M., Tischer, T., Santhanam, B., & Schuh, M. (2015). Live imaging RNAi screen reveals genes essential for meiosis in mammalian oocytes. *Nature*, *524*(7564), 239–242. http://doi.org/10.1038/nature14568.

Pfender, S., Kuznetsov, V., Pleiser, S., Kerkhoff, E., & Schuh, M. (2011). Spire-type actin nucleators cooperate with Formin-2 to drive asymmetric oocyte division. *Current Biology: CB*, *21*(11), 955–960. http://doi.org/10.1016/j.cub.2011.04.029.

Piel, M., & Joanny, J.-F. (2011). The rounding of mitotic cells. *Médecine Sciences: M/S*, 590–592. Éditions EDK, Groupe EDP Sciences. http://doi.org/10.1051/medsci/2011276010.

Pinyol, R., Scrofani, J., & Vernos, I. (2013). The role of NEDD1 phosphorylation by Aurora A in chromosomal microtubule nucleation and spindle function. *Current Biology: CB*, *23*(2), 143–149. http://doi.org/10.1016/j.cub.2012.11.046.

Pirino, G., Wescott, M. P., & Donovan, P. J. (2009). Protein kinase A regulates resumption of meiosis by phosphorylation of Cdc25B in mammalian oocytes. *Cell Cycle (Georgetown, Tex.)*, *8*(4), 665–670.

Polanski, Z., Ledan, E., Brunet, S., Louvet, S., Verlhac, M. H., Kubiak, J. Z., et al. (1998). Cyclin synthesis controls the progression of meiotic maturation in mouse oocytes. *Development*, *125*(24), 4989–4997.

Rabitsch, K. P., Gregan, J., Schleiffer, A., Javerzat, J.-P., Eisenhaber, F., & Nasmyth, K. (2004). Two fission yeast homologs of Drosophila Mei-S332 are required for chromosome segregation during meiosis I and II. *Current Biology: CB*, *14*(4), 287–301. http://doi.org/10.1016/j.cub.2004.01.051.

Rangone, H., Wegel, E., Gatt, M. K., Yeung, E., Flowers, A., Debski, J., et al. (2011). Suppression of scant identifies Endos as a substrate of greatwall kinase and a negative regulator of protein phosphatase 2A in mitosis. *PLoS Genetics*, 7(8), e1002225. http://doi.org/10.1371/journal.pgen.1002225.

Rattani, A., Wolna, M., Ploquin, M., Helmhart, W., Morrone, S., Mayer, B., et al. (2013). Sgol2 provides a regulatory platform that coordinates essential cell cycle processes during meiosis I in oocytes. *eLife*, *2*, e01133. http://doi.org/10.7554/eLife.01133.

Rauh, N. R., Schmidt, A., Bormann, J., Nigg, E. A., & Mayer, T. U. (2005). Calcium triggers exit from meiosis II by targeting the APC/C inhibitor XErp1 for degradation. *Nature*, *437*(7061), 1048–1052. http://doi.org/10.1038/nature04093.

Reis, A., Chang, H.-Y., Levasseur, M., & Jones, K. T. (2006). APCcdh1 activity in mouse oocytes prevents entry into the first meiotic division. *Nature Cell Biology*, *8*(5), 539–540.

Reis, A., Madgwick, S., Chang, H.-Y., Nabti, I., Levasseur, M., & Jones, K. T. (2007). Prometaphase APCcdh1 activity prevents non-disjunction in mammalian oocytes. *Nature Cell Biology*, *9*(10), 1192–1198. http://doi.org/10.1038/ncb1640.

Revenkova, E., Herrmann, K., Adelfalk, C., & Jessberger, R. (2010). Oocyte cohesin expression restricted to predictyate stages provides full fertility and prevents aneuploidy. *Current Biology: CB*, *20*(17), 1529–1533. http://doi.org/10.1016/j.cub.2010.08.024.

Rodrigues-Martins, A., Riparbelli, M., Callaini, G., Glover, D. M., & Bettencourt-Dias, M. (2007). Revisiting the role of the mother centriole in centriole biogenesis. *Science*, *316*(5827), 1046–1050. http://doi.org/10.1126/science.1142950.

Sakakibara, Y., Hashimoto, S., Nakaoka, Y., Kouznetsova, A., Hoog, C., & Kitajima, T. S. (2015). Bivalent separation into univalents precedes age-related meiosis I errors in oocytes. *Nature Communications*, *6*, 7550. http://doi.org/10.1038/ncomms8550.

Sampath, S. C., Ohi, R., Leismann, O., Salic, A., Pozniakovski, A., & Funabiki, H. (2004). The chromosomal passenger complex is required for chromatin-induced microtubule stabilization and spindle assembly. *Cell*, *118*(2), 187–202. http://doi.org/10.1016/j.cell.2004.06.026.

Sánchez, F., & Smitz, J. (2012). Molecular control of oogenesis. *Biochimica Et Biophysica Acta (BBA)—Molecular Basis of Disease*, *1822*(12), 1896–1912. http://doi.org/10.1016/j.bbadis.2012.05.013.

Santaguida, S., Vernieri, C., Villa, F., Ciliberto, A., & Musacchio, A. (2011). Evidence that Aurora B is implicated in spindle checkpoint signalling independently of error correction. *The EMBO Journal*, *30*(8), 1508–1519. http://doi.org/10.1038/emboj.2011.70.

Sardet, C., Prodon, F., Dumollard, R., Chang, P., & Chenevert, J. (2002). Structure and function of the egg cortex from oogenesis through fertilization. *Developmental Biology*, *241*(1), 1–23. http://doi.org/10.1006/dbio.2001.0474.

Sasai, K., Katayama, H., Stenoien, D. L., Fujii, S., Honda, R., Kimura, M., et al. (2004). Aurora-C kinase is a novel chromosomal passenger protein that can complement Aurora-B kinase function in mitotic cells. *Cell Motility and the Cytoskeleton*, *59*(4), 249–263. http://doi.org/10.1002/cm.20039.

Saskova, A., Solc, P., Baran, V., Kubelka, M., Schultz, R. M., & Motlik, J. (2008). Aurora kinase A controls meiosis I progression in mouse oocytes. *Cell Cycle (Georgetown, Tex.)*, 7(15), 2368–2376.

Saunders, C. M., Larman, M. G., Parrington, J., Cox, L. J., Royse, J., Blayney, L. M., et al. (2002). PLC zeta: A sperm-specific trigger of Ca(2+) oscillations in eggs and embryo development. *Development*, *129*(15), 3533–3544.

Schatten, G. (1994). The centrosome and its mode of inheritance: The reduction of the centrosome during gametogenesis and its restoration during fertilization. *Developmental Biology*, *165*(2), 299–335. http://doi.org/10.1006/dbio.1994.1256.

Schatten, G., Simerly, C., & Schatten, H. (1985). Microtubule configurations during fertilization, mitosis, and early development in the mouse and the requirement for egg microtubule-mediated motility during mammalian fertilization. *Proceedings of the National Academy of Sciences of the United States of America*, *82*(12), 4152–4156.

Schindler, K., & Schultz, R. M. (2009a). CDC14B acts through FZR1 (CDH1) to prevent meiotic maturation of mouse oocytes. *Biology of Reproduction*, *80*(4), 795–803. http://doi.org/10.1095/biolreprod.108.074906.

Schindler, K., & Schultz, R. M. (2009b). The CDC14A phosphatase regulates oocyte maturation in mouse. *Cell Cycle (Georgetown, Tex.)*, *8*(7), 1090–1098.

Schmid, M., Steinlein, C., Tian, Q., Hanlon Newell, A. E., Gessler, M., Olson, S. B., et al. (2014). Mosaic variegated aneuploidy in mouse BubR1 deficient embryos and pregnancy loss in human. *Chromosome Research*, *22*(3), 375–392. http://doi.org/10.1007/s10577-014-9432-x.

Schmidt, A., Rauh, N. R., Nigg, E. A., & Mayer, T. U. (2006). Cytostatic factor: An activity that puts the cell cycle on hold. *Quarterly Journal of Microscopical Science*, *119*(Pt. 7), 1213–1218. http://doi.org/10.1242/jcs.02919.

Schuh, M., & Ellenberg, J. (2007). Self-organization of MTOCs replaces centrosome function during acentrosomal spindle assembly in live mouse oocytes. *Cell*, *130*(3), 484–498. http://doi.org/10.1016/j.cell.2007.06.025.

Schuh, M., & Ellenberg, J. (2008). A new model for asymmetric spindle positioning in mouse oocytes. *Current Biology: CB*, *18*(24), 1986–1992. http://doi.org/10.1016/j.cub.2008.11.022.

Schultz, R. M., & Kopf, G. S. (1995). Molecular basis of mammalian egg activation. *Current Topics in Developmental Biology*, *30*, 21–62.

Schultz, R. M., Montgomery, R. R., & Belanoff, J. R. (1983). Regulation of mouse oocyte meiotic maturation: Implication of a decrease in oocyte cAMP and protein dephosphorylation in commitment to resume meiosis. *Developmental Biology*, *97*(2), 264–273.

Sela-Abramovich, S., Galiani, D., Nevo, N., & Dekel, N. (2008). Inhibition of rat oocyte maturation and ovulation by nitric oxide: Mechanism of action. *Biology of Reproduction*, *78*(6), 1111–1118. http://doi.org/10.1095/biolreprod.107.065490.

Sharif, B., Na, J., Lykke-Hartmann, K., McLaughlin, S. H., Laue, E., Glover, D. M., et al. (2010). The chromosome passenger complex is required for fidelity of chromosome transmission and cytokinesis in meiosis of mouse oocytes. *Journal of Cell Science*, *123*(Pt. 24), 4292–4300. http://doi.org/10.1242/jcs.067447.

Shoji, S.,Yoshida,N.,Amanai,M.,Ohgishi,M.,Fukui,T.,Fujimoto,S.,etal.(2006).Mammalian Emi2mediatescytostaticarrestandtransducesthesignalformeioticexitviaCdc20. *TheEMBO Journal*,*25*(4),834–845.http://doi.org/10.1038/sj.emboj.7600953.

Solc, P., Kitajima, T. S., Yoshida, S., Brzakova, A., Kaido, M., Baran, V., et al. (2015). Multiple requirements of PLK1 during mouse oocyte maturation. *PloS One*, *10*(2), e0116783. http://doi.org/10.1371/journal.pone.0116783.

Solc, P., Saskova, A., Baran, V., Kubelka, M., Schultz, R. M., & Motlik, J. (2008). CDC25A phosphatase controls meiosis I progression in mouse oocytes. *Developmental Biology*, *317*(1), 260–269. http://doi.org/10.1016/j.ydbio.2008.02.028.

Stricker, S. A. (1999). Comparative biology of calcium signaling during fertilization and egg activation in animals. *Developmental Biology*, *211*(2), 157–176. http://doi.org/10.1006/dbio.1999.9340.

Szollosi, D., Calarco, P., & Donahue, R. P. (1972). Absence of centrioles in the first and second meiotic spindles of mouse oocytes. *Quarterly Journal of Microscopical Science*, *11*(2), 521–541.

Tachibana-Konwalski, K., Godwin, J., van der Weyden, L., Champion, L., Kudo, N. R., Adams, D. J., et al. (2010). Rec8-containing cohesin maintains bivalents without turnover during the growing phase of mouse oocytes. *Genes & Development*, *24*(22), 2505–2516. http://doi.org/10.1101/gad.605910.

Tanenbaum, M. E., Macůrek, L., Janssen, A., Geers, E. F., Alvarez-Fernández, M., & Medema, R. H. (2009). Kif15 cooperates with eg5 to promote bipolar spindle assembly. *Current Biology: CB*, *19*(20), 1703–1711. http://doi.org/10.1016/j.cub.2009.08.027.

Taylor, S. S., Scott, M. I. F., & Holland, A. J. (2004). The spindle checkpoint: A quality control mechanism which ensures accurate chromosome segregation. *Chromosome Research*, *12*(6), 599–616. http://doi.org/10.1023/B:CHRO.0000036610.78380.51.

Thery, M., Jiménez-Dalmaroni, A., Racine, V., Bornens, M., & JUlicher, F. (2007). Experimental and theoretical study of mitotic spindle orientation. *Nature*, *447*(7143), 493–496. http://doi.org/10.1038/nature05786.

Thery, M., Racine, V., Pépin, A., Piel, M., Chen, Y., Sibarita, J.-B., et al. (2005). The extracellular matrix guides the orientation of the cell division axis. *Nature Cell Biology*, 7(10), 947–953. http://doi.org/10.1038/ncb1307.

Tong, C., Fan, H.-Y., Lian, L., Li, S.-W., Chen, D.-Y., Schatten, H., et al. (2002). Polo-like kinase-1 is a pivotal regulator of microtubule assembly during mouse oocyte meiotic maturation, fertilization, and early embryonic mitosis. *Biology of Reproduction*, *67*(2), 546–554.

Tóth, A., Rabitsch, K. P., Gálová, M., Schleiffer, A., Buonomo, S. B., & Nasmyth, K. (2000). Functional genomics identifies monopolin: A kinetochore protein required for segregation of homologs during meiosis i. *Cell*, *103*(7), 1155–1168.

Touati, S. A., Buffin, E., Cladière, D., Hached, K., Rachez, C., van Deursen, J. M., et al. (2015). Mouse oocytes depend on BubR1 for proper chromosome segregation but not for prophase I arrest. *Nature Communications*, *6*, 6946. http://doi.org/10.1038/ncomms7946.

Tsurumi, C., Hoffmann, S., Geley, S., Graeser, R., & Polanski, Z. (2004). The spindle assembly checkpoint is not essential for CSF arrest of mouse oocytes. *The Journal of Cell Biology*, *167*(6), 1037–1050. http://doi.org/10.1083/jcb.200405165.

Tung, J. J., Hansen, D. V., Ban, K. H., Loktev, A. V., Summers, M. K., Adler, J. R., et al. (2005). A role for the anaphase-promoting complex inhibitor Emi2/XErp1, a homolog of early mitotic inhibitor 1, in cytostatic factor arrest of Xenopus eggs. *Proceedings of the National Academy of Sciences of the United States of America*, *102*(12), 4318–4323. http://doi.org/10.1073/pnas.0501108102.

Uehara, R., Nozawa, R.-S., Tomioka, A., Petry, S., Vale, R. D., Obuse, C., et al. (2009). The augmin complex plays a critical role in spindle microtubule generation for mitotic progression and cytokinesis in human cells. *Proceedings of the National Academy of Sciences*, *106*(17), 6998–7003. http://doi.org/10.1073/pnas.0901587106.

Vaccari, S., Weeks, J. L., Hsieh, M., Menniti, F. S., & Conti, M. (2009). Cyclic GMP signaling is involved in the luteinizing hormone-dependent meiotic maturation of mouse oocytes. *Biology of Reproduction*, *81*(3), 595–604. http://doi.org/10.1095/biolreprod.109.077768.

Varmark, H., Llamazares, S., Rebollo, E., Lange, B., Reina, J., Schwarz, H., et al. (2007). Asterless is a centriolar protein required for centrosome function and embryo development in Drosophila. *Current Biology: CB*, *17*(20), 1735–1745. http://doi.org/10.1016/j.cub.2007.09.031.

Verlhac, M. H., Kubiak, J. Z., Clarke, H. J., & Maro, B. (1994). Microtubule and chromatin behavior follow MAP kinase activity but not MPF activity during meiosis in mouse oocytes. *Development*, *120*(4), 1017–1025.

Verlhac, M. H., Kubiak, J. Z., Weber, M., Géraud, G., Colledge, W. H., Evans, M. J., et al. (1996). Mos is required for MAP kinase activation and is involved in microtubule organization during meiotic maturation in the mouse. *Development*, *122*(3), 815–822.

Verlhac, M. H., Lefebvre, C., Guillaud, P., Rassinier, P., & Maro, B. (2000). Asymmetric division in mouse oocytes: With or without Mos. *Current Biology: CB*, *10*(20), 1303–1306.

Verlhac, M.-H., Terret, M.-E., & Pintard, L. (2010). Control of the oocyte-to-embryo transition by the ubiquitin-proteolytic system in mouse and C. elegans. *Current Opinion in Cell Biology*, *22*(6), 758–763. http://doi.org/10.1016/j.ceb.2010.09.003.

Vivarelli, E., Conti, M., De Felici, M., & Siracusa, G. (1983). Meiotic resumption and intracellular cAMP levels in mouse oocytes treated with compounds which act on cAMP metabolism. *Cell Differentiation, 12*(5), 271–276.

Wainman, A., Buster, D. W., Duncan, T., Metz, J., Ma, A., Sharp, D., et al. (2009). A new Augmin subunit, Msd1, demonstrates the importance of mitotic spindle-templated microtubule nucleation in the absence of functioning centrosomes. *Genes & Development, 23*(16), 1876–1881. http://doi.org/10.1101/gad.532209.

Wassarman, P. M., Josefowicz, W. J., & Letourneau, G. E. (1976). Meiotic maturation of mouse oocytes in vitro: Inhibition of maturation at specific stages of nuclear progression. *Quarterly Journal of Microscopical Science, 22*(3), 531–545.

Watanabe, Y., & Nurse, P. (1999). Cohesin Rec8 is required for reductional chromosome segregation at meiosis. *Nature, 400*(6743), 461–464. http://doi.org/10.1038/22774.

Wei, Y., Multi, S., Yang, C.-R., Ma, J., Zhang, Q.-H., Wang, Z.-B., et al. (2011). Spindle assembly checkpoint regulates mitotic cell cycle progression during preimplantation embryo development. *PloS One, 6*(6), e21557. http://doi.org/10.1371/journal.pone.0021557.

Welburn, J. P. I., Vleugel, M., Liu, D., Yates, J. R., Lampson, M. A., Fukagawa, T., et al. (2010). Aurora B phosphorylates spatially distinct targets to differentially regulate the kinetochore-microtubule interface. *Molecular Cell, 38*(3), 383–392. http://doi.org/10.1016/j.molcel.2010.02.034.

Wianny, F., Tavares, A., Evans, M. J., Glover, D. M., & Zernicka-Goetz, M. (1998). Mouse polo-like kinase 1 associates with the acentriolar spindle poles, meiotic chromosomes and spindle midzone during oocyte maturation. *Chromosoma, 107*(6-7), 430–439. http://doi.org/10.1007/s004120050327.

Wianny, F., & Zernicka-Goetz, M. (2000). Specific interference with gene function by double-stranded RNA in early mouse development. *Nature Cell Biology, 2*(2), 70–75. http://doi.org/10.1038/35000016.

Winston, N. J. (1997). Stability of cyclin B protein during meiotic maturation and the first mitotic cell division in mouse oocytes. *Biology of the Cell/Under the Auspices of the European Cell Biology Organization, 89*(3), 211–219.

Woolley, D. M., & Fawcett, D. W. (1973). The degeneration and disappearance of the centrioles during the development of the rat spermatozoon. *The Anatomical Record, 177*(2), 289–301. http://doi.org/10.1002/ar.1091770209.

Wu, Q., Guo, Y., Yamada, A., Perry, J. A., Wang, M. Z., Araki, M., et al. (2007). A role for Cdc2- and PP2A-mediated regulation of Emi2 in the maintenance of CSF arrest. *Current Biology: CB, 17*(3), 213–224. http://doi.org/10.1016/j.cub.2006.12.045.

Wu, J. Q., & Kornbluth, S. (2008). Across the meiotic divide—CSF activity in the post-Emi2/XErp1 era. *Quarterly Journal of Microscopical Science, 121*(Pt. 21), 3509–3514. http://doi.org/10.1242/jcs.036855.

Yang, K.-T., Li, S.-K., Chang, C.-C., Tang, C.-J. C., Lin, Y.-N., Lee, S.-C., et al. (2010). Aurora-C kinase deficiency causes cytokinesis failure in meiosis I and production of large polyploid oocytes in mice. *Molecular Biology of the Cell, 21*(14), 2371–2383. http://doi.org/10.1091/mbc.E10-02-0170.

Yi, K., & Li, R. (2012). Actin cytoskeleton in cell polarity and asymmetric division during mouse oocyte maturation. *Cytoskeleton (Hoboken, N.J.), 69*(10), 727–737. http://doi.org/10.1002/cm.21048.

Yi, K., Unruh, J. R., Deng, M., Slaughter, B. D., Rubinstein, B., & Li, R. (2011). Dynamic maintenance of asymmetric meiotic spindle position through Arp2/3-complex-driven cytoplasmic streaming in mouse oocytes. *Nature Cell Biology, 13*(10), 1252–1258. http://doi.org/10.1038/ncb2320.

Yokobayashi, S., & Watanabe, Y. (2005). The kinetochore protein Moa1 enables cohesion-mediated monopolar attachment at meiosis I. *Cell, 123*(5), 803–817. http://doi.org/10.1016/j.cell.2005.09.013.

Yoshida, S., Kaido, M., & Kitajima, T. S. (2015). Inherent instability of correct kinetochore-microtubule attachments during meiosis I in oocytes. *Developmental Cell*, *33*(5), 589–602. http://doi.org/10.1016/j.devcel.2015.04.020.

Yuan, K., Hu, H., Guo, Z., Fu, G., Shaw, A. P., Hu, R., et al. (2007). Phospho-regulation of HsCdc14A By Polo-like kinase 1 is essential for mitotic progression. *Journal of Biological Chemistry*, *282*(37), 27414–27423. http://doi.org/10.1074/jbc.M703555200.

Zamora, I., & Marshall, W. F. (2005). A mutation in the centriole-associated protein centrin causes genomic instability via increased chromosome loss in Chlamydomonas reinhardtii. *BMC Biology*, *3*(1), 15. http://doi.org/10.1186/1741-7007-3-15.

Zhang, Y., Zhang, Z., Xu, X.-Y., Li, X.-S., Yu, M., Yu, A.-M., et al. (2008). Protein kinase A modulates Cdc25B activity during meiotic resumption of mouse oocytes. *Developmental Dynamics*, *237*(12), 3777–3786. http://doi.org/10.1002/dvdy.21799.

Zhu, H., Coppinger, J. A., Jang, C.-Y., Yates, J. R., & Fang, G. (2008). FAM29A promotes microtubule amplification via recruitment of the NEDD1-gamma-tubulin complex to the mitotic spindle. *The Journal of Cell Biology*, *183*(5), 835–848. http://doi.org/10.1083/jcb.200807046.

Zhu, H., Fang, K., & Fang, G. (2009). Mechanism, function and regulation of microtubule-dependent microtubule amplification in mitosis. *Molecules and Cells*, *27*(1), 1–3. http://doi.org/10.1007/s10059-009-0014-2.

Zielinska, A. P., Holubcová, Z., Blayney, M., Elder, K., & Schuh, M. (2015). Sister kinetochore splitting and precocious disintegration of bivalents could explain the maternal age effect. *eLife*, *4*, 794. http://doi.org/10.7554/eLife.11389.

Zyss, D., & Gergely, F. (2009). Centrosome function in cancer: Guilty or innocent? *Trends in Cell Biology*, *19*(7), 334–346. http://doi.org/10.1016/j.tcb.2009.04.001.

CHAPTER FIVE

The Genetic Regulation of Cell Fate During Preimplantation Mouse Development

A.A. Lokken, A. Ralston[1]
Michigan State University, East Lansing, MI, United States
[1]Corresponding author: e-mail address: aralston@msu.edu

Contents

Abstract

The adult body is estimated to contain several hundred distinct cell types, each with a specialized physiological function. Failure to maintain cell fate can lead to devastating diseases and cancer, but understanding how cell fates are assigned and maintained during animal development provides new opportunities for human health intervention. The mouse is a premier model for evaluating the genetic regulation of cell fate during development because of the wide variety of tools for measuring and manipulating gene expression levels, the ability to access embryos at desired developmental stages, and

Current Topics in Developmental Biology, Volume 120
ISSN 0070-2153
http://dx.doi.org/10.1016/bs.ctdb.2016.04.006

the similarities between mouse and human development, particularly during the early stages of development. During the first 3 days of mouse development, the preimplantation embryo sets aside cells that will contribute to the extraembryonic tissues. The extraembryonic tissues are essential for establishing pregnancy and ensuring normal fetal development in both mice and humans. Genetic analyses of mouse preimplantation development have permitted identification of genes that are essential for specification of the extraembryonic lineages. In this chapter, we review the tools and concepts of mouse preimplantation development. We describe genes that are essential for cell fate specification during preimplantation stages, and we describe diverse models proposed to account for the mechanisms of cell fate specification during early development.

1. INTRODUCTION

1.1 Cell Fate Specification in the Mouse Embryo

The extraembryonic membranes include placenta and yolk sac (Fig. 1) and are the first lineages established following fertilization. Failure to generate normal extraembryonic tissues can lead to devastating outcomes, including infertility, birth defects, gestational diseases, and reproductive cancers. Environmental and genetic factors may increase the risk of these disorders, and systematic identification of the underlying genetic causes of reproductive disorders is facilitated by the study of experimental models. The mouse is an ideal model in which to study the genetic regulation of cell fate because mice share many developmental features with humans, but genes can be readily manipulated in the mouse. The multitude of genetic tools available for manipulating gene function in mice facilitates the identification and

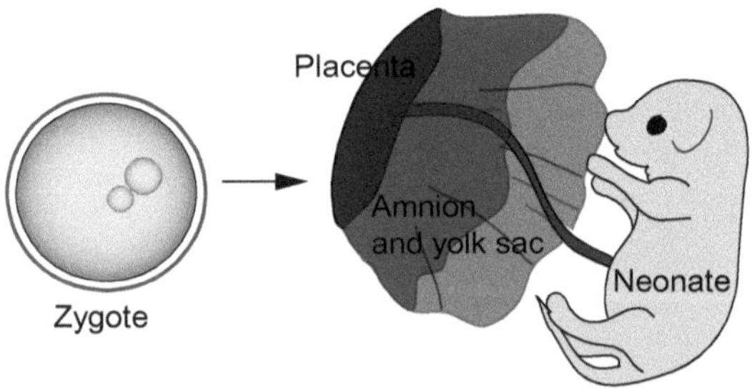

Fig. 1 The genetic regulation of cell fate specification. The 1-cell zygote is totipotent since it has the potential to form fetal and extraembryonic membranes, such as placenta and yolk sac. Yet the genomes of every cell, from zygote to birth, are essentially identical. Understanding how genomes are regulated to create this cellular diversity is an important goal in biomedical research.

study of disease-causing mutations. In this chapter, we first review the principles and tools of mouse genetics, and then describe how genetic approaches have illuminated our understanding of the specification of extraembryonic and fetal lineages.

2. THE PRINCIPLES OF MOUSE GENETICS

2.1 What Is the Genetic Regulation of Cell Fate?

A major goal in developmental, reproductive, and stem cell biology is to understand how genes regulate cell fate. With few exceptions, every cell in the adult body contains the same DNA as the 1-cell zygote from which all cells in the body developed. Yet hundreds of different cell types, each with specialized functions, exist in the adult body. Functional evidence that each of these cell types contains equivalent DNA was provided by experiments recognized by the 2012 Nobel Prize in Physiology or Medicine, which was awarded to John Gurdon and Shinya Yamanaka for showing that adult cell nuclei can be reprogrammed to express embryonic genes and reenter development to create diverse mature cell types (Colman, 2013). If all cell types in the adult body contain the same DNA, they must therefore use the DNA differently—a phenomenon known as regulated gene expression. Regulated gene expression produces cellular diversity by enabling cells with identical *genomes* or *genotypes* (Table 1) to differentiate their *phenotypes*, manifest at the level of *transcriptome*, *proteome*, and cell morphology and function. Since RNAs and proteins are the functional end products of most genes in the genome, mechanisms regulating the differential expression of RNAs and proteins are fundamental to generating cellular diversity—starting in the embryo, and continuing throughout adult life.

2.2 Why Should We Study the Genetic Regulation of Cell Fate?

Proper assignment and maintenance of cell fate is essential for embryogenesis and adult homeostasis. Failure to assign and maintain cell fate is the cause of many human diseases. For example, many cancers arise because cells lose their identity or aberrantly adopt a different identity. Therefore, identifying the genes that are essential for defining cell fate is the first step in devising diagnoses and treatments for human diseases. Identifying genes that regulate cell fates during embryogenesis is the first step toward treating a range of reproductive problems, including infertility, birth defects, gestational diseases, and reproductive cancers. In this goal, the mouse model is a worthy partner.

Table 1 Definitions and Terminology

Term	Definition
Allele	The specific sequence of a gene, given that each gene can carry naturally occurring or experimentally induced sequence variation.
Blastocyst	The embryo after formation of the blastocoel (32 cells to >100 cells and E3.25–E4.5).
Blastomere	A cell within the preimplantation embryo.
Cleavage	Specialized cell division that does not involve cell growth.
Compaction	A developmental stage and the process by which cells of the 8-cell embryo become polarized and tightly adhered to one another.
Conditional allele	An allele that can be activated or deleted in a spatially or temporally regulated manner.
Downstream	In the context of genetic hierarchies, this refers to gene products that act later, or lower in the hierarchy.
Epiblast (EPI)	Cells present in the ICM of the blastocyst that are pluripotent and committed to the fetal lineage.
Epigenetic	Modifications of genotype or phenotype that do not affect the DNA nucleotide sequence.
Forward genetics	An experimental approach relying first on identifying a mutant phenotyps of interest and then identifying the gene that is disrupted to produce that phenotype.
Genetic epistasis	The interaction between two or more genes. How multiple mutations affect a phenotype can enable inference about how the gene products relate to each other.
Genetic redundancy	Cooperative function of two or more genes, usually paralogues.
Genome	All the nuclear DNA within a cell.
Genotype	The alleles within the genome.
Implantation	The process of embryo burrowing into the uterine wall. Also a developmental stage and a stage of pregnancy.
Inner cell mass	The cells inside the blastocyst that will contribute primarily to fetal and yolk sac lineages.
Morula	A morphological description of embryos between the 8-cell and blastocyst stages (8–20 cells).

Table 1 Definitions and Terminology—cont'd

Term	Definition
Null allele	An allele that encodes a nonfunctional protein.
Orthologue	Genes with sequence similarity between species.
Paralogue	Genes with high degree of sequence similarity within a species.
Periimplantation	The stage of development/pregnancy around the time that the embryo implants into the uterus.
Phenocopy	Identical phenotypes resulting from differing genotypes.
Phenotype	The final readout of the genotype, either molecularly or morphologically.
Preimplantation	The period of development between fertilization and implantation into the uterus.
Primitive endoderm (PE)	Cells present in the ICM of the blastocyst that are committed to the extraembryonic endoderm lineage.
Proteome	All proteins expressed within a cell or collection of cells.
Reverse genetics	An experimental approach relying on first altering the function of a specific gene and then evaluating the resulting phenotype.
Transcriptome	All the RNAs expressed within a cell or collection of cells.
Trophectoderm (TE)	Cells present in the blastocyst that surround the ICM. Most will contribute to the trophoblast portion of the placenta, but some will divide to contribute to the ICM.
Upstream	In the context of genetic hierarchies, this refers to gene products that act earlier, or higher up in the hierarchy.
Zona pellucida	The thick protein shell surrounding the preimplantation embryo, the purpose of which is to facilitate fertilization of the oocyte initially, and prevent embryos from precocious implantation subsequently.

3. THE TOOLS OF MOUSE GENETICS

3.1 How Do We Study the Genetic Regulation of Cell Fate?

Investigations of the genetic regulation of cell fate have three main goals. The first goal is to identify genes that are essential for embryonic development (gene discovery). The second goal is to understand how genes are expressed in the right time and place (gene regulation). The third goal is to understand how gene products (RNAs and proteins) cause cells to differentiate from each other (gene roles). Each of these is discussed next.

3.2 Discovering Genes Essential for Cell Fate

Discovery of genes that are essential for the regulation of cell fate is challenging in humans because the human genome is large and naturally variable among individuals. By contrast, experimental animal genomes are often more compact and less variable, but contain *orthologues* of genes found in the human genome. Therefore, initial efforts to identify genes important for animal cell fate specification used the fruit fly *Drosophila melanogaster*. The first trailblazing effort by Nüsslein-Volhard and Wieschaus was recognized by the 1995 Nobel Prize in Physiology or Medicine and relied on screening fruit fly embryos to identify genes essential for embryogenesis, after randomly mutagenizing fruit fly genomes (Roush, 1995). This work provided a paradigm, now known as *forward genetics*. Forward genetic approaches have subsequently been used in mouse models (Liu & Eggenschwiler, 2014), but are not as common as in fly, since smaller numbers of progeny and internally developing embryos diminish the throughput of mouse forward genetic screens, relative to invertebrate models.

Forward genetic screens in fruit flies led to identification of the sequences of essential genes. Since many essential fruit fly genes are also essential for mammalian development, forward genetic screens in *Drosophila* provided some of the first candidates for genetic analysis in mammalian models. Additional candidates have been provided by identifying mouse genes that are specifically expressed in a cell type or tissue of interest. *Reverse genetics* is practical in mice because multiple genome editing approaches are available. For example, reverse genetics is facilitated by the existence of mouse embryonic stem (ES) cells, which can be genetically engineered and then used to establish offspring with mutant *alleles* of genes of interest, technology that was awarded the 2007 Nobel Prize in Physiology or Medicine (Vogel, 2007). Currently, upward of half of the genes in the mouse genome have been knocked out, and efforts to knock out all of the genes are underway (Hall, Limaye, & Kulkarni, 2009). More recently, highly efficient genome editing tools using Cas9/CRISPR have been developed, and these enable editing of zygotic genomes directly (Aida, Imahashi, & Tanaka, 2014), providing a much more rapid path to manipulating the mouse genome.

The essential roles of the majority of the genes described in this chapter were discovered using *null* alleles, created using mouse knockout technology. However, gain-of-function assays are also a well-established genetic approach in mice. For example, the gene *Nanog* was identified using loss- (Mitsui et al., 2003) and gain- (Chambers et al., 2003) of-function approaches. Thus, the mouse model provides an exceptional platform for discovery of genes that are essential for cell fate in mammals.

3.3 Discovering the Roles and Regulation of New Genes in the Regulation of Cell Fate

In addition to the approaches described earlier, additional tools exist to discover the roles and regulation of genes during cell fate specification (Table 2). One widely used approach for discovering the roles of genes in cell fate specification are systems such as Cre/Lox that enable use of *conditional alleles* to investigate the consequences of gene activation or inactivation in precisely defined spatial and temporal domains. The Cre/Lox system has also been adapted to enable genetic lineage tracing, in which fluorescent reporters are activated or inactivated in a group of cells and all of their descendants (Jensen & Dymecki, 2014).

Insight into gene roles is provided by phenotype analysis. Traditionally, phenotype analysis included analysis of embryo morphology and the expression level(s) of one or two downstream target genes. However, since the dawn of affordable deep sequencing and other high-throughput gene expression analysis methods, the field has shifted to analysis of hundreds to thousands of genes simultaneously. This innovation has enabled researchers to understand the genetics of cell fate specification at level of networks and pathways (Parfitt & Shen, 2014). Finally, gene expression reporters provide additional genetic tools and are commonly used in mice. These are created by introducing a fluorescent protein into the gene so that the fluorescent protein reports that gene's expression in real time. Fluorescent reporters are a great way to visualize dynamic changes in gene expression without disrupting important spatial information (Nowotschin & Hadjantonakis, 2014).

4. THE GENETIC REGULATION OF THE FIRST CELL FATE DECISION

4.1 The Genetics of Preimplantation Embryos

The extraembryonic membranes are the first lineages to be established during mammalian development, because they are essential for establishing and maintaining pregnancy. Because the extraembryonic lineages are already specified by the time of *implantation*, many additional genetic approaches

Table 2 Approaches to Evaluating the Genetic Regulation of Cell Fate in Mouse Preimplantation Embryo

Approach	Purpose	Advantage	Disadvantage
siRNA injection	Loss-of-gene function	Rapid, can be highly penetrant, enabling embryos to be pooled for downstream analysis.	Off-target effects possible, incomplete loss of function possible. Specialized equipment and training needed.
mRNA injection	Gain-of-gene function		Specialized equipment and training needed.
Gene targeting by homologous recombination in ES cells	Loss- or gain-of-gene function	Null alleles provide definitive test of complete loss-of-gene function.	Slow to generate new alleles. Low penetrance of desired genotype. Genotype must be determined by PCR.
CRISPR/Cas9 gene targeting in ES cells or zygotes	Loss- or gain-of-gene function		Rapid, once tools are established in the lab.
Immunofluorescence and confocal microscopy	Phenotype analysis (protein)	Rapid, provides spatial information about gene expression.	Antibody may not be available, or specificity may be unknown.
Quantitative PCR	Phenotype analysis (mRNA)	Rapid, any gene can be analyzed.	Spatial information usually lost.
RNA sequencing		All genes are analyzed.	Slow and costly, spatial information usually lost.
Cre/Lox	Conditional ablation or activation of mouse genes.	Restricts functional analysis to genetically defined temporal and spatial window.	May require multiple generations' breeding to bring all alleles into the same genetic background.
	Conditional activation of introduced genes, such as fluorescent proteins.	Used for lineage tracing.	
Reporters	Evaluating gene expression.	Enable live imaging of dynamic changes in gene expression.	Introduced into mouse genome transgenically, which can be slow.

are uniquely possible in preimplantation mouse embryos (Table 2). This is because preimplantation embryos can be cultivated in the tissue culture incubator, which provides access to all developmental stages and allows for genetic interventions that are less straightforward after the embryo's development becomes dependent on connections within the mother. For example, growth factors and signaling pathway inhibitors can be added directly to the preimplantation embryo culture media, fluorescent gene reporters can be imaged in real time (Saiz, Plusa, & Hadjantonakis, 2015), and expression constructs, mRNAs, or small-interfering RNAs (siRNAs) can be introduced into preimplantation embryos by viral infection (Cross, 2007; Dietrich et al., 2015) or microinjection. These approaches are not as straightforward at later developmental stages because the cells become much smaller than in preimplantation embryos, and living embryos more challenging to access. Finally, many of these genetic manipulations can be combined—for example, embryos carrying null alleles can be cultured in the presence of growth factors or signaling pathway inhibitors, facilitating *genetic epistasis* analysis. Therefore, the genetic tools for evaluating gain- and loss-of-gene function prior to implantation are exceptional. This genetic toolbox has provided exciting insight into how cell fates are specified during mammalian development that has not been possible in humans.

4.2 Mammalian Developmental Biology Before Genetics

Prior to the current genetic era, the mouse model was already firmly established as a premiere model for understanding mammalian preimplantation development. Classical, pregenetic experimental approaches led to the identification of (1) optimal conditions for culturing preimplantation embryos, (2) approaches for manipulating the spatial arrangement or number of individual *blastomeres* within the preimplantation embryo, (3) methods for tracking the cleavage patterns and lineages of individual blastomeres, and (4) descriptions of the morphological changes in blastomeres, such as changes in cell polarization and position. These approaches were integrated with analyses of cell fate and these, now classic, studies provided a road map of the major developmental milestones occurring during preimplantation development, and comparison against the developmental strategies understood in nonmammalian animal models. A multitude of studies have provided evidence that mouse embryogenesis does not rely on a prepatterned oocyte (Solter, 2016), arguing that emergent properties of the embryo dynamically regulate cell fate decisions, rather than cytoplasmic determinants. Prior to zygotic genome activation at the 2-cell stage, embryogenesis relies on maternal gene products, loaded into the oocyte during oogenesis (Zheng & Liu, 2012); however, the role of maternal gene products in overseeing cell fate decisions has been found to be minimal (Frum & Ralston, 2015; Zheng & Liu, 2012). Therefore, the majority of models of preimplantation development focus on the causal relationship between the emergent properties of the blastomeres, such as cell position or polarity, and cell fate. The earlier models, including Inside-Out Model and Cell Polarity Model, reviewed elsewhere (Lanner, 2014; Schrode et al., 2013; Yamanaka, Ralston, Stephenson, & Rossant, 2006), provided the initial framework for studies of cell fate specification in the genetic era. This chapter focuses mainly on transcription factors and signaling pathways because they are important for specifying cell fates. Other mechanisms, such as *epigenetic* mechanisms, are important for maintaining cell fates (Marcho, Cui, & Mager, 2015), and are therefore not the focus of this chapter.

4.3 Milestones of Preimplantation

To understand the genetic regulation of cell fate specification during preimplantation mouse development, it is important to first understand the morphological milestones. Morphological observation has revealed stereotyped stages of mouse preimplantation development, which can categorized by cell number, and predicted by the number of days from the time of fertilization (embryonic days, E; Fig. 2). During preimplantation development, mouse and human embryos proceed through remarkably similar morphological changes, first segregating *trophectoderm* (TE, future placenta) and *inner cell mass* (ICM) as the *blastocyst* forms. Within the blastocyst, the ICM further differentiates into epiblast (EPI, future fetus, allantois, yolk sac mesoderm, and amnion) and *primitive endoderm* (PE, future yolk sac endoderm). The morphological events that occur prior to formation of TE, EPI, and PE have been scrutinized in order to understand how an initially uniform ball of cells self-organizes to produce three distinct cell types.

During the first few days of mouse development (E0.0–E2.5), cells of the mouse embryo cleave asynchronously to produce morphologically identical blastomeres. During the late 8-cell stage (E2.75), the blastomeres become polarized along the radial (inside/outside) axis of the embryo. This polarization event, known as *compaction*, establishes apical–basal polarity within each blastomere, with the apical surface exposed to the outside of the embryo, and basolateral surfaces making contact among cells. The apical

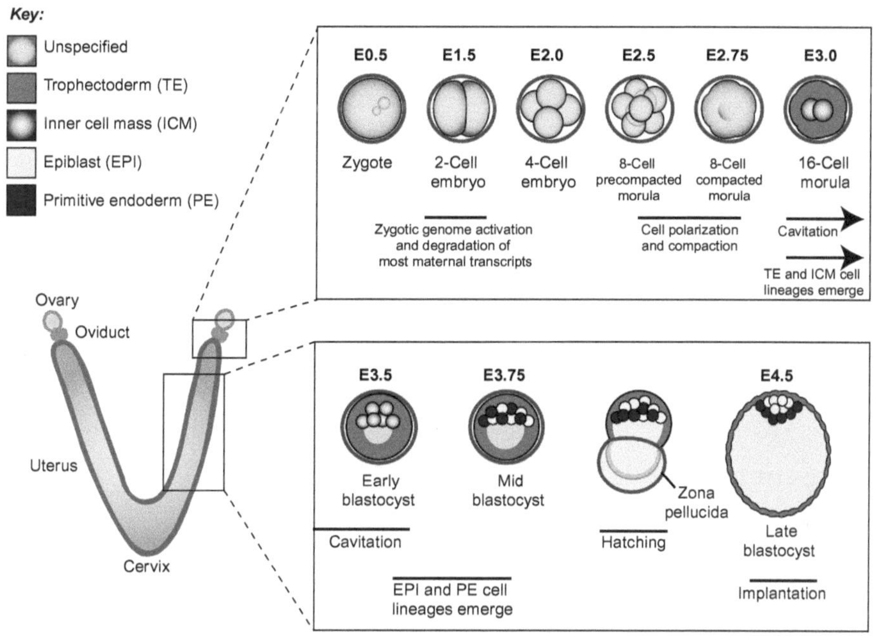

Fig. 2 Morphological events and embryo staging. Embryos proceed through stereotyped stages and are essentially synchronized within a litter. Fertilization occurs within the oviduct, and embryogenesis proceeds as embryos move toward the uterus, entering around E3.5. (See the color plate.)

domain is defined by localization of cell polarity proteins and tight junctions, and the basolateral surface is defined by the localization of adherens junction proteins. During the subsequent *cleavage* divisions, cells that remain on the outer surface of the *morula* tend to maintain polarity, while cells that move to the inside become apolar. Cell polarity and cell position are both known to influence expression of markers of cell fate, and expression of cell fate markers can influence cell position and expression of cell polarity proteins. Therefore, it has been difficult to discern which event comes first in development. For example, lineage tracing has shown that position influences cell fate, since cells remaining exposed to the outside surface tend to contribute to TE (placenta lineage) and inside cells tend to become ICM (fetal and yolk sac lineages; Fleming, 1987). In addition, cell polarity proteins are essential for maintaining positional information in the blastocyst (Plusa et al., 2005; Stephenson, Yamanaka, & Rossant, 2010).

However, cell movement between inside and outside domains is common, especially during cell division (McDole, Xiong, Iglesias, & Zheng, 2011; Watanabe, Biggins, Tannan, & Srinivas, 2014), indicating that cell polarization can be dynamically altered, downstream of spatial position. Cell polarity proteins are essential for maintaining expression of cell fate markers (Alarcon, 2010; Hirate et al., 2013; Kono, Tamashiro, & Alarcon, 2014), and have been proposed to oversee cell fates in a deterministic manner. However, the angle of cleavage has been shown not to predict cell fate (Watanabe et al., 2014), arguing that polarity proteins may participate in cell fate specification through a more dynamically adaptable, regulated mechanism. Indeed, feedback between polarity/position/fate makes it difficult to discern cause and effect at later stages, but cells polarize at the 8-cell stage (Johnson & McConnell, 2004), which precedes formation of topologically distinct inside and outside positions and expression of cell fate markers.

4.4 Genetic Regulation of TE and ICM Fates

Genetic analysis of preimplantation mouse development has illuminated genes and pathways regulating the first cell fates. These studies laid the foundation for efforts to describe the genetic pathways overseeing human preimplantation development, and numerous similarities have been noted (Blakeley et al., 2015; Niakan, Han, Pedersen, Simon, & Pera, 2012). One of the first genetic differences in the early mouse embryo is expression of the genes encoding the transcription factors SOX2 and CDX2, which exhibit a mutually exclusive inside/outside expression pattern at the 16-cell stage (Guo et al., 2010; Strumpf et al., 2005; Wicklow et al., 2014). Apart from being interesting regulators of ICM and TE cell fate, SOX2 and CDX2 also provide readouts for identifying how the emergent properties of the mouse embryo, such as cell cleavage, position, and polarity, establish cell fates. For instance, biased blastomeres' inheritance of *Cdx2* mRNA has been observed (Shi et al., 2015; Skamagki, Wicher, Jedrusik, Ganguly, & Zernicka-Goetz, 2013), suggesting that CDX2 could act as a cell fate determinant. However, live imaging of a *Cdx2* expression reporter showed that *Cdx2* expression dynamically adapts to changes in cell position during formation of the blastocyst (McDole & Zheng, 2012), and that *Cdx2* expression is therefore *downstream* of cell position. At later embryonic stages, CDX2 helps to reinforce expression of polarity pathway components (Jedrusik et al., 2008; Ralston & Rossant, 2008). However, neither the initial outside positioning of blastomeres nor the expression of polarity components depends on *Cdx2* (Blij, Frum, Akyol, Fearon, & Ralston, 2012; Ralston & Rossant, 2008). These observations point to the existence of a position or polarity-sensing mechanism in the 16-cell morula, which is *upstream* of the patterned expression of cell fate markers such as *Cdx2* and

Sox2. The molecular mechanisms acting downstream of cell polarity and position to regulate expression and patterning of TE and ICM cell fate markers are becoming more clear.

4.4.1 HIPPO Signaling Pathway Regulates Patterning of TE Genes

Although the mechanistic details are not yet entirely clear, it is now widely accepted that the HIPPO signaling pathway interprets positional and polarity information to pattern expression of lineage-specific genes prior to blastocyst formation. The HIPPO pathway is highly conserved across animal species, and plays diverse roles in tissue homeostasis (Halder & Johnson, 2011; Meng, Moroishi, & Guan, 2016). In the context of the preimplantation embryo, the HIPPO signaling pathway is thought to be active in inside, apolar cells as early as the 16-cell stage (Fig. 3), and HIPPO signaling is repressed in outside cells by cell polarity pathway components, including protein products of *Angiomotin paralogues* (Anani, Bhat, Honma-Yamanaka, Krawchuk, & Yamanaka, 2014; Hirate et al., 2015, 2013; Hirate & Sasaki, 2014; Kono et al., 2014; Leung & Zernicka-Goetz, 2013; Lorthongpanich, Doris, Limviphuvadh, Knowles, & Solter, 2012), as well as the transcription factor TCFAP2C (Cao et al., 2015). In inside cells, HIPPO signaling pathway targets include the transcription factors YAP and WWTR1, whose activities are repressed by HIPPO signaling (Nishioka et al., 2009). Thus, YAP/WWTR1 are active only in outside cells, where they partner with TEAD4 to induce expression of TE genes including *Cdx2* and *Gata3* (Nishioka et al., 2009; Ralston et al., 2010). Several unusual aspects of HIPPO signaling in the preimplantation embryo have been noted. For example, the kinases MST1/2 have been found not to function in this context (Cockburn, Biechele, Garner, & Rossant, 2013), NOTCH signaling modestly potentiates HIPPO signaling (Rayon et al., 2014), and *Tead4* is dispensable for cell fate specification in embryos raised in hypoxic conditions (Kaneko & DePamphilis, 2013). These observations underscore the importance of invertebrate genetic models to identify new pathway members that may later turn out to play pivotal, if unique, roles in the context of mammalian early development.

4.4.2 Roles of CDX2 and GATA3

The expression pattern of *Cdx2* and *Gata3* is significant because CDX2, and to a lesser extent GATA3, regulates TE maturation (Fig. 3). In this regard, CDX2 appears to be a more potent regulator of TE fate because *Cdx2* null embryos exhibit a severe phenotype including (1) ectopic expression of ICM genes such as *Oct4* and *Nanog* in the TE, (2) loss of expression of TE

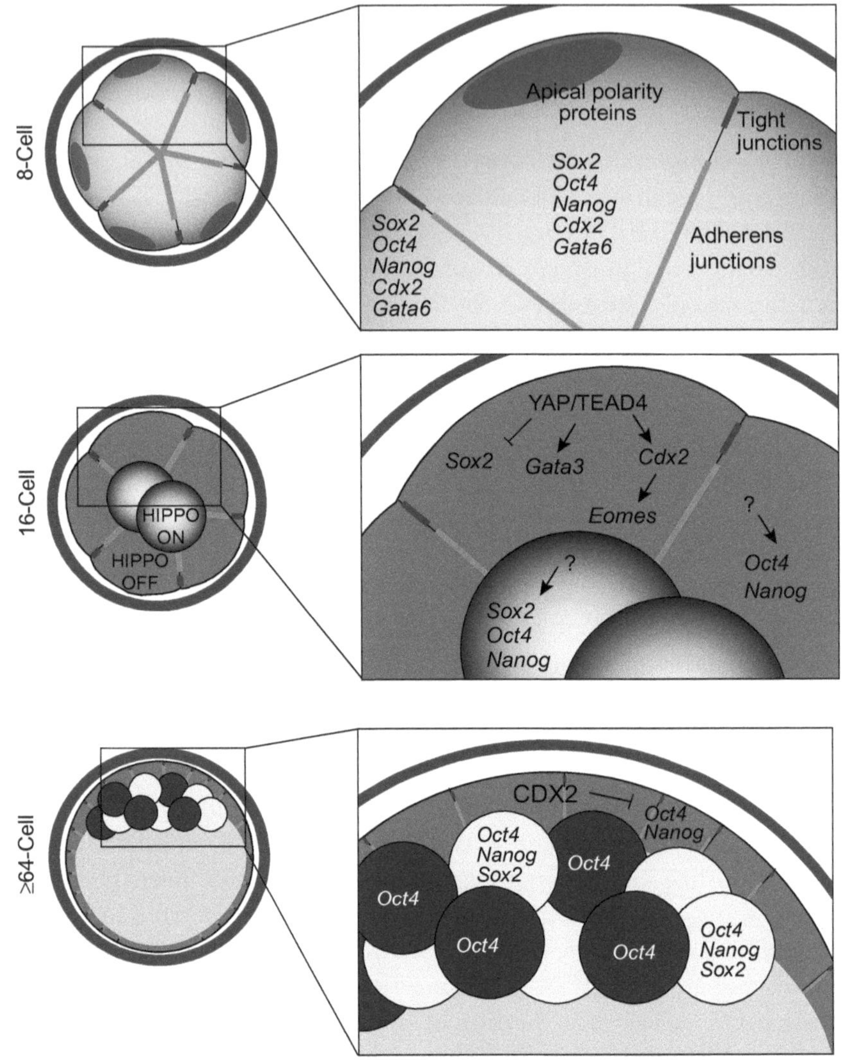

Fig. 3 Segregation of ICM and TE lineages. At the late 8-cell stage, cells polarize and the embryo compacts. At this stage, many genes that will later identify unique cell populations are expressed throughout the embryo. By the 16-cell stage, the HIPPO signaling pathway is active within inside cells, and inactivated by cell polarity components in outside cells. Thus proteins that are negatively regulated by HIPPO, such as YAP and WWTR1, are active in outside cells where they partner with TEAD4 to regulate expression of ICM and TE genes, *Sox2* and *Cdx2/Gata3*, respectively. At this stage, *Oct4* and *Nanog* are expressed ubiquitously, indicating that these genes are not regulated by HIPPO signaling. By the late blastocyst stage, CDX2 represses expression of *Oct4* and *Nanog* within TE cells, thereby restricting the expression of these genes to the ICM. (See the color plate.)

maturation genes, such as *Eomes* and *Krt8*, and (3) preimplantation lethality (Ralston & Rossant, 2008; Strumpf et al., 2005). By contrast, the *Gata3* null allele produces a less severe phenotype, since *Gata3* null embryos survive until midgestation (Ma et al., 1997). *Genetic redundancy* between the GATA2 and GATA3 paralogues has been proposed to mask an earlier *Gata3* null phenotype, although current evidence for this is indirect (Home et al., 2009) and awaits testing with null alleles in double knockout embryos. While CDX2 is likely to regulate expression of hundreds of genes important for TE maturation, only a few targets are known. For example, EOMES is important for implantation and TE proliferation, and *Eomes* null alleles lead to *periimplantation* lethality (Russ et al., 2000; Strumpf et al., 2005). Identification of CDX2, and possibly GATA2/3, targets will therefore provide a wealth of information regarding the earliest events in the establishment of the placenta lineage.

4.4.3 HIPPO Signaling Pathway Regulates ICM Fate

At the same time that HIPPO signaling represses expression of TE genes in inside cells, the HIPPO signaling pathway also promotes expression of *Sox2* in inside cells (Fig. 3; Wicklow et al., 2014). At this stage, *Oct4* and *Nanog* are still expressed ubiquitously throughout the embryo (Palmieri, Peter, Hess, & Schöler, 1994; Strumpf et al., 2005), and loss of the HIPPO-regulated transcription factor TEAD4 fails to disrupt expression of *Oct4* or *Nanog* (Nishioka et al., 2008; Yagi et al., 2007). Thus both the timing and the mechanisms of *Sox2* patterning are distinct from those of *Oct4* and *Nanog* (Fig. 3). Expression of *Oct4* and *Nanog* is restricted to the ICM by CDX2 (Strumpf et al., 2005), which does not affect the expression of *Sox2* (Wicklow et al., 2014). The mechanism by which CDX2 represses OCT4 in TE cells is thought to be indirect; CDX2 is proposed to bind to OCT4 and prevent OCT4 from acting as a transcriptional activator to drive its own expression (Niwa et al., 2005), and possibly the expression of *Nanog*. While *Nanog* expression in the ICM is independent of *Oct4* (Frum et al., 2013; Wu et al., 2013), *Nanog* expression in *Cdx2* null TE could require *Oct4*. This model has not been tested, but could be tested by evaluating embryos lacking both *Cdx2* and *Oct4*. At later stages, disruption of HIPPO signaling disrupts ICM differentiation (Cockburn et al., 2013; Lorthongpanich et al., 2013; Mihajlović, Thamodaran, & Bruce, 2015), indicating that HIPPO targets besides YAP/WWTR1 must regulate ICM maturation and the second lineage decision, described later.

5. THE GENETIC REGULATION OF THE SECOND CELL FATE DECISION

5.1 Genetic Regulation of EPI and PE Fates

While the first lineage decision allocates cells to the TE and the ICM, the second lineage decision resolves the identity of the ICM cells to either PE or pluripotent EPI fates. Therefore, pluripotency must be established within blastomeres that are otherwise differentiating to extraembryonic cell types. Thus the second lineage decision provides insight into the establishment of pluripotency that may have important implications for regenerative medicine (Wu & Schöler, 2016). Several models have been proposed to account for formation of EPI and PE fates within the ICM. It was originally proposed that the position of cells within the ICM influenced cell fate, with cells bordering the blastocoel adopting PE fate, and cells bordering the TE adopting EPI fate. However, more recent genetic approaches using lineage tracing and gene expression analyses have shown that EPI and PE cells are genetically distinct before these lineages sort into two distinct layers, and that cell position does not influence EPI or PE fate (Chazaud, Yamanaka, Pawson, & Rossant, 2006; Kurimoto et al., 2006). By the late blastocyst stage (E3.75), the cells of the ICM express either pluripotency genes *Nanog* and *Sox2* or they express endodermal genes *Gata6*, *Gata4*, and *Sox17* (Fig. 4; Artus, Panthier, & Hadjantonakis, 2010; Chazaud et al., 2006; Niakan et al., 2010; Plusa, Piliszek, Frankenberg, Artus, & Hadjantonakis, 2008; Wicklow et al., 2014). This salt and pepper expression pattern is resolved by the time of implantation (by E4.5), when ICM cells sort into two distinct layers by migration and apoptosis (Meilhac et al., 2009; Plusa et al., 2008). Therefore, during the second cell fate decision, differential gene expression of EPI and PE genes precedes their spatial positioning. Several studies have focused on identifying how the salt and pepper pattern of EPI and PE gene expression is first established.

5.1.1 Models for Salt and Peppering of the ICM

Identifying the mechanisms that initiate the heterogeneity within the ICM has been a major focus of genetic investigation. These studies have yielded two main models to explain how the apparently homogeneous cells of the morula/early blastocyst transition to two distinct populations, with cells biased toward EPI or PE fates. One model to account for the origins of the salt and peppered ICM is the *cleavage history-dependent model* (Fig. 5A). The cleavage history-dependent model posits that ICM cells are biased

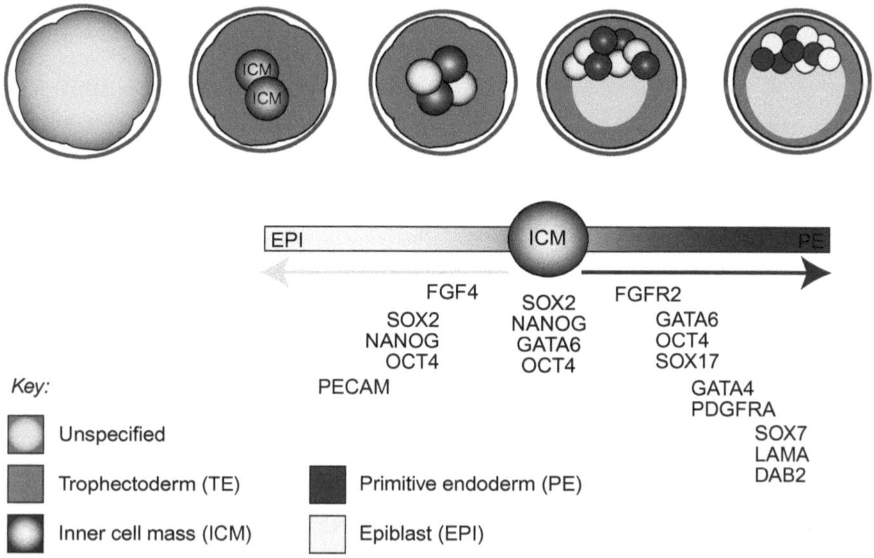

Fig. 4 The salt and pepper ICM. In the immature ICM, *Fgf4* and *Fgfr2* are the earliest to show complementary expression patterns in the ICM. Subsequently, EPI and PE genes become upregulated in *Fgf4*- and *Fgfr2*-expressing cells, respectively. Later, EPI and PE populations begin to sort into discrete layers and express markers of mature EPI and PE cells. (See the color plate.)

toward EPI or PE fate depending on which cell division produced them (Chisholm & Houliston, 1987; Rossant, Chazaud, & Yamanaka, 2003). Since ICM cells can be descended from either TE or ICM cells, lineage could influence cell fate, especially if TE and ICM cells contained cell fate determinants. To add to the mechanistic complexity, ICM cells are born at the 8- to 16-cell, the 16- to 32-cell, and the 32- to 64-cell transitions, suggesting that changes in the molecular composition of the TE over time could also contribute to cellular diversity, according to this model.

To test this model, studies have examined the developmental fates of ICM cells generated during the morula-to-blastocyst transition. For example, live imaging of embryos expressing a membrane-tethered GFP provided a dynamic history of individual ICM cells from the time they were created (8–16 cell stages) until their final positioning in the blastocyst (end of the 32- to 64-cell stages; Morris et al., 2010). The analysis showed that ICM cells generated during the first cell division (8- to 16-cell transition) are biased to acquire EPI cell fate, whereas cells generated later (16- to 32-cell or 32- to 64-cell transitions) are biased to acquire PE fate. Thus, the particular "wave" of division biases daughter cells to EPI or PE fate, with daughter cells arising in the first wave from developmentally younger, less differentiated mother cells biased toward EPI, and progenitors arising from the later, more differentiated mother cell biased toward PE. The molecular mechanism proposed to account for different fates involves Fibroblast Growth Factor Receptor 2 (FGFR2), which is proposed to be inherited from the TE during the second and third waves of division and to promote PE fate (Mihajlović et al., 2015; Morris, Graham, Jedrusik, & Zernicka-Goetz, 2013).

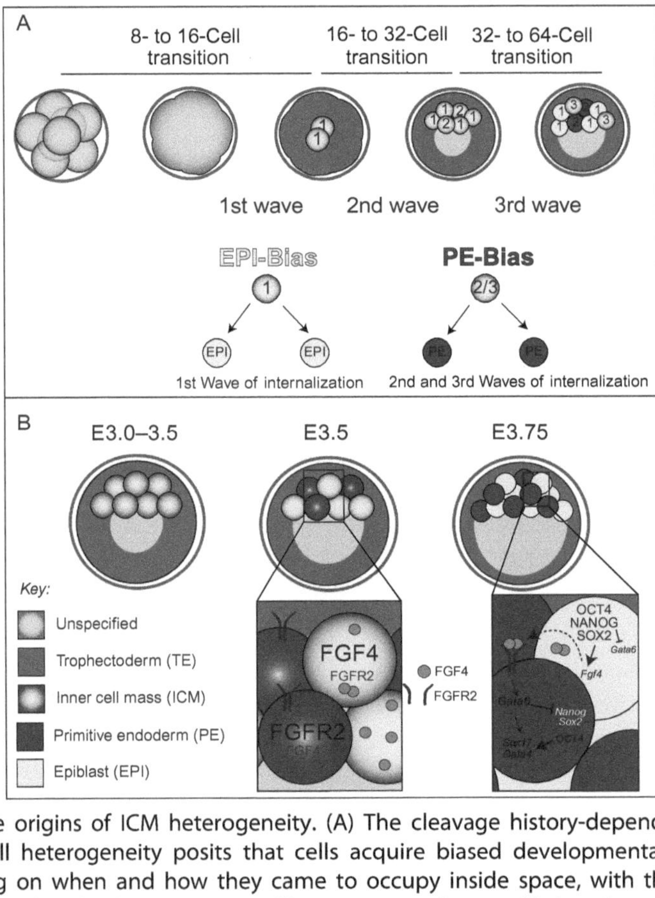

Fig. 5 The origins of ICM heterogeneity. (A) The cleavage history-dependent model of ICM cell heterogeneity posits that cells acquire biased developmental potential, depending on when and how they came to occupy inside space, with the first ICM cells more inclined to become EPI, and later-arising cells more likely to become PE, having descended from TE cells, and inherited PE determinants such as FGFR2. (B) The stochastic gene expression model of ICM cell heterogeneity posits that the initially identical cells of the ICM spontaneously express slightly different levels of FGF4 or FGFR2, and this signaling is amplified in the process of specifying EPI and PE progenitors. (See the color plate.)

However, another study that tested the cleavage history-dependent model did not support the model. In this study, lineage tracing was used to show that cells have an equivalent chance of becoming EPI or PE, regardless of when they occupied inside space (Yamanaka, Lanner, & Rossant, 2010). To arrive at this conclusion, a Cre/Lox lineage tracing approach was used and ICM cells generated from either the first or second waves of cell division were found to be equally capable of contributing to EPI or PE lineages. These observations do not support the cleavage history-dependent model. Possible reasons for disagreement have been discussed (Morris, 2011; Yamanaka, 2011), but other lines of investigation have shown that ICM cells maintain plasticity throughout the waves of cell division, retaining the ability to change their minds about committing to EPI or PE cell fate (Grabarek et al., 2012). Therefore, any model must be sufficiently dynamic to account for observations such as these.

Another model of EPI and PE origins is the *stochastic gene activation model* (Fig. 5B; Chazaud et al., 2006). In this model, gene expression differences arise randomly within cells, independently of cleavage history. The molecular mechanism underlying this model invokes a subset of cells stochastically express differing levels of either the FGF ligand (FGF4) or its receptor (FGFR2), and this biases the cells toward either EPI or PE fate. This model is supported by single-cell transcriptome studies of embryogenesis, which have shown that the earliest difference in gene expression is the differential expression of *Fgf4* and *Fgfr2* as early as the 32-cell stage (Guo et al., 2010), which is prior to salt and peppering of EPI and PE genes. Importantly, both *Fgf4* and *Fgfr2* are essential for PE development (Arman, Haffner-Krausz, Chen, Heath, & Lonai, 1998; Feldman, Poueymirou, Papaioannou, DeChiara, & Goldfarb, 1995; Kang, Piliszek, Artus, & Hadjantonakis, 2013; Wilder et al., 1997). Thus, the salt and pepper expression of *Fgf4* and *Ffgr2* would establish subsequent salt and pepper patterns of the EPI and PE genes. One prediction of this model is that cell fate should be sensitive to FGF4 dose, with higher doses increasing the proportion of PE cells and lower doses increasing the proportion of EPI cells. Consistent with this prediction, *Fgf4* gene copy number influences the proportion cells expressing EPI and PE genes (Krawchuk, Honma-Yamanaka, Anani, & Yamanaka, 2013). To determine whether there are gene expression differences upstream or parallel to FGF signaling, another study examined gene expression patterns in *Fgf4* null embryos (Ohnishi et al., 2014). However, this analysis found no evidence for gene expression differences among ICM cells in the absence of *Fgf4*. These observations suggest that

EPI and PE cell identities are not predetermined by obvious gene expression differences upstream of FGF4/FGFR2 signaling.

5.1.2 Downstream of FGF4

Genetic studies have helped elucidate several of the pathway members acting downstream of FGF4/FGFR2. For example, knockout of *Grb2*, which encodes an SH2 domain-containing adaptor protein that mediates signaling downstream of receptor tyrosine kinases (RTKs), such as FGFR2, was reported to disrupt formation of extraembryonic endoderm (Cheng et al., 1998). Consistent with this observation, PE cell fate was disrupted at the molecular level in the absence of *Grb2*, since all ICM cells expressed *Nanog*, at the expense of *Gata6* (Chazaud et al., 2006). Finally, pharmacological inhibition of FGFR2 and MEK *phenocopies Grb2* null embryos (Chazaud et al., 2006; Nichols, Silva, Roode, & Smith, 2009; Yamanaka et al., 2010), suggesting that MAPK acts toward the bottom of the signaling cascade to promote PE genes and repress EPI genes. Interestingly, FGF4/MAPK signaling is also necessary and sufficient to repress expression of *Sox2*, as well as *Nanog*, in PE cells (Wicklow et al., 2014), suggesting that *Nanog* and *Sox2* are coregulated by FGF4/MAPK signaling. The mechanisms by which MAPK influences expression of EPI and PE genes are not entirely resolved. Genetic epistasis experiments have shown that OCT4 promotes expression of several PE genes, including *Sox17* and *Gata4*, downstream of MAPK (Frum et al., 2013). However, MAPK promotes expression of *Gata6* and represses expression of *Nanog* in an OCT4-independent manner (Frum et al., 2013). Therefore, the pathway appears to branch downstream of MAPK, with *Oct4*-dependent and *Oct4*-independent arms (Fig. 5B). Identification of the mechanisms regulating EPI and PE genes downstream of MAPK will provide key elements of the model. The roles and regulation of other ICM genes downstream of FGF4/MAPK signaling are described next.

5.1.3 The Regulation and Roles of EPI Genes

As described earlier, FGF4/MAPK signaling is responsible for promoting expression of PE genes and repressing expression of EPI genes within PE cells. It is not currently clear how EPI genes are first expressed in ICM and EPI cells. Nor is it yet clear how pluripotency is first established during embryogenesis (Wu & Schöler, 2016). In ES cells, OCT4 and SOX2 maintain each other's expression, as well as expression of *Nanog* and other pluripotency genes (Ralston & Rossant, 2010). However, in the embryo, none of these three pluripotency factors is required for the initial expression of any of the others (Frankenberg et al., 2011; Frum et al., 2013; Silva et al., 2009; Wicklow et al., 2014; Wu et al., 2013). These observations indicate striking differences in the genetic regulation of the initiation (embryo) and maintenance (ES cells) of pluripotency, consistent with evidence that ES cells are transcriptionally divergent from early ICM cells, and more similar to very late ICM cells (Boroviak, Loos, Bertone, Smith, & Nichols, 2014). In addition, these observations indicate that exciting mysteries remain to be solved regarding how pluripotency is first established in vivo.

Some insight into the dynamics of the establishment of pluripotent gene expression during preimplantation development has been provided by detailed time course analyses of gene expression. Expression of OCT4 and NANOG is detected randomly throughout the embryo, from about 8-cell to early blastocyst stages (Dietrich & Hiiragi, 2007; Komatsu & Fujimori, 2015; Palmieri et al., 1994; Xenopoulos, Kang, Puliafito, Di Talia, & Hadjantonakis, 2015), a pattern that mirrors that of other lineage-specific transcription factors such as CDX2 and GATA6 (Dietrich & Hiiragi, 2007; Ralston & Rossant, 2008). Therefore these lineage factors are coexpressed prior to the blastocyst stage, and should not be considered lineage markers per se prior to this stage. By contrast, SOX2 expression is unique because SOX2 is not detected until the 16-cell stage when it is specifically expressed in inside cells of the morula (Wicklow et al., 2014). Pluripotency targets of SOX2 and NANOG in EPI cells remain to be elucidated, but *Oct4*-dependent genes have been identified by RNA sequencing *Oct4* null blastocysts, and these include pluripotency markers such as *Essrb*, *Klf2*, and *Zscan10* (Frum et al., 2013). In addition, all three pluripotency proteins help promote the expression of *Fgf4* in EPI cells (Frankenberg et al., 2011; Frum et al., 2013; Wicklow et al., 2014). Accordingly, expression of PE genes is reduced or eliminated in *Sox2*, *Oct4*, and *Nanog* null embryos (Frankenberg et al., 2011; Frum et al., 2013; Le Bin et al., 2014; Messerschmidt & Kemler, 2010; Wicklow et al., 2014). While exogenous FGF4 can rescue expression of PE genes in *Sox2* and *Nanog* null embryos (Frankenberg et al., 2011; Wicklow et al., 2014), exogenous FGF4 does not rescue PE gene expression in *Oct4* null embryos (Frum et al., 2013), consistent with a role for OCT4 in FGF4-receiving PE cells. NANOG has an additional role in EPI cells, which is to repress expression of *Gata6* (Frankenberg et al., 2011). Thus these three pluripotency factors have differing, as well as overlapping, roles in cell fate specification.

5.1.4 The Regulation and Roles of PE Genes

Gata6 encodes a transcription factor and the loss-of-function phenotype disrupts PE development earlier than other known transcription factors (Bessonnard et al., 2014; Cai, Capo-Chichi, Rula, Yang, & Xu, 2008; Koutsourakis, Langeveld, Patient, Beddington, & Grosveld, 1999; Morrisey, Ip, Lu, & Parmacek, 1996; Schrode et al., 2013). *Gata6* is one of the first genes to be specifically expressed in PE cells at E3.75 (Artus, Piliszek, & Hadjantonakis, 2011; Chazaud et al., 2006). From 8-cell to E3.75, *Gata6* is expressed ubiquitously, indicating that GATA6 is not a PE marker prior to E3.75. Consistent with this, *Gata6* expression prior to E3.75 is independent of *Fgf4*, while the later, PE-specific expression of *Gata6* depends on FGF signaling (Chazaud et al., 2006; Kang et al., 2013; Krawchuk et al., 2013; Nichols et al., 2009; Ohnishi et al., 2014; Yamanaka et al., 2010). FGF signaling is thought to boost levels of GATA6 in PE cells, above the level at which GATA6 is initially detected throughout the embryo, because decreases in FGF signaling pathway components lead to expression of lower levels of GATA6 in presumptive PE cells (Frum et al., 2013; Wicklow et al., 2014). Interestingly, GATA6 does not reinforce its own expression (Freyer et al., 2015).

The loss of *Gata6* is embryonic lethal at E6.5–E7.5 (Morrisey et al., 1996). At the blastocyst stage, *Gata6* null embryos *phenocopy Fgf4* null embryos, since *Gata6* null ICM cells all express *Nanog*, at the expense of PE genes such as *Sox17* and *Gata4* (Bessonnard et al., 2014; Schrode, Saiz, Di Talia, & Hadjantonakis, 2014), indicating that GATA6 acts in parallel to, or downstream of, FGF4. However, addition of exogenous FGF4 to *Gata6* null embryos does not rescue loss of PE gene expression (Schrode et al., 2014), indicating that GATA6 mediates expression of PE genes downstream of FGF4 (Fig. 5B). In addition, FGF4 treatment fails to repress expression of *Nanog* in *Gata6* null blastocysts (Schrode et al., 2014), suggesting that GATA6 also mediates repression of EPI genes downstream of FGF4. Notably, interfering with FGFR and ERK signaling (which normally increases expression of *Nanog* in wild-type ICMs) in *Gata6* null blastocysts leads to increased expression levels of *Nanog* (Schrode et al., 2014). This observation suggests that additional factors act alongside GATA6 to mediate FGFR/ERK-dependent repression of *Nanog*. For example, the loss of PE gene expression in *Gata6* null embryos could be due to persistent expression of *Nanog* in PE cells, where NANOG may repress PE genes. Whether GATA6 promotes PE gene expression directly or indirectly could be addressed by analyzing embryos lacking both *Nanog* and *Gata6*.

While *Gata6* is expressed at the 8-cell stage, other PE genes are not expressed until the salt and pepper pattern has been established in the blastocyst. Studies examining the sequential activation of PE genes during specification of the ICM have shown that GATA6 is the initial pro-PE transcription factor, which precedes expression of *Sox17*, followed by *Pdgfra*, *Gata4*, and finally *Sox7* (Artus et al., 2011). Embryos lacking *Gata4* fail to develop beyond E7.0–E9.5, due to heart defects (Kuo et al., 1997; Molkentin, Lin, Duncan, & Olson, 1997). *Sox17* null embryos still make a proper PE and survive through midgestation, but are growth retarded and eventually die as a result of gut endoderm defects (Kanai-Azuma et al., 2002; Kim, 2007; Niakan et al., 2010). The relatively mild phenotype of *Sox17* null embryos may be a result of compensation by paralogue SOX7. *Sox7* is expressed periimplantation coincident with other PE-associated transcription factors, and *Sox7* expression is maintained in *Sox17*-deficient embryos (Artus et al., 2011; Kanai-Azuma et al., 2002). SOX17 is first detected around E3.5, and its expression overlaps with *Nanog*, *Pou5f1*, and *Gata6* (Morris et al., 2010; Niakan et al., 2010). By the mid-to-late blastocyst stage, *Sox17* expression is restricted to PDGFRA-expressing cells, and SOX17 is thus one of the earliest markers of specified PE cells (Artus et al., 2010; Niakan et al., 2010; Plusa et al., 2008). Similar to SOX2, SOX17 also interacts with OCT4 to bind to the regulatory region of PE genes and activate their expression. Specifically, the OCT4/SOX17 complex has been shown to directly activate the expression of *Gata6* and *Gata4* to reinforce the PE transcriptional network in stem cell lines (Aksoy et al., 2013). An exciting alternative model is that the OCT4 nuclear dwell-time differs between extraembryonic and pluripotent cell types (Plachta, Bollenbach, Pease, Fraser, & Pantazis, 2011).

5.2 Wrapping Up the ICM

A host of additional genes, pathways, and mechanisms have been shown to functionally participate in sorting and maintaining the cell types of the ICM. For example, DAB2 (Morris, Tallquist, Rock, & Cooper, 2002), Laminins (Smyth et al., 1999; Yang et al., 2002), TEAD4 (Mihajlović et al., 2015), Leukemia Inhibitory Factor (Morgani et al., 2013), STAT3-dependent IL6 signaling (Do et al., 2013), cell polarity proteins (Saiz, Grabarek, Sabherwal, Papalopulu, & Plusa, 2013), BMP signaling (Graham et al., 2014; Reyes de Mochel et al., 2015), SALL4 (Elling, Klasen, Eisenberger, Anlag, & Treier, 2006), KLF5 (Lin, Wani, Whitsett, & Wells, 2010), and many help refine and reinforce the cell fates established through the mechanisms described in this chapter. These studies have been fundamental in identifying the key similarities and differences between mouse and human preimplantation embryos.

ACKNOWLEDGMENT

This work is supported by National Institutes of Health R01 GM104009 to A.R.

REFERENCES

Aida, T., Imahashi, R., & Tanaka, K. (2014). Translating human genetics into mouse: The impact of ultra-rapid in vivo genome editing. *Development, Growth & Differentiation, 56*(1), 34–45.

Aksoy, I., Jauch, R., Chen, J., Dyla, M., Divakar, U., Bogu, G. K., et al. (2013). Oct4 switches partnering from Sox2 to Sox17 to reinterpret the enhancer code and specify endoderm. *The EMBO Journal, 32*(7), 938–953.

Alarcon, V. B. (2010). Cell polarity regulator PARD6B is essential for trophectoderm formation in the preimplantation mouse embryo. *Biology of Reproduction, 83*(3), 347–358.

Anani, S., Bhat, S., Honma-Yamanaka, N., Krawchuk, D., & Yamanaka, Y. (2014). Initiation of Hippo signaling is linked to polarity rather than to cell position in the preimplantation mouse embryo. *Development, 141*(14), 2813–2824.

Arman, E., Haffner-Krausz, R., Chen, Y., Heath, J., & Lonai, P. (1998). Targeted disruption of fibroblast growth factor (FGF) receptor 2 suggests a role for FGF signaling in pregastrulation mammalian development. *Proceedings of the National Academy of Sciences of the United States of America, 95*(9), 5082–5087.

Artus, J., Panthier, J. J., & Hadjantonakis, A. K. (2010). A role for PDGF signaling in expansion of the extra-embryonic endoderm lineage of the mouse blastocyst. *Development, 137*(20), 3361–3372.

Artus, J., Piliszek, A., & Hadjantonakis, A. K. (2011). The primitive endoderm lineage of the mouse blastocyst: Sequential transcription factor activation and regulation of differentiation by Sox17. *Developmental Biology, 350*(2), 393–404.

Bessonnard, S., De Mot, L., Gonze, D., Barriol, M., Dennis, C., Goldbeter, A., et al. (2014). Gata6, Nanog and Erk signaling control cell fate in the inner cell mass through a tristable regulatory network. *Development, 141*(19), 3637–3648.

Blakeley, P., Fogarty, N. M., Del Valle, I., Wamaitha, S. E., Hu, T. X., Elder, K., et al. (2015). Defining the three cell lineages of the human blastocyst by single-cell RNA-seq. *Development, 142*(20), 3613.

Blij, S., Frum, T., Akyol, A., Fearon, E., & Ralston, A. (2012). Maternal Cdx2 is dispensable for mouse development. *Development, 139*(21), 3969–3972.

Boroviak, T., Loos, R., Bertone, P., Smith, A., & Nichols, J. (2014). The ability of inner-cell-mass cells to self-renew as embryonic stem cells is acquired following epiblast specification. *Nature Cell Biology, 16*(6), 516–528.

Cai, K. Q., Capo-Chichi, C. D., Rula, M. E., Yang, D. H., & Xu, X. X. (2008). Dynamic GATA6 expression in primitive endoderm formation and maturation in early mouse embryogenesis. *Developmental Dynamics, 237*(10), 2820–2829.

Cao, Z., Carey, T. S., Ganguly, A., Wilson, C. A., Paul, S., & Knott, J. G. (2015). Transcription factor AP-2gamma incudes early Cdx2 expression and represses HIPPO signaling to specify trophectoderm lineage. *Development, 142*(9), 1606–1615. http://www.ncbi.nlm.nih.gov/pubmed/25858457.

Chambers, I., Colby, D., Robertson, M., Nichols, J., Lee, S., Tweedie, S., et al. (2003). Functional expression cloning of Nanog, a pluripotency sustaining factor in embryonic stem cells. *Cell, 113*(5), 643–655.

Chazaud, C., Yamanaka, Y., Pawson, T., & Rossant, J. (2006). Early lineage segregation between epiblast and primitive endoderm in mouse blastocysts through the Grb2-MAPK pathway. *Developmental Cell, 10*(5), 615–624.

Cheng, A., Saxton, T., Sakai, R., Kulkarni, S., Mbamalu, G., Vogel, W., et al. (1998). Mammalian Grb2 regulates multiple steps in embryonic development and malignant transformation. *Cell, 95*(6), 793–803.

Chisholm, J., & Houliston, E. (1987). Cytokeratin filament assembly in the preimplantation mouse embryo. *Development, 101*(3), 565–582.

Cockburn, K., Biechele, S., Garner, J., & Rossant, J. (2013). The Hippo pathway member Nf2 is required for inner cell mass specification. *Current Biology, 23*(13), 1195–1201.

Colman, A. (2013). Profile of John Gurdon and Shinya Yamanaka, 2012 Nobel laureates in medicine or physiology. *Proceedings of the National Academy of Sciences of the United States of America, 110*(15), 5740–5741.

Cross, J. C. (2007). Lentiviruses to the placental rescue. *Nature Biotechnology, 25*(2), 190–191.

Dietrich, J. E., & Hiiragi, T. (2007). Stochastic patterning in the mouse pre-implantation embryo. *Development, 134*(23), 4219–4231.

Dietrich, J. E., Panavaite, L., Gunther, S., Wennekamp, S., Groner, A. C., Pigge, A., et al. (2015). Venus trap in the mouse embryo reveals distinct molecular dynamics underlying specification of first embryonic lineages. *EMBO Reports, 16*(8), 1005–1021.

Do, D. V., Ueda, J., Messerschmidt, D. M., Lorthongpanich, C., Zhou, Y., Feng, B., et al. (2013). A genetic and developmental pathway from STAT3 to the OCT4-NANOG circuit is essential for maintenance of ICM lineages in vivo. *Genes & Development, 27*(12), 1378–1390.

Elling, U., Klasen, C., Eisenberger, T., Anlag, K., & Treier, M. (2006). Murine inner cell mass-derived lineages depend on Sall4 function. *Proceedings of the National Academy of Sciences of the United States of America, 103*(44), 16319–16324.

Feldman, B., Poueymirou, W., Papaioannou, V., DeChiara, T., & Goldfarb, M. (1995). Requirement of FGF-4 for postimplantation mouse development. *Science, 267*(5195), 246–249.

Fleming, T. (1987). A quantitative analysis of cell allocation to trophectoderm and inner cell mass in the mouse blastocyst. *Developmental Biology, 119*(2), 520–531.

Frankenberg, S., Gerbe, F., Bessonnard, S., Belville, C., Pouchin, P., Bardot, O., et al. (2011). Primitive endoderm differentiates via a three-step mechanism involving Nanog and RTK signaling. *Developmental Cell, 21*(6), 1005–1013.

Freyer, L., Schröter, C., Saiz, N., Schrode, N., Nowotschin, S., Martinez-Arias, A., et al. (2015). A loss-of-function and H2B-Venus transcriptional reporter allele for Gata6 in mice. *BMC Developmental Biology, 15*, 38.

Frum, T., Halbisen, M. A., Wang, C., Amiri, H., Robson, P., & Ralston, A. (2013). Oct4 cell-autonomously promotes primitive endoderm development in the mouse blastocyst. *Developmental Cell, 25*(6), 610–622.

Frum, T., & Ralston, A. (2015). Cell signaling and transcription factors regulating cell fate during formation of the mouse blastocyst. *Trends in Genetics, 31*(7), 402–410.

Grabarek, J. B., Zyzynska, K., Saiz, N., Piliszek, A., Frankenberg, S., Nichols, J., et al. (2012). Differential plasticity of epiblast and primitive endoderm precursors within the ICM of the early mouse embryo. *Development, 139*(1), 129–139.

Graham, S. J., Wicher, K. B., Jedrusik, A., Guo, G., Herath, W., Robson, P., et al. (2014). BMP signalling regulates the pre-implantation development of extra-embryonic cell lineages in the mouse embryo. *Nature Communications, 5*, 5667.

Guo, G., Huss, M., Tong, G., Wang, C., Li Sun, L., Clarke, N., et al. (2010). Resolution of cell fate decisions revealed by single-cell gene expression analysis from zygote to blastocyst. *Developmental Cell, 18*(4), 675–685.

Halder, G., & Johnson, R. L. (2011). Hippo signaling: Growth control and beyond. *Development, 138*(1), 9–22.

Hall B., Limaye A., & Kulkarni A.B. (2009). Overview: generation of gene knockout mice. *Current Protocols in Cell Biology/Editorial Board, Juan S. Bonifacino ... [et al.], Chapter*, Unit—19.1217. http://doi.org/10.1002/0471143030.cb1912s44.

Hirate, Y., Hirahara, S., Inoue, K., Kiyonari, H., Niwa, H., & Sasaki, H. (2015). Par-aPKC-dependent and -independent mechanisms cooperatively control cell polarity, Hippo signaling, and cell positioning in 16-cell stage mouse embryos. *Development, Growth & Differentiation, 57*(8), 544–556.

Hirate, Y., Hirahara, S., Inoue, K., Suzuki, A., Alarcon, V. B., Akimoto, K., et al. (2013). Polarity-dependent distribution of angiomotin localizes Hippo signaling in preimplantation embryos. *Current Biology, 23*(13), 1181–1194.

Hirate, Y., & Sasaki, H. (2014). The role of angiomotin phosphorylation in the Hippo pathway during preimplantation mouse development. *Tissue Barriers, 2*(1), e28127.

Home, P., Ray, S., Dutta, D., Bronshteyn, I., Larson, M., & Paul, S. (2009). GATA3 is selectively expressed in the trophectoderm of peri-implantation embryo and directly regulates Cdx2 gene expression. *The Journal of Biological Chemistry, 284*(42), 28729–28737.

Jedrusik, A., Parfitt, D., Guo, G., Skamagki, M., Grabarek, J., Johnson, M., et al. (2008). Role of Cdx2 and cell polarity in cell allocation and specification of trophectoderm and inner cell mass in the mouse embryo. *Genes & Development, 22*(19), 2692–2706.

Jensen, P., & Dymecki, S. M. (2014). Essentials of recombinase-based genetic fate mapping in mice. *Methods in Molecular Biology, 1092*, 437–454.

Johnson, M., & McConnell, J. (2004). Lineage allocation and cell polarity during mouse embryogenesis. *Seminars in Cell & Developmental Biology, 15*(5), 583–597.

Kanai-Azuma, M., Kanai, Y., Gad, J. M., Tajima, Y., Taya, C., Kurohmaru, M., et al. (2002). Depletion of definitive gut endoderm in Sox17-null mutant mice. *Development, 129*(10), 2367–2379.

Kaneko, K. J., & DePamphilis, M. L. (2013). TEAD4 establishes the energy homeostasis essential for blastocoel formation. *Development, 140*(17), 3680–3690.

Kang, M., Piliszek, A., Artus, J., & Hadjantonakis, A. K. (2013). FGF4 is required for lineage restriction and salt-and-pepper distribution of primitive endoderm factors but not their initial expression in the mouse. *Development, 140*(2), 267–279.

Kim, S. K. (2007). Common aging pathways in worms, flies, mice and humans. *The Journal of Experimental Biology, 210*(Pt. 9), 1607–1612.

Komatsu, K., & Fujimori, T. (2015). Multiple phases in regulation of Nanog expression during pre-implantation development. *Development, Growth & Differentiation, 57*(9), 648–656.

Kono, K., Tamashiro, D. A., & Alarcon, V. B. (2014). Inhibition of RHO-ROCK signaling enhances ICM and suppresses TE characteristics through activation of Hippo signaling in the mouse blastocyst. *Developmental Biology, 394*(1), 142–155.

Koutsourakis, M., Langeveld, A., Patient, R., Beddington, R., & Grosveld, F. (1999). The transcription factor GATA6 is essential for early extraembryonic development. *Development, 126*(4), 723–732.

Krawchuk, D., Honma-Yamanaka, N., Anani, S., & Yamanaka, Y. (2013). FGF4 is a limiting factor controlling the proportions of primitive endoderm and epiblast in the ICM of the mouse blastocyst. *Developmental Biology, 384*(1), 65–71.

Kuo, C. T., Morrisey, E. E., Anandappa, R., Sigrist, K., Lu, M. M., Parmacek, M. S., et al. (1997). GATA4 transcription factor is required for ventral morphogenesis and heart tube formation. *Genes & Development, 11*(8), 1048–1060.

Kurimoto, K., Yabuta, Y., Ohinata, Y., Ono, Y., Uno, K., Yamada, R., et al. (2006). An improved single-cell cDNA amplification method for efficient high-density oligonucleotide microarray analysis. *Nucleic Acids Research*, *34*(5), e42.

Lanner, F. (2014). Lineage specification in the early mouse embryo. *Experimental Cell Research*, *321*(1), 32–39.

Le Bin, G. C., Muñoz-Descalzo, S., Kurowski, A., Leitch, H., Lou, X., Mansfield, W., et al. (2014). Oct4 is required for lineage priming in the developing inner cell mass of the mouse blastocyst. *Development*, *141*(5), 1001–1010.

Leung, C. Y., & Zernicka-Goetz, M. (2013). Angiomotin prevents pluripotent lineage differentiation in mouse embryos via Hippo pathway-dependent and -independent mechanisms. *Nature Communications*, *4*, 2251.

Lin, S. C., Wani, M. A., Whitsett, J. A., & Wells, J. M. (2010). Klf5 regulates lineage formation in the pre-implantation mouse embryo. *Development*, *137*(23), 3953–3963.

Liu, A., & Eggenschwiler, J. (2014). Identifying essential genes in mouse development via an ENU-based forward genetic approach. *Methods in Molecular Biology*, *1092*, 95–118.

Lorthongpanich, C., Doris, T. P., Limviphuvadh, V., Knowles, B. B., & Solter, D. (2012). Developmental fate and lineage commitment of singled mouse blastomeres. *Development*, *139*(20), 3722–3731.

Lorthongpanich, C., Messerschmidt, D. M., Chan, S. W., Hong, W., Knowles, B. B., & Solter, D. (2013). Temporal reduction of LATS kinases in the early preimplantation embryo prevents ICM lineage differentiation. *Genes & Development*, *27*(13), 1441–1446.

Ma, G., Roth, M., Groskopf, J., Tsai, F., Orkin, S., Grosveld, F., et al. (1997). GATA-2 and GATA-3 regulate trophoblast-specific gene expression in vivo. *Development*, *124*(4), 907–914.

Marcho, C., Cui, W., & Mager, J. (2015). Epigenetic dynamics during preimplantation development. *Reproduction*, *150*(3), R109–R120.

McDole, K., Xiong, Y., Iglesias, P. A., & Zheng, Y. (2011). Lineage mapping the preimplantation mouse embryo by two-photon microscopy, new insights into the segregation of cell fates. *Developmental Biology*, *355*(2), 239–249.

McDole, K., & Zheng, Y. (2012). Generation and live imaging of an endogenous Cdx2 reporter mouse line. *Genesis*, *50*(10), 775–782.

Meilhac, S., Adams, R., Morris, S., Danckaert, A., Le Garrec, J., & Zernicka-Goetz, M. (2009). Active cell movements coupled to positional induction are involved in lineage segregation in the mouse blastocyst. *Developmental Biology*, *331*(2), 210–221.

Meng, Z., Moroishi, T., & Guan, K. L. (2016). Mechanisms of Hippo pathway regulation. *Genes & Development*, *30*(1), 1–17.

Messerschmidt, D. M., & Kemler, R. (2010). Nanog is required for primitive endoderm formation through a non-cell autonomous mechanism. *Developmental Biology*, *344*(1), 129–137.

Mihajlović, A. I., Thamodaran, V., & Bruce, A. W. (2015). The first two cell-fate decisions of preimplantation mouse embryo development are not functionally independent. *Scientific Reports*, *5*, 15034.

Mitsui, K., Tokuzawa, Y., Itoh, H., Segawa, K., Murakami, M., Takahashi, K., et al. (2003). The homeoprotein Nanog is required for maintenance of pluripotency in mouse epiblast and ES cells. *Cell*, *113*(5), 631–642.

Molkentin, J. D., Lin, Q., Duncan, S. A., & Olson, E. N. (1997). Requirement of the transcription factor GATA4 for heart tube formation and ventral morphogenesis. *Genes & Development*, *11*(8), 1061–1072.

Morgani, S. M., Canham, M. A., Nichols, J., Sharov, A. A., Migueles, R. P., Ko, M. S., et al. (2013). Totipotent embryonic stem cells arise in ground-state culture conditions. *Cell Reports*, *3*(6), 1945–1957.

Morris, S. A. (2011). Cell fate in the early mouse embryo: Sorting out the influence of developmental history on lineage choice. *Reproductive Biomedicine Online*, *22*(6), 521–524.

Morris, S. A., Graham, S. J., Jedrusik, A., & Zernicka-Goetz, M. (2013). The differential response to Fgf signalling in cells internalized at different times influences lineage segregation in preimplantation mouse embryos. *Open Biology*, *3*(11), 130104.

Morris, S., Tallquist, M., Rock, C., & Cooper, J. (2002). Dual roles for the Dab2 adaptor protein in embryonic development and kidney transport. *The EMBO Journal*, *21*(7), 1555–1564.

Morris, S. A., Teo, R. T., Li, H., Robson, P., Glover, D. M., & Zernicka-Goetz, M. (2010). Origin and formation of the first two distinct cell types of the inner cell mass in the mouse embryo. *Proceedings of the National Academy of Sciences of the United States of America*, *107*(14), 6364–6369.

Morrisey, E., Ip, H., Lu, M., & Parmacek, M. (1996). GATA-6: A zinc finger transcription factor that is expressed in multiple cell lineages derived from lateral mesoderm. *Developmental Biology*, *177*(1), 309–322.

Niakan, K. K., Han, J., Pedersen, R. A., Simon, C., & Pera, R. A. (2012). Human preimplantation embryo development. *Development*, *139*(5), 829–841.

Niakan, K. K., Ji, H., Maehr, R., Vokes, S. A., Rodolfa, K. T., Sherwood, R. I., et al. (2010). Sox17 promotes differentiation in mouse embryonic stem cells by directly regulating extraembryonic gene expression and indirectly antagonizing self-renewal. *Genes & Development*, *24*(3), 312–326.

Nichols, J., Silva, J., Roode, M., & Smith, A. (2009). Suppression of Erk signalling promotes ground state pluripotency in the mouse embryo. *Development*, *136*(19), 3215–3222.

Nishioka, N., Inoue, K., Adachi, K., Kiyonari, H., Ota, M., Ralston, A., et al. (2009). The Hippo signaling pathway components Lats and Yap pattern Tead4 activity to distinguish mouse trophectoderm from inner cell mass. *Developmental Cell*, *16*(3), 398–410.

Nishioka, N., Yamamoto, S., Kiyonari, H., Sato, H., Sawada, A., Ota, M., et al. (2008). Tead4 is required for specification of trophectoderm in pre-implantation mouse embryos. *Mechanisms of Development*, *125*(3–4), 270–283.

Niwa, H., Toyooka, T., Shimosato, D., Strumpf, D., Takahashi, K., Yagi, R., et al. (2005). Interaction between Oct3/4 and Cdx2 determines trophectoderm differentiation. *Cell*, *123*(5), 917–929.

Nowotschin, S., & Hadjantonakis, A. K. (2014). Live imaging mouse embryonic development: Seeing is believing and revealing. *Methods in Molecular Biology*, *1092*, 405–420.

Ohnishi, Y., Huber, W., Tsumura, A., Kang, M., Xenopoulos, P., Kurimoto, K., et al. (2014). Cell-to-cell expression variability followed by signal reinforcement progressively segregates early mouse lineages. *Nature Cell Biology*, *16*(1), 27–37.

Palmieri, S., Peter, W., Hess, H., & Schöler, H. (1994). Oct-4 transcription factor is differentially expressed in the mouse embryo during establishment of the first two extraembryonic cell lineages involved in implantation. *Developmental Biology*, *166*(1), 259–267.

Parfitt, D. E., & Shen, M. M. (2014). From blastocyst to gastrula: Gene regulatory networks of embryonic stem cells and early mouse embryogenesis. *Philosophical Transactions of the Royal Society of London. Series B, Biological Sciences*, *369*(1657), pii: 20130542. http://www.ncbi.nlm.nih.gov/pubmed/?term=PMC4216465.

Plachta, N., Bollenbach, T., Pease, S., Fraser, S. E., & Pantazis, P. (2011). Oct4 kinetics predict cell lineage patterning in the early mammalian embryo. *Nature Cell Biology*, *13*(2), 117–123.

Plusa, B., Frankenberg, S., Chalmers, A., Hadjantonakis, A., Moore, C., Papalopulu, N., et al. (2005). Downregulation of Par3 and aPKC function directs cells towards the ICM in the preimplantation mouse embryo. *Journal of Cell Science*, *118*(Pt. 3), 505–515.

Plusa, B., Piliszek, A., Frankenberg, S., Artus, J., & Hadjantonakis, A. (2008). Distinct sequential cell behaviours direct primitive endoderm formation in the mouse blastocyst. *Development*, *135*(18), 3081–3091.

Ralston, A., Cox, B., Nishioka, N., Sasaki, H., Chea, E., Rugg-Gunn, P., et al. (2010). Gata3 regulates trophoblast development downstream of Tead4 and in parallel to Cdx2. *Development*, *137*(3), 395–403.

Ralston, A., & Rossant, J. (2008). Cdx2 acts downstream of cell polarization to cell-autonomously promote trophectoderm fate in the early mouse embryo. *Developmental Biology*, *313*(2), 614–629.

Ralston, A., & Rossant, J. (2010). The genetics of induced pluripotency. *Reproduction*, *139*(1), 35–44.

Rayon, T., Menchero, S., Nieto, A., Xenopoulos, P., Crespo, M., Cockburn, K., et al. (2014). Notch and hippo converge on cdx2 to specify the trophectoderm lineage in the mouse blastocyst. *Developmental Cell*, *30*(4), 410–422.

Reyes de Mochel, N. S., Luong, M., Chiang, M., Javier, A. L., Luu, E., Toshihiko, F., et al. (2015). BMP signaling is required for cell cleavage in preimplantation-mouse embryos. *Developmental Biology*, *397*(1), 45–55.

Rossant, J., Chazaud, C., & Yamanaka, Y. (2003). Lineage allocation and asymmetries in the early mouse embryo. *Philosophical Transactions of the Royal Society of London. Series B, Biological Sciences*, *358*(1436), 1341–1348 (discussion 1349).

Roush, W. (1995). Nobel prizes: Fly development work bears prize-winning fruit. *Science*, *270*(5235), 380–381.

Russ, A., Wattler, S., Colledge, W., Aparicio, S., Carlton, M., Pearce, J., et al. (2000). Eomesodermin is required for mouse trophoblast development and mesoderm formation. *Nature*, *404*(6773), 95–99.

Saiz, N., Grabarek, J. B., Sabherwal, N., Papalopulu, N., & Plusa, B. (2013). Atypical protein kinase C couples cell sorting with primitive endoderm maturation in the mouse blastocyst. *Development*, *140*(21), 4311–4322.

Saiz, N., Plusa, B., & Hadjantonakis, A. K. (2015). Single cells get together: High-resolution approaches to study the dynamics of early mouse development. *Seminars in Cell & Developmental Biology*, *47–48*, 92–100.

Schrode, N., Saiz, N., Di Talia, S., & Hadjantonakis, A. K. (2014). GATA6 levels modulate primitive endoderm cell fate choice and timing in the mouse blastocyst. *Developmental Cell*, *29*(4), 454–467.

Schrode, N., Xenopoulos, P., Piliszek, A., Frankenberg, S., Plusa, B., & Hadjantonakis, A. K. (2013). Anatomy of a blastocyst: Cell behaviors driving cell fate choice and morphogenesis in the early mouse embryo. *Genesis*, *51*(4), 219–233.

Shi, J., Chen, Q., Li, X., Zheng, X., Zhang, Y., Qiao, J., et al. (2015). Dynamic transcriptional symmetry-breaking in pre-implantation mammalian embryo development revealed by single-cell RNA-seq. *Development*, *142*(20), 3468–3477.

Silva, J., Nichols, J., Theunissen, T., Guo, G., van Oosten, A., Barrandon, O., et al. (2009). Nanog is the gateway to the pluripotent ground state. *Cell*, *138*(4), 722–737.

Skamagki, M., Wicher, K. B., Jedrusik, A., Ganguly, S., & Zernicka-Goetz, M. (2013). Asymmetric localization of Cdx2 mRNA during the first cell-fate decision in early mouse development. *Cell Reports*, *3*(2), 442–457.

Smyth, N., Vatansever, H. S., Murray, P., Meyer, M., Frie, C., Paulsson, M., et al. (1999). Absence of basement membranes after targeting the LAMC1 gene results in embryonic lethality due to failure of endoderm differentiation. *The Journal of Cell Biology*, *144*(1), 151–160.

Solter, D. (2016). Preformation versus epigenesis in early mammalian development. *Current Topics in Developmental Biology*, *117*, 337–391. http://www.ncbi.nlm.nih.gov/pubmed/26969990.

Stephenson, R. O., Yamanaka, Y., & Rossant, J. (2010). Disorganized epithelial polarity and excess trophectoderm cell fate in preimplantation embryos lacking E-cadherin. *Development*, *137*(20), 3383–3391.

Strumpf, D., Mao, C. A., Yamanaka, Y., Ralston, A., Chawengsaksophak, K., Beck, F., et al. (2005). Cdx2 is required for correct cell fate specification and differentiation of trophectoderm in the mouse blastocyst. *Development*, *132*(9), 2093–2102.

Vogel, G. (2007). Nobel Prizes. A knockout award in medicine. *Science*, *318*(5848), 178–179.

Watanabe, T., Biggins, J. S., Tannan, N. B., & Srinivas, S. (2014). Limited predictive value of blastomere angle of division in trophectoderm and inner cell mass specification. *Development*, *141*(11), 2279–2288.

Wicklow, E., Blij, S., Frum, T., Hirate, Y., Lang, R. A., Sasaki, H., et al. (2014). HIPPO pathway members restrict SOX2 to the inner cell mass where it promotes ICM fates in the mouse blastocyst. *PLoS Genetics*, *10*(10), e1004618.

Wilder, P., Kelly, D., Brigman, K., Peterson, C., Nowling, T., Gao, Q., et al. (1997). Inactivation of the FGF-4 gene in embryonic stem cells alters the growth and/or the survival of their early differentiated progeny. *Developmental Biology*, *192*(2), 614–629.

Wu, G., Han, D., Gong, Y., Sebastiano, V., Gentile, L., Singhal, N., et al. (2013). Establishment of totipotency does not depend on Oct4A. *Nature Cell Biology*, *15*(9), 1089–1097.

Wu, G., & Schöler, H. R. (2016). Lineage segregation in the totipotent embryo. *Current Topics in Developmental Biology*, *117*, 301–317. http://www.ncbi.nlm.nih.gov/pubmed/?term=26969985.

Xenopoulos, P., Kang, M., Puliafito, A., Di Talia, S., & Hadjantonakis, A. K. (2015). Heterogeneities in Nanog expression drive stable commitment to pluripotency in the mouse blastocyst. *Cell Reports*, *10*, 1508–1520, pii: S2211-1247(15)00138-2. http://www.ncbi.nlm.nih.gov/pubmed/25753417.

Yagi, R., Kohn, M., Karavanova, I., Kaneko, K., Vullhorst, D., DePamphilis, M., et al. (2007). Transcription factor TEAD4 specifies the trophectoderm lineage at the beginning of mammalian development. *Development*, *134*(21), 3827–3836.

Yamanaka, Y. (2011). Response: Cell fate in the early mouse embryo—Sorting out the influence of developmental history on lineage choice. *Reproductive Biomedicine Online*, *22*(6), 525–527 (discussion 528).

Yamanaka, Y., Lanner, F., & Rossant, J. (2010). FGF signal-dependent segregation of primitive endoderm and epiblast in the mouse blastocyst. *Development*, *137*(5), 715–724.

Yamanaka, Y., Ralston, A., Stephenson, R., & Rossant, J. (2006). Cell and molecular regulation of the mouse blastocyst. *Developmental Dynamics*, *235*(9), 2301–2314.

Yang, D. H., Smith, E. R., Roland, I. H., Sheng, Z., He, J., Martin, W. D., et al. (2002). Disabled-2 is essential for endodermal cell positioning and structure formation during mouse embryogenesis. *Developmental Biology*, *251*(1), 27–44.

Zheng, W., & Liu, K. (2012). Maternal control of mouse preimplantation development. *Results and Problems in Cell Differentiation*, *55*, 115–139.

CHAPTER SIX

Polarity in Cell-Fate Acquisition in the Early Mouse Embryo

C.Y. Leung[1,2], M. Zhu[2], M. Zernicka-Goetz[3]
University of Cambridge, Cambridge, United Kingdom
[3]Corresponding author: e-mail address: mz205@cam.ac.uk

Contents

Abstract

Establishing polarity is a fundamental part of embryogenesis and can be traced back to the earliest developmental stages. It can be achieved in one of two ways: through the preexisting polarization of germ cells before fertilization or via symmetry breaking after fertilization. In mammals, it seems to be the latter, and we will discuss the various cytological and molecular events that lead up to this event, its mechanisms and the consequences. In mammals, the first polarization event occurs in the preimplantation period, when the embryo is but a cluster of cells, free-floating in the oviduct. This provides a unique, autonomous system to study the de novo polarization that is essential to

[1] Current address: Integrated Cardio Metabolic Centre, Novum, Hälsovägen 7, Karolinska Institutet, Huddinge, 141 57 Sweden.
[2] These authors contributed equally to this work.

Current Topics in Developmental Biology, Volume 120
ISSN 0070-2153
http://dx.doi.org/10.1016/bs.ctdb.2016.04.008

life. In this review, we will cover modern and past studies on the polarization of the early embryo, using the mouse as a model system, as well as hypothesizing the potential implications and functions of the biological events involved.

1. INTRODUCTION

A new generation of mammalian life begins with fertilization when an oocyte fuses with a sperm to form the zygote. The developmental window between the zygote and implantation into the uterus is termed preimplantation development (Fig. 1A). The main purpose of this period of embryogenesis is the partitioning of cellular lineages and to form a structure for embryo implantation. A pluripotent, embryonic lineage is preserved while extra-embryonic lineages differentiate from totipotent cells that would become specialized in processes such as implantation. Early divisions in the embryo are cleavage divisions, where cells divide without increasing in mass. The zygote undergoes cleavage while lacking conventional epithelial polarity, until an event known as compaction takes place. In the mouse embryo compaction occurs at roughly two and half days after fertilization,

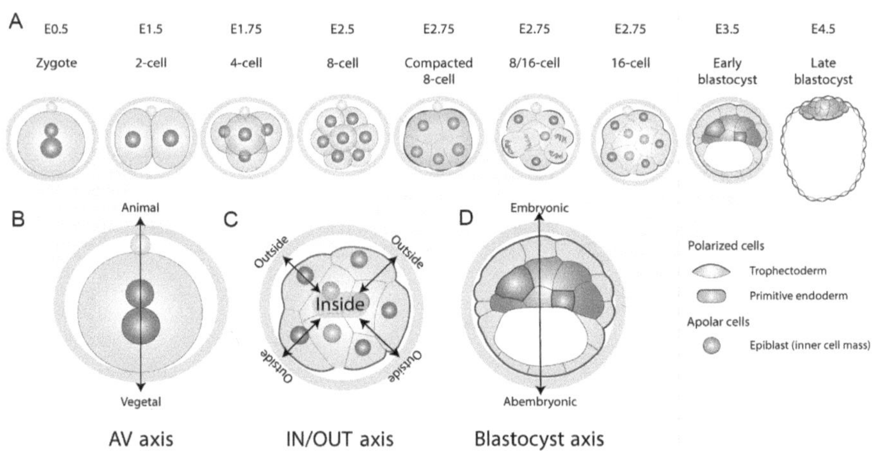

Fig. 1 An overview of the polarity axes of preimplantation development. (A) Schematic view of preimplantation development. (B) From the zygote stage onward the animal–vegetal axis is defined by the polar body marking the animal pole. (C) After compaction and asymmetric divisions, cells can divide along the IN/OUT axis into the "inside" of the embryo, making them "inside cells." (D) By the blastocyst stage an inner cell mass is formed which marks the embryonic pole of the embryo. There are two types of polarized cells in the E4.5 blastocyst, the trophectoderm and the primitive endoderm, while the epiblast cells that later form the embryo proper remain apolar. (See the color plate.)

and simultaneously acquires some features of canonical epithelial polarity. This establishes a radial polarity in the embryo where cells can now distinguish its physical position within the embryo. The physical position, in combination with cellular polarization plays a critical role in establishing cell fate. Polarized cells on the outside of the embryo differentiate to become an epithelial-like lineage called the trophectoderm (TE). The TE consists of progenitors for lineages such as trophoblast giant cells that mediate embryo implantation and generate an extra-embryonic ectoderm, essential for signaling into the developing embryo. Meanwhile, apolar inside cells retain their potency and ability to produce every single lineage of the embryo proper. While epithelial polarity becomes apparent at the 8-cell stage, it is not the only type of polarity that can affect cell fate. Polarity can take many forms and it is evident when symmetry is broken or when heterogeneities come to lie along an axis. The mouse embryo is certainly not a perfectly symmetrical system, and heterogeneity is abundant. Some aspects of this heterogeneity can influence cell fate and others may be neutral to development. Here we will proceed chronologically through early mouse embryo development and discuss all forms of polarity with a specific emphasis on their impact on cell-fate choices.

2. POLARITY AXES IN PREIMPLANTATION EMBRYOS

The early embryo exhibits consistent patterns of polarization at successive stages that demarcate various embryonic axes. The first embryonic axis can be observed as early as the 1-cell stage. The fertilized mouse egg is a large spherical cell with two smaller cells, called polar bodies, one loosely and one more firmly attached at one end. These polar bodies are the by-products of asymmetric meiotic divisions. The first polar body is extruded by the oocyte during oogenesis, and the second polar body is produced immediately after fertilization. The majority of the first polar bodies undergo apoptosis as early as the zygote stage but the second polar body persists longer into development, as it remains attached by an actin cable (Fabian, Cikos, Rehak, & Koppel, 2014; Gardner, 1997; Plusa, Grabarek, Piotrowska, Glover, & Zernicka-Goetz, 2002). The dramatic difference in size between the polar body and the zygote is the result of meiotic spindle migration and asymmetric cytokinesis, demonstrating the highly polarized nature of the intracellular machinery. The polar bodies define the first embryonic axis, the animal–vegetal (AV) axis, where the site of polar body extrusion defines the animal pole (Fig. 1B). The terms "animal" and "vegetal" are borrowed terms from nonmammalian embryos. In many nonmammalian model organisms the embryo is clearly divided into two halves where one half contains cells that divide quickly (hence animal), and the other contains yolky cells that divide slowly (hence vegetal). In these embryos, cellular determinants are already prepositioned at the 1-cell stage and any perturbations such as destruction of one pole will affect development, in some cases severely. This is not the case for mammalian embryos, and removal of either the animal or vegetal pole of the zygote still results in development of fertile mice (Zernicka-Goetz, 1998). Furthermore, there is no known polarization of a molecular determinant of cell fate in the mouse zygote. The mammalian embryo is unique in that it appears to lack pre-established polarity components that would become critical for development or cell fate. Visually polarization become noticeable after the embryo completes three rounds of cleavage at the 8-cell stage (Alford, Ng, & Burgess, 2009; Bowerman & Shelton, 1999; Guethhallonet & Maro, 1992; Muller & Hausen, 1995). At this time, compaction takes place, when blastomeres that were once loosely aggregated become a tightly compacted ball of cells (Ducibella & Anderson, 1975; Winkel, Ferguson, Takeichi, & Nuccitelli, 1990). A dramatic reorganization of the cytoskeleton and cytoplasmic organelles takes place during compaction, after which a clear IN–OUT polarity axis is generated: an apical domain facing the outside, and a basal domain facing the center of the embryo (Fleming & Johnson, 1988) (Fig. 1C). After compaction and the simultaneous reorganization of cytoplasmic components, the blastomeres are primed for asymmetric cell divisions. Blastomeres can divide along the IN–OUT axis, asymmetrically giving rise to inside cells and outside cells, or the division can be perpendicular to the IN–OUT axis, giving rise to two outside cells (Johnson & Ziomek, 1981a). The polarity state is hugely different between outside cells and inside cells. Outside cells have some typical characteristics of a polarized epithelial cell while inside cells are apolar (Anani, Bhat, Honma-Yamanaka, Krawchuk, & Yamanaka, 2014; Fleming & Johnson, 1988). Outside cells will then differentiate to form the very first embryonic epithelium-like lineage: TE, while the inside cells will contribute to the embryonic lineage. Around the third day after fertilization, an aqueous cavity appears within the embryo, bound by the one-cell thick TE. This pushes the inside cells to one side and creates the inner cell mass (ICM). Even though preimplantation embryos are free of any physical attachments that may signal as a polarity reference, by the time the embryo is ready to implant into the uterus it has a clearly polarized structure: a hollow ball of cells with an ICM positioned at one end of the embryo. At this point the ICM marks another embryonic axis: the blastocyst axis, also referred to as the embryonic–abembryonic axis, with the pluripotent ICM defines the embryonic pole (Fig. 1D). The TE is not the only lineage that develops epithelial polarity. At the late blastocyst stage, an epithelium-like layer develops over the ICM, representing the second polarized epithelium-like layer, the primitive endoderm (PE) (Fig. 1A).

3. PRECOMPACTION POLARITY

3.1 Zygote to 2-Cell Embryos

Early mammalian embryos are remarkably regulative. Removing a blastomere from any cleavage-stage embryo does not result in developmental abnormalities (Tarkowski, 1959; Tsunoda & McLaren, 1983). Indeed preimplantation genetic diagnosis is routinely performed on human embryos by removing and analyzing an 8-cell blastomere. This plasticity could be evidence for lack of polarization, or an ability of the embryo to compensate, in the face of damage or manipulation, by restructuring polarity. To date this is still under debate but there is increasing evidence that early cleavage blastomeres show more heterogeneity and bias to their development than expected. Till now, only few proteins are known to have a polarized distribution from the first cleavage stage onward. Most of them belong to the subcortical maternal complex (SCMC) or are postulated to be part of or interact with the SCMC. The SCMC is a protein complex that is exclusively produced by the oocyte but persists throughout preimplantation development. What makes it remarkable is that it localizes adjacent to the plasma membrane in zygotes and after cleavage divisions is excluded from sites of cell–cell contact (Fig. 2A). This creates a radial asymmetry where the center of the embryo lacks the SCMC. The first protein of the SCMC identified to play a role in embryogenesis is *Nlrp5* (also known as *Mater*). Its expression is oocyte-specific, the transcript degrades following fertilization but the protein persists until the blastocyst stage (Tong et al., 2000, 2004). *Nlrp5* mutant embryos arrest at the 2-cell stage in the mouse (Tong et al., 2000). Although its mechanism of function is not fully elucidated, evidence points to a role in zygotic genome activation, mitochondrial function, endoplasmic reticulum distribution, and calcium homeostasis (Fernandes et al., 2012; Kim et al., 2014; Tong et al., 2000). Nlrp5 was found to be a major component of a SCMC complex, with binding partners Filia, FLOPED, and TLE6 (Ohsugi, Zheng, Baibakov, Li, & Dean, 2008; To et al., 2008). Throughout

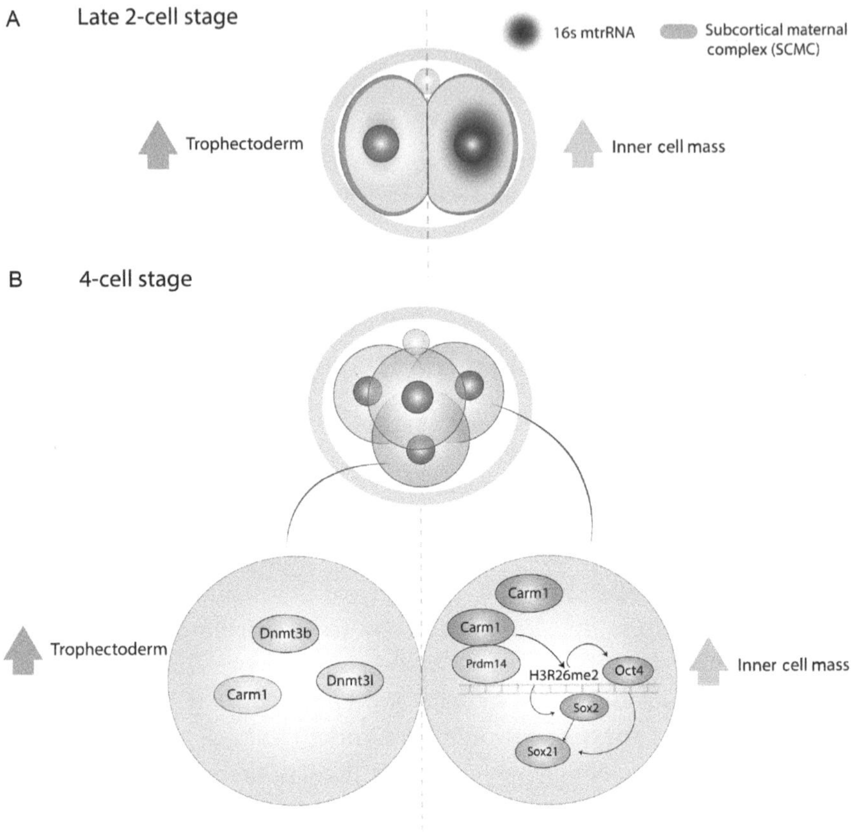

Fig. 2 Molecular hetereogeneities in the 2- and 4-cell embryo. (A) There are two molecular hetereogeneities in the 2-cell embryo, the subcortical maternal complex (SCMC) which is excluded from the cell–cell contact regions of the embryo, and the presence of 16S mtrRNA. 16S mtrRNA is the only heterogeneous molecule at the 2-cell stage known to be a bona fide cell-fate regulator. (B) A hypothetical model on the molecular differences of cell-fate regulators between 4-cell blastomeres. The defining difference in the levels of Prdm14 starts off a cascade that involves Carm1 and possibly Sox21 to promote the pluripotent lineage. (See the color plate.)

preimplantation development, all of the SCMC components are excluded from cell–cell contact regions, creating a striking polarized localization pattern as early as the 2-cell stage embryo (To et al., 2008; Tong et al., 2004). This polarized localization persists until the blastocyst stage, where only the TE cells inherit this maternal complex (To et al., 2008; Tong et al., 2004). SCMC polarization is driven by topology rather than cell fate, as in the TE specifier *Tead4* mutant embryos, the SCMC remains polarized in the outside cells. Other proteins with a similar distribution pattern to the SCMC include maternally derived leptin (Schulz & Roberts, 2011), Tcl1 (Narducci et al., 2002), Uch-l1 (Sekiguchi et al., 2006), and Padi6 (Yurttas et al., 2008), a relative of Padi4, which is a pluripotency regulator (Christophorou et al., 2014). There is no direct evidence that these proteins are bona fide cell-fate regulators, with efforts to investigate them complicated by early phenotypes and very stable proteins. They are, however, some of the few known polarized proteins in early mammalian embryos.

Do 2-cell blastomeres contain heterogeneous amounts of cell-fate guiding factors that significantly alter their fate? Interrogation of early 2-cell blastomeres using single-cell RNA sequencing revealed little difference in the inheritance bias of transcripts (Shi et al., 2015). Despite this, a very recent study reported that 2-cell stage blastomeres have an unequal distribution of mitochondrial ribosomal RNAs (mtrRNAs) (Zheng, Li, Zhang, Yang, & Qi, 2016). This is the earliest molecular asymmetry reported in the mammalian embryo. Specifically, the 16S mtrRNA, which can be detected from the zygote onward, becomes progressively heterogeneous at the 2-cell stage. The early 2-cell stage embryo has equal amounts of 16S mtrRNA but by the late 2-cell stage, this was lost in one of the two blastomeres (Fig. 2A). From the 4-cell stage onward, 16S mtrRNA is restricted to the basal–lateral region of the cytoplasm, and later is only detected in the ICM of the blastocyst. In addition, 16S mtrRNA seems to play a functional role in the ICM fate, where its downregulation in one of two blastomeres result in a significantly higher contribution to the TE lineage (Zheng et al., 2016). It would be interesting to investigate its mechanism of function, and in particular, whether it relates to other axes and physical cues. For example, do 16S mtrRNA levels in the 2-cell blastomere affect its subsequent division plane, which has been shown to affect cell fate (Piotrowska-Nitsche, Perea-Gomez, Haraguchi, & Zernicka-Goetz, 2005; Piotrowska-Nitsche & Zernicka-Goetz, 2005)? Moreover, does this heterogeneity relate to the sperm entry point (Piotrowska & Zernicka-Goetz, 2001)? Since paternal mitochondria are not inherited, there is a possibility that mtrRNA is exclusively maternal, however 16S mtrRNA is not localized within mitochondria, and does not rule out paternal transmission. There has been precedent of sperm-mediated transmission of RNA molecules that causes intergenerational phenotypes (Chen et al., 2016) and this would be an exciting prospect for mtrRNAs.

3.2 4-Cell Embryos

The 4-cell stage is the earliest stage where increasingly growing evidence points to different cell-lineage potency between the blastomeres. This implies the existence of polarity gradients or heterogeneous expression/ inheritance of cell-fate guiding factors at the 2- to 4-cell transition. First observations were made on the probabilistic relationship between the order and orientation of the 2- to 4-cell division plane and the blastomere's subsequent cell fate (Piotrowska-Nitsche et al., 2005; Piotrowska-Nitsche & Zernicka-Goetz, 2005). The molecular mechanism that mediates this effect is still not absolutely known but two recent studies demonstrate that blastomeres generated from different cell division patterns do indeed present distinct transcriptional states that guide their fate (Goolam et al., 2016). This heterogeneity in gene expression patterns between cells is particular significant for targets of the key pluripotency master genes, *Oct4* and *Sox2*, and was shown to guide cell fate. Despite the fact that a 4-cell blastomere is not totipotent on its own, its potency relative to the AV axis can be tested by reconstructing embryos using only blastomeres from either the animal or the vegetal pole. It was found that embryos reconstructed with 4-cell blastomeres from the animal pole produced viable offspring but not embryos reconstructed solely from vegetal 4-cell blastomeres (Morris, Guo, & Zernicka-Goetz, 2012; Piotrowska-Nitsche et al., 2005). Although a number of studies did not detect any cell-fate differences between the progeny of 4-cell blastomeres (Alarcon & Marikawa, 2003; Chroscicka, Komorowski, & Maleszewski, 2004; Hiiragi & Solter, 2004; Louvet-Vallee, Vinot, & Maro, 2005; Motosugi, Bauer, Polanski, Solter, & Hiiragi, 2005), a more conclusive study using genetic tracing demonstrated individual 4-cell blastomeres indeed has a differential bias in cell fate (Tabansky et al., 2013). This provides an important clue that there may be some form of early polarity subtly biasing cell fate. It is also important to note that these biases in cell fate are probabilistic and not deterministic, a major difference to other nonmammalian embryos that have been studied.

The first molecules to be identified to bias cell fate at the 4-cell stage are epigenetic regulators. Carm1 is a histone arginine methyltransferase that is able to affect cell-fate allocation via histone arginine modifications (Torres-Padilla, Parfitt, Kouzarides, & Zernicka-Goetz, 2007). Carm1 itself does not display a polarized distribution in the early preimplantation embryo (Torres-Padilla et al., 2007), but its interacting partner, Prdm14, is found to be present at highly variable levels between 4-cell blastomeres (Burton et al., 2013). Importantly, both Prdm14 and Carm1 can actively direct cell fate: if either one is overexpressed in a 2-cell blastomere, its progeny will have higher levels of H3R26me2 and become more likely to take on an ICM fate (Burton et al., 2013). Prdm14 is a zinc-finger DNA-binding protein, and the zinc-finger motif is responsible for its methylation and cell-fate specification activity. Since Prdm14 directly interacts with Carm1, it was suggested that Prdm14 binds specific regions in the DNA, then recruits Carm1 and its histone methyltransferase activity. Since Carm1 is expressed at a relatively even level between 4-cell blastomeres, the heterogeneous levels of Prdm14 could be the key asymmetry that directs Carm1 activity to bias cell fate (Burton et al., 2013). Furthermore, Prdm14 is not the only heterogeneously expressed epigenetic modifier. The DNA methyltransferases Dnmt3b and Dnmt3l are expressed in an opposing pattern to Prdm14 in the 4-cell stage, similar to an inverse correlation found in the blastocyst where Prdm14 is expressed in the ICM and the DNA methyltransferases are found in the TE (Burton et al., 2013) (Fig. 2B).

Advances in single-cell sequencing technologies have provided the opportunity to link observations made at a morphogenetic level to individual effector molecules. Is there a systemic heterogeneity in the early embryo that contributes to a polarized embryonic state? Most recently, comparison of transcriptomes of individual blastomeres at the 4-cell stage highlighted targets of Sox2 and Oct4 such as Sox21 as a potential mediator of early cell-fate bias and polarity (Goolam et al., 2016). Sox21 expression was highly variable between the four blastomeres, and at least one cell in each embryo consistently had very low to no Sox21 transcripts. The function of Sox21 has not been previously studied in the preimplantation embryo but there have been hints to an important role in cell-fate regulation. In embryonic stem (ES) cells and induced pluripotent stem (iPS) cells Sox21 is a known protagonist of the pluripotent lineage. It plays an important role in the reprogramming process of iPS cells and in ES cells it inhibits the expression of Cdx2, a TE master regulator (Kuzmichev et al., 2012). It also interacts with key pluripotency factors, such as Oct4 and Sox2. Sox2 directly regulates the expression of its cousin, Sox21 (Chakravarthy, Ormsbee, Mallanna, & Rizzino, 2011; Kuzmichev et al., 2012; Mallanna et al., 2010), and Oct4 can interact with the regulatory region of Sox21 (Chakravarthy et al., 2011; Goke et al., 2011; Kuzmichev et al., 2012; Mallanna et al., 2010). Direct in vivo analysis of Sox21 in the embryo confirmed that Sox21 indeed promotes the pluripotent lineage. Decreasing Sox21 expression led to a premature upregulation of Cdx2 and the cell would be heavily biased to differentiate into the TE. Furthermore, Sox21 may only be one of many early, polarized, cell-fate regulators. Within the highly variable genes at the 4-cell stage, Oct4 and Sox2 targets are remarkably overrepresented, with a subset correlating with the expression pattern of Sox21. Notably these include other well-known pluripotent factors such as Nanog and Esrrb. This implicates that even as early as the 4-cell stage, polarized transcriptomic states already exists, which biases the cell-fate outcome of progenies. These transcriptomic states are not independent of the epigenetic regulations mentioned previously. Carm1 activity is closely associated with Sox21 expression. Inhibition of Carm1 results in a near complete loss of Sox21 expression at the 4-cell stage, and the opposite is true when Carm1 is overexpressed (Goolam et al., 2016).

It is increasingly evident that cells of the 4-cell embryo are not identical. The new evidence gathered enables a hypothetical model for the underlying mechanism behind the different transcriptional states to be pieced together. At the top of the cascade, Prdm14 is asymmetrically distributed within the 4-cell embryo via a yet unknown mechanism. Prdm14 binds specific regions in the genome, and recruits Carm1 and its histone methylation activity to its vicinity. Methylation marks on H3R26 potentially lead to the expression of downstream targets of Oct4 and Sox2, such as Sox21 (Fig. 2B). This is a possible mechanism for symmetry breaking, the earliest yet known to involve transcriptional cell-fate determinants in any mammalian embryo.

It is a tempting model, but there are still many gaps to be filled in. First of all, is the level of variability in the endogenous levels of Prdm14 and Sox21 enough to bias cell fate? This is a technically difficult question to answer, as it requires determination of transcript levels and live-cell tracking on the same embryo. Thus far, only experimental knockdown or overexpression has been performed, and these manipulations are likely to influence gene expression on a much greater scale than the inherent variability of transcripts between blastomeres. Another pending question is whether the "Sox21 axis" relates to other axes. For example, does Sox21 heterogeneity correlate with Prdm14 heterogeneity, or to any physical axes such as the AV axis? And perhaps the most interesting question of all is why does this heterogeneity exist, is it simply a result of transcriptional noise and feedback loops (Shi et al., 2015) or more than this? Does it serve a purpose in development? Sox21 could play a role in preventing premature differentiation and preserving some cells for the embryonic lineage—indeed Sox21 expression prevents premature activation of Cdx2 that leads to lose of pluripotency (Goolam et al., 2016). But knockout models of *Carm1* (Kim, Lee, et al., 2004; Yadav et al., 2003, 2008), *Prdm14* (Yamaji et al., 2008), and *Sox21*

(Kiso et al., 2009) do not have an early arrest phenotype and embryo development can progress beyond the cleavage stages. However, such knockouts do not remove any maternal contribution and so there remains the possibility for a role for maternal transcripts. It is also possible that any of these proteins are part of redundant pathways that are able to compensate either for the lack of one gene or that while heterogeneity presents a preferred pathway that gives some advantages, it is possible for development to take place without it. Such flexibility could be a unique evolutionary feature of mammalian embryos as it provides a stronger safeguard to development seen in other metazoans. In the development of many non-mammalian embryos following a predetermined body plan, death, or abnormality in any single cell in early development would have serious ramifications for the development of the embryo as a whole. In amphibians and fish, this could be compensated by having large numbers of offspring. In mammals, small litter sizes have made each embryo more "valuable," and any abnormality in one cell within a predetermined embryo would become very costly to the animal. Furthermore, mammals have a far longer lifespan that imposes the possibility of age-related errors in chromosome segregation that can have a direct impact on the quality of development. By having a flexible polarity in the early cleavage stages, the mammalian embryo ensures no single blastomere would be essential for the development of the whole embryo.

4. 8-CELL EMBRYOS: FROM COMPACTION ONWARD

The early 8-cell stage embryo is a physically symmetrical ball of cells in which all the cells are in a physically equivalent position within the embryo. Within 12 h of the 4- to 8-cell transition a process called compaction occurs where all of the blastomeres flatten against each other, eliminating the spaces between the cells. This is also the time when blastomeres acquire conventional epithelial polarity. As a result, the next cell division can be either symmetric where outside cells give rise to two daughter cells that also remain outside; or asymmetric where outside cells divide along the IN/OUT axis and give rise to one inside cell and one outside cell (Johnson & Ziomek, 1981a). The defining difference between an inside cell and an outside cell is the asymmetry in its cell–cell contact. An inside cell experiences symmetrical cell–cell contact on all sides, whereas an outside cell has its apical surface exposed. This difference plays an instructive role in polarity and cell fate, as demonstrated by the compaction-defective *E-cadherin* mutant (Stephenson,

Yamanaka, & Rossant, 2010). These embryos continue to divide in the absence of compaction, creating a ball of loosely attached cells with partially exposed surfaces. This asymmetric cell–cell contact results in cells becoming polarized and taking on an outside TE cell fate. Apart from the cell–cell contact regions, other cellular features are also remodeled during compaction. The plasma membrane, cytoskeleton as well as cytoplasmic organelles become clearly segregated into apical and basal–lateral compartments (Fleming, Pickering, Qasim, & Maro, 1986; Handyside, 1980). Two relatively old terms, surface polarity and cytoplasmic polarity, were used to describe the polarization state of cell before the widespread use of molecular markers. Surface polarity describes the partition of the plasma membrane and the cytocortex, while cytoplasmic polarity describes the polarization of the cytoskeleton and cytoplasmic organelles (Reeve, 1981). Although both types of polarity are generated simultaneously, they require and are regulated by different cellular mechanisms.

4.1 Surface Polarity

Surface polarity is usually defined by the presence of microvilli and binding of Concavalin A, a leptin, to the apical surface. As the embryo compacts, the E-cadherin-based adherens junctions (AJ) become elongated as cell–cell contacts are maximized. This contact region delineates the basal–lateral domain of the cell and the contact free area marks the apical domain. Apart from AJs, other junctional components such as gap junctions and tight junctions (TJs), as well as ion channel components, are also recruited to the cell–cell contact site (Ducibella, Albertini, Anderson, & Biggers, 1975; Winston, Johnson, McConnell, Cook, & Day, 2004). Meanwhile, cytocortical elements that include the actin cytoskeleton, microtubules, and their associated proteins, are reassembled so that they accumulate under the apical cytocortex, away from the cell–cell contact regions (Houliston, Pickering, & Maro, 1987; Johnson & Maro, 1984; Lehtonen & Reima, 1986; Reima & Lehtonen, 1985) (Fig. 3). As a final step, cytoskeletal microfilaments form a recognizable ring pattern underneath the apical region of the plasma membrane, as observed by F-actin localization (Anani et al., 2014; Thomas et al., 2004). Reassembly of the actin cytoskeleton also results in the restriction of microvilli, the actin-based membrane protrusions, to the apical pole, where previously they would have been distributed uniformly (Reeve & Ziomek, 1981). The restriction of microvilli to the apical pole after compaction probably reflects the realignment of the actin cytoskeleton

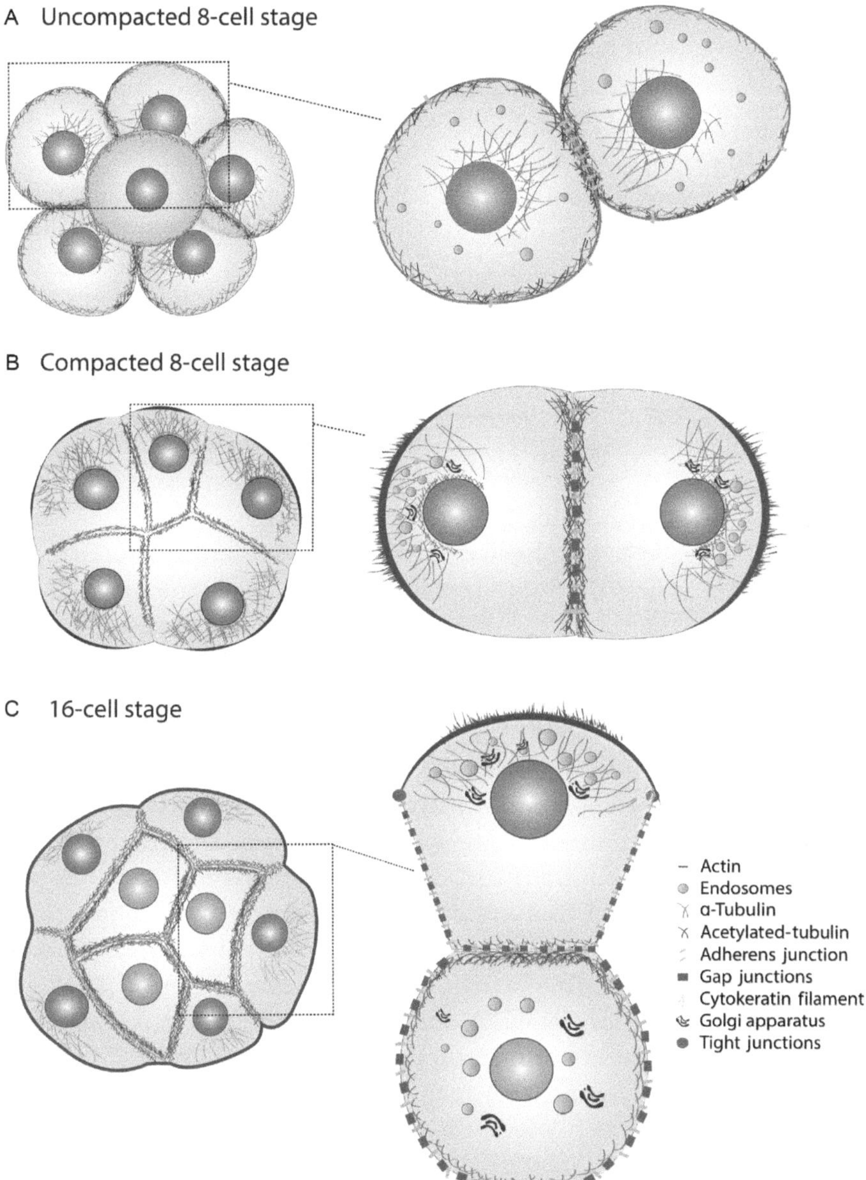

Fig. 3 The cytoplasmic and surface polarity of the (A) uncompacted 8-cell stage embryo, (B) compacted 8-cell stage embryo, and (C) the 8–16 cell stage embryo. (See the color plate.)

along the IN/OUT axis. It is not clear why cleavage-stage embryos develop microvilli as these are usually found on cells specialized in absorption and secretion, functions that are not apparent in the early embryo. It is therefore possible that the presence of microvilli represents a side effect of a highly dynamic actin cytoskeleton.

Once established, the surface pole is robust and resistant to external disruption such as cell isolation or cytoskeleton inhibitor treatment (Johnson & Maro, 1985; Johnson & Ziomek, 1981b). Although the establishment of surface polarity is marked by changes in actin and tubulin-based cytoskeleton dynamics, pharmacological disruption of either cytoskeleton does not prevent surface pole formation or maintenance. Surface polarity is only significantly disrupted when both microtubule and actin polymerization are inhibited (Johnson & Maro, 1985). The robustness of the surface pole alludes to the idea that it can act as a memory of epithelial polarity (Johnson, Pickering, Dhiman, Radcliffe, & Maro, 1988). By anchoring a set of factors at the apical pole, it can be inherited in symmetric divisions but not in asymmetric division, where only outside cells inherit polarity components in the surface pole.

4.2 Cytoplasmic Polarity

Concomitant with surface polarity, the cytoplasmic contents of blastomeres also polarize during compaction, which is especially apparent for organelles associated with microtubule filaments. Before compaction, microtubules extend from the entire nuclear surface to the cytocortex and projects perpendicularly to cell–cell contact sites. After compaction is complete, the microtubule cytoskeleton becomes realigned so that polymerization occurs mainly on the nuclear surface facing the apical zone and the filaments extend in parallel to the contact regions (Houliston et al., 1987) (Fig. 3). Subsets of microtubules can also be distinguished by posttranslational modifications. Acetylated tubulin is enriched basally in outside cells and preferentially accumulates in inside cells (Houliston & Maro, 1989). Acetylated microtubules are known to be more stable than their nonacetylated counterparts, in that they can withstand mild treatments of microtubule depolymerizing drugs (De Brabander, Van de Veire, Aerts, Borgers, & Janssen, 1976; LeDizet & Piperno, 1986; Piperno, LeDizet, & Chang, 1987). This may give outside polarized cells a more dynamic microtubule cytoskeleton, however in vitro data suggests tubulin acetylation does not affect polymerization kinetics (Maruta, Greer, & Rosenbaum, 1986). Embryos with hyperacetylated tubulin can develop normally (Zhang et al., 2008) and it is not known whether tubulin acetylation confers a function in inside cells.

Along with microtubule polarization, the pericentriolar material, a component of the centrosome that mainly comprises of nucleated-tubulin, redistributes from being dispersed in the cytoplasm to being tethered and aggregated onto the apical surface of the nucleus (Houliston et al., 1987). The nucleus itself is subject to repositioning during compaction. The majority of nuclei migrates toward the basal half of the cell, via kinesin-mediated microtubule transport (Ajduk, Biswas Shivhare, & Zernicka-Goetz, 2014; Reeve & Kelly, 1983). Interestingly this can be controlled by the transcription or polarization state of the cell, where in embryos lacking a key differentiation factor, Cdx2, or apical polarity, the incidence of basal nuclei increases. The position of the nuclei is also strongly correlated to the orientation of the division the cell would undergo. An apical nucleus means that the cell almost exclusively undergoes symmetric division, which generates outside progenitors cells for the TE. Cells with baso-central nuclei can divide either way, both symmetrically and asymmetrically (Ajduk et al., 2014). Therefore, Cdx2 and polarity can perhaps promote symmetric divisions over asymmetric ones by influencing nuclear position. It is not clear whether nuclear position is causal to the division plane or what kind of mechanism controls the two.

As the outside polarized cells differentiate, they acquire many features typical of an epithelial cell. This includes the expression of cytokeratin filaments that provide mechanical support as well as other additional functions. After the start of asymmetric divisions at the 8- to 16-cell stage, polarized outside cells quantitatively express higher levels of cytokeratin (Chisholm & Houliston, 1987) and by the blastocyst stage, cytokeratin 8 is exclusively expressed in the TE lineage. Membrane trafficking systems also reassemble in outside cells as they polarize. Clathrin-coated endosomes become transiently clustered under the apical region after compaction (Fleming, Cannon, & Pickering, 1986; Reeve, 1981), and the Golgi apparatus and lysosomes show clear basal distribution in outside cells at the 16-cell stage (Fleming, Cannon, et al., 1986) (Fig. 3). As opposed to surface polarity, cytoplasmic polarity requires both the actin and tubulin cytoskeletons (Fleming & Pickering, 1985). It has therefore been proposed that surface polarity is a prerequisite for cytoplasmic polarity (Houliston et al., 1987; Johnson & Maro, 1985). One way to interpret this is that the signal for polarization comes from the cell surface, which is then transmitted throughout the cell. The origin of this signal likely comes from cell–cell contact sites, which will be discussed further on. Overall, cytoplasmic and surface polarity are descriptive concepts, they mainly describe macroscopic events at the organelle and cellular level, and the specific function of these events are still largely unknown. Despite that, they play an important role in segmenting cytoplasmic contents to establish the correct cell fate in their descendants. One such example is the segregation of mRNA transcripts. Polarizing mRNA transcripts of cell-fate determinants is a highly conserved mechanism in asymmetric divisions during development. However in mammals, at least thus far, there is only one known mRNA that is polarized, the mRNA of *Cdx2* (Skamagki, Wicher, Jedrusik, Ganguly, & Zernicka-Goetz, 2013). *Cdx2* mRNA becomes polarized toward the apical from compaction onward, ensuring that outside cells inherit more *Cdx2* transcripts. The mechanism to anchor *Cdx2* mRNA apically is dependent on both the actin and microtubule cytoskeleton, the kinesin motor protein and the polarity protein atypical protein kinase C (aPKC) (Skamagki et al., 2013). This example highlights a common feature among many polarized components of the cell: multiple cellular processes need to act in synchrony to ensure a correct cell fate is established.

5. THE MOLECULAR LANDSCAPE OF POLARIZATION AFTER COMPACTION

The epithelial polarization state of the embryo is established and maintained by the molecular regulation between different protein complexes. These complexes localize to their respective regions and locally antagonize their counterparts to delineate the apical and basal–lateral regions. They are highly conserved throughout evolution and perhaps the most well-known apical polarity complex is the Par complex. The evolutionarily conserved Par complex always marks the apical domain and is excluded cell–cell contact regions as a result of the mutual antagonistic interactions with the basal–lateral components. The Par complex consists of three components: Par6, Par3, and aPKC, all of which were discovered in *C. elegans* embryo through genetic screening (Etemad-Moghadam, Guo, & Kemphues, 1995). Simultaneously with the construction of the apical domain, many conserved basal–lateral polarity regulators can also be detected colocalizing with E-cadherin basal-laterally. Scribble (Scrib) and Lgl from Lgl/Dlg/Scribble complexes colocalize with E-cadherin at cell–cell contacts and demarcate the basal–lateral region of the blastomeres (Kono, Tamashiro, & Alarcon, 2014; Vinot et al., 2005). The Par complex restricts Lgl and Scrib to the basal–lateral domain, a feature conserved in *Drosophila* epithelial cells. Knockdown of Par complex components or overexpression of dominant-negative aPKC, causes Scrib to expand over the whole cell membrane and the Lgl homologue Llgl1 to be totally lost from the cell membrane (Hirate et al., 2013). Thus, the epithelial polarity in the mouse embryo is maintained by the mutual antagonization of opposing protein complexes (Fig. 4).

5.1 The Par Complex and Associated Proteins

Par proteins have been found to be present and functional in a wide range of contexts, including both embryonic and somatic tissues (Goldstein & Macara, 2007). Par6 and Par3 are scaffolds and aPKC is a signaling kinase. aPKC belongs to the protein kinase C (PKC) family, but unlike other PKC members, its activation is independent of both diacylglycerol and inositol triphosphate (IP_3)-triggered calcium signals. Mammals have several homologues of Par3 and Par6, and two isozymes of aPKC: aPKCζ and aPKCι/λ. When recruited to the apical domain, aPKC is able to phosphorylate many apical proteins (Suzuki & Ohno, 2006). During compaction at the 8-cell stage, Par6 and aPKC are strongly enriched at the apical pole and are also colocalized with F-actin (Anani et al., 2014; Vinot et al., 2005). The third PAR complex member, Par3, initially polarizes apically at the late 8-cell stage (Plusa et al., 2005), but by the blastocyst stage it becomes colocalized with TJs in outside cells and can be found diffused on the membrane of inside cells (Vinot et al., 2005). The Par complex components are important cell-fate regulators. RNAi knockdown of Par3 or aPKC leads to blastomeres losing their apical domain and becoming incorporated into the inside of the embryo (Plusa et al., 2005). This suggests the polarity state of a blastomere can actively instruct its positional allocation within the embryo. Pard6b (a mouse Par6 homologue) is equally important in maintaining the polarization state, it is required to recruit aPKC to the apical domain, without which multiple pathways required to set up cell fate will be affected (Alarcon, 2010).

How does the apical Par complex control transcription from the cell cortex? In recent years, the signaling pathways linking polarity to cell fate have been gradually uncovered. From the 16-cell stage onward, lineage specification is controlled by the transcriptional coactivator Yap (Nishioka et al., 2009). Yap binds the transcription factor Tead4 and activates transcription of its targets, which include many TE genes such as *Cdx2*. *Tead4* is a TE master

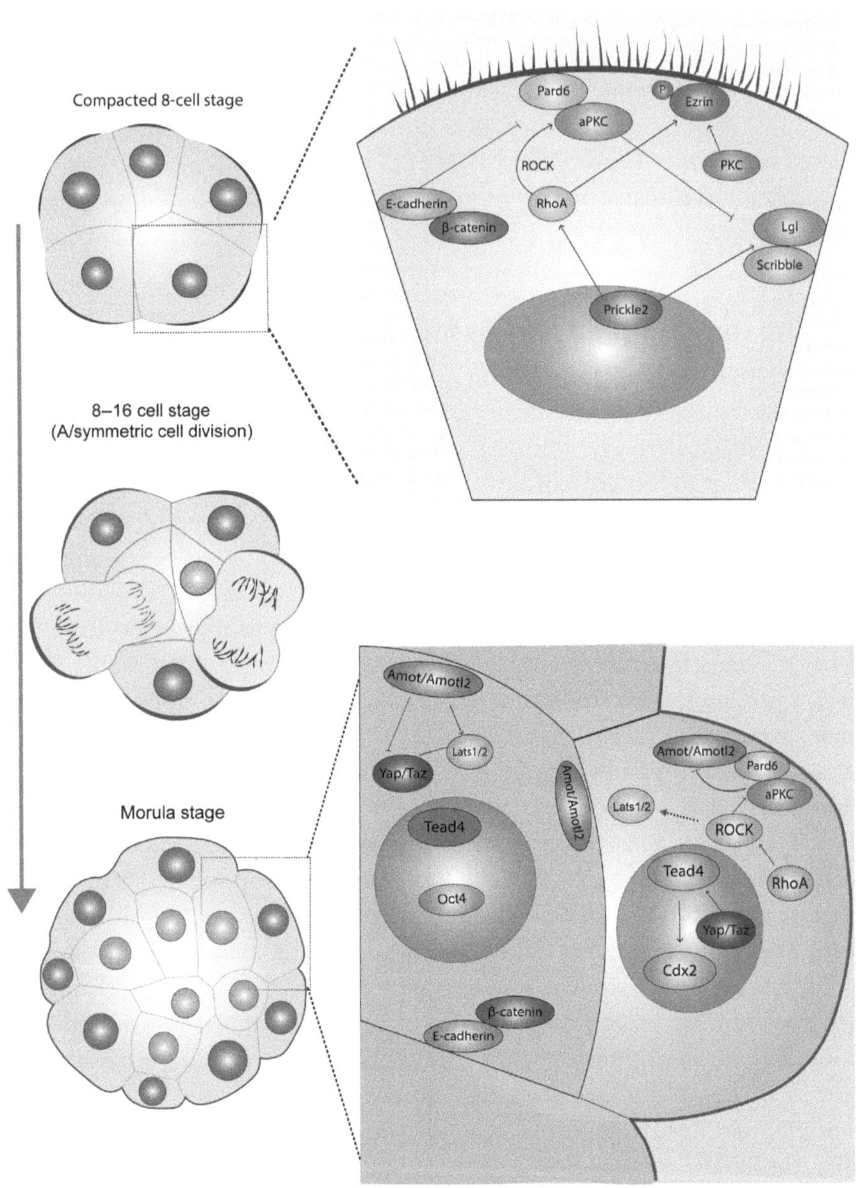

Fig. 4 The molecular landscape of 8-cell stage polarization formation and its role in pre-implantation cell-fate decision. In the late 8-cell stage, Par complex and phosphorylated Ezrin enriched at the apical domain. The Par complex apical localization was antagonized by E-cadherin. RhoA activity is also required for the apical domain formation. Other conserved polarity regulators are also present in the mouse embryo. Polarity is essential in segregating trophectoderm (TE) and inner cell mass (ICM) lineage through regulating the activity of the core hippo pathway regulators. (See the color plate.)

gene that specifies the TE fate (Nishioka et al., 2008; Yagi et al., 2007), but there is evidence that suggest an alternative mechanism for TE fate specification at low oxygen levels (Kaneko & DePamphilis, 2013). Under atmospheric oxygen, polarized outside cells have nuclear Yap and acquire a TE fate, apolar inside cells have cytoplasmic Yap and acquire an ICM fate. In the ICM lineage, Yap is inactivated by the Hippo signaling pathway. The Hippo pathway kinase Lats phosphorylates Yap and leads to its cytoplasmic sequestration (Nishioka et al., 2009). As a result, Tead4 remains inactive and ICM genes are derepressed. Disruption of apical polarity via *Pard6b* knockdown led to an ectopic accumulation of Yap in the cytoplasm in the entire embryo (Hirate et al., 2013). Thus the Hippo pathway, and subsequently, transcription can be controlled by the polarization state. How does the Hippo pathway sense polarity? The missing link was identified to be Angiomotin (Amot). In outside cells, Amot is polarized to the apical regions but in inside cells Amot is present throughout cell–cell contact regions (Hirate et al., 2013; Leung & Zernicka-Goetz, 2013). Functionally, Amot can activate the Hippo pathway and suppress TE fate specification by inhibiting the nuclear translocation of Yap. In the absence of Amot, the Hippo pathway is rendered inactive, Yap shuttles into the nucleus and activate the TE gene program via Tead4 (Hirate et al., 2013; Leung & Zernicka-Goetz, 2013). In summary, asymmetric cell–cell contact leads to an accumulation of the Par complex on the apical membrane and this polarization recruits and disables Amot. Without Amot activity, the Hippo pathway is not activated, Yap enters the nucleus and the TE fate is specified (Fig. 4).

The Par complex also plays a role in PE development. In the mid-blastocyst stage, PE precursors arise from the ICM cells with the expression of early PE marker Gata4. They are initially mixed with the Nanog expressing Epiblast progenitors in a "salt and pepper" pattern (Chazaud, Yamanaka, Pawson, & Rossant, 2006; Guo et al., 2010; Plusa, Piliszek, Frankenberg, Artus, & Hadjantonakis, 2008; Rossant, Chazaud, & Yamanaka, 2003). This heterogeneity is resolved by the migration of PE progenitor cells toward the blastocyst cavity, forming a layer over the ICM (Artus, Piliszek, & Hadjantonakis, 2011; Chazaud et al., 2006; Frankenberg et al., 2011; Meilhac et al., 2009; Morris et al., 2010; Plusa et al., 2008). During the sorting process, aPKC localizes on the membrane of only the Gata4 positive cells and tethered on the surface area of PE when the sorting completes. aPKC is also important for PE sorting and survival. Downregulating aPKC activity largely prevents PE cells from reaching the ICM surface, with some PE cells undergoing apoptosis (Saiz, Grabarek,

Sabherwal, Papalopulu, & Plusa, 2013). Formation of the TE and PE are the first two incidences where the Par complex is critical in establishing polarity and subsequently cell fate. Over the course of development the Par complex plays recurring, essential roles in diverse processes in multiple tissues.

5.2 Junction Assembly and Cell Fate

Formation of mature junctions is a key step in acquiring a polarized epithelial phenotype. Desmosome components, which make up specialized adhesive junctions found between cells, begin to be expressed at the morula stage, forming plaques on intracellular junctions and linking the cytokeratin filaments between neighboring cells (Ducibella et al., 1975; Fleming, Garrod, & Elsmore, 1991). At the same time sodium channels are recruited to the intracellular junctions of the outer TE cells, resulting in the accumulation of liquid within the embryo and the formation of the blastocyst cavity (MacPhee et al., 2000). To contain the blastocyst cavity, TJs create a permeability seal between TE cells (Ducibella et al., 1975; Eckert & Fleming, 2008; Moriwaki, Tsukita, & Furuse, 2007). TJ biogenesis, another hallmark of differentiated epithelial cells, occurs exclusively in the TE and its maturation occurs throughout preimplantation development. TJ assembly is thought to be a manifestation of cell-fate specification and epithelial biogenesis rather than a causal agent (Fleming, McConnell, Johnson, & Stevenson, 1989). However, some TJ components may be required for the maintenance of correct apical–basal polarity and then subsequently cell fate. Occludin is a 4-pass transmembrane protein that is assembled into TJs at the 32-cell stage (Sheth, Moran, Anderson, & Fleming, 2000). Incubating embryos with an antibody targeting occludin disrupts paracellular permeability and inhibits blastocyst cavity formation. Apart from cavity formation, expression of the key pluripotency factor, Oct4, increases while the TE marker H19 becomes barely detectable (Kim, Gye, & Kim, 2004). JAM-1 is another transmembrane TJ component. It is first expressed at the uncompacted 8-cell stage where it is localized basal–laterally at cell–cell contact regions. As with other TJ components, disrupting JAM-1 function affects blastocyst cavity formation (Thomas et al., 2004). Interestingly, JAM-1 localization from becoming apical, colocalizing with the apical microvillus pole, reverts back to basal–lateral at the 16- to 32-cell stage (Thomas et al., 2004). This transient apical localization could point to a role for JAM-1 in recruiting polarity factors to the apical pole, and indeed JAM-1 has been reported to bind Par3 and aPKC in epithelial cells (Ebnet et al., 2001; Itoh et al., 2001).

5.3 Ezrin

Another player in apical polarity is the ERM family protein Ezrin. The ERM protein family consists of ezrin, radixin, and moesin. Together, this family functions as a linkage between the plasma membrane and the actin cytoskeleton, and thus is especially important for cytoskeletal–membrane interactions (Neisch & Fehon, 2011). Like other ERM members, its activation requires the binding of phosphatidylinositol bisphosphate (PiP_2) and also C-terminal phosphorylation by a variety of protein kinases, such as PKC, MEK, and Rho-associated protein kinases (ROCKs) (Zhang et al., 2014). In a variety of epithelial cells including the polarized cells of the early mouse embryo, Ezrin is concentrated at the microvillus pole. Before compaction Ezrin is distributed around the cell cortex but afterward, together with the microvilli, Ezrin becomes restricted to the apical pole (Louvet, Aghion, Santa-Maria, Mangeat, & Maro, 1996). Ezrin polarization actually precedes that of the microvilli and thus may be involved in their formation (Louvet-Vallee, Dard, Santa-Maria, Aghion, & Maro, 2001). Indeed, aPKC-mediated phosphorylation of Ezrin is required for apical microvilli pole formation (Liu et al., 2013). Furthermore, a constitutively active Ezrin leads to a higher number of abnormally long microvilli that are no longer restricted to the apical pole (Dard, Louvet-Vallee, Santa-Maria, & Maro, 2004). When microvilli are no longer restricted, normal cell adhesion and polarization are affected (Dard et al., 2004). This highlights a potential developmental role for Ezrin, where it diverts the highly dynamic microvilli away from cell–cell contact sites, so that mature junctions can form in a stable manner.

5.4 Planar Cell Polarity

Planar cell polarity (PCP) is a common phenomenon that is well studied in many epithelial tissues. Canonical PCP probably does not exist in preimplantation embryo, with the epithelium-like TE constantly differentiating, unlike classic PCP models such as the fully differentiated mammalian epidermis. Despite this, a PCP gene, Prickle2 (Pk2), was found to be an important apical–basal polarity regulator in early mouse embryos. *Pk2* maternal-zygotic mutant embryos die shortly after reaching the early blastocyst stage without forming a blastocyst cavity (Tao et al., 2012). In these mutant embryos, microtubules become short and punctuated around the cell membrane, the actin cytoskeleton is similarly perturbed and polarity components are mislocalized. aPKC and Scrib localization is also perturbed, becoming diffuse in the cytoplasm instead of concentrated on the cell cortex, whereas Llgl1 remains unaffected. With this disturbed polarity, the Hippo pathway is likely to be overactivated, and indeed the embryo takes on an ICM fate and Cdx2 expression is suppressed (Tao et al., 2012). Curiously, Pk2 needs to be nuclear to exert its function (Tao et al., 2012) and it is not clear whether it affects transcription. There is still a lot unknown about how Pk2 affects polarity and cell fate and whether other PCP components also play a role in the early mouse embryo.

5.5 E-Cadherin

E-cadherin is a highly conserved calcium-dependent transmembrane adhesion molecule that plays recurring roles in development. It can be detected on the cell membrane throughout preimplantation. During compaction, E-cadherin is progressively removed from the apical domain and becomes enriched at the site of cell–cell contact. After an asymmetric cell division, E-cadherin is evenly distributed on the cell membrane of inside cells but found only on the basal–lateral domains of the outside cells (Vestweber, Gossler, Boller, & Kemler, 1987). E-cadherin plays a role in compaction, polarization, blastocyst cavity formation, and cell-fate specification (Ao & Erickson, 1992; Larue, Ohsugi, Hirchenhain, & Kemler, 1994; Stephenson et al., 2010). Interestingly, exchanging endogenous E-cadherin with a similar adhesion molecule, N-cadherin, does not affect adhesion and the embryo is able to compact but it fails to form a polarized TE. In contrast, the ICM lineage seems to be unaffected as pluripotent ES cells can be derived (Kan et al., 2007). This genetic replacement experiment demonstrates E-cadherin function spans more than simply mediating adhesion between blastomeres, and it has a unique function in the TE lineage that cannot be substituted by other adhesion molecules.

E-cadherins are an integral part of AJs, which are important in cellular adhesion, and also consist of β-catenin, α-catenin, and p120 linked to the actin cytoskeleton (Hirano, Nose, Hatta, Kawakami, & Takeichi, 1987). E-cadherin maternal-zygotic null embryos do not compact, pointing to its nonredundant role in adhesion (Stephenson et al., 2010). Although these embryos do not form a blastocyst, cell division is unaffected up until the equivalent of the blastocyst stage when the embryo becomes a loose cluster of blastomeres, with aPKC taking over the cortex, the majority of blastomeres are interpreted as "outside cells." Indeed these cells have nuclear Yap, express high levels of Cdx2 and low levels of Oct4 (Stephenson et al., 2010). Therefore, E-cadherin is crucial in establishing and maintaining mature cell junctions, creating a permissive context for the ICM fate and counteracting the polarizing effects of the Par complex. The relationship between cell–cell contact, polarity and cell fate can be further investigated by experimentally isolating 8-cell blastomeres. In these singled out cells, gene expression is characteristic of neither ICM nor TE, but somewhat in between and tending to be more TE-like (Lorthongpanich, Doris, Limviphuvadh, Knowles, & Solter, 2012). Furthermore, singled out blastomeres are not able to polarize correctly and only after division can they regain apical polarity through its cell–cell contact sites. This phenotype is more severe than the *E-cadherin* maternal-zygotic mutants as embryos completely lacking E-cadherin still recruit the Par complex to the membrane, which is not the case for physically isolated blastomeres. Thus, E-cadherin is not the only molecule mediating adhesion in the embryo, and it has been suggested that cadherin 3 could be mediating some residual weak adhesion that could allow the *E-cadherin* mutant to polarize (Lorthongpanich et al., 2012).

5.6 RhoA-ROCK Signaling

The Rho family of GTPases are well-known, important regulators for diverse cellular processes, including cellular polarization and morphogenesis. In the mouse embryo, inhibition of the Rho subclass of GTPases (RhoA, RhoB, and RhoC) disrupted both compaction and concomitantly polarization at the 8-cell stage. Various polarity factors become mislocalized in these embryos, including aPKC, Pard6b, Scrib, Cdh1 (E-cadherin), and Llgl1 (Clayton, Hall, & Johnson, 1999; Kono et al., 2014; Liu et al., 2013). These factors were no longer restricted to their respective domains, both apical and basal factors polarization is lost and their localization became overlapped with each other. This may suggest Rho is required for the antagonistic relationships between the polarity factors. Despite this striking phenotype, it is unclear how Rho family members regulate compaction and polarization. Rho could mediate its effect partly through its downstream effector, the ROCK. Treating embryos with a ROCK inhibitor did not cause any compaction defect but resulted in a very similar, but less severe, polarization phenotype (Kono et al., 2014; Liu et al., 2013). There were no clear differences in the actin and microtubule cytoskeletal structures in embryos treated with ROCK inhibitor, suggesting that the polarity defect may be independent of the cytoskeletons (Kono et al., 2014). The lack of correctly established apical/basal–lateral domains also coincided in a defect in cell fate. ROCK inhibition led to an upregulation of Nanog, a pluripotent gene expressed in the ICM, and the downregulation of Cdx2. Similarly for embryos treated with a Rho inhibitor, Cdx2 expression was decreased, as a result of an overly active Hippo signaling, and the cell-fate phenotype could be reversed by the knockdown of LATS kinase (Kono et al., 2014). These embryos failed to form blastocysts due to defective TJs (Kawagishi et al., 2004; Kono et al., 2014). Depending on the concentration of the Rho inhibitor, it could also prevent cleavage divisions (Duan et al., 2014). Although *Rock1* and *Rock2* mutant embryos form intact blastocysts (Kamijo et al., 2011), the inhibitor treatment phenotype was postulated to be a result of the inhibition of maternally deposited *Rock* mRNA (Kono et al., 2014). Overall these experiments suggest the possibility of a Rho-ROCK signaling cascade for epithelial polarity establishment in the 8-cell embryo. In this hypothetical model, Rho activity triggers ROCK to maintain or setup the correct localization and the antagonistic activity of the Par complex with basal–lateral factors. This allows the Par complex to selectively inhibit the Hippo pathway in outside cells via Amot, thus setting up the TE fate in outside cells and the ICM fate in inside cells (Fig. 4).

6. SUMMARY AND DISCUSSION

The early mouse embryo exhibits two main phases of polarity. One is the early heterogeneity that biases cell fate from the 2- to 4-cell stage, the other is an epithelial-like polarity that is acquired during compaction at the 8-cell stage. Traditionally these polarization events have been investigated and perceived separately, whereas in reality they must be interrelated, as both have an effect on cell fate. Thus far, there is little known about the relationship between early heterogeneities and the later epithelial-like polarity. This is, mostly because the discovery of the molecular nature of early heterogeneity is still new and the practical difficulty of unraveling molecular mechanisms in the early cleavage embryo is very high, compounded by their regulative nature. The mechanism through which early embryos acquire heterogeneity remains still to be discovered; possibilities include transcriptional noise, differential parental inheritance and differential inheritance during cell division, all of which may not be necessarily mutually exclusive. Much more is known about the establishment of epithelial polarity at the 8-cell stage, mainly because it utilizes a similar system to the fully differentiated epithelia and therefore is much easier to study. We now have a roadmap of the main molecular players that connect apical polarity to transcription factors that instruct cell fate. Although this map is far from complete, it delineates a hierarchical signaling cascade that travels from the cell membrane to the DNA in the nucleus. There are many other players not connected to this network, and efforts directed to unify the different molecular regulators to pinpoint their role within this complex process would be very fruitful.

One of the most interesting unsolved question is what triggers the de novo polarization of the mouse embryo at the 8-cell stage. There appears to be two essential prerequisites: cell–cell contact and a timing cue triggered specifically at the 8-cell stage. As described earlier, it is well known that cell–cell contact is required for polarization at the 8-cell stage. Yet cell–cell contact is likely a permissive signaling, as it is insufficient to polarize the cytocortex until the 8-cell stage. Therefore, there must be a more important timing cue at the 8-cell stage that activates the cell–cell contact and triggers polarity. The source of such a timing cue is intriguing and would be one of the key mysteries to resolve in future studies. Indeed, the mouse embryo provides a unique, system-autonomous model to study the mechanism and the timing of de novo polarity establishment.

REFERENCES

Ajduk, A., Biswas Shivhare, S., & Zernicka-Goetz, M. (2014). The basal position of nuclei is one pre-requisite for asymmetric cell divisions in the early mouse embryo. *Developmental Biology*, *392*, 133–140.

Alarcon, V. B. (2010). Cell polarity regulator PARD6B is essential for trophectoderm formation in the preimplantation mouse embryo. *Biology of Reproduction*, *83*, 347–358.

Alarcon, V. B., & Marikawa, Y. (2003). Deviation of the blastocyst axis from the first cleavage plane does not affect the quality of mouse postimplantation development. *Biology of Reproduction*, *69*, 1208–1212.

Alford, L. M., Ng, M. M., & Burgess, D. R. (2009). Cell polarity emerges at first cleavage in sea urchin embryos. *Developmental Biology*, *330*, 12–20.

Anani, S., Bhat, S., Honma-Yamanaka, N., Krawchuk, D., & Yamanaka, Y. (2014). Initiation of Hippo signaling is linked to polarity rather than to cell position in the preimplantation mouse embryo. *Development (Cambridge, England)*, *141*, 2813–2824.

Ao, A., & Erickson, R. P. (1992). Injection of antisense RNA specific for E-cadherin demonstrates that E-cadherin facilitates compaction, the first differentiative step of the mammalian embryo. *Antisense Research and Development*, *2*, 153–163.

Artus, J., Piliszek, A., & Hadjantonakis, A. K. (2011). The primitive endoderm lineage of the mouse blastocyst: Sequential transcription factor activation and regulation of differentiation by Sox17. *Developmental Biology*, *350*, 393–404.

Bowerman, B., & Shelton, C. A. (1999). Cell polarity in the early Caenorhabditis elegans embryo. *Current Opinion in Genetics & Development*, *9*, 390–395.

Burton, A., Muller, J., Tu, S., Padilla-Longoria, P., Guccione, E., & Torres-Padilla, M. E. (2013). Single-cell profiling of epigenetic modifiers identifies PRDM14 as an inducer of cell fate in the mammalian embryo. *Cell Reports*, *5*, 687–701.

Chakravarthy, H., Ormsbee, B. D., Mallanna, S. K., & Rizzino, A. (2011). Rapid activation of the bivalent gene Sox21 requires displacement of multiple layers of gene-silencing machinery. *FASEB Journal*, *25*, 206–218.

Chazaud, C., Yamanaka, Y., Pawson, T., & Rossant, J. (2006). Early lineage segregation between epiblast and primitive endoderm in mouse blastocysts through the Grb2-MAPK pathway. *Developmental Cell*, *10*, 615–624.

Chen, Q., Yan, M., Cao, Z., Li, X., Zhang, Y., Shi, J., et al. (2016). Sperm tsRNAs contribute to intergenerational inheritance of an acquired metabolic disorder. *Science*, *351*, 397–400.

Chisholm, J. C., & Houliston, E. (1987). Cytokeratin filament assembly in the preimplantation mouse embryo. *Development*, *101*, 565–582.

Christophorou, M. A., Castelo-Branco, G., Halley-Stott, R. P., Oliveira, C. S., Loos, R., Radzisheuskaya, A., et al. (2014). Citrullination regulates pluripotency and histone H1 binding to chromatin. *Nature*, *507*, 104–108.

Chroscicka, A., Komorowski, S., & Maleszewski, M. (2004). Both blastomeres of the mouse 2-cell embryo contribute to the embryonic portion of the blastocyst. *Molecular Reproduction and Development*, *68*, 308–312.

Clayton, L., Hall, A., & Johnson, M. H. (1999). A role for Rho-like GTPases in the polarisation of mouse eight-cell blastomeres. *Developmental Biology*, *205*, 322–331.

Dard, N., Louvet-Vallee, S., Santa-Maria, A., & Maro, B. (2004). Phosphorylation of ezrin on threonine T567 plays a crucial role during compaction in the mouse early embryo. *Developmental Biology*, *271*, 87–97.

De Brabander, M. J., Van De Veire, R. M., Aerts, F. E., Borgers, M., & Janssen, P. A. (1976). The effects of methyl (5-(2-thienylcarbonyl)-1H-benzimidazol-2-yl) carbamate, (R 17934; NSC 238159), a new synthetic antitumoral drug interfering with microtubules, on mammalian cells cultured in vitro. *Cancer Research*, *36*, 905–916.

Duan, X., Chen, K. L., Zhang, Y., Cui, X. S., Kim, N. H., & Sun, S. C. (2014). ROCK inhibition prevents early mouse embryo development. *Histochemistry and Cell Biology*, *142*, 227–233.

Ducibella, T., Albertini, D. F., Anderson, E., & Biggers, J. D. (1975). The preimplantation mammalian embryo: Characterization of intercellular junctions and their appearance during development. *Developmental Biology*, *45*, 231–250.

Ducibella, T., & Anderson, E. (1975). Cell shape and membrane changes in the eight-cell mouse embryo: Prerequisites for morphogenesis of the blastocyst. *Developmental Biology*, *47*, 45–58.

Ebnet, K., Suzuki, A., Horikoshi, Y., Hirose, T., Meyer Zu Brickwedde, M. K., Ohno, S., et al. (2001). The cell polarity protein ASIP/PAR-3 directly associates with junctional adhesion molecule (JAM). *EMBO Journal*, *20*, 3738–3748.

Eckert, J. J., & Fleming, T. P. (2008). Tight junction biogenesis during early development. *Biochimica et Biophysica Acta*, *1778*, 717–728.

Etemad-Moghadam, B., Guo, S., & Kemphues, K. J. (1995). Asymmetrically distributed PAR-3 protein contributes to cell polarity and spindle alignment in early C. elegans embryos. *Cell*, *83*, 743–752.

Fabian, D., Cikos, S., Rehak, P., & Koppel, J. (2014). Do embryonic polar bodies commit suicide? *Zygote*, *22*, 10–17.

Fernandes, R., Tsuda, C., Perumalsamy, A. L., Naranian, T., Chong, J., Acton, B. M., et al. (2012). NLRP5 mediates mitochondrial function in mouse oocytes and embryos. *Biology of Reproduction*, *86*(138), 1–10.

Fleming, T. P., Cannon, P. M., & Pickering, S. J. (1986). The cytoskeleton, endocytosis and cell polarity in the mouse preimplantation embryo. *Developmental Biology*, *113*, 406–419.

Fleming, T. P., Garrod, D. R., & Elsmore, A. J. (1991). Desmosome biogenesis in the mouse preimplantation embryo. *Development*, *112*, 527–539.

Fleming, T. P., & Johnson, M. H. (1988). From egg to epithelium. *Annual Review of Cell Biology, 4*, 459–485.
Fleming, T. P., McConnell, J., Johnson, M. H., & Stevenson, B. R. (1989). Development of tight junctions de novo in the mouse early embryo: Control of assembly of the tight junction-specific protein, ZO-1. *Journal of Cell Biology, 108*, 1407–1418.
Fleming, T. P., & Pickering, S. J. (1985). Maturation and polarization of the endocytotic system in outside blastomeres during mouse preimplantation development. *Journal of Embryology and Experimental Morphology, 89*, 175–208.
Fleming, T. P., Pickering, S. J., Qasim, F., & Maro, B. (1986). The generation of cell surface polarity in mouse 8-cell blastomeres: The role of cortical microfilaments analysed using cytochalasin D. *Journal of Embryology and Experimental Morphology, 95*, 169–191.
Frankenberg, S., Gerbe, F., Bessonnard, S., Belville, C., Pouchin, P., Bardot, O., et al. (2011). Primitive endoderm differentiates via a three-step mechanism involving Nanog and RTK signaling. *Developmental Cell, 21*, 1005–1013.
Gardner, R. L. (1997). The early blastocyst is bilaterally symmetrical and its axis of symmetry is aligned with the animal-vegetal axis of the zygote in the mouse. *Development, 124*, 289–301.
Goke, J., Jung, M., Behrens, S., Chavez, L., O'Keeffe, S., Timmermann, B., et al. (2011). Combinatorial binding in human and mouse embryonic stem cells identifies conserved enhancers active in early embryonic development. *PLoS Computational Biology, 7*, e1002304.
Goldstein, B., & Macara, I. G. (2007). The PAR proteins: Fundamental players in animal cell polarization. *Developmental Cell, 13*, 609–622.
Goolam, M., Scialdone, A., Graham, S. J., Macaulay, I. C., Jedrusik, A., Hupalowska, A., et al. (2016). Heterogeneity in Oct4 and Sox2 targets biases cell fate in 4-cell mouse embryos. *Cell, 165*, 61–74.
Guethhallonet, C., & Maro, B. (1992). Cell polarity and cell diversification during early mouse embryogenesis. *Trends in Genetics, 8*, 274–279.
Guo, G., Huss, M., Tong, G. Q., Wang, C., Li Sun, L., Clarke, N. D., et al. (2010). Resolution of cell fate decisions revealed by single-cell gene expression analysis from zygote to blastocyst. *Developmental Cell, 18*, 675–685.
Handyside, A. H. (1980). Distribution of antibody- and lectin-binding sites on dissociated blastomeres from mouse morulae: Evidence for polarization at compaction. *Journal of Embryology and Experimental Morphology, 60*, 99–116.
Hiiragi, T., & Solter, D. (2004). First cleavage plane of the mouse egg is not predetermined but defined by the topology of the two apposing pronuclei. *Nature, 430*, 360–364.
Hirano, S., Nose, A., Hatta, K., Kawakami, A., & Takeichi, M. (1987). Calcium-dependent cell-cell adhesion molecules (cadherins): Subclass specificities and possible involvement of actin bundles. *Journal of Cell Biology, 105*, 2501–2510.
Hirate, Y., Hirahara, S., Inoue, K., Suzuki, A., Alarcon, V. B., Akimoto, K., et al. (2013). Polarity-dependent distribution of angiomotin localizes Hippo signaling in preimplantation embryos. *Current Biology, 23*, 1181–1194.
Houliston, E., & Maro, B. (1989). Posttranslational modification of distinct microtubule subpopulations during cell polarization and differentiation in the mouse preimplantation embryo. *Journal of Cell Biology, 108*, 543–551.
Houliston, E., Pickering, S. J., & Maro, B. (1987). Redistribution of microtubules and pericentriolar material during the development of polarity in mouse blastomeres. *Journal of Cell Biology, 104*, 1299–1308.
Itoh, M., Sasaki, H., Furuse, M., Ozaki, H., Kita, T., & Tsukita, S. (2001). Junctional adhesion molecule (JAM) binds to PAR-3: A possible mechanism for the recruitment of PAR-3 to tight junctions. *Journal of Cell Biology, 154*, 491–497.
Johnson, M. H., & Maro, B. (1984). The distribution of cytoplasmic actin in mouse 8-cell blastomeres. *Journal of Embryology and Experimental Morphology, 82*, 97–117.

Johnson, M. H., & Maro, B. (1985). A dissection of the mechanisms generating and stabilizing polarity in mouse 8- and 16-cell blastomeres: The role of cytoskeletal elements. *Journal of Embryology and Experimental Morphology*, *90*, 311–334.
Johnson, M. H., Pickering, S. J., Dhiman, A., Radcliffe, G. S., & Maro, B. (1988). Cytocortical organization during natural and prolonged mitosis of mouse 8-cell blastomeres. *Development*, *102*, 143–158.
Johnson, M. H., & Ziomek, C. A. (1981a). The foundation of two distinct cell lineages within the mouse morula. *Cell*, *24*, 71–80.
Johnson, M. H., & Ziomek, C. A. (1981b). Induction of polarity in mouse 8-cell blastomeres: Specificity, geometry, and stability. *Journal of Cell Biology*, *91*, 303–308.
Kamijo, H., Matsumura, Y., Thumkeo, D., Koike, S., Masu, M., Shimizu, Y., et al. (2011). Impaired vascular remodeling in the yolk sac of embryos deficient in ROCK-I and ROCK-II. *Genes to Cells*, *16*, 1012–1021.
Kan, N. G., Stemmler, M. P., Junghans, D., Kanzler, B., De Vries, W. N., Dominis, M., et al. (2007). Gene replacement reveals a specific role for E-cadherin in the formation of a functional trophectoderm. *Development*, *134*, 31–41.
Kaneko, K. J., & Depamphilis, M. L. (2013). TEAD4 establishes the energy homeostasis essential for blastocoel formation. *Development*, *140*, 3680–3690.
Kawagishi, R., Tahara, M., Sawada, K., Ikebuchi, Y., Morishige, K., Sakata, M., et al. (2004). Rho-kinase is involved in mouse blastocyst cavity formation. *Biochemical and Biophysical Research Communications*, *319*, 643–648.
Kim, J., Gye, M. C., & Kim, M. K. (2004). Role of occludin, a tight junction protein, in blastocoel formation, and in the paracellular permeability and differentiation of trophectoderm in preimplantation mouse embryos. *Molecules and Cells*, *17*, 248–254.
Kim, J., Lee, J., Yadav, N., Wu, Q., Carter, C., Richard, S., et al. (2004). Loss of CARM1 results in hypomethylation of thymocyte cyclic AMP-regulated phosphoprotein and deregulated early T cell development. *Journal of Biological Chemistry*, *279*, 25339–25344.
Kim, B., Zhang, X., Kan, R., Cohen, R., Mukai, C., Travis, A. J., et al. (2014). The role of MATER in endoplasmic reticulum distribution and calcium homeostasis in mouse oocytes. *Developmental Biology*, *386*, 331–339.
Kiso, M., Tanaka, S., Saba, R., Matsuda, S., Shimizu, A., Ohyama, M., et al. (2009). The disruption of Sox21-mediated hair shaft cuticle differentiation causes cyclic alopecia in mice. *Proceedings of the National Academy of Sciences of the United States of America*, *106*, 9292–9297.
Kono, K., Tamashiro, D. A., & Alarcon, V. B. (2014). Inhibition of RHO-ROCK signaling enhances ICM and suppresses TE characteristics through activation of Hippo signaling in the mouse blastocyst. *Developmental Biology*, *394*, 142–155.
Kuzmichev, A. N., Kim, S. K., D'Alessio, A. C., Chenoweth, J. G., Wittko, I. M., Campanati, L., et al. (2012). Sox2 acts through Sox21 to regulate transcription in pluripotent and differentiated cells. *Current Biology*, *22*, 1705–1710.
Larue, L., Ohsugi, M., Hirchenhain, J., & Kemler, R. (1994). E-cadherin null mutant embryos fail to form a trophectoderm epithelium. *Proceedings of the National Academy of Sciences of the United States of America*, *91*, 8263–8267.
Ledizet, M., & Piperno, G. (1986). Cytoplasmic microtubules containing acetylated alpha-tubulin in Chlamydomonas reinhardtii: Spatial arrangement and properties. *Journal of Cell Biology*, *103*, 13–22.
Lehtonen, E., & Reima, I. (1986). Changes in the distribution of vinculin during preimplantation mouse development. *Differentiation*, *32*, 125–134.
Leung, C. Y., & Zernicka-Goetz, M. (2013). Angiomotin prevents pluripotent lineage differentiation in mouse embryos via Hippo pathway-dependent and -independent mechanisms. *Nature Communications*, *4*, 2251.

Liu, H., Wu, Z., Shi, X., Li, W., Liu, C., Wang, D., et al. (2013). Atypical PKC, regulated by Rho GTPases and Mek/Erk, phosphorylates Ezrin during eight-cell embryocompaction. *Developmental Biology*, *375*, 13–22.

Lorthongpanich, C., Doris, T. P., Limviphuvadh, V., Knowles, B. B., & Solter, D. (2012). Developmental fate and lineage commitment of singled mouse blastomeres. *Development*, *139*, 3722–3731.

Louvet, S., Aghion, J., Santa-Maria, A., Mangeat, P., & Maro, B. (1996). Ezrin becomes restricted to outer cells following asymmetrical division in the preimplantation mouse embryo. *Developmental Biology*, *177*, 568–579.

Louvet-Vallee, S., Dard, N., Santa-Maria, A., Aghion, J., & Maro, B. (2001). A major post-translational modification of ezrin takes place during epithelial differentiation in the early mouse embryo. *Developmental Biology*, *231*, 190–200.

Louvet-Vallee, S., Vinot, S., & Maro, B. (2005). Mitotic spindles and cleavage planes are oriented randomly in the two-cell mouse embryo. *Current Biology*, *15*, 464–469.

Macphee, D. J., Jones, D. H., Barr, K. J., Betts, D. H., Watson, A. J., & Kidder, G. M. (2000). Differential involvement of Na(+), K(+)-ATPase isozymes in preimplantation development of the mouse. *Developmental Biology*, *222*, 486–498.

Mallanna, S. K., Ormsbee, B. D., Iacovino, M., Gilmore, J. M., Cox, J. L., Kyba, M., et al. (2010). Proteomic analysis of Sox2-associated proteins during early stages of mouse embryonic stem cell differentiation identifies Sox21 as a novel regulator of stem cell fate. *Stem Cells*, *28*, 1715–1727.

Maruta, H., Greer, K., & Rosenbaum, J. L. (1986). The acetylation of alpha-tubulin and its relationship to the assembly and disassembly of microtubules. *Journal of Cell Biology*, *103*, 571–579.

Meilhac, S. M., Adams, R. J., Morris, S. A., Danckaert, A., Le Garrec, J. F., & Zernicka-Goetz, M. (2009). Active cell movements coupled to positional induction are involved in lineage segregation in the mouse blastocyst. *Developmental Biology*, *331*, 210–221.

Moriwaki, K., Tsukita, S., & Furuse, M. (2007). Tight junctions containing claudin 4 and 6 are essential for blastocyst formation in preimplantation mouse embryos. *Developmental Biology*, *312*, 509–522.

Morris, S. A., Guo, Y., & Zernicka-Goetz, M. (2012). Developmental plasticity is bound by pluripotency and the Fgf and Wnt signaling pathways. *Cell Reports*, *2*, 756–765.

Morris, S. A., Teo, R. T., Li, H., Robson, P., Glover, D. M., & Zernicka-Goetz, M. (2010). Origin and formation of the first two distinct cell types of the inner cell mass in the mouse embryo. *Proceedings of the National Academy of Sciences of the United States of America*, *107*, 6364–6369.

Motosugi, N., Bauer, T., Polanski, Z., Solter, D., & Hiiragi, T. (2005). Polarity of the mouse embryo is established at blastocyst and is not prepatterned. *Genes & Development*, *19*, 1081–1092.

Muller, H. A. J., & Hausen, P. (1995). Epithelial-cell polarity in early Xenopus development. *Developmental Dynamics*, *202*, 405–420.

Narducci, M. G., Fiorenza, M. T., Kang, S. M., Bevilacqua, A., Di Giacomo, M., Remotti, D., et al. (2002). TCL1 participates in early embryonic development and is overexpressed in human seminomas. *Proceedings of the National Academy of Sciences of the United States of America*, *99*, 11712–11717.

Neisch, A. L., & Fehon, R. G. (2011). Ezrin, Radixin and Moesin: Key regulators of membrane-cortex interactions and signaling. *Current Opinion in Cell Biology*, *23*, 377–382.

Nishioka, N., Inoue, K., Adachi, K., Kiyonari, H., Ota, M., Ralston, A., et al. (2009). The Hippo signaling pathway components Lats and Yap pattern Tead4 activity to distinguish mouse trophectoderm from inner cell mass. *Developmental Cell*, *16*, 398–410.

Nishioka, N., Yamamoto, S., Kiyonari, H., Sato, H., Sawada, A., Ota, M., et al. (2008). Tead4 is required for specification of trophectoderm in pre-implantation mouse embryos. *Mechanisms of Development, 125*, 270–283.

Ohsugi, M., Zheng, P., Baibakov, B., Li, L., & Dean, J. (2008). Maternally derived FILIA-MATER complex localizes asymmetrically in cleavage-stage mouse embryos. *Development, 135*, 259–269.

Piotrowska, K., & Zernicka-Goetz, M. (2001). Role for sperm in spatial patterning of the early mouse embryo. *Nature, 409*, 517–521.

Piotrowska-Nitsche, K., Perea-Gomez, A., Haraguchi, S., & Zernicka-Goetz, M. (2005). Four-cell stage mouse blastomeres have different developmental properties. *Development, 132*, 479–490.

Piotrowska-Nitsche, K., & Zernicka-Goetz, M. (2005). Spatial arrangement of individual 4-cell stage blastomeres and the order in which they are generated correlate with blastocyst pattern in the mouse embryo. *Mechanisms of Development, 122*, 487–500.

Piperno, G., Ledizet, M., & Chang, X. J. (1987). Microtubules containing acetylated alpha-tubulin in mammalian cells in culture. *Journal of Cell Biology, 104*, 289–302.

Plusa, B., Frankenberg, S., Chalmers, A., Hadjantonakis, A. K., Moore, C. A., Papalopulu, N., et al. (2005). Downregulation of Par3 and aPKC function directs cells towards the ICM in the preimplantation mouse embryo. *Journal of Cell Science, 118*, 505–515.

Plusa, B., Grabarek, J. B., Piotrowska, K., Glover, D. M., & Zernicka-Goetz, M. (2002). Site of the previous meiotic division defines cleavage orientation in the mouse embryo. *Nature Cell Biology, 4*, 811–815.

Plusa, B., Piliszek, A., Frankenberg, S., Artus, J., & Hadjantonakis, A. K. (2008). Distinct sequential cell behaviours direct primitive endoderm formation in the mouse blastocyst. *Development, 135*, 3081–3091.

Reeve, W. J. (1981). Cytoplasmic polarity develops at compaction in rat and mouse embryos. *Journal of Embryology and Experimental Morphology, 62*, 351–367.

Reeve, W. J., & Kelly, F. P. (1983). Nuclear position in the cells of the mouse early embryo. *Journal of Embryology and Experimental Morphology, 75*, 117–139.

Reeve, W. J., & Ziomek, C. A. (1981). Distribution of microvilli on dissociated blastomeres from mouse embryos: Evidence for surface polarization at compaction. *Journal of Embryology and Experimental Morphology, 62*, 339–350.

Reima, I., & Lehtonen, E. (1985). Localization of nonerythroid spectrin and actin in mouse oocytes and preimplantation embryos. *Differentiation, 30*, 68–75.

Rossant, J., Chazaud, C., & Yamanaka, Y. (2003). Lineage allocation and asymmetries in the early mouse embryo. *Philosophical Transactions of the Royal Society of London. Series B, Biological Sciences, 358*, 1341–1348. discussion 1349.

Saiz, N., Grabarek, J. B., Sabherwal, N., Papalopulu, N., & Plusa, B. (2013). Atypical protein kinase C couples cell sorting with primitive endoderm maturation in the mouse blastocyst. *Development, 140*, 4311–4322.

Schulz, L. C., & Roberts, R. M. (2011). Dynamic changes in leptin distribution in the progression from ovum to blastocyst of the pre-implantation mouse embryo. *Reproduction, 141*, 767–777.

Sekiguchi, S., Kwon, J., Yoshida, E., Hamasaki, H., Ichinose, S., Hideshima, M., et al. (2006). Localization of ubiquitin C-terminal hydrolase L1 in mouse ova and its function in the plasma membrane to block polyspermy. *American Journal of Pathology, 169*, 1722–1729.

Sheth, B., Moran, B., Anderson, J. M., & Fleming, T. P. (2000). Post-translational control of occludin membrane assembly in mouse trophectoderm: A mechanism to regulate timing of tight junction biogenesis and blastocyst formation. *Development, 127*, 831–840.

Shi, J., Chen, Q., Li, X., Zheng, X., Zhang, Y., Qiao, J., et al. (2015). Dynamic transcriptional symmetry-breaking in pre-implantation mammalian embryo development revealed by single-cell RNA-seq. *Development*, *142*, 3468–3477.

Skamagki, M., Wicher, K. B., Jedrusik, A., Ganguly, S., & Zernicka-Goetz, M. (2013). Asymmetric localization of Cdx2 mRNA during the first cell-fate decision in early mouse development. *Cell Reports*, *3*, 442–457.

Stephenson, R. O., Yamanaka, Y., & Rossant, J. (2010). Disorganized epithelial polarity and excess trophectoderm cell fate in preimplantation embryos lacking E-cadherin. *Development*, *137*, 3383–3391.

Suzuki, A., & Ohno, S. (2006). The PAR-aPKC system: Lessons in polarity. *Journal of Cell Science*, *119*, 979–987.

Tabansky, I., Lenarcic, A., Draft, R. W., Loulier, K., Keskin, D. B., Rosains, J., et al. (2013). Developmental bias in cleavage-stage mouse blastomeres. *Current Biology*, *23*, 21–31.

Tao, H., Inoue, K., Kiyonari, H., Bassuk, A. G., Axelrod, J. D., Sasaki, H., et al. (2012). Nuclear localization of Prickle2 is required to establish cell polarity during early mouse embryogenesis. *Developmental Biology*, *364*, 138–148.

Tarkowski, A. K. (1959). Experiments on the development of isolated blastomers of mouse eggs. *Nature*, *184*, 1286–1287.

Thomas, F. C., Sheth, B., Eckert, J. J., Bazzoni, G., Dejana, E., & Fleming, T. P. (2004). Contribution of JAM-1 to epithelial differentiation and tight-junction biogenesis in the mouse preimplantation embryo. *Journal of Cell Science*, *117*, 5599–5608.

To, C., Kulkarni, S., Pawson, T., Honda, T., Gribble, G. W., Sporn, M. B., et al. (2008). The synthetic triterpenoid 2-cyano-3,12-dioxooleana-1,9-dien-28-oic acid-imidazolide alters transforming growth factor beta-dependent signaling and cell migration by affecting the cytoskeleton and the polarity complex. *Journal of Biological Chemistry*, *283*, 11700–11713.

Tong, Z. B., Gold, L., De Pol, A., Vanevski, K., Dorward, H., Sena, P., et al. (2004). Developmental expression and subcellular localization of mouse MATER, an oocyte-specific protein essential for early development. *Endocrinology*, *145*, 1427–1434.

Tong, Z. B., Gold, L., Pfeifer, K. E., Dorward, H., Lee, E., Bondy, C. A., et al. (2000). Mater, a maternal effect gene required for early embryonic development in mice. *Nature Genetics*, *26*, 267–268.

Torres-Padilla, M. E., Parfitt, D. E., Kouzarides, T., & Zernicka-Goetz, M. (2007). Histone arginine methylation regulates pluripotency in the early mouse embryo. *Nature*, *445*, 214–218.

Tsunoda, Y., & McLaren, A. (1983). Effect of various procedures on the viability of mouse embryos containing half the normal number of blastomeres. *Journal of Reproduction and Fertility*, *69*, 315–322.

Vestweber, D., Gossler, A., Boller, K., & Kemler, R. (1987). Expression and distribution of cell adhesion molecule uvomorulin in mouse preimplantation embryos. *Developmental Biology*, *124*, 451–456.

Vinot, S., Le, T., Ohno, S., Pawson, T., Maro, B., & Louvet-Vallee, S. (2005). Asymmetric distribution of PAR proteins in the mouse embryo begins at the 8-cell stage during compaction. *Developmental Biology*, *282*, 307–319.

Winkel, G. K., Ferguson, J. E., Takeichi, M., & Nuccitelli, R. (1990). Activation of protein kinase-C triggers premature compaction in the 4-cell stage mouse embryo. *Developmental Biology*, *138*, 1–15.

Winston, N. J., Johnson, M. H., McConnell, J. M., Cook, D. I., & Day, M. L. (2004). Expression and role of the ether-a-go-go-related (MERG1A) potassium-channel protein during preimplantation mouse development. *Biology of Reproduction*, *70*, 1070–1079.

Yadav, N., Cheng, D., Richard, S., Morel, M., Iyer, V. R., Aldaz, C. M., et al. (2008). CARM1 promotes adipocyte differentiation by coactivating PPARgamma. *EMBO Reports*, *9*, 193–198.

Yadav, N., Lee, J., Kim, J., Shen, J., Hu, M. C., Aldaz, C. M., et al. (2003). Specific protein methylation defects and gene expression perturbations in coactivator-associated arginine methyltransferase 1-deficient mice. *Proceedings of the National Academy of Sciences of the United States of America*, *100*, 6464–6468.

Yagi, R., Kohn, M. J., Karavanova, I., Kaneko, K. J., Vullhorst, D., Depamphilis, M. L., et al. (2007). Transcription factor TEAD4 specifies the trophectoderm lineage at the beginning of mammalian development. *Development*, *134*, 3827–3836.

Yamaji, M., Seki, Y., Kurimoto, K., Yabuta, Y., Yuasa, M., Shigeta, M., et al. (2008). Critical function of Prdm14 for the establishment of the germ cell lineage in mice. *Nature Genetics*, *40*, 1016–1022.

Yurttas, P., Vitale, A. M., Fitzhenry, R. J., Cohen-Gould, L., Wu, W., Gossen, J. A., et al. (2008). Role for PADI6 and the cytoplasmic lattices in ribosomal storage in oocytes and translational control in the early mouse embryo. *Development*, *135*, 2627–2636.

Zernicka-Goetz, M. (1998). Fertile offspring derived from mammalian eggs lacking either animal or vegetal poles. *Development*, *125*, 4803–4808.

Zhang, Y., Kwon, S., Yamaguchi, T., Cubizolles, F., Rousseaux, S., Kneissel, M., et al. (2008). Mice lacking histone deacetylase 6 have hyperacetylated tubulin but are viable and develop normally. *Molecular and Cellular Biology*, *28*, 1688–1701.

Zhang, C., Wu, Y., Xuan, Z., Zhang, S., Wang, X., Hao, Y., et al. (2014). p38MAPK, Rho/ROCK and PKC pathways are involved in influenza-induced cytoskeletal rearrangement and hyperpermeability in PMVEC via phosphorylating ERM. *Virus Research*, *192*, 6–15.

Zheng, Z., Li, H., Zhang, Q., Yang, L., & Qi, H. (2016). Unequal distribution of 16S mtrRNA at the 2-cell stage regulates cell lineage allocations in mouse embryos. *Reproduction*, *151*, 351–367.

CHAPTER SEVEN

Mouse Embryo Compaction

M.D. White*, S. Bissiere*, Y.D. Alvarez†, N. Plachta*,[1]
*Institute of Molecular and Cell Biology, A*STAR, Singapore, Singapore
†Facultad de Ciencias Exactas y Naturales, Universidad de Buenos Aires, CONICET, Buenos Aires, Argentina
[1]Corresponding author: e-mail address: plachtan@imcb.a-star.edu.sg

Contents

Abstract

Compaction is a critical first morphological event in the preimplantation development of the mammalian embryo. Characterized by the transformation of the embryo from a loose cluster of spherical cells into a tightly packed mass, compaction is a key step in the establishment of the first tissue-like structures of the embryo. Although early investigation of the mechanisms driving compaction implicated changes in cell–cell adhesion, recent work has identified essential roles for cortical tension and a compaction-specific class of filopodia. During the transition from 8 to 16 cells, as the embryo is compacting, it must also make fundamental decisions regarding cell position, polarity, and fate. Understanding how these and other processes are integrated with compaction requires further investigation. Emerging imaging-based techniques that enable quantitative analysis from the level of cell–cell interactions down to the level of individual regulatory molecules will provide a greater understanding of how compaction shapes the early mammalian embryo.

Current Topics in Developmental Biology, Volume 120
ISSN 0070-2153
http://dx.doi.org/10.1016/bs.ctdb.2016.04.005

1. THE FIRST CHANGE IN CELL SHAPE DURING DEVELOPMENT

Compaction is the first identifiable morphogenetic process during mammalian embryogenesis and it is critical for the divergence of cell lineages and subsequent development. During the transition from 8 to 16 cells, there is a striking morphological change as cells flatten against each other, increasing their contact areas and making the embryo more spherical (Calarco & Brown, 1969; Ducibella, Ukena, Karnovsky, & Anderson, 1977). The change in cell shape is so evident that compacted embryos can be easily distinguished from noncompacted ones using a simple light microscope (Fig. 1).

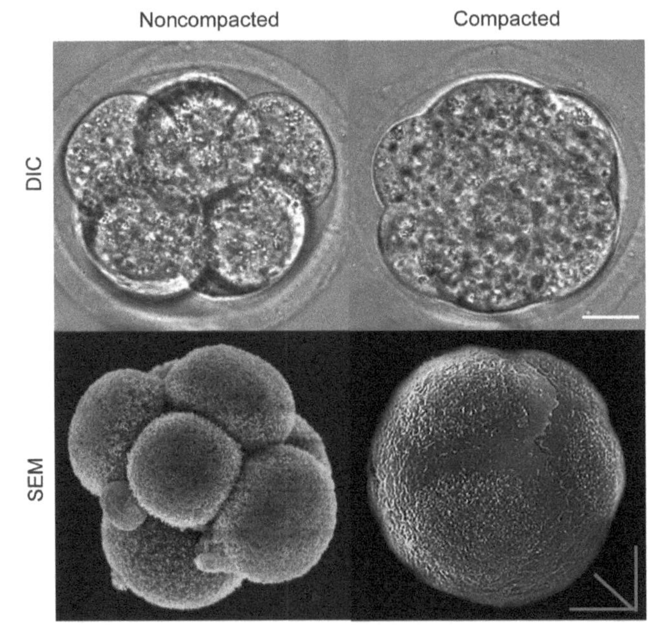

Fig. 1 Morphological changes in preimplantation mouse embryos undergoing compaction. Membranes of individual blastomeres are clearly delineated under DIC optics in a noncompacted 8-cell mouse embryo (*top left panel*). Scanning electron microscopy reveals a uniform distribution of microvilli across all cell surfaces and blastomeres are relatively spherical (*bottom left panel*). Individual cell membranes are no longer discernible by DIC in the compacted embryo (*top right panel*). The blastomeres have flattened and the microvilli are localized to apical zones surrounded by smooth membrane at cell–cell junctions (*bottom right panel*). Scale bars: 10 μm in *x*, *y* and 15 μm in *x*, *y*, *z*.

Embryos that fail to compact will eventually arrest and the degree of compaction is positively correlated with increased success of in vitro fertilization (IVF) treatments (Le Cruguel et al., 2013; Tao et al., 2002). Since the first live birth from a human embryo generated using IVF in 1978 (Steptoe & Edwards, 1978), more than five million babies have been born as a result of this technology. Use of assisted reproductive technologies is increasing worldwide and 80% of today's IVF babies have been born since 2000 (Adamson, Tabangin, Macaluso, & de Mouzon, 2013). Transferring multiple embryos during IVF is associated with adverse pregnancy outcomes (Luke et al., 2015; Practice Committee of American Society for Reproductive Medicine, 2012) so it is crucial to identify and understand the mechanisms that produce the most viable embryo. In addition, beyond the biomedical implications, compaction also provides a unique window into how mammalian cells polarize and interact with each other to form tissue-like structures in vivo.

Mammalian life begins with a newly fertilized zygote, which is round and enclosed within a protective coat of glycoproteins called the zona pellucida. The zygote then undergoes three rounds of cleavage division, from one to two cells, two to four cells, and four to eight cells, producing progressively smaller cells, or blastomeres, after each division. The blastomeres are relatively round and identical in appearance until the late 8-cell stage. Compaction then begins as cell membranes and organelles separate into apical and basolateral domains to establish cell polarity (Fleming & Pickering, 1985; Handyside, 1980; Johnson & McConnell, 2004; Johnson & Ziomek, 1981a; Maro, Johnson, Pickering, & Louvard, 1985; Reeve, 1981; Reeve & Ziomek, 1981). Cell contacts extend outward toward the surface of the embryo minimizing intercellular spaces and placing the cell membranes in close apposition (Goodall & Johnson, 1984; Lo & Gilula, 1979). The blastomeres deform and flatten their apical surfaces forming a tightly grouped mass of cells with indistinct cell boundaries (Fig. 1).

Embryo compaction is critical for blastocyst formation and subsequent development. As the embryo transitions from 8 to 16 cells, the first spatial segregation of cells occurs, with most cells positioned on the exterior of the embryo and a few interior cells completely enclosed. Blastocyst formation is characterized by cavitation and further differentiation of inner and outer cells. The close membrane apposition that arises during compaction likely facilitates the assembly of tight junctions that form a permeability seal between adjacent outer cells. During the 16- to 32-cell stage, the outer cells pump fluid into the intercellular spaces, which coalesce to form a fluid-filled blastocoel. Blastomeres in the outer layer remain polarized and differentiate into the trophectoderm (TE), eventually giving rise to the fetal part of the placenta. The enclosed blastomeres lose their apical features (Johnson & Ziomek, 1983) and form the pluripotent inner cell mass (ICM), from which the embryo and remaining supportive tissues are derived (Dyce, George, Goodall, & Fleming, 1987).

Due to their similarities in preimplantation development, the mouse has been used as a model for the early human embryo for over 45 years (Cockburn & Rossant, 2010). Although preimplantation development naturally occurs within the oviduct, it can be recapitulated in vitro without adversely affecting the developmental potential of embryos (Summers & Biggers, 2003). Mouse embryos can be easily removed from the maternal oviducts and cultured in simple media conditions. Under these ex utero conditions the embryos develop almost as rapidly as they do in utero and if transferred back to the uterus they can implant and produce viable offspring.

In addition, the relatively large cells of the mouse embryo allow imaging of subcellular processes. Pronuclear microinjection of mRNA or DNA is a well-established technique for expression of exogenous proteins and mouse embryos can endure this process with high efficiency. Furthermore, many genetic tools are available for manipulation of proteins of interest in the mouse. Thousands of genetically modified animals carrying targeted endogenous genes or expressing various transgenic constructs are also now available. This combination of factors makes the mouse an ideal model system for studying compaction in early mammalian development.

2. PROPOSED MECHANISMS FOR COMPACTION

2.1 Cell Adhesion

A key driver of tissue morphogenesis is the formation and rearrangement of cell–cell contacts mediated by cell adhesion molecules (CAMs). Pioneering studies demonstrated that when cells from different embryonic germ layers are artificially mixed, they spontaneously sort into separate populations (Townes & Holtfreter, 1955). It was predicted, and later demonstrated (Nose, Nagafuchi, & Takeichi, 1988), that this cell sorting was based on the differential expression of CAMs. These experiments prompted the differential adhesion hypothesis (DAH), in which tissues are equated to immiscible liquids. The main principles of the DAH are that (1) cells adhere to each other, (2) adhesion properties vary between different cells, and (3) aggregations of cells tend to arrange themselves in a way that minimizes the free energy of the system (Steinberg, 1970). Many developmental morphogenetic processes including epiboly (Kane, McFarland, & Warga, 2005), myotome (Cortes et al., 2003) and rhombomere formation in the zebrafish (Cooke, Kemp, & Moens, 2005), cell sorting in the *Drosophila* imaginal wing disc (Dahmann & Basler, 2000), and patterning of the mouse cerebellum (Gliem et al., 2006) can be explained, at least to some extent, by the DAH. Additionally, the malignant invasion of cancerous cells is consistent with the DAH and alterations in many CAMs have been associated with the development and progression of cancer (Okegawa, Pong, Li, & Hsieh, 2004).

Early studies revealed that calcium-dependent adhesion is critical for embryo compaction (Fleming, Sheth, & Fesenko, 2001). Interfering with adhesion by chelating calcium ions or using antibodies targeting a cell surface glycoprotein decompacted embryos and prevented development to the blastocyst stage (Ducibella & Anderson, 1975; Wales, 1970; Whitten, 1971). Although some members of the calcium-dependent integrin family of CAMs are also expressed throughout early mouse preimplantation development (Sutherland, Calarco, & Damsky, 1993), their function does not appear to be required during compaction (Richa, Damsky, Buck, Knowles, & Solter, 1985). The cell surface glycoprotein essential for compaction was later identified as E-cadherin (uvomorulin) (Hyafil, Babinet, & Jacob, 1981), a member of the multigene cadherin family of homophilic CAMs found in most animal cells (Meng & Takeichi, 2009; Yoshida & Takeichi, 1982). Cadherins function by forming transmembrane macromolecular complexes. They mediate adhesion by ligating their extracellular domain to that of cadherins presented on neighboring cells. Interactions with cytoplasmic and cytoskeletal proteins via their intracellular domain serve to connect the cell membrane to the cortex (Hoffman & Yap, 2015).

Evidence for a leading role for E-cadherin in the DAH is provided by studies demonstrating that manipulating the expression level of E-cadherin directly affects how cells from different tissue layers sort (Foty & Steinberg, 2005; Godt & Tepass, 1998; Schotz et al., 2008). Although named for its expression in epithelial cell layers, E-cadherin is also present from the very early stages of development. It is maternally derived in the oocyte and de novo E-cadherin zygotic synthesis starts at the 2-cell stage (Vestweber, Gossler, Boller, & Kemler, 1987). In blastomeres, E-cadherin is enriched basolaterally, forming intercellular adherens junctions and connecting to the actin cytoskeleton via catenin proteins (Ozawa, Ringwald, & Kemler, 1990). Embryos lacking zygotic E-cadherin die before implantation. They are able to compact due to residual maternal E-cadherin but fail to form normal blastocysts (Larue, Ohsugi, Hirchenhain, & Kemler, 1994). Embryos lacking both maternal and zygotic E-cadherin cannot compact or form a blastocyst and instead remain as loose clusters of cells (Stephenson, Yamanaka, & Rossant, 2010).

E-cadherin is uniformly distributed in the cell membrane until the 8-cell stage. Then it begins to accumulate in cell–cell junctions and is predominantly localized to basolateral membrane regions by the 16-cell stage (Fig. 2).

Cell adhesion is generally proposed to be proportional to cadherin expression levels (Foty & Steinberg, 2005; Krieg et al., 2008). Hence, if compaction were driven by intercellular differences in adhesion, it would be reasonable to expect variation in the expression of E-cadherin between blastomeres. However, it has recently been demonstrated that there are no marked differences in expression levels or mobility of E-cadherin between cells of the embryo during compaction (Samarage et al., 2015). Furthermore, it is uncertain whether the decrease in surface energy resulting from cadherin ligation could even generate sufficient forces to deform tissues (Maitre et al., 2012). More microscopic differences, or a potential role of other yet unidentified molecules cannot be excluded, but there is currently

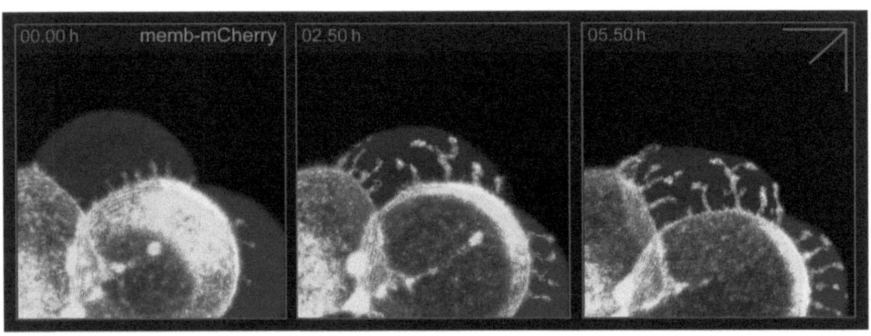

Fig. 2 Filopodia extended by some blastomeres facilitate changes in cell shape that compact the entire embryo. One cell of the 2-cell embryo was injected with membrane-targeted mCherry (memb-mCherry) to label half of the embryo. This allows visualization of memb-mCherry positive filopodia that extend from some blastomeres onto their unlabeled neighbors (DIC image pseudo-colored *blue* and *green*) during compaction. Cells are initially relatively spherical (*left panel*) but both filopodia-extending cells and their neighbors flatten and draw closer together as filopodia are extended (*middle, right panel*). Scale bars: 10 μm in *x, y, z*. (See the color plate.)

little evidence that compaction is mediated by a change in adhesion at basolateral cell regions.

The redistribution of E-cadherin at the onset of compaction coincides with the assembly of another junctional complex, the gap junction (McLachlin, Caveney, & Kidder, 1983). As the regulation of connexin-based intercellular communication is controlled by E-cadherin in mouse cells in vitro (Jongen et al., 1991), it is plausible that this relocalization of E-cadherin acts as a trigger for the assembly of gap junctions, possibly through calcium signaling. Functional communication between cells at the 8-cell stage might also mark the end of solitary blastomeres and the establishment of a syncytium that may be necessary for the maintenance of compaction. Antibodies targeting the gap junction component connexin 43 (Cx43), block intercellular dye transfer and cause decompaction and extrusion of affected blastomeres from the rest of the embryo (Becker & Davies, 1995). However, the successful progression of Cx43 null embryos through compaction suggests compensatory mechanisms exist, as previously demonstrated by the failure of connexin 36 knockout mice to show the cognitive deficits attributed to acute blockade of neuronal gap junctions (Bissiere et al., 2011). Further investigation is required to dissect the contribution of gap junctions to the process of compaction.

2.2 Cortex Tension

Although widely accepted to explain cell sorting (Grubb, 2006) and elegant in its simplicity, the DAH cannot fully explain many experimental observations and simulations of sorting behavior (Brodland & Chen, 2000). Accounting for the effects of cortical tension prompted development of an alternative model called the differential interfacial tension hypothesis (DITH) (Brodland, 2002). Cortical tension is generated by the contractility of actomyosin networks (Pasternak, Spudich, & Elson, 1989) and is now recognized to be a key determinant of the shape of both individual cells and tissues (Heisenberg & Bellaiche, 2013; Lecuit & Lenne, 2007). A cellular cortex lies under the plasma membrane of most eukaryotic cells and is comprised of a dynamic network of actin filaments connected to the cell membrane. Myosin motor proteins hydrolyse ATP to slide pairs of these actin filaments with respect to each other, generating contractility or expansion of the cortex. The DITH postulates that cells sort to minimize the interfacial tension of the system, which results from both adhesive and cortical tensions. Adhesive tension mediated by cadherins increases contact areas between cells, but cortical tension generated by actomyosin contractility acts in opposition to decrease intercellular contacts. According to the DITH, cells sort in a manner that maximizes their adhesive tension, while minimizing their cortical tension and the balance of these two forces determines cell shape. Morphogenetic processes such as boundary formation (Monier, Pelissier-Monier, Brand, & Sanson, 2010), cell intercalation (Bertet, Sulak, & Lecuit, 2004), and tissue invagination (Chandrasekaran & Beckendorf, 2005; Lee et al., 2006) can all be described in terms of changes in the balance between adhesive and cortical tensions as defined by the DITH.

Recent work examining the function of cortical tension in the preimplantation embryo has demonstrated that it plays a role in driving the compaction process (Maitre, Niwayama, Turlier, Nedelec, & Hiiragi, 2015) and the first spatial segregation of mammalian cells (Samarage et al., 2015). Actomyosin has been shown to clear from cell–cell contacts and accumulate at the surface during compaction (Maitre et al., 2015). This has been proposed to form a contractile shell on the surface of the embryo, to which about 75% of the cell shape changes during compaction are attributed. In this model, E-cadherin does not directly generate forces, but acts to exclude actomyosin from cell–cell contacts, facilitating cortical contractility. E-cadherin is required, however, for the function of a newly discovered class of filopodia proposed to provide an additional mechanism to draw neighboring cells closer together during compaction (Fierro-Gonzalez, White, Silva, & Plachta, 2013).

2.3 Filopodia

Recently, it was demonstrated that the onset of compaction is accompanied by the extension of long membrane protrusions from some cells in the mouse embryo (Fig. 3). These protrusions were identified as filopodia and shown to stretch across the apical membrane of neighboring cells. They differ in both length and molecular composition from previously observed microvilli present at the apical pole of the mouse blastomeres (Calarco & Epstein, 1973; Ducibella et al., 1977). The key in observing the formation of these filopodia during embryonic compaction was to perform pronuclear injection in only one of the cells at the 2-cell stage.

The filopodia contain F-actin and express both E-cadherin and proteins that link it to the actin cytoskeleton, such as α- and β-catenin. Furthermore, these filopodia also express the unconventional myosin protein, myosin-10. This myosin has previously been shown to promote the formation of

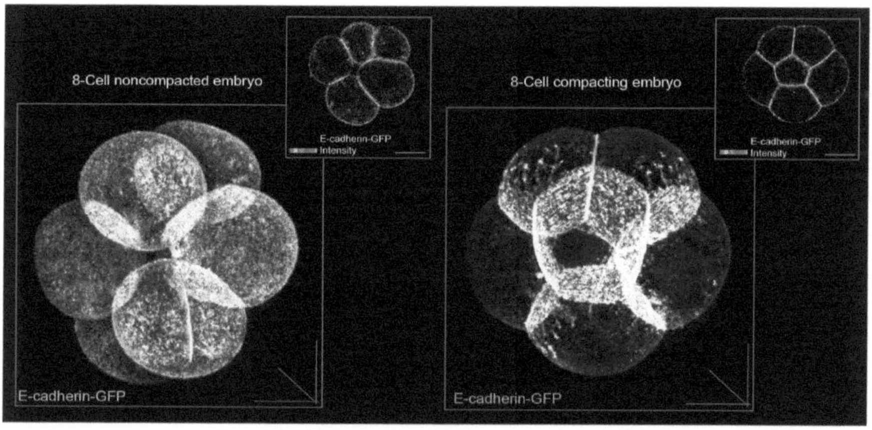

Fig. 3 E-cadherin redistributes to basolateral cell membranes during compaction. E-cadherin-GFP expression in an embryo injected at the 1-cell stage. At the early 8-cell stage, prior to the onset of compaction, E-cadherin-GFP is distributed across the entire cell membrane (*left panel*). As compaction begins, E-cadherin-GFP becomes predominantly localized to the basolateral cell membrane and compaction-specific filopodia (*right panel*). Insets show intensity of E-cadherin-GFP expression in single 2D planes. Scale bars: 10 μm in *x*, *y* and *x*, *y*, *z*. (See the color plate.)

filopodia-like structures in multiple cultured cell lines (Kerber & Cheney, 2011). In the preimplantation mouse embryo, myosin-10 levels increase during compaction and its expression is both required and sufficient to trigger compaction. These recently discovered filopodia only appear at the 8-cell stage as compaction commences, and disappear around the 16-cell stage when the embryo is fully compacted. Trans interactions between E-cadherin molecules most likely anchor the filopodia to neighboring cell membranes, where they remain for several hours before retracting immediately prior to the division of the filopodia-forming cell. Importantly, disrupting the filopodia using laser-based ablations causes rapid deformation of the cell membrane indicating that filopodia provide structural support for the alterations of cell shape that are required during compaction.

These findings provide a new mechanism controlling compaction relying on the action of long cellular protrusions. However, several open questions remain. For example, it is yet to be determined why only some cells of the embryo extend filopodia while others do not. Indeed, it was observed that only about 60% of cells of the embryo become filopodia-forming cells and that, while they can project filopodia onto up to three other cells simultaneously, they never receive reciprocal filopodia on their own apical membrane. It is also striking how filopodia extend to very similar lengths and their bases are almost equally spaced. How the number and length of these structures is controlled is currently unknown. This tight regulation of the distribution and timing of filopodia indicates that they may also have an unidentified signaling role. Investigation of these questions will likely reveal further details about how the process of compaction is controlled in the mouse embryo.

3. KEY CELL DECISIONS DURING COMPACTION

Accompanying compaction are several other major morphogenetic processes. The blastomeres develop polarity, reposition to form inner and outer cells, and start to display differential expression of cell fate markers. Understanding how all of these processes are integrated remains an open challenge.

3.1 Cell Position Changes: Division and Internalization

The process of embryo compaction is synchronized with two more key morphogenetic events: cell division and cell internalization.

3.1.1 Cell Division

During compaction, each cell of the 8-cell embryo divides once, thereby producing a 16-cell embryo. Cell division requires the existence of important regulatory mechanisms that must control the timing of each cell division and the positioning of resultant daughter cells within the embryo. Although few mechanistic insights have been elucidated, it is of note that, unlike in many nonmammalian embryos, neighboring cells in the mouse embryo do not divide simultaneously. Instead, a dividing cell typically completes its division minutes to hours before any of its neighbors divide (Fierro-Gonzalez et al., 2013; Samarage et al., 2015). It is plausible that such an orchestration of cell division times ensures that the global embryo architecture is not compromised. As a mouse blastomere undergoes division, it loses its compacted shape and becomes highly spherical, before cleaving into two daughter cells. These daughter cells are also initially spherical and must then flatten and compact to incorporate into the rest of the embryo. Unlike in more tightly packed tissues, if two neighboring cells were to divide at the same time it may become difficult to simultaneously reintegrate the four resulting daughter cells into the embryo. Therefore, elucidating mechanisms that control cell division timing in the embryo and understanding how these might impact on embryo architecture remains an interesting future challenge. In particular, it will be important to understand how molecular mechanisms regulating the assembly of the mitotic and cytokinesis structures relate to the regulation of cytoskeletal components supporting cell shape and position. It will also be of interest to understand what type of communication exists between a dividing cell and its neighbors in the embryo. For example, can cells receive information regarding the division status of a neighbor cell and change their own division behaviors accordingly? A key goal will be determining how the control of cell division in the early mouse embryo, which is composed of few cells, compares to the control of cell divisions in much more coherent embryonic structures, such as the early *Drosophila* embryo in which mitoses occur in a highly synchronous fashion.

3.1.2 Cell Position

Additionally, during compaction cells must not only divide but also relocate to different positions. Compaction coincides with the spatial separation of embryonic cells into the first two distinct cell lineages of the conceptus. During the 8- to 16-cell stage some cells become internalized to form the pluripotent inner mass of the embryo. By the 16-cell stage, most embryos contain an inner mass composed of three of these pioneer cells. These cells then go on to divide internally to expand the ICM at blastocyst stages, accompanied by a small population of new cells that are internalized during the 16- to 32-cell stage. Understanding how the initial pioneer cells become internalized is of great interest as they form the first separate embryonic compartments. Moreover, repositioning cells to the interior of the embryo generates a unique niche in which for the first time, some cells of the embryo may be exposed to different signals than others.

Currently, little is known about how the mechanisms controlling embryo compaction integrate with those regulating the internalization of inner cells. Until recently, even the morphogenetic events founding the inner mass remained poorly defined. Traditional views had assumed that the inner mass formed via highly orientated cell divisions, referred to as asymmetric divisions (Yamanaka, Ralston, Stephenson, & Rossant, 2006; Zernicka-Goetz, Morris, & Bruce, 2009). It was generally accepted that some cells of the early embryo would divide with a highly orientated cleavage plane and push one of their resulting daughter cells into the inner mass as a direct result of the scission. However, these asymmetric cell division events had not been visualized in real time. Most observations, including very recent studies have relied on tracking fluorescently labeled cell nuclei using markers such as histone-GFP (Strnad et al., 2015). However, tracking cell nuclei is inaccurate because the nuclear spatial coordinates cannot be directly translated into position and morphology of the entire cell. For instance, some cells in the embryo have their nuclei located more basolaterally while others are located more apically (Ajduk, Biswas Shivhare, & Zernicka-Goetz, 2014), so nuclear position does not necessarily inform about the inner/outer status of the entire cell membrane. Instead, tracking cells by their membrane offers substantially more information about the overall position of each cell in the embryo with relation to its neighbors. Membrane segmentation has previously proven useful to investigate morphogenetic processes in *Drosophila* (Gelbart et al., 2012), zebrafish (Xiong et al., 2014), and plant (Yoshida et al., 2014) embryos. Recently, this approach has also been used to follow the process of inner mass formation in living mouse embryos (Samarage et al., 2015). Instead of highly orientated asymmetric cell divisions, most cells founding the inner mass originate from a symmetric division and are subsequently allocated inside the embryo via a process displaying all of the classical features of apical constriction (Sawyer et al., 2010). Subcellular heterogeneities in tensile forces, generated by actomyosin cortical networks were shown to drive a decrease in apical surface area, an increase in basolateral area, and gradual repositioning of cells to the interior of the embryo. Interestingly, on average the first internalization event occurs at the 12-cell stage, exactly as the embryo is compacting. The degree of overlap in the embryo between the processes of compaction and cell internalization remains an important open question.

3.2 Cell Polarity

Unlike cells in nonmammalian embryos, the blastomeres of the early mouse embryo initially show fewer features of cell polarity. At the onset of compaction, cues arising from cell–cell interactions direct the establishment and orientation of polarity. The contact-free surface of each blastomere forms a recognizable apical domain enriched in microvilli, actin, and actin-binding proteins (Ducibella et al., 1977; Louvet, Aghion, Santa-Maria, Mangeat, & Maro, 1996; Reeve & Ziomek, 1981). Cues from the cell membrane induce the asymmetrical localization of apical and basolateral polarity proteins which then reinforce formation of discrete domains through mutually antagonistic interactions. E-cadherin is one of the first proteins to polarize in the mouse embryo, becoming enriched in cell–cell contacts as compaction commences (Fig. 2) (Vestweber et al., 1987). Other classical cell polarity proteins such as Ezrin, Pard6b, and the aPKCs (PKCζ and PKCλ) then localize to the apical domain while Par-1, Jam-1, and Na/K ATPase accumulate at basolateral cell–cell contacts (Barcroft, Moseley, Lingrel, & Watson, 2004; Louvet et al., 1996; Pauken & Capco, 2000; Thomas et al., 2004; Vinot et al., 2005; Wang, Ojakian, & Nelson, 1990). Asymmetric cell–cell contacts are required and sufficient for blastomeres to polarize (Ziomek & Johnson, 1980), but are not necessary to maintain polarization once it has been established (Johnson & Ziomek, 1981b). Although the precise trigger for the establishment of polarity in the mouse embryo has not been elucidated, it is likely to involve E-cadherin. Embryos lacking both maternal and zygotic *E-cadherin* reveal that it is required to restrict the area of the apical domain and confine basolateral proteins, ensuring appropriate segregation of apical and basolateral domains (Stephenson et al., 2010). However, due to its requirement for cell–cell adhesion, it remains difficult to dissect the exact contribution of E-cadherin to polarity. It is possible that E-cadherin is simply necessary for cells to make sufficient contact for an unidentified cadherin-independent polarity cue to take effect.

The establishment of discrete domains of E-cadherin-mediated adhesion and apical polarity during compaction has important downstream consequences for the acquisition of cell fate. Blastomeres use their specific combination of adhesion and apical polarity to convert information about their position within the embryo into cell fate decisions. This is achieved by regulating the subcellular localization of members of the Hippo signaling pathway through interactions with adhesion complexes or apical polarity domains (Cockburn, Biechele, Garner, & Rossant, 2013; Hirate et al., 2013).

3.3 Cell Fate Decisions

A final, critical event occurring concurrent with compaction is the embarkation of some cells along the pathway to establishing the first cell lineages. Although segregation of most lineage-specific cell markers is not established until the blastocyst stage, the allocation of some cells to the interior of the embryo during compaction is the foundational morphogenetic process forming the future ICM. The internalized cells can be said to have adopted a distinct fate because they will remain inside and only contribute to the expansion of the inner mass (Samarage et al., 2015). This is in contrast to the cells positioned in the outer layer, which still have the potential to contribute to TE, or provide new inner cells during later developmental stages.

Whether blastomeres acquire molecular heterogeneities associated with cell fate before, during, or after compaction is currently still debated

(Burton & Torres-Padilla, 2014; Rossant & Tam, 2009; Zernicka-Goetz et al., 2009). Unlike in some nonmammalian species, identifying molecular heterogeneities that predict cell fate in the early mouse and human embryo has proven challenging (Rossant & Tam, 2009; Zernicka-Goetz et al., 2009). Original observations of morphological uniformity and a lack of evidence for uneven distribution of fate determinants in mouse blastomeres led to a primarily stochastic view of early fate determination (Rossant, 1976; Tarkowski, 1959, 1961; Tarkowski & Wroblewska, 1967). Some studies have concluded that the separation of pluripotent and extraembryonic cell fates does not occur until after the embryo becomes a compacted morula (Dietrich & Hiiragi, 2007; Kurotaki, Hatta, Nakao, Nabeshima, & Fujimori, 2007; Louvet-Vallee, Vinot, & Maro, 2005; Motosugi, Bauer, Polanski, Solter, & Hiiragi, 2005). By contrast, others propose that mammalian cell fates might be predictable as early as the 2- to 8-cell stage (Gardner, 2001, 2002a, 2002b; Piotrowska, Wianny, Pedersen, & Zernicka-Goetz, 2001; Piotrowska-Nitsche, Perea-Gomez, Haraguchi, & Zernicka-Goetz, 2005; Piotrowska-Nitsche & Zernicka-Goetz, 2005; Plachta, Bollenbach, Pease, Fraser, & Pantazis, 2011; Shi et al., 2015). Either way, recent studies have shed further light on how processes occurring during compaction impact on cell fate decisions.

Differentiation of the outer cells of the morula into TE requires the expression of Cdx2 and Gata3, which are driven by the transcription factor Tead4 and its coactivator, Yap1 (Ralston et al., 2010). The Hippo signaling pathway kinase, Lats1/2, phosphorylates Yap1 to control its subcellular localization (Nishioka et al., 2009). Angiomotin (Amot) is required to activate Lats1/2 and switch on the Hippo signaling pathway (Hirate et al., 2013). In outer cells, Amot is sequestered by components of the apical polarity complex and localized to the apical domain (Hirate et al., 2013). Here it is bound to actin and held in an inactive state, preventing activation of the Hippo signaling pathway. Unphosphorylated Yap1 can enter the nucleus and induce transcription of the TE-specific genes Cdx2 and Gata3. In inner cells lacking apical polarity, Amot can interact with Lats1/2 and the E-cadherin adhesion complex at adherens junctions via Nf2 (Hirate et al., 2013). This interaction activates Amot and stabilizes it at adherens junctions where it switches Hippo signaling on. Yap1 is phosphorylated and excluded from the nucleus, promoting the transcription of ICM-specific genes. In this way, the reorganization of proteins into apical and basolateral domains that occurs during compaction directs the subsequent fate of the cell.

Hippo pathway signaling has also been shown in vitro to be responsive to actin rearrangement caused by mechanical cues such as cell shape and geometry (Dupont et al., 2011; Wada, Itoga, Okano, Yonemura, & Sasaki, 2011). It is, therefore, possible that the changes in cell morphology and tensile forces that develop during compaction may act in parallel to regulate the Hippo signaling pathway, although this remains to be demonstrated in the living embryo.

4. OPEN QUESTIONS ABOUT COMPACTION

Although the process of embryo compaction has been studied for decades, important questions remain unresolved regarding how it occurs and how it is regulated. For example, the chain of events that triggers this critical first change in embryo morphology is undetermined. PKC-α-mediated phosphorylation of β-catenin is known to be part of the molecular signal that prompts the onset of compaction (Pauken & Capco, 1999). Inhibiting PKC-α blocks compaction and activating it can induce early compaction (Ohsugi, Ohsawa, & Semba, 1993; Winkel, Ferguson, Takeichi, & Nuccitelli, 1990); however, the upstream event activating PKC-α at this time in the embryo is unknown.

In fact, little is known about how the timing of compaction is controlled. Why do mouse embryos begin to compact at the 8-cell stage and not earlier or later? The timing of compaction is independent of cell number so must be regulated through another mechanism (Fernandez & Izquierdo, 1980). The total cytoplasmic volume of the embryo remains constant during cleavage division of the 1-cell embryo to the 8-cell embryo. This results in an exponential increase in nucleocytoplasmic ratio with the largest change occurring during the first three to four cleavage divisions (Aiken, Swoboda, Skepper, & Johnson, 2004). Artificially increasing the nucleocytoplasmic ratio by extracting cytoplasmic material from the 1-cell embryo induces early compaction at the 4-cell stage (Lee, Lee, Yoon, Roh, & Kim, 2001). Although this may indicate a direct link between the onset of compaction and the cell's nucleocytoplasmic ratio, the answer is more likely to lie in the cytoplasmic composition. Restoring the volume of a cytoplasmically depleted 1-cell embryo has varying effects on the timing of compaction depending on the stage of the embryo that donor cytoplasm is derived from. It is probable that degradation or dilution of an inhibitory cytoplasmic factor present in the 1-cell embryo regulates the timing of compaction, but the identity of this factor remains unknown.

And finally, how is compaction integrated with other processes occurring during this critical developmental window, such as changes in cell polarity, the start of cell differentiation, and the morphological reorganization of the entire embryo?

5. EMERGING TECHNIQUES AND FUTURE WORK

New techniques based on imaging will enable the discovery of further mechanisms regulating compaction. The field of developmental biology has benefited greatly from the establishment of live-imaging methods. Yet, a key future requirement will be the utilization of more quantitative analytical tools to study compaction at multiple levels, ranging from macroscopic cell–cell interactions to the dynamics of regulatory molecules (Fig. 4).

Although it is now possible to label cells in the embryo with various fluorescently tagged proteins, new studies should focus on the development of computational image segmentation approaches to quantify the main changes in cell shape and position during compaction (Fig. 4A). A greater challenge is the development of techniques to track the dynamics of subcellular features such as the cytoskeleton, including actin, microtubules, and intermediate filaments, as well as various intracellular organelles, including the nucleus, endoplasmic reticulum, and mitochondria and filopodia (Fig. 4A). Revealing the dynamic behaviors of these structures may facilitate the discovery of other cell processes involved in compaction. Computational image segmentation approaches can also provide quantitative information about the subcellular forces acting in the embryo (Fig. 4B). Measuring the contact angles between cells reveals the local balance of adhesion and tension and changes in the distribution of these forces can be tracked throughout compaction.

In addition to techniques for studying whole cells or intracellular structures, methods for studying the dynamics of the key regulatory proteins controlling compaction also need to be applied. For example, compaction provides an excellent system to probe the dynamics of actomyosin networks. Previous work has performed measurements of adhesion forces and cortex tension using dual pipette aspiration assays. However, this approach is very invasive and cannot reveal the subcellular organization of tensile forces acting during compaction. It will be important to study actomyosin dynamics at a more microscopic level to understand the forces driving compaction. Laser ablations offer a good entry point to probe the relative magnitude and directionality of tensile forces, as they can be targeted to defined subcellular regions and their noninvasiveness allows repeat measurements during development.

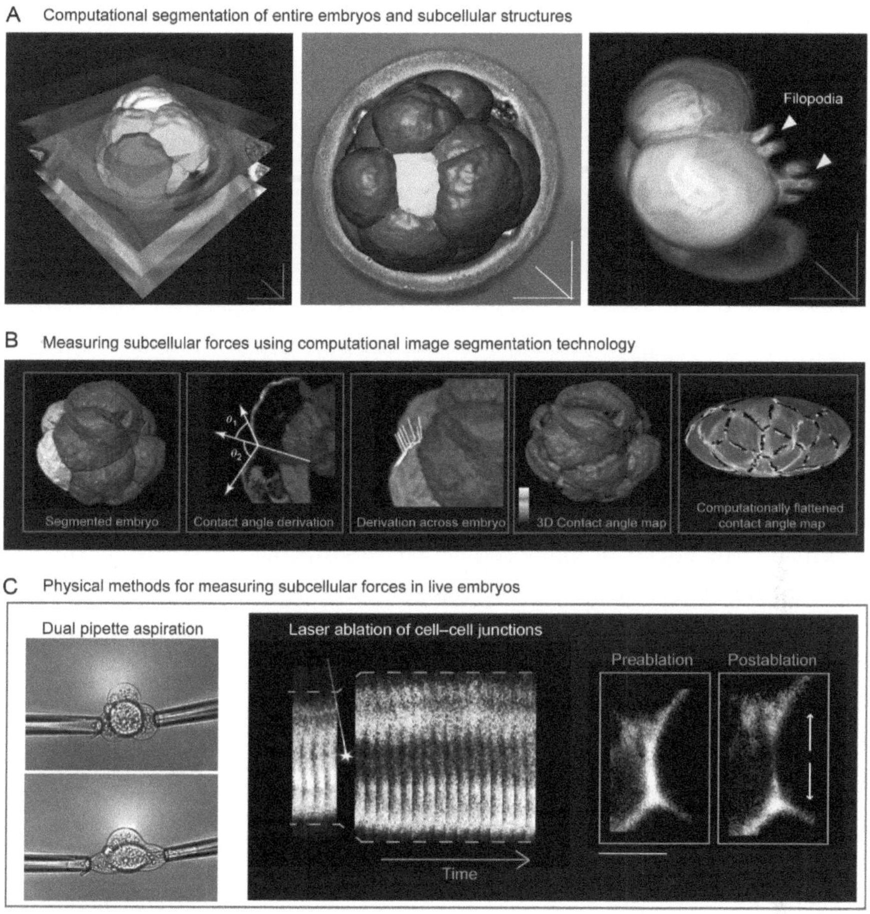

Fig. 4 Emerging technologies for quantitative investigations of mechanisms controlling compaction. New quantitative imaging-based technologies are facilitating the investigation of compaction. (A) Computational segmentation of entire embryos (*left, middle panel*) or selected cells and filopodia (*arrowhead, right panel*). Embryos expressing fluorescently labeled proteins are imaged in 4D using two-photon microscopy, and individual features are computationally segmented. (B) Image segmentation technologies enable calculation of the subcellular distribution of biomechanical forces. Contact angles at cell–cell junctions result from the balance of subcellular forces and can be measured at multiple points along the junction. This provides a 3D map of contact angles, which can be projected to a 2D map of the apical surface of the embryo. (C) Dual pipette aspiration is a physical method for measuring subcellular forces in the embryo. An embryo is held between two micropipettes (*top left panel*) and aspiration is increased (*bottom left panel*) until the cells separate. Although this method provides useful information about the force of cell–cell adhesion, it is not compatible with ongoing development of the embryo. A less invasive physical technique for measuring subcellular forces is laser ablation. Targeting high laser energy into a localized region of cell cortex ablates cortical structure causing rapid recoil about the ablation that reflects the size and directionality of subcellular forces in that region. When cell–cell junctions are laser ablated (*right panel*) the degree by which adjacent vertices separate (*white arrows, right panel*) reveals the residual tensile forces. Scale bars: 5 μm in *x*, *y* and 10 μm in *x*, *y*, *z*. (See the color plate.)

The mobility of E-cadherin and many of its intracellular interacting proteins could be probed by combining the use of fluorescence recovery after photobleaching (FRAP), fluorescence correlation spectroscopy (FCS), and image correlation techniques such as raster image correlation spectroscopy (RICS) (Digman & Gratton, 2011; Kaur et al., 2013). Moreover, new photoactivatable and photoconvertible proteins are excellent tools for labeling defined subpopulations of regulatory proteins and following their behaviors with high temporal resolution imaging.

The engineering of more light-activatable regulatory proteins would be a very valuable tool to manipulate compaction at the subcellular level. So far, most of these tools have been used to study cell migration in culture conditions, yet the accessibility of the mouse embryo during compaction should make future studies in this direction possible.

Finally, the effects of many molecular and mechanical perturbations can now be tested at the level of gene expression even in single cells. Recently, the mouse embryo has been analyzed at the single-cell level using RNAseq (Shi et al., 2015; Tang et al., 2011), thus these methods could be rapidly applied to investigate gene expression changes occurring during normal or manipulated compaction.

6. CONCLUSIONS

Compaction offers an excellent experimental system to study how cells interact with each other in a whole embryo to form the first tissue-like structures during mammalian development. Yet, most studies so far have relied on the use of static analysis and our understanding of this morphogenetic process is limited to a few cellular components. A key future challenge is to perform more integrative dynamic analyses to reveal how various cellular and molecular processes regulate these early changes in cell shape. As is the case for other fields, new cross-disciplinary collaborations among labs experienced in embryology, imaging, genetics, and mechanobiology will be of great benefit to our understanding of this critical process in mammalian development.

REFERENCES

Adamson, G. D., Tabangin, M., Macaluso, M., & de Mouzon, J. (2013). The number of babies born globally after treatment with the assisted reproductive technologies (ART). *Fertility and Sterility*, *100*, S42.

Aiken, C. E., Swoboda, P. P., Skepper, J. N., & Johnson, M. H. (2004). The direct measurement of embryogenic volume and nucleo-cytoplasmic ratio during mouse preimplantation development. *Reproduction*, *128*, 527–535.

Ajduk, A., Biswas Shivhare, S., & Zernicka-Goetz, M. (2014). The basal position of nuclei is one pre-requisite for asymmetric cell divisions in the early mouse embryo. *Developmental Biology*, *392*, 133–140.

Barcroft, L. C., Moseley, A. E., Lingrel, J. B., & Watson, A. J. (2004). Deletion of the Na/K-ATPase alpha1-subunit gene (Atp1a1) does not prevent cavitation of the preimplantation mouse embryo. *Mechanisms of Development*, *121*, 417–426.

Becker, D. L., & Davies, C. S. (1995). Role of gap junctions in the development of the preimplantation mouse embryo. *Microscopy Research and Technique*, *31*, 364–374.

Bertet, C., Sulak, L., & Lecuit, T. (2004). Myosin-dependent junction remodelling controls planar cell intercalation and axis elongation. *Nature*, *429*, 667–671.

Bissiere, S., Zelikowsky, M., Ponnusamy, R., Jacobs, N. S., Blair, H. T., & Fanselow, M. S. (2011). Electrical synapses control hippocampal contributions to fear learning and memory. *Science*, *331*, 87–91.

Brodland, G. W. (2002). The Differential Interfacial Tension Hypothesis (DITH): A comprehensive theory for the self-rearrangement of embryonic cells and tissues. *Journal of Biomechanical Engineering*, *124*, 188–197.

Brodland, G. W., & Chen, H. H. (2000). The mechanics of cell sorting and envelopment. *Journal of Biomechanics*, *33*, 845–851.

Burton, A., & Torres-Padilla, M. E. (2014). Chromatin dynamics in the regulation of cell fate allocation during early embryogenesis. *Nature Reviews. Molecular Cell Biology*, *15*, 723–735.

Calarco, P. G., & Brown, E. H. (1969). An ultrastructural and cytological study of preimplantation development of the mouse. *The Journal of Experimental Zoology*, *171*, 253–283.

Calarco, P. G., & Epstein, C. J. (1973). Cell surface changes during preimplantation development in the mouse. *Developmental Biology*, *32*, 208–213.

Chandrasekaran, V., & Beckendorf, S. K. (2005). Tec29 controls actin remodeling and endoreplication during invagination of the Drosophila embryonic salivary glands. *Development*, *132*, 3515–3524.

Cockburn, K., Biechele, S., Garner, J., & Rossant, J. (2013). The Hippo pathway member Nf2 is required for inner cell mass specification. *Current Biology*, *23*, 1195–1201.

Cockburn, K., & Rossant, J. (2010). Making the blastocyst: Lessons from the mouse. *The Journal of Clinical Investigation*, *120*, 995–1003.

Cooke, J. E., Kemp, H. A., & Moens, C. B. (2005). EphA4 is required for cell adhesion and rhombomere-boundary formation in the zebrafish. *Current Biology*, *15*, 536–542.

Cortes, F., Daggett, D., Bryson-Richardson, R. J., Neyt, C., Maule, J., Gautier, P., et al. (2003). Cadherin-mediated differential cell adhesion controls slow muscle cell migration in the developing zebrafish myotome. *Developmental Cell*, *5*, 865–876.

Dahmann, C., & Basler, K. (2000). Opposing transcriptional outputs of Hedgehog signaling and engrailed control compartmental cell sorting at the Drosophila A/P boundary. *Cell*, *100*, 411–422.

Dietrich, J. E., & Hiiragi, T. (2007). Stochastic patterning in the mouse pre-implantation embryo. *Development*, *134*, 4219–4231.

Digman, M. A., & Gratton, E. (2011). Lessons in fluctuation correlation spectroscopy. *Annual Review of Physical Chemistry*, *62*, 645–668.

Ducibella, T., & Anderson, E. (1975). Cell shape and membrane changes in the eight-cell mouse embryo: Prerequisites for morphogenesis of the blastocyst. *Developmental Biology*, *47*, 45–58.

Ducibella, T., Ukena, T., Karnovsky, M., & Anderson, E. (1977). Changes in cell surface and cortical cytoplasmic organization during early embryogenesis in the preimplantation mouse embryo. *The Journal of Cell Biology*, *74*, 153–167.

Dupont, S., Morsut, L., Aragona, M., Enzo, E., Giulitti, S., Cordenonsi, M., et al. (2011). Role of YAP/TAZ in mechanotransduction. *Nature*, *474*, 179–183.

Dyce, J., George, M., Goodall, H., & Fleming, T. P. (1987). Do trophectoderm and inner cell mass cells in the mouse blastocyst maintain discrete lineages? *Development*, *100*, 685–698.

Fernandez, M. S., & Izquierdo, L. (1980). Blastocoel formation in half and double mouse embryos. *Anatomy Embryology*, *160*, 77–81.

Fierro-Gonzalez, J. C., White, M. D., Silva, J. C., & Plachta, N. (2013). Cadherin-dependent filopodia control preimplantation embryo compaction. *Nature Cell Biology*, *15*, 1424–1433.

Fleming, T. P., & Pickering, S. J. (1985). Maturation and polarization of the endocytotic system in outside blastomeres during mouse preimplantation development. *Journal of Embryology and Experimental Morphology*, *89*, 175–208.

Fleming, T. P., Sheth, B., & Fesenko, I. (2001). Cell adhesion in the preimplantation mammalian embryo and its role in trophectoderm differentiation and blastocyst morphogenesis. *Frontiers in Bioscience*, *6*, D1000–D1007.

Foty, R. A., & Steinberg, M. S. (2005). The differential adhesion hypothesis: A direct evaluation. *Developmental Biology*, *278*, 255–263.

Gardner, R. L. (2001). Specification of embryonic axes begins before cleavage in normal mouse development. *Development*, *128*, 839–847.

Gardner, R. L. (2002a). Experimental analysis of second cleavage in the mouse. *Human Reproduction*, *17*, 3178–3189.

Gardner, R. L. (2002b). Patterning is initiated before cleavage in the mouse. *Annals of Anatomy*, *184*, 577–581.

Gelbart, M. A., He, B., Martin, A. C., Thiberge, S. Y., Wieschaus, E. F., & Kaschube, M. (2012). Volume conservation principle involved in cell lengthening and nucleus movement during tissue morphogenesis. *Proceedings of the National Academy of Sciences of the United States of America*, *109*, 19298–19303.

Gliem, M., Weisheit, G., Mertz, K. D., Endl, E., Oberdick, J., & Schilling, K. (2006). Expression of classical cadherins in the cerebellar anlage: Quantitative and functional aspects. *Molecular and Cellular Neurosciences*, *33*, 447–458.

Godt, D., & Tepass, U. (1998). Drosophila oocyte localization is mediated by differential cadherin-based adhesion. *Nature*, *395*, 387–391.

Goodall, H., & Johnson, M. H. (1984). The nature of intercellular coupling within the preimplantation mouse embryo. *Journal of Embryology and Experimental Morphology*, *79*, 53–76.

Grubb, B. J. (2006). Developmental Biology, Eighth Edition. Scott F. Gilbert, editor. *Integrative and Comparative Biology*, *46*, 652–653.

Handyside, A. H. (1980). Distribution of antibody- and lectin-binding sites on dissociated blastomeres from mouse morulae: Evidence for polarization at compaction. *Journal of Embryology and Experimental Morphology*, *60*, 99–116.

Heisenberg, C. P., & Bellaiche, Y. (2013). Forces in tissue morphogenesis and patterning. *Cell*, *153*, 948–962.

Hirate, Y., Hirahara, S., Inoue, K., Suzuki, A., Alarcon, V. B., Akimoto, K., et al. (2013). Polarity-dependent distribution of angiomotin localizes Hippo signaling in preimplantation embryos. *Current Biology*, *23*, 1181–1194.

Hoffman, B. D., & Yap, A. S. (2015). Towards a dynamic understanding of cadherin-based mechanobiology. *Trends in Cell Biology*, *25*, 803–814.

Hyafil, F., Babinet, C., & Jacob, F. (1981). Cell-cell interactions in early embryogenesis: A molecular approach to the role of calcium. *Cell*, *26*, 447–454.

Johnson, M. H., & McConnell, J. M. L. (2004). Lineage allocation and cell polarity during mouse embryogenesis. *Seminars in Cell & Developmental Biology*, *15*, 583–597.

Johnson, M. H., & Ziomek, C. A. (1981a). The foundation of two distinct cell lineages within the mouse morula. *Cell*, *24*, 71–80.

Johnson, M. H., & Ziomek, C. A. (1981b). Induction of polarity in mouse 8-cell blastomeres: Specificity, geometry, and stability. *The Journal of Cell Biology, 91*, 303–308.
Johnson, M. H., & Ziomek, C. A. (1983). Cell interactions influence the fate of mouse blastomeres undergoing the transition from the 16- to the 32-cell stage. *Developmental Biology, 95*, 211–218.
Jongen, W. M., Fitzgerald, D. J., Asamoto, M., Piccoli, C., Slaga, T. J., Gros, D., et al. (1991). Regulation of connexin 43-mediated gap junctional intercellular communication by Ca2+ in mouse epidermal cells is controlled by E-cadherin. *The Journal of Cell Biology, 114*, 545–555.
Kane, D. A., McFarland, K. N., & Warga, R. M. (2005). Mutations in half baked/E-cadherin block cell behaviors that are necessary for teleost epiboly. *Development, 132*, 1105–1116.
Kaur, G., Costa, M. W., Nefzger, C. M., Silva, J., Fierro-Gonzalez, J. C., Polo, J. M., et al. (2013). Probing transcription factor diffusion dynamics in the living mammalian embryo with photoactivatable fluorescence correlation spectroscopy. *Nature Communications, 4*, 1637.
Kerber, M. L., & Cheney, R. E. (2011). Myosin-X: A MyTH-FERM myosin at the tips of filopodia. *Journal of Cell Science, 124*, 3733–3741.
Krieg, M., Arboleda-Estudillo, Y., Puech, P. H., Kafer, J., Graner, F., Muller, D. J., et al. (2008). Tensile forces govern germ-layer organization in zebrafish. *Nature Cell Biology, 10*, 429–436.
Kurotaki, Y., Hatta, K., Nakao, K., Nabeshima, Y., & Fujimori, T. (2007). Blastocyst axis is specified independently of early cell lineage but aligns with the ZP shape. *Science, 316*, 719–723.
Larue, L., Ohsugi, M., Hirchenhain, J., & Kemler, R. (1994). E-cadherin null mutant embryos fail to form a trophectoderm epithelium. *Proceedings of the National Academy of Sciences of the United States of America, 91*, 8263–8267.
Le Cruguel, S., Ferre-L'Hotellier, V., Moriniere, C., Lemerle, S., Reynier, P., Descamps, P., et al. (2013). Early compaction at day 3 may be a useful additional criterion for embryo transfer. *Journal of Assisted Reproduction and Genetics, 30*, 683–690.
Lecuit, T., & Lenne, P. F. (2007). Cell surface mechanics and the control of cell shape, tissue patterns and morphogenesis. *Nature Reviews. Molecular Cell Biology, 8*, 633–644.
Lee, D. R., Lee, J. E., Yoon, H. S., Roh, S. I., & Kim, M. K. (2001). Compaction in preimplantation mouse embryos is regulated by a cytoplasmic regulatory factor that alters between 1- and 2-cell stages in a concentration-dependent manner. *The Journal of Experimental Zoology, 290*, 61–71.
Lee, J. Y., Marston, D. J., Walston, T., Hardin, J., Halberstadt, A., & Goldstein, B. (2006). Wnt/Frizzled signaling controls C. elegans gastrulation by activating actomyosin contractility. *Current Biology, 16*, 1986–1997.
Lo, C. W., & Gilula, N. B. (1979). Gap junctional communication in the preimplantation mouse embryo. *Cell, 18*, 399–409.
Louvet, S., Aghion, J., Santa-Maria, A., Mangeat, P., & Maro, B. (1996). Ezrin becomes restricted to outer cells following asymmetrical division in the preimplantation mouse embryo. *Developmental Biology, 177*, 568–579.
Louvet-Vallee, S., Vinot, S., & Maro, B. (2005). Mitotic spindles and cleavage planes are oriented randomly in the two-cell mouse embryo. *Current Biology, 15*, 464–469.
Luke, B., Stern, J. E., Kotelchuck, M., Declercq, E. R., Hornstein, M. D., Gopal, D., et al. (2015). Adverse pregnancy outcomes after in vitro fertilization: Effect of number of embryos transferred and plurality at conception. *Fertility and Sterility, 104*, 79–86.
Maitre, J. L., Berthoumieux, H., Krens, S. F., Salbreux, G., Julicher, F., Paluch, E., et al. (2012). Adhesion functions in cell sorting by mechanically coupling the cortices of adhering cells. *Science, 338*, 253–256.

Maitre, J. L., Niwayama, R., Turlier, H., Nedelec, F., & Hiiragi, T. (2015). Pulsatile cell-autonomous contractility drives compaction in the mouse embryo. *Nature Cell Biology*, *17*, 849–855.

Maro, B., Johnson, M. H., Pickering, S. J., & Louvard, D. (1985). Changes in the distribution of membranous organelles during mouse early development. *Journal of Embryology and Experimental Morphology*, *90*, 287–309.

McLachlin, J. R., Caveney, S., & Kidder, G. M. (1983). Control of gap junction formation in early mouse embryos. *Developmental Biology*, *98*, 155–164.

Meng, W., & Takeichi, M. (2009). Adherens junction: Molecular architecture and regulation. *Cold Spring Harbor Perspectives in Biology*, *1*, a002899.

Monier, B., Pelissier-Monier, A., Brand, A. H., & Sanson, B. (2010). An actomyosin-based barrier inhibits cell mixing at compartmental boundaries in Drosophila embryos. *Nature Cell Biology*, *12*, 60–65 (Suppl. pp. 1–9).

Motosugi, N., Bauer, T., Polanski, Z., Solter, D., & Hiiragi, T. (2005). Polarity of the mouse embryo is established at blastocyst and is not prepatterned. *Genes & Development*, *19*, 1081–1092.

Nishioka, N., Inoue, K., Adachi, K., Kiyonari, H., Ota, M., Ralston, A., et al. (2009). The Hippo signaling pathway components Lats and Yap pattern Tead4 activity to distinguish mouse trophectoderm from inner cell mass. *Developmental Cell*, *16*, 398–410.

Nose, A., Nagafuchi, A., & Takeichi, M. (1988). Expressed recombinant cadherins mediate cell sorting in model systems. *Cell*, *54*, 993–1001.

Ohsugi, M., Ohsawa, T., & Semba, R. (1993). Similar responses to pharmacological agents of 1,2-OAG-induced compaction-like adhesion of two-cell mouse embryo to physiological compaction. *The Journal of Experimental Zoology*, *265*, 604–608.

Okegawa, T., Pong, R. C., Li, Y., & Hsieh, J. T. (2004). The role of cell adhesion molecule in cancer progression and its application in cancer therapy. *Acta Biochimica Polonica*, *51*, 445–457.

Ozawa, M., Ringwald, M., & Kemler, R. (1990). Uvomorulin-catenin complex formation is regulated by a specific domain in the cytoplasmic region of the cell adhesion molecule. *Proceedings of the National Academy of Sciences of the United States of America*, *87*, 4246–4250.

Pasternak, C., Spudich, J. A., & Elson, E. L. (1989). Capping of surface receptors and concomitant cortical tension are generated by conventional myosin. *Nature*, *341*, 549–551.

Pauken, C. M., & Capco, D. G. (1999). Regulation of cell adhesion during embryonic compaction of mammalian embryos: Roles for PKC and beta-catenin. *Molecular Reproduction and Development*, *54*, 135–144.

Pauken, C. M., & Capco, D. G. (2000). The expression and stage-specific localization of protein kinase C isotypes during mouse preimplantation development. *Developmental Biology*, *223*, 411–421.

Piotrowska, K., Wianny, F., Pedersen, R. A., & Zernicka-Goetz, M. (2001). Blastomeres arising from the first cleavage division have distinguishable fates in normal mouse development. *Development*, *128*, 3739–3748.

Piotrowska-Nitsche, K., Perea-Gomez, A., Haraguchi, S., & Zernicka-Goetz, M. (2005). Four-cell stage mouse blastomeres have different developmental properties. *Development*, *132*, 479–490.

Piotrowska-Nitsche, K., & Zernicka-Goetz, M. (2005). Spatial arrangement of individual 4-cell stage blastomeres and the order in which they are generated correlate with blastocyst pattern in the mouse embryo. *Mechanisms of Development*, *122*, 487–500.

Plachta, N., Bollenbach, T., Pease, S., Fraser, S. E., & Pantazis, P. (2011). Oct4 kinetics predict cell lineage patterning in the early mammalian embryo. *Nature Cell Biology*, *13*, 117–123.

Practice Committee of American Society for Reproductive Medicine. (2012). Multiple gestation associated with infertility therapy: An American Society for Reproductive Medicine Practice Committee opinion. *Fertility and Sterility, 97*, 825–834.

Ralston, A., Cox, B. J., Nishioka, N., Sasaki, H., Chea, E., Rugg-Gunn, P., et al. (2010). Gata3 regulates trophoblast development downstream of Tead4 and in parallel to Cdx2. *Development, 137*, 395–403.

Reeve, W. J. (1981). Cytoplasmic polarity develops at compaction in rat and mouse embryos. *Journal of Embryology and Experimental Morphology, 62*, 351–367.

Reeve, W. J., & Ziomek, C. A. (1981). Distribution of microvilli on dissociated blastomeres from mouse embryos: Evidence for surface polarization at compaction. *Journal of Embryology and Experimental Morphology, 62*, 339–350.

Richa, J., Damsky, C. H., Buck, C. A., Knowles, B. B., & Solter, D. (1985). Cell surface glycoproteins mediate compaction, trophoblast attachment, and endoderm formation during early mouse development. *Developmental Biology, 108*, 513–521.

Rossant, J. (1976). Postimplantation development of blastomeres isolated from 4- and 8-cell mouse eggs. *Journal of Embryology and Experimental Morphology, 36*, 283–290.

Rossant, J., & Tam, P. P. (2009). Blastocyst lineage formation, early embryonic asymmetries and axis patterning in the mouse. *Development, 136*, 701–713.

Samarage, C. R., White, M. D., Alvarez, Y. D., Fierro-Gonzalez, J. C., Henon, Y., Jesudason, E. C., et al. (2015). Cortical tension allocates the first inner cells of the mammalian embryo. *Developmental Cell, 34*, 435–447.

Sawyer, J. M., Harrell, J. R., Shemer, G., Sullivan-Brown, J., Roh-Johnson, M., & Goldstein, B. (2010). Apical constriction: A cell shape change that can drive morphogenesis. *Developmental Biology, 341*, 5–19.

Schotz, E. M., Burdine, R. D., Julicher, F., Steinberg, M. S., Heisenberg, C. P., & Foty, R. A. (2008). Quantitative differences in tissue surface tension influence zebrafish germ layer positioning. *Human Frontier Science Program, 2*, 42–56.

Shi, J., Chen, Q., Li, X., Zheng, X., Zhang, Y., Qiao, J., et al. (2015). Dynamic transcriptional symmetry-breaking in pre-implantation mammalian embryo development revealed by single-cell RNA-seq. *Development, 142*, 3468–3477.

Steinberg, M. S. (1970). Does differential adhesion govern self-assembly processes in histogenesis? Equilibrium configurations and the emergence of a hierarchy among populations of embryonic cells. *The Journal of Experimental Zoology, 173*, 395–433.

Stephenson, R. O., Yamanaka, Y., & Rossant, J. (2010). Disorganized epithelial polarity and excess trophectoderm cell fate in preimplantation embryos lacking E-cadherin. *Development, 137*, 3383–3391.

Steptoe, P. C., & Edwards, R. G. (1978). Birth after the reimplantation of a human embryo. *Lancet, 2*, 366.

Strnad, P., Gunther, S., Reichmann, J., Krzic, U., Balazs, B., de Medeiros, G., et al. (2015). Inverted light-sheet microscope for imaging mouse pre-implantation development. *Nature Methods, 13*, 139–142.

Summers, M. C., & Biggers, J. D. (2003). Chemically defined media and the culture of mammalian preimplantation embryos: Historical perspective and current issues. *Human Reproduction Update, 9*, 557–582.

Sutherland, A. E., Calarco, P. G., & Damsky, C. H. (1993). Developmental regulation of integrin expression at the time of implantation in the mouse embryo. *Development, 119*, 1175–1186.

Tang, F., Barbacioru, C., Nordman, E., Bao, S., Lee, C., Wang, X., et al. (2011). Deterministic and stochastic allele specific gene expression in single mouse blastomeres. *PLoS One, 6*, e21208.

Tao, J., Tamis, R., Fink, K., Williams, B., Nelson-White, T., & Craig, R. (2002). The neglected morula/compact stage embryo transfer. *Human Reproduction, 17*, 1513–1518.

Tarkowski, A. K. (1959). Experiments on the development of isolated blastomers of mouse eggs. *Nature, 184*, 1286–1287.
Tarkowski, A. K. (1961). Mouse chimaeras developed from fused eggs. *Nature, 190*, 857–860.
Tarkowski, A. K., & Wroblewska, J. (1967). Development of blastomeres of mouse eggs isolated at the 4- and 8-cell stage. *Journal of Embryology and Experimental Morphology, 18*, 155–180.
Thomas, F. C., Sheth, B., Eckert, J. J., Bazzoni, G., Dejana, E., & Fleming, T. P. (2004). Contribution of JAM-1 to epithelial differentiation and tight-junction biogenesis in the mouse preimplantation embryo. *Journal of Cell Science, 117*, 5599–5608.
Townes, P. L., & Holtfreter, J. (1955). Directed movements and selective adhesion of embryonic amphibian cells. *Journal of Experimental Zoology, 128*, 53–120.
Vestweber, D., Gossler, A., Boller, K., & Kemler, R. (1987). Expression and distribution of cell adhesion molecule uvomorulin in mouse preimplantation embryos. *Developmental Biology, 124*, 451–456.
Vinot, S., Le, T., Ohno, S., Pawson, T., Maro, B., & Louvet-Vallee, S. (2005). Asymmetric distribution of PAR proteins in the mouse embryo begins at the 8-cell stage during compaction. *Developmental Biology, 282*, 307–319.
Wada, K., Itoga, K., Okano, T., Yonemura, S., & Sasaki, H. (2011). Hippo pathway regulation by cell morphology and stress fibers. *Development, 138*, 3907–3914.
Wales, R. G. (1970). Effects of ions on the development of the preimplantation mouse embryo in vitro. *Australian Journal of Biological Sciences, 23*, 421–429.
Wang, A. Z., Ojakian, G. K., & Nelson, W. J. (1990). Steps in the morphogenesis of a polarized epithelium. I. Uncoupling the roles of cell-cell and cell-substratum contact in establishing plasma membrane polarity in multicellular epithelial (MDCK) cysts. *Journal of Cell Science, 95*(Pt. 1), 137–151.
Whitten, W. K. (1971). Nutritional requirements for the culture of preimplantation embryos in vitro. *Advances in the Biosciences, 6*, 129–139.
Winkel, G. K., Ferguson, J. E., Takeichi, M., & Nuccitelli, R. (1990). Activation of protein kinase C triggers premature compaction in the four-cell stage mouse embryo. *Developmental Biology, 138*, 1–15.
Xiong, F., Ma, W., Hiscock, T. W., Mosaliganti, K. R., Tentner, A. R., Brakke, K. A., et al. (2014). Interplay of cell shape and division orientation promotes robust morphogenesis of developing epithelia. *Cell, 159*, 415–427.
Yamanaka, Y., Ralston, A., Stephenson, R. O., & Rossant, J. (2006). Cell and molecular regulation of the mouse blastocyst. *Developmental Dynamics, 235*, 2301–2314.
Yoshida, S., Barbier de Reuille, P., Lane, B., Bassel, G. W., Prusinkiewicz, P., Smith, R. S., et al. (2014). Genetic control of plant development by overriding a geometric division rule. *Developmental Cell, 29*, 75–87.
Yoshida, C., & Takeichi, M. (1982). Teratocarcinoma cell adhesion: Identification of a cell-surface protein involved in calcium-dependent cell aggregation. *Cell, 28*, 217–224.
Zernicka-Goetz, M., Morris, S. A., & Bruce, A. W. (2009). Making a firm decision: Multifaceted regulation of cell fate in the early mouse embryo. *Nature Reviews. Genetics, 10*, 467–477.
Ziomek, C. A., & Johnson, M. H. (1980). Cell surface interaction induces polarization of mouse 8-cell blastomeres at compaction. *Cell, 21*, 935–942.

CHAPTER EIGHT

Metabolism of Preimplantation Embryo Development: A Bystander or an Active Participant?☆

K.J. Kaneko[1]
Center for Food Safety and Applied Nutrition, U.S. Food and Drug Administration, College Park, MD, United States
[1]Correponding author: e-mail address: kotaro.kaneko@fda.hhs.gov

Contents

☆The opinions expressed in this chapter are the author's personal opinions and do not necessarily reflect those of FDA, DHHS, or the Federal Government.

Current Topics in Developmental Biology, Volume 120
ISSN 0070-2153
http://dx.doi.org/10.1016/bs.ctdb.2016.04.010

Abstract

Unicellular organisms are exquisitely sensitive to nutrient availability in the environment and have evolved elaborate mechanisms to sense the levels and types of nutrients, altering gene expression patterns accordingly to adjust the metabolic activities required to survive. Thus, environmental cues induce adaptive metabolic differentiation through transcriptional and posttranscriptional changes. Similarly, preimplantation embryos are exposed to various environmental cues within the maternal reproductive tract prior to implantation. Because only "simple" culture conditions are required, it is assumed that these embryos are genetically preprogrammed to develop with little influence from the environment, with the exception of few "necessities" provided by the environment. However, a wealth of literature now suggests that the developing embryos are greatly influenced by the maternal environment. Even though the developing embryos have the capacity and plasticity to deal with nutritional imbalance posed by an altered maternal environment, there is often a trade-off to the overall fitness of those embryos later in life. Despite these studies that underline the general importance of the reproductive environment during development, it is thought that the primary driver of mammalian development is strictly genetic and that metabolic adaptation by the preimplantation embryo is secondary to genetic control. In this review, I propose that not only does the maternal environment of developing preimplantation embryos influence developmental potential, pregnancy outcomes, and postnatal disease states, but that it has an active role in induction and potentiation of the first differentiation event, the production of trophectoderm and inner cell mass lineages.

1. INTRODUCTION

For unicellular microorganisms such as yeast, sensing and responding to environmental changes are critical to their survival. Decision to grow, divide, mate, and differentiate is ultimately dictated by the amount and the type of nutrients, including oxygen, temperature, toxicants, and other variables that induce stress response. These microorganisms must achieve an intricate balance between energy intake and cellular processes, such that under favorable conditions, they can grow and divide at will, whereas under unfavorable conditions, they must make wise decisions to optimize growth and survival. In higher multicellular eukaryotic organisms such as mammals, the notion that environmental cues can dictate specific genetic program as a response to these cues is implicit from organismal perspective but often minimized at the cellular level. This is magnified in developmental biology, where a "gene-centric" point of view aims the focus on specific genes, often transcription factors, that direct specific developmental programs. However, it is becoming clear that metabolism plays a significant role in directing developmental programs (Folmes et al., 2011; Onjiko, Moody, & Nemes,

2015; Shyh-Chang, Daley, & Cantley, 2013). For example, maternal diabetes, resulting in hyperglycemic conditions, may alter the gene expression pattern, hence development potential, of the preimplantation embryos (Fraser, Waite, Wood, & Martin, 2007; Jungheim & Moley, 2008). In addition, in many mammalian species, environmental cues and/or metabolic stress can and do elicit developmental pause, such as delayed implantation or embryonic diapause (McGowen, Erez, Romero, & Wildman, 2014; Nieder & Weitlauf, 1984). These and other studies show that the maternal environment strongly influences the developmental potential of the preimplantation embryo (Fleming et al., 2004; Thompson, Mitchell, & Kind, 2007; Watkins, Papenbrock, & Fleming, 2008).

Executing precise metabolic adjustments to achieve "balanced growth" requires complex interplay between sensing what the environment provides to gene expression patterns required to fit the needs of the cell. For example, since the time of Louis Pasteur, it has been known that yeast use respiration/oxidative phosphorylation (OXPHOS) for energy metabolism in the presence of oxygen and anaerobic glycolysis in its absence (Vadlakonda, Dash, Pasupuleti, Anil Kumar, & Reddanna, 2013). However, even under aerobic conditions, yeast can choose respiration or glycolysis depending on the amount and the type of carbon source, such that in the presence of high levels of favorable carbon source such as glucose, yeast will preferably use glycolysis/fermentation for energy metabolism. Under these conditions, yeast will exhibit "glucose repression" whereby a large number of genes not required to ferment glucose are suppressed at the transcriptional level (Conrad et al., 2014; Kayikci & Nielsen, 2015). Aerobic glycolysis and glucose repression can be abrogated simply by modulating glucose uptake; thus, wild-type yeast senses levels of glucose through its metabolism, and this metabolic flux directs transcriptional control. Similarly, yeast can sense the level of nitrogen in the environment upon conditions of nitrogen starvation and respond at the transcriptional level by undergoing differentiation known as pseudophyphal differentiation (Gagiano, Bauer, & Pretorius, 2002).

In higher eukaryotes, whereas it is clear that transcriptional control directs the developmental program, what signal triggers specific cascade of genetic programming is often not known. In the case of preimplantation embryos, a number of transcription factors have been proposed to trigger the onset of differentiation. However, whether or not environmental cues, through metabolic reprogramming, can influence or even trigger preimplantation embryo development has not been adequately explored. In this chapter, I propose that not only metabolic states of preimplantation embryo significantly contribute to its development but that metabolic states, as dictated by its environment, may well be the trigger for induction of the first differentiation event, the production of trophectoderm and inner cell mass (ICM) lineages.

2. PREIMPLANTATION DEVELOPMENT OVERVIEW

Upon fertilization, mammalian zygote begins its developmental journey in the oviduct by first activating its transcription and translation at the late 1-cell/early 2-cell stage in mice[a] (Fig. 1A) (Hamatani et al., 2006; Li, Zheng, & Dean, 2010; Nothias, Majumder, Kaneko, & DePamphilis, 1995; Schultz, 2002). Up to the 8-cell stage, each blastomere of the mouse embryo is equipotent such that any of the blastomeres can give rise to either embryonic or extraembryonic lineages (Kelly, 1977). By the late 8-cell stage, however, the blastomeres become less totipotent and begin showing signs of the first mammalian differentiation event that results in producing the trophectoderm (TE)/placental lineage and the ICM/epiblast lineage (Cockburn & Rossant, 2010; Zernicka-Goetz, Morris, & Bruce, 2009). The blastomeres that reside on the outside will go on to become TE, whereas those residing on the inside eventually give rise to ICM (Tarkowski & Wroblewska, 1967). However, at this early developmental stage, the lineage specification is not set in stone as lineage markers (for example, Cdx2 for TE and Oct4 for ICM) are present in both inside and outside blastomeres (Dietrich & Hiiragi, 2007; Ralston & Rossant, 2008), and the two sets of blastomeres can be manipulated in vitro to produce the other lineage as late as the 32-cell stage (Suwinska, Czolowska, Ozdzenski, & Tarkowski, 2008). At ~8–16-cell stage, the embryos undergo compaction, where individual blastomeres fuse to form a spherical structure called morula. These compacted embryos then transverse into the uterine tract from the oviduct and transform into an embryo structure called the blastocysts in which a "ball" of ICM is surrounded by a layer of TE (Nagy, 2003). TE, being transporting group of epithelial cells, actively establishes an ion gradient, resulting in influx of fluids from the uterine environment to form the blastocoel cavity, a process termed "cavitation" (Kidder, 2002; Watson, 1992). Expanded, fluid-filled blastocysts eventually "hatch" from their outer glycoprotein shell (zona pellucida); the precise mechanism

[a] At the 4-cell stage in humans.

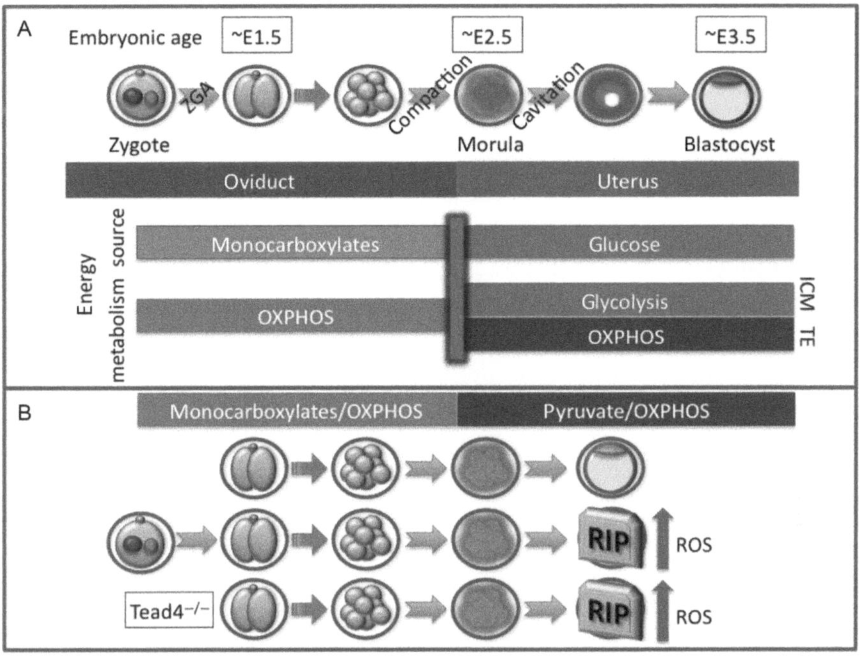

Fig. 1 Overview of mouse preimplantation development. (A) In the oviduct, transcriptionally silent egg is fertilized by the sperm to form the zygote (1-cell embryo), initially with two parental pronuclei, which are chromatin-modified and replicated prior to formation of a 2-cell embryo. Zygotic gene activation (ZGA) begins in the late 1-cell stage, but the major ZGA occurs after the 2-cell embryos are formed at ~E1.5. After the 8-cell stage, Cadherin1-mediated compaction occurs ~E2.5 to form the morula, which now enters the uterus. Outer cells of morula differentiate into trophectoderm (TE), whereas inner cells become the inner cell mass (ICM). Morula-to-blastocyst transition occurs during cavitation, where TE initiates and expands a fluid-filled cavity called the blastocoel. Mouse embryos use monocarboxylates, lactate, and pyruvate, for energy prior to the morula stage. At the morula stage, there is a switch to glucose for energy substrate with corresponding increase in O_2 consumption. Whereas glucose and monocarboxylates are present in both oviduct and the uterus, preimplantation embryos cannot use glucose for energy prior to the morula stage, whereas they preferentially use glucose for energy after the morula stage. Thus glycolytic activity is inhibited prior to the morula stage, presumably due to regulatory mechanisms involving glucose transporters and metabolic inhibition of hexokinase, PFK, and/or other enzymes involved in Embden-Meyerhof-Parnas pathway. Whereas monocarboxylates can only be metabolized as energy through oxidative phosphorylation (OXPHOS), glucose can be metabolized either by glycolysis or OXPHOS. As TE must use energy consuming Na^+/K^+ ATPase for blastocoel expansion, it relies on complete oxidation of glucose through OXPHOS in vivo. ICM, on the other hand, is concerned with replication, maintenance of pluripotency, and anticipation of rapid increase in biomass during early stages of implantation in a hypoxic environment, all suited for glycolytic energy metabolism. (B) Mouse preimplantation embryos have the capacity for metabolic adaptation, as

(Continued)

involved in hatching is not clear and in fact, the process in vivo appears to differ from that of in vitro (Gonzales & Bavister, 1995; Ralston & Rossant, 2008). Clearly, initiation and expansion, as well as maintenance of the blastocoel cavity, are essential for embryo development because it cannot implant and further develop without hatching. At ~4.0 days after fertilization, ICM differentiates to give rise to the primitive endoderm/yolk sac lineage and the epiblast, and shortly thereafter, the blastocysts implant into the uterine wall (Cockburn & Rossant, 2010; Nagy, 2003; Nagy, Gertsenstein, Vintersten, & Behringer, 2003; Rossant & Tam, 2009; Simmons & Cross, 2005; Yamanaka, Ralston, Stephenson, & Rossant, 2006; Zernicka-Goetz et al., 2009). Implantation absolutely relies on functional TE and their differentiated derivative, trophoblast giant cells (Armant, 2005; Cross, 2000; Wang & Dey, 2006). Thus, during preimplantation development, a murine fertilized egg has the unenviable task of dividing 6–7 times to produce 64–128 cells, consisting of 2 extraembryonic lineages in addition to pluripotent embryonic cells, expanding the blastocoel cavity, and hatching from zona pellucida all within ~4.5 days. The rate at which the embryo develops in the female reproductive tract is critical as it must reach a hatched blastocyst stage within the narrow window of uterine receptivity (Ma, Song, Das, Paria, & Dey, 2003).

With the advances in manipulation of gene expression via knockout mouse and gene knockdown methodologies combined with in silico-aided survey of changing gene expression landscape during preimplantation development, molecular mechanisms that control development at this stage are becoming clearer. However, many of these studies involving genetic manipulations mostly adopt a strictly "gene-centric" point of view to explain the observed developmental phenotype. Although morula-to-blastocyst transition designates one of the most important early developmental milestones by generating two cell lineages, a survey of knockout mice with lethality prior to blastocyst formation reveals that relatively few genes, especially transcription factors, cause this phenotype (Kohn, Kaneko, & DePamphilis, 2012).

Fig. 1—Cont'd evidenced by the 2-cell embryos to develop to blastocysts in complete absence of glucose in culture (Jansen, Cashman, Thompson, Pantaleon, & Kaye, 2009; Martin & Leese, 1995). Wild-type embryos can undergo metabolic reprogramming at the morula stage to utilize pyruvate in the absence of glucose. Glucose, however, is essential for 1-cell embryos to develop to blastocysts, presumably because glucose metabolism is needed not for energy but for alleviation of oxidative stress (ROS) during blastocyst formation. Interestingly, *Tead4*$^{-/-}$ 2-cell embryos cannot develop to blastocysts in the absence of glucose, but can form blastocoel by addition of an antioxidant during glucose-free culture (Kaneko & DePamphilis, 2013). (See the color plate.)

For example, ablation of neither TE-specific Cdx2 nor ICM-specific Oct4 causes preblastocyst lethality, even though both genes play a major role in their respective cell lineages after the formation of blastocysts. Whereas the persistence of maternal mRNA or proteins complicates the interpretation of early embryonic phenotype of knockout mouse models, it is now clear that neither the elimination of maternal Oct4 nor Cdx2 results in the prevention of blastocyst formation (Blij, Frum, Akyol, Fearon, & Ralston, 2012; Wu et al., 2013). These observations suggest a need to reexamine the role of these lineage-specific transcription factors with respect to specification and differentiation of the cell lineages produced during preimplantation development.

Underlying the recent elucidation of preimplantation embryo development at the molecular level is the wealth of knowledge obtained during development of protocols for in vitro fertilization and preimplantation embryo culture (Gardner, 1998; Johnson, Mahmood, & Patel, 2003; Lawitts & Biggers, 1993; Leese, Conaghan, Martin, & Hardy, 1993). These studies revealed not only what the minimal culture requirements were but also the changing energy substrates and metabolic reprogramming that occurs during preimplantation development. Furthermore, these studies revealed the inherent adaptability of the developing embryos in a variety of culture conditions. These studies also identified seemingly peculiar metabolic needs by the developing embryos. Furthermore, despite the plasticity of preimplantation embryos to develop under varying conditions, presumably through adaptive metabolic reprogramming, it seems that certain metabolic parameters must be met in order for development to proceed.

The goal of this chapter is to provide one developmental biologist's view on how preimplantation embryo development during blastocyst formation might proceed by interpreting some metabolic peculiarities observed during preimplantation embryo in light of increasing understanding of how metabolism affects gene expression. But in order to make a seemingly coherent argument, a brief review of energy metabolism is first provided. Next, several highlights of how environmental cues dictate yeast differentiation programs are presented to serve as a model to understand how metabolic cues may influence preimplantation embryo development.

3. ENERGY METABOLISM

For the purposes of this discussion, I will define "energy metabolism" as those metabolic pathways involved in generating energy, mainly in the form of ATP and NAD/NADPH, as well as those pathways involved in maintaining redox balance within the cell. Energy "homeostasis" is used here to confer the idea that a cell must produce sufficient ATP to energize a needed metabolic function (ie, maintain osmolarity, pH, etc., or increase biomass) but still retain appropriate redox balance. Sections 3.1–3.6 (Fig. 2) summarize a simplified overview of important metabolic pathways relevant to the discussion of energy metabolism in preimplantation embryos, with the caveat that the operational metabolic system is far more complex than described.

3.1 Glycolysis vs OXPHOS

For most organisms, glucose is the preferred carbon source for energy production. Not surprisingly, glucose metabolism is conserved throughout almost all living organisms with little variation (exceptions exist in prokaryotes), likely owing to the fact that the pathway evolved in a thermodynamically and chemically sensible manner (Bar-Even, Flamholz, Noor, & Milo, 2012). The catabolism of glucose (6-carbon molecule) to 2 molecules each of pyruvate (3-carbon molecule), ATP, NADH, and H_2O using 10 sequential enzymatic reactions (Embden–Meyerhof–Parnas or EMP pathway) is used for both glycolysis and OXPHOS pathways. The first half of the EMP pathway is often termed the "preparatory phase" because ATP is consumed to "prime" glucose, whereas in the second half of the pathway is often referred as the "pay-off phase" because ATP and NADH are produced. In the first step of glycolysis, glucose is first phosphorylated by hexokinase or glucokinase to produce glucose-6-phosphate (G6P). G6P is not only a substrate for an important divergent metabolic pathway (pentose phosphate pathway/PPP—see later) but also a substrate for storage in the form of glycogen. In EMP pathway, G6P is further modified to fructose 1,6-bisphosphate (F1,6BP) by phosphofructokinase (PFK), a rate-limiting step in the EMP pathway. F1,6BP is then cleaved to form two related 3-carbon molecules, dihydroxyacetone-phosphate and its isomer, glyceraldehyde 3-phosphate (G3P). G3P is then converted to glycerate 1,3-bisphosphate by G3P dehydrogenase, which requires NAD^+. Glycerate 1,3-bisphosphate is then used to generate ATP through substrate-level phosphorylation. A second substrate-level phosphorylation occurs at the last step to produce pyruvate via pyruvate kinase (PK), with the overall net gain of 2 ATPs per 1 molecule of glucose. However, much of the chemical energy is still retained within the final by-product of the EMP pathway, pyruvate. EMP pathway reactions in eukaryotes are normally carried out in the cytoplasm.

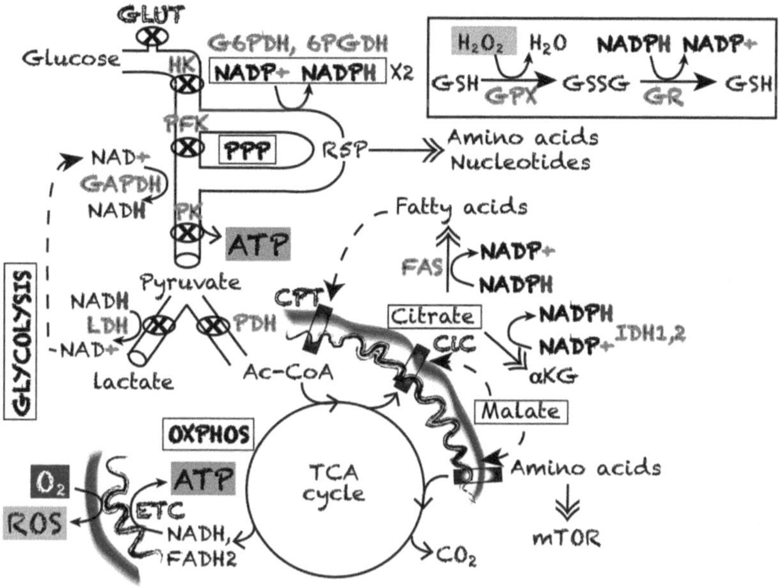

Fig. 2 Overview of relevant energy metabolism. Glucose can be metabolized in a number of ways, which is regulated at multiple points. Glucose entry into cells can be controlled by glucose transporters (GLUT), which is one target of Hypoxia Inducible Factor 1 (HIF1α). Glucose is then phosphorylated by hexose kinase (HK) to glucose 6-phosphate (G6P), which can be directed toward either Embden–Meyerhof–Parnas pathway (EMP) for ATP generation or diverted toward pentose phosphate pathway (PPP) to generate NADPH and ribose 5-phosphate (R5P). G6P is also a substrate for glycogen and glucosamine pathways (not shown). EMP is regulated at multiple points by phosphofructokinase (PFK), pyruvate kinase (PK), and glyceraldehyde phosphate dehydrogenase (GAPDH), which relies on the availability of NAD^+. Whether glucose is metabolized by glycolysis or oxidative phosphorylation (OXPHOS) is determined by the fate of pyruvate. Lactate dehydrogenase (LDH) produces lactate and regenerates NAD^+ for glycolysis, while for OXPHOS, pyruvate is transported to the mitochondria to generate acetyl-CoA (Ac-CoA), which is fed into the tricarboxylic acid (TCA) cycle through the action of pyruvate dehydrogenase complex (PDH). TCA cycle generates reducing equivalents in the form of NADH and $FADH_2$, which is used by the electron transport chain (ETC) in the inner membrane to generate an electrochemical gradient that ultimately drives ATP synthesis. O_2 serves as the final electron acceptor to produce H_2O, but the chemical properties of O_2 make it a promiscuous electron acceptor, resulting in production of reactive oxygen species (ROS). Cells have multiple pathways to neutralize ROS, including superoxide dismutase (not shown) that reacts with oxygen radical to produce hydrogen peroxide (H_2O_2), which is then neutralized by either catalase (not shown) or glutathione peroxidase (GPX) to form H_2O. Reduction of H_2O_2 results in oxidation of glutathione (GSH) to glutathione disulfide (GSSG), which must be reduced by glutathione reductase (GR) and NADPH to regenerate GSH (*inset*). Amino acids can be metabolized for energy through pyruvate or Ac-CoA intermediates which feed into the TCA cycle and OXPHOS. To regenerate TCA components that leave the cycle for

(Continued)

Whether glycolysis or OXPHOS is utilized depends on the fate of pyruvate and the presence of oxygen. In most circumstances, in the absence of oxygen, pyruvate is oxidized to lactate (in mammals) or acetaldehyde (in yeast) by enzymes LDH and pyruvate decarboxylase, respectively. Acetaldehyde is then converted to ethanol by aldehyde dehydrogenase. Lactate and ethanol production by the dehydrogenases allows continuation of the EMP pathway by regenerating oxidized NAD^+ needed for G3P dehydrogenase. Although this pathway does not result in efficient generation of ATP, it does allow efficient flux through the system. Thus, when excess glucose is present in the environment and little oxygen is available, glycolysis can supply ATP at a much faster rate, albeit less efficiently compared to OXPHOS.

In the presence of oxygen, pyruvate is converted to acetyl-CoA (Ac-CoA) in the mitochondrial matrix by pyruvate dehydrogenase complex (PDH), producing NADH and CO_2 in the process. Ac-CoA then enters the tricarboxylic acid (TCA) cycle, which results in generation of CO_2, reducing equivalents, in the forms of NADH and $FADH_2$, GTP, as well as substrates for biosynthetic pathways. NADH and $FADH_2$ are then used as electron donors in the electron transport chain (ETC) in the inner

Fig. 2—Cont'd biosynthetic pathways such as citrate, some amino acids (like glutamine) can be metabolized into TCA components such as α-ketoglutarate (αKG) by a process known as anaplerosis. Amino acids, especially leucine and glutamine, upregulate mammalian target of rapamycin (mTOR) activity, leading to increase in OXPHOS activity. Fatty acids (FAs) are also energy substrates, which enter the mitochondria through regulated transport systems, which include carnitine palmitoyltransferase 1 (CPT1), a rate-limiting step that is irreversibly inhibited by etomoxir. Upon entering the mitochondria, FAs are metabolized to Ac-CoA units which feed into the TCA cycle. As FAs are synthesized in the cytoplasm by fatty acid synthase (FAS), which requires citrate as a substrate, glucose and amino acids can provide carbon sources for citrate synthesis in the mitochondria. However, citrate must be transported out of mitochondria by citrate carrier (CiC), which is an antiporter that requires dicarboxylates such as malate for exchange. FAS also requires NADPH, which is mostly provided by G6P dehydrogenase (G6PD) and 6-phosphogluconate dehydrogenase (6PGDH) in the PPP, as well as by NADPH-dependent isocitrate dehydrogenases 1 and 2 (IDH1, 2) that oxidize isocitrate (which can be derived from cytoplasmic citrate) to αKG. Thus in addition to its role in fatty acid biosynthesis, cytoplasmic citrate has multiple roles in preimplantation embryos as it not only can inhibit PFK (thus regulate glycolysis) but also regenerate NADPH (Dumollard, Carroll, Duchen, Campbell, & Swann, 2009). Cytosolic malate is also an important regulator of redox balance as part of "malate-aspartate" shuttle, which is essential during preimplantation development (Lane & Gardner, 2005), as well as being a substrate for regenerating NADPH via malic enzyme (not shown). (See the color plate.)

mitochondrial membrane, leading to an electrochemical gradient (resulting in proton motive force) that drives ATP synthesis. Molecular oxygen is an absolute essential component in this process as it acts as the electron acceptor. Thus, theoretically, this pathway, often referred as respiration, can generate up to 36 net molecules of ATP per 1 molecule of glucose, compared to 2 net molecules of ATP per glucose in the glycolytic pathway.

3.2 Problem with OXPHOS

Whereas complete oxidation of glucose via OXPHOS results in more efficient ATP synthesis, it does come with a price: a generation of reactive oxygen species (ROS). Molecular oxygen, while being a suitable electron acceptor, has two unpaired electrons and therefore is a "biradical" (Davies, 1995). As such, O_2 serves a useful role acting as an electron sink at the end of the ETC, where cytochrome oxidase (Complex IV) catalyzes the formation of H_2O via the transfer of four electrons. However, the components of the ETC prior to cytochrome oxidase are involved in transfer of one or two electrons, making them susceptible to uncontrolled electron transfer to O_2 for the generation of superoxides, hydrogen peroxides, and hydroxyl radicals (Murphy, 2009). These ROS species can readily react with various macromolecules including lipids, carbohydrates, proteins, and nucleic acids. Therefore, chemistry of O_2 and the biology of ETC necessarily result in production of ROS when glucose is oxidized via OXPHOS (Adelman, Saul, & Ames, 1988; Davies, 1995; Finkel & Holbrook, 2000; Turrens, 2003). It should be noted that OXPHOS-associated ROS is not the only endogenous ROS produced. However, given the central importance of OXPHOS to the metabolic needs of an aerobic organism, it is likely that ROS produced by the mitochondria plays a major role during active metabolic processes that require efficient ATP production (Andreyev, Kushnareva, Murphy, & Starkov, 2015; Balaban, Nemoto, & Finkel, 2005).

Aerobic organisms have evolved numerous strategies to neutralize ROS, including several enzymatic systems, such as superoxide dismutases, which convert superoxide to hydrogen peroxide. H_2O_2 is then neutralized by heme-containing catalases, as well as glutathione peroxidases, and peroxiredoxins. The latter two scavenger systems rely on redox states of glutathione and thioredoxins, respectively. In turn, reduced forms of glutathione and thioredoxins must be regenerated from NADPH (Fig. 2). Therefore, ROS scavenger activities rely on adequate reducing equivalents within the cell, specifically in the form of NADPH. In other words, the cells must harbor sufficient NADPH/$NADP^+$ ratio in order to maintain redox balance to neutralize ROS production during OXPHOS.

3.3 Importance of PPP

Eukaryotes mainly possess three major pathways[b] for generating NADPH from $NADP^+$. In the cytoplasm, enzymes comparable but distinct from those involved in the TCA cycle, isocitrate dehydrogenase and malic enzymes, can produce NADPH (Cairns & Mak, 2013; Dalziel, 1980; Lee et al., 2002). Additionally in the cytosol, NADPH is produced from PPP, a pathway intimately linked to EMP pathway (Fig. 2). PPP is divided into oxidative and nonoxidative branch (Patra & Hay, 2014; Riganti, Gazzano, Polimeni, Aldieri, & Ghigo, 2012). In the nonoxidative branch, EMP by-products, G3P, and fructose-6-phosphate combine in a series of reversible reactions to produce ribose-5-phosphate (R5P) and xylulose-5-phosphate (X5P). R5P is a precursor to nucleotide biosynthesis, whereas X5P can be used to synthesize aromatic amino acids. In the oxidative branch, G6P is converted to ribulose-5-phosphate (Rb5P) producing two equivalents of NADPH. Rb5P can then be converted to both X5P and R5P, linking the two branches of PPP. Whereas reactions in the oxidative branch are not reversible, all reactions within the nonoxidative branch are reversible, essentially allowing glycolysis in reverse under certain circumstances (Patra & Hay, 2014).

The importance of PPP during cell proliferation lies in its ability to synthesize both NADPH and critical biosynthetic precursors, R5P and X5P. In addition to its ability to regenerate redox components within the antioxidant defense, NADPH is also necessary to provide reducing power for various biosynthetic pathways, including fatty acid (FA) biosynthesis. Whereas the oxidative branch alone can provide NADPH, R5P, and X5P, the reversible nonoxidative branch allows exquisite control of glucose metabolism when both reducing powers and ATPs are needed. This is because the nonoxidative branch can convert X5P and R5P back to F5P and G3P, which can then be metabolized by EMP pathway for ATP generation. Thus, PPP allows glucose to be metabolized to generate NADPH, R5P/X5P, and ATP. Because other energy sources, such as lipids, monocarboxylates, and amino acids, cannot directly be metabolized to G6P (a substrate for PPP), glucose becomes a crucial energy substrate in cells that require energy,

[b] Mitochondria also have nicotinamide nucleotide transhydrogenase for NADPH production (Yin, Sancheti, & Cadenas, 2012).

anabolic precursors, and reducing power. It is therefore not surprising that functional PPP is associated with cells undergoing rapid proliferation such as cancer cells (Jiang, Du, & Wu, 2014; Patra & Hay, 2014).

A related but parallel pathway that is essential to production of cytoplasmic NADPH is the beta-oxidation of FAs (Hayes & Dinkova-Kostova, 2014). Whereas NADPH serves as a reducing agent for FA synthesis, which takes place in the cytoplasm, NAD + is the oxidizing agent during oxidation, which takes place in the mitochondria. FAs are oxidized to Ac-CoA, combined with oxaloacetate to generate citrate, which can be incorporated into the TCA cycle for energy production. In addition, citrate produced upon FA oxidation can be transported to the cytoplasm and used to generate NADPH through two pathways: (1) citrate is converted to isocitrate, a substrate for isocitrate dehydrogenase 1 that produces NAPDH and α-ketoglutarate; (2) citrate is converted to Ac-CoA and oxaloacetate by ATP citrate lyase; oxaloacetate is then converted to malate, which is then converted to pyruvate by malic enzyme, producing NADPH. It has also been suggested that FA oxidation may directly or indirectly influence PPP activity (Tyra, Spitz, & Rutkowski, 2012). Therefore, FA oxidation plays an essential role during cancer cell proliferation by providing needed NADPH for redox balance (Carracedo, Cantley, & Pandolfi, 2013; Hayes & Dinkova-Kostova, 2014; Pike, Smift, Croteau, Ferrick, & Wu, 2011).

3.4 Regulation of Glucose Metabolism: Glycolysis vs OXPHOS

Since the days of Pasteur, it has been observed that under anaerobic conditions, glycolysis is the predominant pathway to metabolize glucose, often termed the Pasteur effect, whereas under aerobic conditions, respiration becomes the dominant energy metabolism as O_2 suppresses glycolytic fermentation (Engelhardt, 1974; Vadlakonda et al., 2013). Induction of glycolytic pathway in the absence of O_2 occurs by multiple regulatory pathways, which ultimately result in upregulation of glucose uptake as well as increase in flux through EMP pathway via upregulating the levels and activities of enzymes of EMP as well as LDH or ADH. In simplest terms, decrease in O_2 leads to decrease in ATP yield as well as activation of hypoxia-inducible factor 1 (HIF1). Low ATP/AMP ratio activates PFK as well as AMP-activated protein kinase (AMPK), the latter activating 6-phosphofructo-2-kinase to increase fructose-2,6-bisphosphate, an allosteric activator of PFK. HIF1 upregulates various EMP enzymes as well as LDH to increase glycolysis. In addition, HIF1 upregulated expression of pyruvate dehydrogenase kinase 1, which in turn inhibit PDH, a key enzyme in incorporating pyruvate into the TCA cycle, effectively downregulating OXPHOS (Kim, Tchernyshyov, Semenza, & Dang, 2006).

In differentiated cells, the presence of O_2 results in inhibition of glycolytic metabolism because of increase in ATP/AMP ratio as well as O_2-mediated degradation of HIF1 (Maldonado & Lemasters, 2014; Schonenberger & Kovacs, 2015; Semenza, 1999). Proliferating cells, in contrast, often show high glycolytic activity even in the presence of O_2, phenomenon referred as aerobic glycolysis or Warburg effect (Warburg, 1956). Although Warburg hypothesized that aerobic glycolysis is a result of an aberrant metabolic pathway brought on by a defect in mitochondrial metabolism, it is now clear that highly proliferative "normal" cells, such as embryonic stem cells (ESCs), do show aerobic glycolysis. Although precise mechanism on how normal proliferating cells switch from respiration to aerobic glycolysis is not clear, it is thought that growth factors through protein kinase pathways (such as PI3 or tyrosine kinase) modulate glucose metabolism to shift toward glycolysis rather than the TCA cycle (Feng & Levine, 2010). For example, glucose transporters become upregulated in cancer cells, presumably via upregulated *myc* oncogene, allowing increased uptake of glucose by the cells (Osthus et al., 2000). What is clear is that in the presence of high levels of glucose and a nitrogen source such as glutamine, proliferating cells prefer glycolysis because the pathway allows greater flux through the EMP and PPP pathways that result in increased levels of intermediary substrates as well as NADPH's reducing power to increase biomass (Vander Heiden, Cantley, & Thompson, 2009). A common theme in highly proliferative cells, including cancer cells, is that many biological processes involved in glycolysis are upregulated, whereas OXPHOS processes are downregulated. In fact, in some yeasts and some mammalian cells, glucose concentration inversely correlates with OXPHOS activity, a phenomenon known as Crabtree effect (Crabtree, 1929). A number of explanations have been proposed to explain this effect, including competition for ADP and inorganic phosphate (Pi) between glycolysis and OXPHOS, changes in intracellular pH and $[Ca^{2+}]$, and metabolite (such as F1,6BP) suppression of complexes III and IV and/or modulation of mitochondrial unspecific channel (Diaz-Ruiz et al., 2008; Rodriguez-Enriquez, Juarez, Rodriguez-Zavala, & Moreno-Sanchez, 2001; Rosas-Lemus, Uribe-Alvarez, Chiquete-Felix, & Uribe-Carvajal, 2014; Wojtczak, 1996). No matter the specific mechanism, it seems likely that when glucose is abundant, an overflow of EMP pathway by-products likely led to an evolution of a strategy that resulted in maximizing flux through the EMP pathway (via Warburg/Crabtree effect) in order to increase biomass, and many higher eukaryotes, including humans, utilize a similar strategy (Schuster, Boley, Moller, Stark, & Kaleta, 2015; Vander Heiden et al., 2009).

Aerobic glycolysis does not imply that no O_2 is consumed or that OXPHOS does not take place during cell proliferation. In fact, some OXPHOS is required for proliferation of some cells as ETC and O_2 act as electron sink to support aspartate biosynthesis (Birsoy et al., 2015; Sullivan et al., 2015). Therefore, OXPHOS not only supplies energy but also plays an important role in maintenance of redox balance for biosynthetic pathways critical for increase in biomass. Counterintuitively, reduction in O_2 (ie, anaerobic metabolism) also induces ROS as a signaling molecule (Hamanaka & Chandel, 2009).

Although in the literature, glycolytic pathway is often associated with proliferative cells and OXPHOS associated with differentiated, nondividing cells, in reality, most cells not only have the capacity to adapt their metabolism to fit the need but also often use combination of glycolytic and OXPHOS energy metabolism. In this context, it is important to keep in mind that embryonic cells utilize both types of metabolism in order to accomplish the developmental goal of proliferation and differentiation. In placental mammals, the embryo develops in changing environment that dictate the type of energy metabolism required. For example, during implantation, it is thought that embryos are exposed to low ambient oxygen and thus must accommodate those environmental cues to alter their metabolism; however, exposure to sudden burst in oxygen concentration, expected to occur as the circulatory and placental development are completed, would lead OXPHOS metabolism.

3.5 Mitochondrial Metabolism

Mitochondria are maternally inherited organelle, as paternal mitochondria are eliminated after fertilization (Sato & Sato, 2013). Mitochondria are extensively replicated during oocyte maturation, but after fertilization, the mitochondrial number per embryo remains constant until the blastocyst stage when mitochondrial replication is reinitiated (St John, Facucho-Oliveira, Jiang, Kelly, & Salah, 2010). Mitochondrial dysfunctions induced by increased oxidative stress due to overnutrition or aging during oocyte maturation have been suggested to play a major role in infertility and early pregnancy loss (Grindler & Moley, 2013; Thouas, Trounson, Wolvetang, &

Jones, 2004). In addition, mitochondria play critical roles in maintenance of redox states during oocyte maturation and early preimplantation development (Dumollard et al., 2009; Van Blerkom, 2009). Therefore, mitochondria are a maternally inherited component that plays a major role in preimplantation development.

Although mitochondria are the site of OXPHOS, the organelle also serves as a site of other essential intermediary metabolism, including FA beta-oxidation, amino acid biosynthesis, and cholesterol and phospholipid biosynthesis, as well as being key regulators of apoptosis, calcium, and ROS signaling (Nunnari & Suomalainen, 2012). In addition, emerging discoveries strongly suggest that it also plays critical roles in governing cell fate and differentiation. Mitochondrial genome only encodes 13 polypeptides all involved in OXPHOS machinery along with rRNAs and tRNAs and thus requires nuclear-encoded components to carry out all metabolic and cellular functions (Taanman, 1999). Mitochondria have intricate symbiotic relationship with the nuclear genome as well as with other organelles including peroxisomes and endoplasmic reticulum (Fan & Evans, 2015; Guantes et al., 2015; Kotiadis, Duchen, & Osellame, 2014; Marchi, Patergnani, & Pinton, 2014; Wai & Langer, 2016). In addition, mitochondria exist in dynamic morphological states through fission and fusion processes, whereby they can exist in various forms ranging from punctate moieties to hypercondensed filaments. Under low nutrient or low stress conditions, mitochondrial fusion promotes OXPHOS, whereas under excess nutrient or high stress conditions, mitochondria undergo fission ultimately leading to reduced OXPHOS, mitochondrial quality control of mitophagy, or cell death (Chan, 2012; Stotland & Gottlieb, 2015; Wai & Langer, 2016). Details of how mitochondrial fusion and fission processes are regulated have been reviewed recently and will not be covered here. Multiple players are involved in balancing the rate of mitochondrial fusion and fission to bring about homeostatic balance. Mitochondrial fusion requires GTPase proteins, outer membrane-bound mitofusion proteins (MFN1/2) and inner membrane-bound L-OPA1 protein, whereas mitochondrial fission requires Drp1, a dynamin-related cytoplasmic protein that is recruited to bind to specific receptors on the mitochondrial outer membrane upon appropriate signals (Ishihara, Kohno, & Ishihara, 2015; Wai & Langer, 2016). Not only is balanced mitochondrial fusion/fission essential during development, but they are also critical for preventing the onset of mitochondrial dysfunction diseases in adults. In addition, morphology of individual mitochondria can change depending on the metabolic state of the cells. For example,

OXPHOS activity correlates with changes in the morphology of cristae such that extensive and efficient OXPHOS is associated with elongated and more extensively folded or condensed cristae (Cogliati et al., 2013; Mannella, 2006). L-OPA1 protein functions to regulate the shape of cristae through their interaction with mitochondrial solute carrier protein (SLC25A), independent of its function in the fission/fusion process (Patten et al., 2014). These and other studies strongly suggest that mitochondria do not merely serve a subservient role as a cellular energy factory but plays major roles in sensing and dictating cellular processes that ultimately bring about changes in nuclear gene expression (Guantes, Diaz-Colunga, & Iborra, 2016; Whelan & Zuckerbraun, 2013).

3.6 Contribution of Amino Acids and FAs to Energy Metabolism

Both amino acids and FAs are catabolized in the mitochondria. Amino acids regulate energy production, protein biosynthesis, autophagy, and energy homeostasis through multiple signal transduction pathways that sense nutrient levels (Sengupta, Peterson, & Sabatini, 2010). Thus, amino acids are important components of energy metabolism as well as for anabolic pathways required for cell proliferation. Although the rate of O_2 consumption and OXPHOS is primarily induced by the need for ATP (ie, ATP/ADP ratio), it is thought that amino acids can also modulate OXPHOS activity. For energy production, amino acids enhance OXPHOS by increasing the TCA flux, as metabolites from some amino acids are catabolized to either pyruvate or Ac-CoA which can feed into the TCA cycle (Kingsbury, Sen, & Cardenas, 2015; Nelson, Lehninger, & Cox, 2008). Some amino acids are also metabolized into components of the TCA cycle through a process known as anaplerosis (Owen, Kalhan, & Hanson, 2002). To maintain the anion balance within the TCA cycle, various intermediates are also removed from the cycle through a process known as cataplerosis. Anaplerosis/cataplerosis allows cells to utilize amino acid metabolism for biosynthesis, ultimately leading to increase in biomass for the proliferating cells. This is a critical process during cell proliferation as components of TCA (such as citrate) that leave the mitochondria to be used for FA and other anabolic biosynthesis must be replenished (often through glutamine intake and catabolism).

A critical pathway that integrates amino acid signaling and energy is the "mammalian target of rapamycin" (mTOR) protein kinase complexes (mTORC1/2) that establish energy homeostasis by balancing metabolic demand with supply (Choo et al., 2010; DeBerardinis, Lum, Hatzivassiliou, & Thompson, 2008). Activation of mTOR by amino acids, especially glutamine and branched-chain amino acids such as leucine, leads to an increase in OXPHOS activity as well as in ribosome synthesis for biomass accumulation (Fan et al., 2013; Kingsbury et al., 2015; Laplante & Sabatini, 2012; Schieke et al., 2006; Tokunaga, Yoshino, & Yonezawa, 2004). Glutamine and leucine may both directly activate mTOR through distinct pathways (Jewell et al., 2015), or glutamine may indirectly modulate leucine activation of mTOR through glutamine/leucine antiporter in which glutamine's cellular export is coupled to leucine import (Nicklin et al., 2009). Thus, it is not surprising that cancer cells are "addicted" to glutamine as it serves an important role in anaplerosis as well as in activation of mTOR (Dang, 2012).

FAs also play a central role in energy metabolism. Although FAs can be imported into cells from the environment, cell proliferation often requires de novo FA synthesis. FA synthesis takes place in the cytoplasm via FA synthase, which requires Ac-CoA and NADPH (Wakil, 1989). Ac-CoA is formed from pyruvate oxidation, amino acid catabolism through the TCA cycle, or from FA oxidation, processes that all occur within the mitochondria (Nelson et al., 2008). As Ac-CoA cannot cross the mitochondrial membrane, it must be "transferred" out of mitochondria through intermediate formation of citrate with OAA. Citrate then is transported out of mitochondria through TCA/malate antiporter, upon which citrate lyase converts citrate to cytosolic Ac-CoA with OAA as a by-product. Malate dehydrogenase then converts OAA to malate, which is transported back to mitochondria as a counterion for the TCA/malate antiporter. As FA synthesis requires NADPH, importance of functional operation of the oxidative branch of PPP is revealed (Katz, Landau, & Bartsch, 1966; Patra & Hay, 2014).

When extra ATP is required, FAs serve an important role as energy substrates since they can produce roughly twice as much ATP per mass compared to carbohydrate oxidation (Carracedo et al., 2013; Nelson et al., 2008). FA beta-oxidation is a process whereby FAs of various chain lengths are oxidized to individual Ac-CoA units. As beta-oxidation takes place in the mitochondria, FAs are imported from the cytoplasm using carnitine shuttle system, in which carnitine palmitoyltransferase 1 (CPT1) is a component. FA oxidation is regulated at multiple levels, including peroxisome proliferator-activated receptors (PPARs) and coactivators (PGCs), as well as posttranscriptionally, such as through allosteric control of CPT1 by malonyl-CoA (important building block of FA biosynthesis; Foster, 2012), the levels of which are controlled by PPAR and AMPK (Houten, Violante, Ventura, & Wanders, 2016). Other posttranslational regulations of FA oxidation have been proposed including acetylation/deacetylation by SIRT1 and S-nitrosylation by nitric oxide.

Interestingly, the inhibition of FA beta-oxidation in cancer cells by etomoxir, an inhibitor of CPT1, leads not only to decrease in cellular ATP but also to decrease in NADPH and reduced glutathione with corresponding upregulation of ROS (Pike et al., 2011). Inclusion of an antioxidant in the presence of etomoxir alleviates depletion of NADPH and ATP, strongly suggesting that one metabolic function of FA oxidation is to provide sufficient reducing power to reduce oxidative stress. Although concrete proof is lacking, a link between FA oxidation and NADPH production has been proposed through the activity of mitochondrial NADPH-dependent isocitrate dehydrogenase by FA-derived Ac-CoA. The fact that cancer cells often show upregulation of FA synthase to promote FA synthesis as well as FA oxidation leading to increased cellular respiration suggests that FA metabolism plays a critical role to modulate and prevent ATP depletion (Zaytseva et al., 2015).

3.7 Summary of Basic Energy Metabolism

When glucose is catabolized for energy production, the fate of pyruvate determines whether glycolysis or OXPHOS pathway is used. Generally, in the presence of oxygen, pyruvate is fed into the TCA cycle through Ac-CoA, resulting in reducing equivalents used to generate proton motive force driving ATP synthesis. While overall ATP yield is higher with OXPHOS compared to glycolysis, ROS is necessarily generated due to inefficiency of the ETC in the mitochondria. Because ROS can readily oxidize macromolecules, the levels are strictly regulated through a number of mechanisms, including the presence of glutathione and other scavenger systems. In the absence or in lieu of glucose, monocarboxylates, such as pyruvate and lactate, FAs and amino acids can also be used as energy source through OXPHOS. However, glucose serves additional metabolic roles as various intermediary catabolites within the EMP pathway can be utilized as substrates for biosynthesis of some amino acids and lipids. In addition, diversion of G6P through PPP results in not only substrates for biosynthesis of nucleic acids and aromatic amino acids but also increase in reducing equivalents via generation of NADPH, which in turn can be used to regenerate glutathione as well as promote FA biosynthesis. On the other hand, oxidation of FA can also result in ATP synthesis through the TCA cycle and regenerate NADPH. Amino acids can also be an important source of energy by feeding through the TCA cycle, but they can also provide nitrogen and other building blocks for biosynthesis as well as as well as serve as important regulators of energy metabolism such as mTOR. Thus, for cells undergoing expensive energy expenditures and/or increase in biomass, there is a trade-off between generation of ATP and production of reducing equivalents and other building block substrates. Mitochondria play a central role in sensing and communicating with the nucleus and other organelles to mediate metabolic needs and energy provision. Maintaining this balance, energy homeostasis, is critical for microorganisms and preimplantation embryos.

4. ENERGY METABOLISM AND YEAST DIFFERENTIATION

Most yeast species found in nature have the capacity to exist in two morphologically distinct states: spherical yeast form and multicellular filamentous form (reviewed in Gagiano et al., 2002; Gancedo, 2001; Madhani & Fink, 1998). These two forms of growth are usually distinguished by the mechanism by which daughter cells are produced (budding during yeast growth vs elongation during filamentous growth) and whether or not physical separation occurs between mother and daughter cells. Although some commonly used yeast strains have lost the ability to grow in filamentous form, it is thought that in nature, this dimorphic switch to filamentous form allows yeast to scavenge for nutrients or escape from harmful environments. Multiple types of filamentous forms are observed, depending on the species, the ploidy, and/or the methods of induction. Although differences exist in the types of filamentous forms (for example, haploid invasive growth and diploid pseudohyphal growth), fundamental mechanisms for induction of this "differentiation" pathway are similar, and "filamentous growth" will be used here to describe both processes interchangeably (Gagiano et al., 2002; Roberts & Fink, 1994). A variety of external signals can induce switch from yeast/spherical growth to filamentous growth, including changes in carbon source, depletion of nitrogen source, pH or osmotic stress, temperatures, and other environment cues. All yeasts have extensive sensing mechanisms that allow rapid metabolic and gene expression adaptations to the changing environment. Various signaling pathways interact in a complex manner to ultimately bring about adaptation, including changes in metabolism and cell cycle, leading to expression of various filamentous target genes, many of which are genes required for adhesion, such as FLO11 and FLO8. The purpose of this section is not to review the copious amount of data on regulation of filamentous phenotype in yeasts but to provide several highlights of the literature that pertain to my discussion of the relationship between energy metabolism and preimplantation development. For readers interested in yeast filamentation, a number of detailed reviews have been published in addition to those cited earlier (Cullen & Sprague, 2012; Schneper, Duvel, & Broach, 2004; Sudbery, 2011; Whiteway & Bachewich, 2007). It should be noted that environmental cues utilize the same or similar pathways to induce other "differentiated" programs in yeast, such as sporulation and quiescence.

4.1 Signaling Pathways Involved in Filamentous Growth

A number of signaling pathways regulate yeast differentiation program of filamentous growth. An environmental signal, such as stress, carbon source, and nitrogen starvation, transmits cascading signals that ultimately upregulate genes for filamentation and suppress pathways for yeast growth. The main pathways identified include mitogen-activated protein kinase (MAPK) and cAMP-dependent protein kinase A (PKA) pathways (Gancedo, 2001). These pathways can interact independently and interdependently to regulate genes necessary for filamentous growth (Fig. 3). Two yeast species most well studied and characterized are *Saccharomyces cerevisiae* and *Candida albicans*. Whereas the precise inductive signals and individual players are not the same between the two species, basic regulatory circuitry appears to be conserved between the two species.

A common factor that links MAPK and PKA pathways is Ras2p (a G protein), which is activated by a variety of environmental signals (see Pan, Harashima, & Heitman, 2000 and other reviews cited earlier). For MAPK signaling pathway, Ras2p activates Cdc42p kinase, which activates Ste20p, an evolutionarily conserved serine/threonine kinase (Dan, Watanabe, & Kusumi, 2001; Strange, Denton, & Nehrke, 2006). Activated Ste20p kinase activity in turn initiates the triple MAPK cascade (Ste20p → Ste11p → Ste7p → Kss1p). Kss1p then inactivates inhibitors, Dig1/2p, leading to the activation of Ste12p and Tec1p complex or Tec1p itself, key transcription activators of filamentation.

Whereas yeast MAPK signaling are often associated with response to pheromones, hyperosmolarity, and stress (Chen & Thorner, 2007), PKA signaling pathway is often associated with nutrient sensing pathway, including sensing carbon source such as glucose. In *S. cerevisiae*, nitrogen starvation is coupled to glucose for induction of pseudohyphal growth. Gpr1p is a

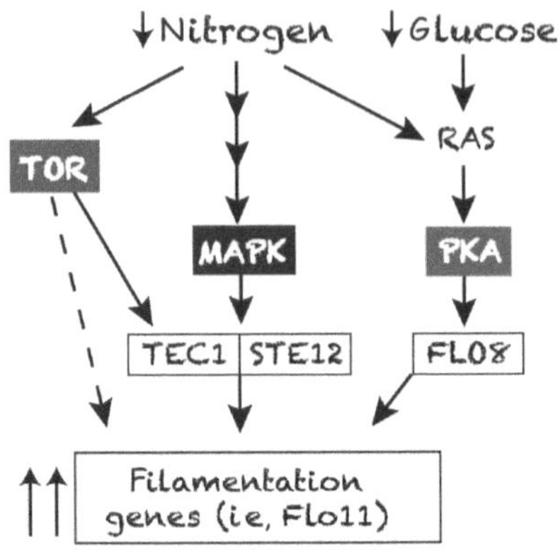

Fig. 3 Overview of induction of filamentation in yeast. Multiple environmental and stress cues can induce filamentation/pseudohyphal differentiation in various yeast species. Yeast can sense quality and quantity of carbon and nitrogen sources, which activate cAMP/protein kinase A (PKA), mitogen-activated protein kinase (MAPK), and target of rapamycin (TOR) pathways (and others) to ultimately activate gene expression specific for filamentation. PKA pathway acts through FLO8 transcription factor, whereas MAPK pathway acts through TEC1 and/or STE12 transcription factors. TOR appears to regulate filamentation in multiple ways including regulation of TEC1 protein stability. *The simplified cartoon is adapted from Bruckner, S., Kern, S., Birke, R., Saugar, I., Ulrich, H. D., & Mosch, H. U. (2011). The TEA transcription factor Tec1 links TOR and MAPK pathways to coordinate yeast development.* Genetics, 189*(2), 479–494. doi: 10.1534/genetics.111.133629; Cullen, P. J., & Sprague, G. F., Jr. (2012). The regulation of filamentous growth in yeast.* Genetics, 190*(1), 23–49. doi: 10.1534/genetics.111.127456; Gancedo, J. M. (2001). Control of pseudohyphae formation in Saccharomyces cerevisiae.* FEMS Microbiology Reviews, 25*(1), 107–123; and Rua, D., Tobe, B. T., & Kron, S. J. (2001). Cell cycle control of yeast filamentous growth.* Current Opinion in Microbiology, 4*(6), 720–727.*

G protein-coupled cell surface receptor that integrates carbon and nitrogen sensing. In fact, nitrogen starvation upregulates GPR1, a receptor for glucose and other fermentable carbon sources, to activate PKA pathway through G proteins, Gpa2p, Gpb1/2p, and Gpg1p (Lorenz et al., 2000; Xue, Batlle, & Hirsch, 1998). To induce filamentation, Gpr1p-coupled G proteins as well as Ras2p activate adenylate cyclase, leading to PKA cascade, which ultimately activates and inhibits FLO8 transcription factor and SFL1 inhibitor, respectively. Flo8p in turn activates FLO11, a cell surface flocculin, essential for cell adhesion and invasive growth. Therefore, both MAPK and PKA pathways converge on common set of filamentous genes (ie, FLO11), although each pathway incurs subtle distinct effects on filamentous phenotype, including changes in cell cycle regulation, budding pattern, elongation, and other morphological changes (Gancedo, 2001; Pan et al., 2000; Rua et al., 2001).

Moreover, pseudohyphal differentiation induced by nitrogen starvation in *S. cerevisiae* was inhibited by rapamycin (Cutler, Pan, Heitman, & Cardenas, 2001). Target of rapamycin (TOR) is a central integrator of nitrogen sensing signals, and TORC1 (a protein complex that contains TOR and other regulatory proteins), but not TORC2, is sensitive to rapamycin. One of the downstream targets of TORC1 kinase activity is phosphorylation of Tap42p, a regulatory subunit for PPA2-like phosphatase Sit4p, which is involved in nitrogen catabolite repression and ribosome biogenesis (Rohde, Bastidas, Puria, & Cardenas, 2008). In addition, rapamycin inhibition of TORC1 leads to Tec1p degradation, thus connecting MAPK and TOR signaling pathways to a common target (Bruckner et al., 2011). However, whereas rapamycin-induced Tec1p degradation directly leads to downregulation of TEC1 target, G1-cyclin/CLN1, no downregulation of FLO11 was observed. Thus, multiple signaling pathways independently and interdependently influence the complex induction of filamentous phenotype.

4.2 Metabolic Control of Filamentation

Induction of filamentation is a result of the ability of yeasts to sense the nutritional status of the environment. In general, yeasts accomplish this by the following general mechanisms: (1) physical interaction of the nutrient via receptor or transporter-like sensing systems; (2) monitoring metabolites via activities of enzymes from specific catabolite pathway. Specifically, yeast can sense quantity and quality of carbon source by using aforementioned Gpr1p, a presumed glucose receptor, or transporter homologues like Rgt2p and Snf3p or by monitoring the activity of glucose/hexose kinase, which participates in the first enzymatic reaction prior to the entry into glycolysis. Gpr1p, as mentioned, activates cAMP/PKA pathway. Glucose-bound Rgt2p and Snf3p induce proteasome degradation of inhibitors that suppress hexose transporter genes, required for efficient glucose uptake. In the presence of high levels of glucose, expression of genes that enhance glycolysis is upregulated and the expression of those genes that enhance catabolism of other types of carbon source, respiratory functions, gluconeogenesis, or FA oxidation is downregulated. This phenotype describes a well-studied but complex phenomenon called glucose repression. Detailed discussion of how glucose repression is achieved is beyond the scope of this chapter. However, one regulatory layer of how glucose repression is partly achieved is discussed later.

Hexokinase 2 (Hxk2p) has a dual function in yeast. Not only does it mediate the enzymatic reaction to phosphorylate glucose at the C6 position prior to its entry into glycolytic or PPP pathway, but it also modulates transcriptional activity in the nucleus. Hxk2p has been proposed to play a key role in glucose repression not only by directly translocating to the nucleus to act as transcriptional repressor along with Mig1p but also to directly regulate the activity of Snf1p, a key AMPK homologue in yeast, that is involved in glucose repression (Conrad et al., 2014; Fernandez-Garcia, Pelaez, Herrero, & Moreno, 2012). Under high glucose environment, Hxk2p is retained in the nucleus, where it associates with Mig1p/Snf1p complex to prevent Mig1p phosphorylation by Snf1p. This allows Mig1p/Hxk2p complex to suppress transcription of genes for respiration and catabolism of alternative carbon source. In a low glucose environment, activated Snf1p phosphorylates Hxk2p at serine 14, resulting in export of Hxk2p out of the nucleus. In the absence of Hxk2p, Mig1p is phosphorylated by Snf1p, which results in derepression of the target genes. Therefore, Snf1p, whose activity is mainly modulated by AMP/ADP/ATP levels, is also a major player in glucose sensing through hexokinase. Activated Snf1p can also directly phosphorylate and activate transcription factors as well as phosphorylate histone H3 to bring about epigenetics changes, resulting in transcription upregulation of hexose transporters.

4.3 Mitochondria and Filamentous Growth

A number of investigations suggested that mitochondria are essential for pseudohyphal differentiation (Jin, Dobry, McCown, & Kumar, 2008; Kang & Jiang, 2005; Starovoytova, Sorokin, Sokolov, Severin, & Knorre, 2013; Wright, Repine, & Repine, 1993), and that specific energy metabolism is associated with yeast vs filamentous growth: respiratory metabolism is associated with yeast growth, whereas glycolytic metabolism is associated with filamentous growth (Land, McDonald, Stjernholm, & Friedman, 1975). Paradoxically, a mutation in NDH51, a subunit of mitochondrial complex I, in *C. albicans* results in complete block of filamentous growth (McDonough, Bhattacherjee, Sadlon, & Hostetter, 2002). This observation is consistent with inhibition of filamentation in wild-type yeasts with rotenone, a Complex I inhibitor (Vellucci, Gygax, & Hostetter, 2007). In addition, mutants in pyruvate dehydrogenase complex protein X (PDX1) and dihydrolipoamide dehydrogenase (LPD1) genes, both components of the pyruvate dehydrogenase complex, also phenocopied NDH51 mutant in their filamentation defects (Kim & Kim, 2010; Vellucci et al., 2007). All three mutants are able to grow as yeast form, and in the case of PDX1 and NDH51 mutants, compensatory mechanism to support altered respiratory metabolism is observed.

Although intact respiratory system is associated with yeast morphology, high O_2 consumption is observed for the first 2.5 min upon induction of filamentous growth prior to suppression of respiration, suggestive of a Crabtree-like effect (Land et al., 1975). However, a low, sublethal dose of H_2O_2 can also induce filamentation (Nasution et al., 2008), and oxygen requirement for pseudohyphal growth did not necessarily involve OXPHOS activity (Wright et al., 1993). In fact, Guedouari et al. found that induction of filamentation was always associated with reduction in reduced/oxidized glutathione ratio (Guedouari et al., 2014). They further found that because PPP is inhibited upon induction of filamentation, yeast–hyphal transition goes through an oxidative stress phase. These observations suggest that the requirement for O_2 and mitochondria for filamentation may be the generation of ROS which contribute to reduction in redox state, although this phenomenon may be a consequence and not the cause of metabolic adaptation during yeast–hyphal transition (Guedouari et al., 2014). Of interest, mitochondrial complex I homologue in mammals is a large complex consisting of ~45 subunits, involved in NADH oxidation/ubiquinone reduction, and is the most upstream complex of the electron transfer chain as well as being a major site of ROS production (Hirst, 2013).

4.4 Concluding Remarks on Yeast Differentiation

Unicellular organisms such as yeast undergo differentiation in the form of morphological changes, accompanied by metabolic adaptation. These organisms can sense the amount and type of nutrients in the environment and alter gene expression patterns to maximize survival. Multiple signaling pathways interact to bring about changes in gene expression. For induction of filamentation, it appears that the metabolic adaptations result in changes in glutathione redox states but do not necessarily correlate with changes in ATP generation. Although the precise relationship is not clear, OXPHOS and functional respiratory chain appears critical for filamentation induction.

5. ENERGY METABOLISM AND PREIMPLANTATION EMBRYO DEVELOPMENT

As preimplantation development proceeds from a zygote to a blastocyst stage, mouse embryos do not substantially increase embryo biomass or volume, as only nucleocytoplasmic ratios increase upon division (Aiken, Swoboda, Skepper, & Johnson, 2004). This phenomenon is reminiscent of the observation in *Xenopus* and other invertebrates, where cleavage stage involves rapid cell division until the mid-blastula transition (MBT), where nucleocytoplasmic ratio and/or nuclear size appear to regulate the delay in the cell cycle by incorporating gap phases and initiate zygotic transcription (Jevtic & Levy, 2015; Newport & Kirschner, 1982). Furthermore, it has been proposed that during the cleavage stage, *Xenopus* embryos rely primarily on OXPHOS using amino acids, especially alanine and glutamine, as energy source, since inhibitors of OXPHOS, but not glycolysis, block development of cleavage-stage embryos (Dworkin & Dworkin-Rastl, 1991; Vastag et al., 2011). Interestingly, injection of [^{32}P] phosphoenolpyruvate (PEP), a substrate for PK, into *Xenopus* oocytes, eggs, and embryos resulted not in [^{32}P]-labeled ATP but in labeled G6P, G1P, and UDP-glucose, the last being a precursor to glycogen synthesis (Dworkin & Dworkin-Rastl, 1989). Thus, EMP pathway appeared to be operational but in reverse. However, when *Xenopus* eggs or embryos were treated with OXPHOS inhibitors, glycolysis could occur, suggesting that the EMP pathway is under metabolic control. In addition, it was further deduced that PEP was being utilized to produce G6P not through the classical gluconeogenesis involving fructose 1,6-bisphosphatase (FBP1), but through the non-oxidative branch of PPP, relying on transaldolase and transketolase enzymes. It was further proposed that the block in glycolysis was at the PK step, which converts PEP to pyruvate during EMP. Embryonic isoform of PK, unlike the liver or muscle isoforms, has a higher K_m for PEP, and its activity is stimulated by FBP1 and inhibited by relatively high ATP concentrations normally found in embryos (Dworkin, Segil, & Dworkin-Rastl, 1987). Thus, glycolytic activity is suppressed during cleavage stage but can be metabolically induced by artificially manipulating ATP/ADP ratio such as by inhibiting OXPHOS (Dworkin & Dworkin-Rastl, 1990). It is likely that activation of glycolytic activity just prior to gastrulation observed in normally developing *Xenopus* embryos may be a result of changes in intermediary metabolism that influences ATP/ADP ratio or other metabolites (Dworkin & Dworkin-Rastl, 1992).

One intriguing possibility is the observation that maternal supply of deoxyribonucleotide phosphates (dNTPs) is exhausted prior to MBT due to increase in DNA replication, resulting in the lengthening of the cell cycle (Vastag et al., 2011). Although Vastag et al. did not observe changes in ATP/ADP or reducing equivalents, a widespread metabolic remodeling occurs during cleavage stage resulting in changes in the levels of various metabolic intermediates including sedoheptulose 7-phosphate, the only PPP intermediate examined in the study. In addition, injection of glycolytic intermediates G6P, G3P, and PEP into oocytes induced apoptosis that is suppressed by upregulation of NADPH. As PPP plays an important role in ribonucleotide synthesis as well as in NADPH production, PPP may play a critical role during early development. These observations all suggest that regulation of energy metabolism in early vertebrate development, and development itself, may be controlled by metabolic flux rather than specific changes in gene expression.

There are a number of key differences between early *Xenopus* and mouse development, among the most notable being that the activation of zygotic gene expression occurs much earlier in mammalian embryos compared to *Xenopus* embryos. In addition, *Xenopus* embryos rely on internal supply of energy until hatching, while mammalian preimplantation embryos have the capacity to import energy substrates from the environment even prior to implantation and the establishment of functional placenta. Finally, unlike in *Xenopus*, mammalian embryos produce a distinct cell type, TE, prior to the production of hypoblast/primitive endoderm and epiblast. However, preimplantation embryos, like *Xenopus* embryos, display substantial changes in energy substrate and metabolism as development proceeds. It is likely that the basic metabolic needs of early *Xenopus* do not substantially differ from those of developing preimplantation embryos. It is also prudent to keep in mind that in mammals, there exists some type of communication between the mother and the preimplantation embryo, as embryo development relies on coordinating its development with physiological changes that occur within the female reproductive tract. Therefore, preimplantation embryos likely are much more sensitive to levels of external nutrients and signaling molecules than cleavage-stage *Xenopus* embryos.

5.1 Changes in Energy Substrate

During development of protocols for in vitro fertilization and preimplantation embryo culture, energy requirements for preimplantation development were elucidated (Gardner, 1998; Johnson et al., 2003; Lawitts & Biggers, 1993; Leese et al., 1993). A zygote can develop into a blastocyst in vitro culture using simple media formulations. Multiple embryo culture media have been formulated, and one of the most popular media used is KSOM developed by Biggers (1998). In addition to some simple salts, KSOM contains monocarboxylates, pyruvate and lactate, and glucose as energy substrates.[c] While monocarboxylates are used as primarily energy sources through low-level OXPHOS activity in early cleavage embryos, at around morula stage, there is an abrupt switch to using glucose for energy (Fig. 1A) (Houghton, Thompson, Kennedy, & Leese, 1996). In fact, prior to this stage, the embryos cannot use exogenous glucose as an energy source, whereas after this stage, the embryos do not use pyruvate as an energy source even though pyruvate is present in the milieu of the developing embryo (Gardner, 1998). This implies that, like the cleavage-stage *Xenopus* embryos, mouse cleavage-stage embryos selectively utilize OXPHOS and not glycolysis as the major energy metabolism. In addition, changes in energy substrate, rather than cell number, cytokinesis, or developmental age, correlate almost precisely to morula–blastocyst transition, strongly suggesting its importance during this developmental stage (Martin & Leese, 1999).

At the morula stage, coincident with changes in energy substrate, an increase in oxygen consumption is observed. Although O_2 consumption increases only two- to threefold between cleavage-stage embryos and blastocysts, a substantial change in OXPHOS activity is observed. During cleavage-stage embryos, whereas only 20–30% of O_2 consumed was sensitive to an OXPHOS blocker, cyanide, ~70% of O_2 consumed was cyanide sensitive at the blastocyst stage (Trimarchi, Liu, Porterfield, Smith, & Keefe, 2000). Furthermore, the switch to glucose metabolism at the morula stage is not due to the embryo's inability to use monocarboxylates as energy source at this stage. Two-cell embryos isolated from in vivo can develop to fully expanded blastocysts in vitro in complete absence of glucose by using pyruvate as energy source (Martin & Leese, 1995). Therefore, mouse morulae have the capacity to utilize monocarboxylates as energy source but metabolically choose glucose as energy source if glucose is available (Fig. 1B).

Glucose, however, is required at the zygote stage to serve a critical, but as yet clearly defined, role that is not revealed until the morula-to-blastocyst stage (Brown & Whittingham, 1991; Martin & Leese, 1995; Pantaleon, Scott, & Kaye, 2008). In contrast to 2-cell embryos, when 1-cell embryos are cultured in the absence of glucose, the embryos arrest development at the morula stage. If, however, 1-cell embryos are briefly exposed to glucose prior to morula stage, these embryos can develop to the blastocyst stage. Although the precise nature of why glucose is required is not known, it

[c] Glucose concentration in KSOM is 0.2 mM rather than 3.4 mM, which more closely approximate concentrations found in the oviduct (Gardner & Leese, 1990).

has been postulated that glucose exposure during cleavage stage mediates expression of a monocarboxylate transporter necessary for compensatory switch to pyruvate energy metabolism after morula formation (Jansen, Esmaeilpour, Pantaleon, & Kaye, 2006). However, the biological significance of this observation is not clear as normal development does not appear to require this compensatory switch. Glucose's critical role may also be through glucosamine signaling and/or glycosylation as it can also be metabolized as a precursor for both pathways through G6P (Pantaleon et al., 2008; Pantaleon, Tan, Kafer, & Kaye, 2010). Another possibility is that glucose must be metabolized through PPP in order to generate sufficient levels of NADPH. As O_2 consumption, as well as OXPHOS activity, increases during blastocoel expansion, ROS production requires ample reducing power to prevent oxidative stress. Consistent with this hypothesis, 1-cell embryos, but not 2-cell embryos, cultured in glucose-free media exhibit oxidative stress through redox imbalance (Jansen et al., 2009). Furthermore, it has been previously shown that PPP activity is high at 2-cell and morula stages (O'Fallon & Wright, 1986).

5.2 Energy Metabolism of ICM and TE

During the morula stage, prior to blastocoel formation, the first differentiation event takes place to produce the ICM and TE (Fig. 1A). It has been shown previously that the two major cell types, TE and ICM, have distinct energy requirements and metabolic pathways suited for their respective functions (Gopichandran & Leese, 2003; Hewitson & Leese, 1993; Houghton, 2006). For example, ICM and its in vitro derivative, ESCs, have high proliferative potential and use "aerobic" glycolysis for increasing biomass (Jang, Yang, Lee, & Cheong, 2015; Vander Heiden et al., 2009). TE, a columnar epithelium, requires copious amount of ATP for energy-expensive Na^+/K^+ ATPase pump to form and expand the blastocoel cavity. Hence, TE has much higher O_2 consumption than ICM presumably due to using OXPHOS to metabolize glucose, the most efficient pathway for producing ATP. In fact, TE has much higher number of mitochondria with morphologies that are associated with active OXPHOS, whereas mitochondria in ICM as well as in ESC exist in more spherical shape with little cristae as compared to those in TE (Cogliati et al., 2013; Houghton, 2006; Stern, Biggers, & Anderson, 1971). Whether changes in TE's mitochondrial morphology are induced after the initiation of TE differentiation or it is integral to the initiation of differentiation is yet to be determined.

A number of studies have suggested that the differences in energy metabolism between TE and ICM also reflect their relative pluripotency states. ICM, as pluripotent group of cells, shows relatively "quiet" metabolism driven mostly by glycolytic pathway, whereas TE displays OXPHOS-driven high energy metabolism, consistent with energy metabolism of other differentiated cells (Houghton, 2006; Teslaa & Teitell, 2015; Vander Heiden et al., 2009). During the induction of induced pluripotent stem cell (iPSC), metabolic reprogramming of OXPHOS-dominated "differentiated" energy metabolism reverts to glycolytic metabolism, consistent with association of OXPHOS with differentiated cells and glycolysis with pluripotent cells (Shyh-Chang et al., 2013; Teslaa & Teitell, 2015; Xu et al., 2013). Intriguingly, during iPSC induction of mouse embryo fibroblasts, changes in metabolic reprogramming seemed to occur prior to the full onset of expression of pluripotency-associated transcription factors such as Oct4, Nanog, and Sox2, suggesting that metabolic reprogramming may not just be a consequence of pluripotency state (Folmes et al., 2011). Furthermore, inhibition of glycolysis and OXPHOS reduced and promoted, respectively, the efficiency of pluripotency induction, supporting the hypothesis that metabolic reprogramming is a critical component of establishing pluripotency. Conversely, in order for pluripotent cells to differentiate, maturation of mitochondrial-mediated OXPHOS is required, consistent with the idea that metabolic reprogramming is essential and may precede the induction of differentiation (Chung et al., 2007; Mandal, Lindgren, Srivastava, Clark, & Banerjee, 2011).

5.3 Effects of Oxygen and Amino Acids in the Culture Media

A number of studies highlighted improvement of in vitro preimplantation embryo development when O_2 concentration/tension of the incubating chamber was reduced from atmospheric O_2 (~21%) to 5%, the latter more closely matching the in vivo environment (Harvey, Kind, & Thompson, 2002; Orsi & Leese, 2001; Rinaudo, Giritharan, Talbi, Dobson, & Schultz, 2006; Wale & Gardner, 2012). The prevailing thought is that reduction in [O_2] during culture results in reduction of ROS from oxidative processes, such as OXPHOS; this is supported by the observations that inclusion of antioxidants in the culture media ameliorates oxidative stress and improves development. However, O_2 itself appears to be required for preimplantation development as inhibition of OXPHOS or in vitro culture at 1% O_2 inhibits development (Kaneko & DePamphilis, 2013; Thomson, 1967), although this sensitivity may depend on species (Brison & Leese, 1994; Kane & Buckley, 1977). As mentioned earlier, the need for low levels of O_2 by the cleavage-stage embryos likely reflects their primary reliance on monocarboxylates, amino acids, and/or FAs for energy, which can only generate ATP through OXPHOS, although other oxidative processes such as demethylation of methylated CpG dinucleotides by ten-eleven translocation (TET) proteins (Li & Zhang, 2014) cannot be excluded.

The boost in energy requirement needed for conversion of morula to fully expanded blastocyst results in increased production of OXPHOS-mediated ROS, which can be partly attenuated by antioxidant defense mechanism present within the embryo and its surroundings (Guerin, El Mouatassim, & Menezo, 2001; Johnson & Nasr-Esfahani, 1994; Orsi & Leese, 2001). The embryo must carefully balance energy and ROS production because imbalanced energy homeostasis will result in oxidative stress, changes in the intracellular redox potential and impaired biosynthetic potential, all detrimental to development (Betts & Madan, 2008; Dumollard et al., 2009; Dumollard, Duchen, & Carroll, 2007; Takahashi, 2012). One potential source of energy that may play a critical role during this process is FAs. As stated previously, FA beta-oxidation produces ATP via OXPHOS through incorporation of Ac-CoA into the TCA cycle. Studies showed that not only do preimplantation embryos can take up exogenous FAs from the environment, but also they show a significant increase in beta-oxidation activity between 8-cell and blastocyst stages (Hillman & Flynn, 1980; Yamada et al., 2012). In addition, inhibition of beta-oxidation by etomoxir as well as genetic ablation of long-chain acyl-CoA dehydrogenase (an enzyme involved in the first step of each oxidation cycle) resulted in decreased blastocoel formation, suggesting that beta-oxidation may contribute to energy production during morula-to-blastocyst transition (Berger & Wood, 2004; Dunning et al., 2010). The fact that FA beta-oxidation can provide both ATP and NADPH (see earlier) during this developmental stage suggests why beta-oxidation may be one of the key players in maintaining energy production and redox balance.

The roles amino acids play in preimplantation embryo development appear complex. Even though all amino acids are present in the female reproductive tract, preimplantation embryos appeared to develop normally in early culture formulations which lacked most exogenously added amino acids (Brinster, 1968; Cholewa & Whitten, 1970). Although original KSOM formulation did not include amino acids other than glutamine, it did contain bovine serum albumin as a potential source of amino acids

(Biggers, McGinnis, & Raffin, 2000). Furthermore, essential and/or nonessential amino acids were shown to improve development in a stage-specific manner with respect to cell number and rate of blastocyst formation resulting in improved developmental potential of the resulting blastocysts (Ho, Wigglesworth, Eppig, & Schultz, 1995; Lane & Gardner, 1997, 1998; Summers, McGinnis, Lawitts, Raffin, & Biggers, 2000). These observations could be explained by a number of potential effects of amino acids on preimplantation development, including their functions as a chelator, osmotic regulator, as well as a buffer for pH regulation (Van Winkle, 2001). Intriguingly, in addition to serving as an energy source, amino acids may have the ability to induce favorable energy metabolism during preimplantation embryo development: in the absence of amino acids, preimplantation embryos display upregulation of glycolysis and downregulation of OXPHOS during blastocyst formation, whereas in the presence of amino acids in the culture media, embryos downregulate glycolysis and upregulate OXPHOS, mimicking the energy metabolism observed in freshly isolated blastocysts (Lane & Gardner, 1994, 1996, 1998). As described in the first section, amino acids have the capacity to modulate OXPHOS in a number of ways including their ability to enter the TCA cycle for energy production, as well as their ability to upregulate OXPHOS through the mTOR pathway (Lu et al., 2015; Ramanathan & Schreiber, 2009). As both glutamine and leucine play crucial roles in activating mTOR (Nicklin et al., 2009), the effects of amino acids in preimplantation development may be mediated partly through mTOR signaling. Thus one contribution of amino acids to preimplantation embryo development is to promote ATP production through OXPHOS, especially during the time of great energy expenditure, such as the expansion of blastocoel cavity. When sufficient levels of amino acids are not present in the environment such as during in vitro culture without amino acids, preimplantation embryo then adapts to alternative energy metabolism (ie, glycolysis) in order to meet the high energy demands during blastocyst formation.

5.4 Preblastocyst Knockout Mouse Phenotype

Despite a large number of knockout mice generated, relatively few mutations give rise to embryonic phenotype that result in arrest in development prior to the formation of blastocysts (or initiation of blastocoel cavity) (Kohn et al., 2012). Many of the genes that display preblastocyst arrest phenotype would be considered housekeeping genes such as those involved in RNA processing, cell cycle, and protein synthesis. Conspicuously absent from the list are "master" regulatory-type of genes, often transcription factors, that initiate gene expression patterns associated with a differentiated phenotype (Baker, 2001), even though blastocoel formation requires differentiation of equipotent blastomeres into TE. In fact, ablation of several key transcription factors associated with ICM (ie, Oct4, Nanog), TE (ie, Cdx2, Eomes), or primitive endoderm (ie, GATA4/6) results in peri-implantation (ie, postblastocyst formation) lethal phenotype. These observations could be due to functional redundancy and mutational robustness (Payne & Wagner, 2015; Sorrells & Johnson, 2015). On the other hand, phenotype of zygotic ablation may be masked by maternally inherited mRNA or protein (Foygel et al., 2008; Keramari et al., 2010). In the case of TE-associated Cdx2, maternally provided Cdx2 are not needed for the blastocyst formation as ablation of both maternal and zygotic Cdx2 still resulted in blastocyst formation (Blij et al., 2012). In addition, although maternal Sox2 was originally hypothesized to be required for blastocyst formation (Keramari et al., 2010), it was subsequently shown that blastocysts with TE and ICM could be produced in the absence of both maternal and zygotic Sox2 (Wicklow et al., 2014). Thus, although lineage-specific transcription factors are clearly required for development past the peri-implantation stage, whether or not they serve indispensable roles in the initial specification of the TE and/or ICM is not clear.

Tead4, a member of evolutionarily conserved TEAD gene family and a TEC1 homologue, is one of handful of transcription factors that show lethality prior to blastocyst formation (Nishioka et al., 2008; Yagi et al., 2007). Embryos lacking *Tead4* arrest development at the morula stage prior to blastocoel cavity formation; these null embryos express ICM marker *Oct4* but not TE marker Cdx2. *Tead4*'s functional role in development appeared restricted to TE lineage because neither conditional ablation in the implanted epiblasts (Yagi et al., 2007) nor differentiation of *Tead4*$^{-/-}$ ESC showed developmental defects (Nishioka et al., 2008; Yagi et al., 2007). Hence, *Tead4* was hypothesized to act upstream of *Cdx2* as a specifier of TE lineage. Furthermore, recent report suggests involvement of Hippo signaling pathway in *Tead4*-mediated TE specification (Nishioka et al., 2009), although exactly how this might occur is not clear (Home et al., 2012; Kaneko & DePamphilis, 2013).

Subsequently, it was observed that manipulation of in vitro culture conditions during *Tead4*$^{-/-}$ embryos culture resulted in formation of a blastocoel cavity, which absolutely requires TE differentiation (Kaneko &

DePamphilis, 2013; Kidder & Watson, 2005). In addition, genes specific for TE lineage, such as *Cdx2* and *AtpB1*, a Na^+/K^+ ATPase pump subunit, were detected in these *Tead4*$^{-/-}$ embryos, including those that did not expand the blastocoel. These observations suggested that *Tead4* was not required for specification and differentiation of TE lineage, but was likely involved in maintenance of TE differentiation. Although the precise role of *Tead4* during morula-to-blastocyst transition is not entirely clear, analyses of *Tead4*$^{-/-}$ embryo development under a variety of culture conditions strongly suggest that it has a fundamental role in reducing oxidative stress. The evidence for how this conclusion came about is presented below.

Although manipulation of culture conditions to assess phenotypic effects has been extensively applied using wild-type preimplantation embryos, the same techniques have not been applied to phenotypic analyses of mutant preimplantation embryos. *Tead4* KO phenotypic analysis is one of the first examples in which manipulation of culture conditions led to reversal of a developmental block. In this case, because its developmental block was morula-to-blastocyst transition, the presence of blastocoel cavity, indicative of specification of functional TE, could be used as a biomarker for the effects of culture conditions. This conveniently allowed assessing what type of changes in culture conditions would or would not allow blastocoel to form. For example, in addition to lowering O_2 tension, manipulation of glucose and amino acid composition of the culture media led to the conclusion that *Tead4* mediates oxidative stress. When 2-cell embryos from *Tead4* heterozygous matings were cultured in the absence of glucose, nullizygous *Tead4* embryos developed to the morula stage but did not form blastocoels (Fig. 1B; Kaneko & DePamphilis, 2013). This observation is in contrast to *Tead4*$^{+/+}$ and *Tead4*$^{+/-}$ embryos or previous observations with wild-type 2-cell embryos, suggesting that *Tead4*$^{-/-}$ embryos could not make a successful switch to utilizing monocarboxylates for energy, phenotype similar to culturing wild-type 1-cell embryos (Kaneko & DePamphilis, 2013; Martin & Leese, 1995). Thus, *Tead4*$^{-/-}$ 2-cell embryos, in the absence of glucose in the media, appeared to lack sufficient antioxidant capacity to support energy metabolism of either monocarboxylates or FAs through OXPHOS, phenotype observed with wild-type 1-cell embryos (Jansen et al., 2009). Consistent with the hypothesis, addition of antioxidant to *Tead4*$^{-/-}$ 2-cell embryos cultured in glucose-free media allowed blastocoel to form.

The same study showed that inclusion of glutamine and Earle's essential amino (EAA) acids prevented blastocoel formation of *Tead4*$^{-/-}$ embryos in an additive manner—that is, addition of either glutamine or EAA did not prevent blastocoel, whereas addition of both did. As amino acids promote OXPHOS over glycolysis, thus oxidative stress, these observations again suggest that *Tead4*$^{-/-}$ embryos have the reduced capacity to deal with oxidative stress. Consistent with this hypothesis, addition of antioxidant to glutamine/EAA containing media ameliorates oxidative stress, allowing *Tead4*$^{-/-}$ embryos to form blastocoels under these conditions. As discussed earlier, amino acids, especially glutamine and leucine (which is part of EAA), promote OXPHOS through activation of mTOR pathway. Thus, inhibition of mTOR is expected to reduce OXPHOS-mediated oxidative stress (Morita et al., 2013; Villa-Cuesta, Holmbeck, & Rand, 2014). When mTOR inhibitor, rapamycin, was included in the culture media containing glutamine and EAA, *Tead4*$^{-/-}$ embryos were able to form a blastocoel, consistent with *Tead4*'s requirement when oxidative stress is induced (Fig. 4, Kaneko & DePamphilis, unpublished results). The results obtained from manipulating culture conditions and the corresponding effects on antioxidant addition all strongly point to the conclusion that *Tead4* is not required for TE specification per se, but its requirement is manifested during the metabolic reprogramming that must occur just prior to or during TE differentiation.

An obvious question that arose from these studies was: if *Tead4*$^{-/-}$ embryos were blocked at the morula-to-blastocyst transition in vivo, what is the biological significance of in vitro manipulations that allowed the reversal of an in vivo phenotype? We know that preimplantation embryos have metabolic peculiarities that fit the developmental needs in vivo. In the case of *Tead4* nullizygous embryos, in vitro culture manipulations allowed these embryos to reprogram their metabolism so that their development could proceed. Normal development within the maternal reproductive tract is designed such that during the morula-to-blastocyst transition in which TE lineage requires large amount of ATP, TEs metabolic reprogramming allows the cells to utilize the most efficient energy metabolism, OXPHOS. In the maternal reproductive tract, morulae are normally exposed to amino acids, glucose, and O_2. Thus, these in vivo conditions allow morulae to upregulate glucose consumption and complete glucose oxidation via OXPHOS, which would provide the necessary ATPs for blastocoel formation and expansion. This metabolic reprogramming mainly by the TE must be offset by the oxidative stress load, which requires efficient regeneration of glutathione to keep the redox balance. By manipulation the culture conditions that suppress OXPHOS or alleviate oxidative stress, the embryos do

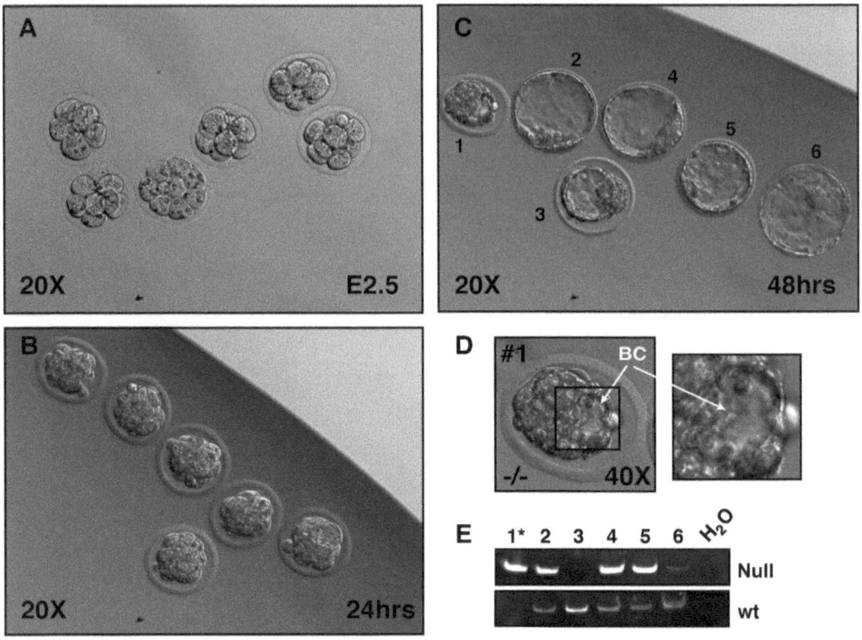

Fig. 4 Inhibition of mTOR rescues *Tead4*$^{-/-}$ embryos. Under culture conditions containing amino acids that blocked blastocoel formation, leading to an abnormal morphology in *Tead4*$^{-/-}$ embryos (Kaneko & DePamphilis, 2013), rapamycin allowed initiation of blastocoel in *Tead4*$^{-/-}$ embryos. (A) Eight-cell embryos were isolated from *Tead4* heterozygous matings and then cultured for 24 h (B) or 48 h (C) in KSOM media with 3.4 mM glucose, supplemented with glutamine, essential amino acids and 25 μM rapamycin, a potent and specific inhibitor of mTOR kinase. Embryo 1 (C, 20 ×) contained a blastocoel cavity ("BC" D, 40 ×) and was genotyped as *Tead4*$^{-/-}$ (E).

not require *Tead4* for blastocyst formation. Thus, the simplest and most likely explanation of *Tead4*'s essential role during morula-to-blastocyst transition in vivo is that of a mediator to reduce oxidative stress specifically in the TE lineages.

Although the phenotypic analyses of *Tead4* knockout mice were extensively discussed, there are a number of other genetic mutations of interest which give preblastocyst arrest phenotype. For example, ablation of nonoxidative PPP pathway enzyme, transketolase, results in preblastocyst developmental arrest (Xu, Wawrousek, & Piatigorsky, 2002). Although transketolase is often upregulated in cancer cells and is thought to play a key role in amino acid-derived purine biosynthesis (Saha et al., 2014), it may also provide an equally important role in suppressing oxidative stress during increased glucose uptake (Hammes et al., 2003). In addition, ablation of glutamine synthetase results in block in blastocyst formation (He, Hakvoort, Vermeulen, Lamers, & Van Roon, 2007). Although ammonia detoxification was provided as a potential explanation for the phenotype, the enzyme may play a critical role in supplying glutamine for mTOR activation (Jewell et al., 2015; Nicklin et al., 2009). Finally, although ablation of X-linked G6P dehydrogenase (G6PD) does not result in preimplantation lethality, ESC lacking G6PD are susceptible to oxidative stress (Filosa et al., 2003; Pandolfi et al., 1995).

6. PERSPECTIVE AND FUTURE CONSIDERATIONS

How the environment and the corresponding metabolic response induce changes in gene expression at the molecular level has been an ongoing topic of research with microorganisms since the days of Louis Pasteur. In mammals, however, the phenomenon has only recently been "rediscovered" as a research topic, roughly 60 years since Otto Warburg published his *Science* article, "On respiratory impairment in cancer cells" (Warburg, 1956). In the same year published in Nature, Wesley Whitten reported a successful development of 8-cell mouse embryos to blastocysts in culture (Whitten, 1956). Since then, the pioneering works of Whitten, Biggers, Brinster, Bavister, Leese, Gardner, and others have led to successful in vitro fertilization and culture technologies. But the knowledge gained about preimplantation embryo metabolism during those efforts has been underappreciated by many researchers studying preimplantation development at the molecular level. Despite the renewed attention of the relationship among gene expression, cell differentiation, and energy metabolism in cancer biology (Wise & Thompson, 2010), it has been proposed that most, if not all, low-hanging fruits for research have already been picked (McKnight, 2010). But biologists studying mouse preimplantation embryos have unique advantages over cancer biologists: (1) preimplantation embryos, despite (or in spite of, some might say) in vitro manipulations, can go on to develop into normal adults, unlike cancer cell lines that have been propagated, selected, and adapted for growth in culture, clouding the biological significance of studies that use them; (2) preimplantation embryos can be cultured under "simple" conditions with defined media that can easily be adapted, manipulated, and modified to look for phenotypic or genotypic effects, much like how yeasts can be studied; (3) there are hundreds of genetic mutations constructed in mice, many of which have not been subjected to analyses of preimplantation embryo culture manipulations. Time has come to utilize and combine technological advances on studying metabolism, manipulations of embryo culture methods, and gene ablation and suppression technologies to better understand the relationship among gene expression, metabolism, and development. I suggest that we reexamine the phenotypes of various knockout mice using manipulations of embryo culture to assess metabolic effects of genetic ablation, which the understanding that the observations made in vitro be rationally interpreted that make metabolic sense in vivo.

One area that is ripe for fruitful research is the role of mitochondria on TE differentiation. Even though we have known for decades that the mitochondrial structure and function undergo profound alterations during TE differentiation, an important question yet to be answered is whether this change precedes or follows the initiation of TE differentiation. As we continue to uncover how mitochondrial morphology (Wai & Langer, 2016) as well as mitochondrial–nuclear reciprocal interactions (Ryan & Hoogenraad, 2007) are regulated, the relationship between mitochondrial morphology and TE differentiation should become clearer.

If mitochondria play a substantial role in initiation of TE differentiation, how might this occur? Upon formation of 8-cell embryos, the spherical geometry places inner blastomeres in a differential oxygen environment compared to outer blastomeres, as previously postulated (Harvey et al., 2002). It is anticipated that compaction will further differentiate the two environments such that the inner cells find themselves in a more hypoxic environment. It is becoming clear that mammalian cells sense hypoxia through mitochondria's ETCs' generation of ROS, which acts as a signaling molecule to initiate the hypoxic response (Waypa, Smith, & Schumacker, 2016). Could one key to TE differentiation be ROS signaling by more hypoxic inner cells, which serves as a trigger to activate a TE differentiation pathway in the outer cells, leading to changes in mitochondrial morphology in the outer cells? This would mean that changes in ATP and redox levels per se are not necessarily the inducer of differentiation but consequence of ROS signaling. The fact that 32-cell stage outer and inner cells can dedifferentiate and take on the other's phenotype upon rearrangement (Suwinska et al., 2008) suggests the possibility that differential, regional [O_2] may be the critical factor to induce TE/ICM lineage specifications. It would then follow that transcription factors associated with TE or ICM lineage may be needed to support the metabolic needs of the two lineages to "lock in" the differentiation pathway.

ACKNOWLEDGMENTS

I would like to apologize to all researchers whose work I did not mention or cite due to space and time constraints. I would like to thank Melvin DePamphilis for his long-standing support in allowing me to explore multiple avenues of research ideas and interests, always reminding me to "look at the bright side" in all experimental failures. I also would like to thank Ronald Chanderbhan and Supratim Choudhuri in their support for writing this manuscript.

REFERENCES

Adelman, R., Saul, R. L., & Ames, B. N. (1988). Oxidative damage to DNA: Relation to species metabolic rate and life span. *Proceedings of the National Academy of Sciences of the United States of America*, *85*(8), 2706–2708.

Aiken, C. E., Swoboda, P. P., Skepper, J. N., & Johnson, M. H. (2004). The direct measurement of embryogenic volume and nucleo-cytoplasmic ratio during mouse preimplantation development. *Reproduction*, *128*(5), 527–535. http://dx.doi.org/10.1530/rep.1.00281.

Andreyev, A. Y., Kushnareva, Y. E., Murphy, A. N., & Starkov, A. A. (2015). Mitochondrial ROS metabolism: 10 years later. *Biochemistry (Mosc)*, *80*(5), 517–531. http://dx.doi.org/10.1134/S0006297915050028.

Armant, D. R. (2005). Blastocysts don't go it alone. Extrinsic signals fine-tune the intrinsic developmental program of trophoblast cells. *Developmental Biology*, *280*(2), 260–280. http://dx.doi.org/10.1016/j.ydbio.2005.02.009.

Baker, N. E. (2001). Master regulatory genes; telling them what to do. *Bioessays*, *23*(9), 763–766. http://dx.doi.org/10.1002/bies.1110.

Balaban, R. S., Nemoto, S., & Finkel, T. (2005). Mitochondria, oxidants, and aging. *Cell*, *120*(4), 483–495.

Bar-Even, A., Flamholz, A., Noor, E., & Milo, R. (2012). Rethinking glycolysis: On the biochemical logic of metabolic pathways. *Nature Chemical Biology*, *8*(6), 509–517. http://dx.doi.org/10.1038/nchembio.971.

Berger, P. S., & Wood, P. A. (2004). Disrupted blastocoele formation reveals a critical developmental role for long-chain acyl-CoA dehydrogenase. *Molecular Genetics and Metabolism*, *82*(4), 266–272. http://dx.doi.org/10.1016/j.ymgme.2004.06.001.

Betts, D. H., & Madan, P. (2008). Permanent embryo arrest: Molecular and cellular concepts. *Molecular Human Reproduction*, *14*(8), 445–453. http://dx.doi.org/10.1093/molehr/gan035.

Biggers, J. D. (1998). Reflections on the culture of the preimplantation embryo. *The International Journal of Developmental Biology*, *42*(7), 879–884.

Biggers, J. D., McGinnis, L. K., & Raffin, M. (2000). Amino acids and preimplantation development of the mouse in protein-free potassium simplex optimized medium. *Biology of Reproduction*, *63*(1), 281–293.

Birsoy, K., Wang, T., Chen, W. W., Freinkman, E., Abu-Remaileh, M., & Sabatini, D. M. (2015). An essential role of the mitochondrial electron transport chain in cell proliferation is to enable aspartate synthesis. *Cell*, *162*(3), 540–551. http://dx.doi.org/10.1016/j.cell.2015.07.016.

Blij, S., Frum, T., Akyol, A., Fearon, E., & Ralston, A. (2012). Maternal Cdx2 is dispensable for mouse development. *Development*, *139*(21), 3969–3972. http://dx.doi.org/10.1242/dev.086025.

Brinster, R. L. (1968). Effect of glutathione on the development of two-cell mouse embryos in vitro. *Journal of Reproduction and Fertility*, *17*(3), 521–525.

Brison, D. R., & Leese, H. J. (1994). Blastocoel cavity formation by preimplantation rat embryos in the presence of cyanide and other inhibitors of oxidative phosphorylation. *Journal of Reproduction and Fertility*, *101*(2), 305–309.

Brown, J. J., & Whittingham, D. G. (1991). The roles of pyruvate, lactate and glucose during preimplantation development of embryos from F1 hybrid mice in vitro. *Development*, *112*(1), 99–105.

Bruckner, S., Kern, S., Birke, R., Saugar, I., Ulrich, H. D., & Mosch, H. U. (2011). The TEA transcription factor Tec1 links TOR and MAPK pathways to coordinate yeast development. *Genetics*, *189*(2), 479–494. http://dx.doi.org/10.1534/genetics.111.133629.

Cairns, R. A., & Mak, T. W. (2013). Oncogenic isocitrate dehydrogenase mutations: Mechanisms, models, and clinical opportunities. *Cancer Discovery*, *3*(7), 730–741. http://dx.doi.org/10.1158/2159-8290.CD-13-0083.

Carracedo, A., Cantley, L. C., & Pandolfi, P. P. (2013). Cancer metabolism: Fatty acid oxidation in the limelight. *Nature Reviews. Cancer*, *13*(4), 227–232. http://dx.doi.org/10.1038/nrc3483.

Chan, D. C. (2012). Fusion and fission: Interlinked processes critical for mitochondrial health. *Annual Review of Genetics*, *46*, 265–287. http://dx.doi.org/10.1146/annurev-genet-110410-132529.

Chen, R. E., & Thorner, J. (2007). Function and regulation in MAPK signaling pathways: Lessons learned from the yeast Saccharomyces cerevisiae. *Biochimica et Biophysica Acta*, *1773*(8), 1311–1340. http://dx.doi.org/10.1016/j.bbamcr.2007.05.003.

Cholewa, J. A., & Whitten, W. K. (1970). Development of two-cell mouse embryos in the absence of a fixed-nitrogen source. *Journal of Reproduction and Fertility*, *22*(3), 553–555.

Choo, A. Y., Kim, S. G., Vander Heiden, M. G., Mahoney, S. J., Vu, H., Yoon, S. O., … Blenis, J. (2010). Glucose addiction of TSC null cells is caused by failed mTORC1-dependent balancing of metabolic demand with supply. *Molecular Cell*, *38*(4), 487–499. http://dx.doi.org/10.1016/j.molcel.2010.05.007. S1097-2765(10)00369-2 [pii].

Chung, S., Dzeja, P. P., Faustino, R. S., Perez-Terzic, C., Behfar, A., & Terzic, A. (2007). Mitochondrial oxidative metabolism is required for the cardiac differentiation of stem cells. *Nature Clinical Practice. Cardiovascular Medicine*, *4*(Suppl. 1), S60–S67. http://dx.doi.org/10.1038/ncpcardio0766.

Cockburn, K., & Rossant, J. (2010). Making the blastocyst: Lessons from the mouse. *The Journal of Clinical Investigation*, *120*(4), 995–1003. http://dx.doi.org/10.1172/JCI41229.

Cogliati, S., Frezza, C., Soriano, M. E., Varanita, T., Quintana-Cabrera, R., Corrado, M., … Scorrano, L. (2013). Mitochondrial cristae shape determines respiratory chain supercomplexes assembly and respiratory efficiency. *Cell*, *155*(1), 160–171. http://dx.doi.org/10.1016/j.cell.2013.08.032.

Conrad, M., Schothorst, J., Kankipati, H. N., Van Zeebroeck, G., Rubio-Texeira, M., & Thevelein, J. M. (2014). Nutrient sensing and signaling in the yeast Saccharomyces cerevisiae. *FEMS Microbiology Reviews*, *38*(2), 254–299. http://dx.doi.org/10.1111/1574-6976.12065.

Crabtree, H. G. (1929). Observations on the carbohydrate metabolism of tumours. *The Biochemical Journal*, *23*(3), 536–545.

Cross, J. C. (2000). Genetic insights into trophoblast differentiation and placental morphogenesis. *Seminars in Cell & Developmental Biology*, *11*(2), 105–113. http://dx.doi.org/10.1006/scdb.2000.0156.

Cullen, P. J., & Sprague, G. F., Jr. (2012). The regulation of filamentous growth in yeast. *Genetics*, *190*(1), 23–49. http://dx.doi.org/10.1534/genetics.111.127456.

Cutler, N. S., Pan, X., Heitman, J., & Cardenas, M. E. (2001). The TOR signal transduction cascade controls cellular differentiation in response to nutrients. *Molecular Biology of the Cell*, *12*(12), 4103–4113.

Dalziel, K. (1980). Isocitrate dehydrogenase and related oxidative decarboxylases. *FEBS Letters*, *117*(Suppl.), K45–K55.
Dan, I., Watanabe, N. M., & Kusumi, A. (2001). The Ste20 group kinases as regulators of MAP kinase cascades. *Trends in Cell Biology*, *11*(5), 220–230.
Dang, C. V. (2012). Links between metabolism and cancer. *Genes & Development*, *26*(9), 877–890. http://dx.doi.org/10.1101/gad.189365.112.
Davies, K. J. (1995). Oxidative stress: The paradox of aerobic life. *Biochemical Society Symposium*, *61*, 1–31.
DeBerardinis, R. J., Lum, J. J., Hatzivassiliou, G., & Thompson, C. B. (2008). The biology of cancer: Metabolic reprogramming fuels cell growth and proliferation. *Cell Metabolism*, 7(1), 11–20. http://dx.doi.org/10.1016/j.cmet.2007.10.002. S1550-4131(07)00295-1 [pii].
Diaz-Ruiz, R., Averet, N., Araiza, D., Pinson, B., Uribe-Carvajal, S., Devin, A., & Rigoulet, M. (2008). Mitochondrial oxidative phosphorylation is regulated by fructose 1,6-bisphosphate. A possible role in Crabtree effect induction? *The Journal of Biological Chemistry*, *283*(40), 26948–26955. http://dx.doi.org/10.1074/jbc.M800408200.
Dietrich, J. E., & Hiiragi, T. (2007). Stochastic patterning in the mouse pre-implantation embryo. *Development*, *134*(23), 4219–4231. http://dx.doi.org/10.1242/dev.003798.
Dumollard, R., Carroll, J., Duchen, M. R., Campbell, K., & Swann, K. (2009). Mitochondrial function and redox state in mammalian embryos. *Seminars in Cell & Developmental Biology*, *20*(3), 346–353.
Dumollard, R., Duchen, M., & Carroll, J. (2007). The role of mitochondrial function in the oocyte and embryo. *Current Topics in Developmental Biology*, 77, 21–49. http://dx.doi.org/10.1016/S0070-2153(06)77002-8. S0070-2153(06)77002-8 [pii].
Dunning, K. R., Cashman, K., Russell, D. L., Thompson, J. G., Norman, R. J., & Robker, R. L. (2010). Beta-oxidation is essential for mouse oocyte developmental competence and early embryo development. *Biology of Reproduction*, *83*(6), 909–918. http://dx.doi.org/10.1095/biolreprod.110.084145.
Dworkin, M. B., & Dworkin-Rastl, E. (1989). Metabolic regulation during early frog development: Glycogenic flux in Xenopus oocytes, eggs, and embryos. *Developmental Biology*, *132*(2), 512–523.
Dworkin, M. B., & Dworkin-Rastl, E. (1990). Regulation of carbon flux from amino acids into sugar phosphates in Xenopus embryos. *Developmental Biology*, *138*(1), 177–187.
Dworkin, M. B., & Dworkin-Rastl, E. (1991). Carbon metabolism in early amphibian embryos. *Trends in Biochemical Sciences*, *16*(6), 229–234.
Dworkin, M. B., & Dworkin-Rastl, E. (1992). Glycogen breakdown in cleaving Xenopus embryos is limited by ADP. *Molecular Reproduction and Development*, *32*(4), 354–362. http://dx.doi.org/10.1002/mrd.1080320408.
Dworkin, M. B., Segil, N., & Dworkin-Rastl, E. (1987). Pyruvate kinase isozymes in oocytes and embryos from the frog Xenopus laevis. *Comparative Biochemistry and Physiology—Part B*, *88*(3), 743–749.
Engelhardt, W. A. (1974). On the dual role of respiration. *Molecular and Cellular Biochemistry*, *5*(1–2), 25–33.
Fan, W., & Evans, R. (2015). PPARs and ERRs: Molecular mediators of mitochondrial metabolism. *Current Opinion in Cell Biology*, *33*, 49–54. http://dx.doi.org/10.1016/j.ceb.2014.11.002.
Fan, J., Kamphorst, J. J., Mathew, R., Chung, M. K., White, E., Shlomi, T., & Rabinowitz, J. D. (2013). Glutamine-driven oxidative phosphorylation is a major ATP source in transformed mammalian cells in both normoxia and hypoxia. *Molecular Systems Biology*, *9*, 712. http://dx.doi.org/10.1038/msb.2013.65.
Feng, Z., & Levine, A. J. (2010). The regulation of energy metabolism and the IGF-1/mTOR pathways by the p53 protein. *Trends in Cell Biology*, *20*(7), 427–434. http://dx.doi.org/10.1016/j.tcb.2010.03.004.

Fernandez-Garcia, P., Pelaez, R., Herrero, P., & Moreno, F. (2012). Phosphorylation of yeast hexokinase 2 regulates its nucleocytoplasmic shuttling. *The Journal of Biological Chemistry*, *287*(50), 42151–42164. http://dx.doi.org/10.1074/jbc.M112.401679.

Filosa, S., Fico, A., Paglialunga, F., Balestrieri, M., Crooke, A., Verde, P., … Martini, G. (2003). Failure to increase glucose consumption through the pentose-phosphate pathway results in the death of glucose-6-phosphate dehydrogenase gene-deleted mouse embryonic stem cells subjected to oxidative stress. *The Biochemical Journal*, *370*(Pt. 3), 935–943. http://dx.doi.org/10.1042/BJ20021614.

Finkel, T., & Holbrook, N. J. (2000). Oxidants, oxidative stress and the biology of ageing. *Nature*, *408*(6809), 239–247. http://dx.doi.org/10.1038/35041687.

Fleming, T. P., Kwong, W. Y., Porter, R., Ursell, E., Fesenko, I., Wilkins, A., … Eckert, J. J. (2004). The embryo and its future. *Biology of Reproduction*, *71*(4), 1046–1054. http://dx.doi.org/10.1095/biolreprod.104.030957. biolreprod.104.030957 [pii].

Folmes, C. D., Nelson, T. J., Martinez-Fernandez, A., Arrell, D. K., Lindor, J. Z., Dzeja, P. P., … Terzic, A. (2011). Somatic oxidative bioenergetics transitions into pluripotency-dependent glycolysis to facilitate nuclear reprogramming. *Cell Metabolism*, *14*(2), 264–271. http://dx.doi.org/10.1016/j.cmet.2011.06.011.

Foster, D. W. (2012). Malonyl-CoA: The regulator of fatty acid synthesis and oxidation. *The Journal of Clinical Investigation*, *122*(6), 1958–1959.

Foygel, K., Choi, B., Jun, S., Leong, D. E., Lee, A., Wong, C. C., … Yao, M. W. (2008). A novel and critical role for Oct4 as a regulator of the maternal-embryonic transition. *PLoS One*, *3*(12), e4109. http://dx.doi.org/10.1371/journal.pone.0004109.

Fraser, R. B., Waite, S. L., Wood, K. A., & Martin, K. L. (2007). Impact of hyperglycemia on early embryo development and embryopathy: In vitro experiments using a mouse model. *Human Reproduction*, *22*(12), 3059–3068. http://dx.doi.org/10.1093/humrep/dem318. dem318 [pii].

Gagiano, M., Bauer, F. F., & Pretorius, I. S. (2002). The sensing of nutritional status and the relationship to filamentous growth in Saccharomyces cerevisiae. *FEMS Yeast Research*, *2*(4), 433–470.

Gancedo, J. M. (2001). Control of pseudohyphae formation in Saccharomyces cerevisiae. *FEMS Microbiology Reviews*, *25*(1), 107–123.

Gardner, D. K. (1998). Changes in requirements and utilization of nutrients during mammalian preimplantation embryo development and their significance in embryo culture. *Theriogenology*, *49*(1), 83–102.

Gardner, D. K., & Leese, H. J. (1990). Concentrations of nutrients in mouse oviduct fluid and their effects on embryo development and metabolism in vitro. *Journal of Reproduction and Fertility*, *88*(1), 361–368.

Gonzales, D. S., & Bavister, B. D. (1995). Zona pellucida escape by hamster blastocysts in vitro is delayed and morphologically different compared with zona escape in vivo. *Biology of Reproduction*, *52*(2), 470–480.

Gopichandran, N., & Leese, H. J. (2003). Metabolic characterization of the bovine blastocyst, inner cell mass, trophectoderm and blastocoel fluid. *Reproduction*, *126*(3), 299–308.

Grindler, N. M., & Moley, K. H. (2013). Maternal obesity, infertility and mitochondrial dysfunction: Potential mechanisms emerging from mouse model systems. *Molecular Human Reproduction*, *19*(8), 486–494. http://dx.doi.org/10.1093/molehr/gat026.

Guantes, R., Diaz-Colunga, J., & Iborra, F. J. (2016). Mitochondria and the non-genetic origins of cell-to-cell variability: More is different. *Bioessays*, *38*(1), 64–76. http://dx.doi.org/10.1002/bies.201500082.

Guantes, R., Rastrojo, A., Neves, R., Lima, A., Aguado, B., & Iborra, F. J. (2015). Global variability in gene expression and alternative splicing is modulated by mitochondrial content. *Genome Research*, *25*(5), 633–644. http://dx.doi.org/10.1101/gr.178426.114.

Guedouari, H., Gergondey, R., Bourdais, A., Vanparis, O., Bulteau, A. L., Camadro, J. M., & Auchere, F. (2014). Changes in glutathione-dependent redox status and mitochondrial energetic strategies are part of the adaptive response during the filamentation process in Candida albicans. *Biochimica et Biophysica Acta, 1842*(9), 1855–1869. http://dx.doi.org/10.1016/j.bbadis.2014.07.006.

Guerin, P., El Mouatassim, S., & Menezo, Y. (2001). Oxidative stress and protection against reactive oxygen species in the pre-implantation embryo and its surroundings. *Human Reproduction Update*, 7(2), 175–189.

Hamanaka, R. B., & Chandel, N. S. (2009). Mitochondrial reactive oxygen species regulate hypoxic signaling. *Current Opinion in Cell Biology, 21*(6), 894–899. http://dx.doi.org/10.1016/j.ceb.2009.08.005.

Hamatani, T., Ko, M., Yamada, M., Kuji, N., Mizusawa, Y., Shoji, M., ... Yoshimura, Y. (2006). Global gene expression profiling of preimplantation embryos. *Human Cell, 19*(3), 98–117. http://dx.doi.org/10.1111/j.1749-0774.2006.00018.x.

Hammes, H. P., Du, X., Edelstein, D., Taguchi, T., Matsumura, T., Ju, Q., ... Brownlee, M. (2003). Benfotiamine blocks three major pathways of hyperglycemic damage and prevents experimental diabetic retinopathy. *Nature Medicine, 9*(3), 294–299. http://dx.doi.org/10.1038/nm834. nm834 [pii].

Harvey, A. J., Kind, K. L., & Thompson, J. G. (2002). REDOX regulation of early embryo development. *Reproduction, 123*(4), 479–486.

Hayes, J. D., & Dinkova-Kostova, A. T. (2014). The Nrf2 regulatory network provides an interface between redox and intermediary metabolism. *Trends in Biochemical Sciences, 39*(4), 199–218. http://dx.doi.org/10.1016/j.tibs.2014.02.002.

He, Y., Hakvoort, T. B., Vermeulen, J. L., Lamers, W. H., & Van Roon, M. A. (2007). Glutamine synthetase is essential in early mouse embryogenesis. *Developmental Dynamics, 236*(7), 1865–1875. http://dx.doi.org/10.1002/dvdy.21185.

Hewitson, L. C., & Leese, H. J. (1993). Energy metabolism of the trophectoderm and inner cell mass of the mouse blastocyst. *The Journal of Experimental Zoology, 267*(3), 337–343. http://dx.doi.org/10.1002/jez.1402670310.

Hillman, N., & Flynn, T. J. (1980). The metabolism of exogenous fatty acids by preimplantation mouse embryos developing in vitro. *Journal of Embryology and Experimental Morphology, 56*, 157–168.

Hirst, J. (2013). Mitochondrial complex I. *Annual Review of Biochemistry, 82*, 551–575. http://dx.doi.org/10.1146/annurev-biochem-070511-103700.

Ho, Y., Wigglesworth, K., Eppig, J. J., & Schultz, R. M. (1995). Preimplantation development of mouse embryos in KSOM: Augmentation by amino acids and analysis of gene expression. *Molecular Reproduction and Development, 41*(2), 232–238. http://dx.doi.org/10.1002/mrd.1080410214.

Home, P., Saha, B., Ray, S., Dutta, D., Gunewardena, S., Yoo, B., ... Paul, S. (2012). Altered subcellular localization of transcription factor TEAD4 regulates first mammalian cell lineage commitment. *Proceedings of the National Academy of Sciences of the United States of America, 109*(19), 7362–7367. http://dx.doi.org/10.1073/pnas.1201595109.

Houghton, F. D. (2006). Energy metabolism of the inner cell mass and trophectoderm of the mouse blastocyst. *Differentiation, 74*(1), 11–18. http://dx.doi.org/10.1111/j.1432-0436.2006.00052.x. S0301-4681(09)60192-4 [pii].

Houghton, F. D., Thompson, J. G., Kennedy, C. J., & Leese, H. J. (1996). Oxygen consumption and energy metabolism of the early mouse embryo. *Molecular Reproduction and Development, 44*(4), 476–485. http://dx.doi.org/10.1002/(SICI)1098-2795(199608)44:4<476::AID-MRD7>3.0.CO;2-I.

Houten, S. M., Violante, S., Ventura, F. V., & Wanders, R. J. (2016). The biochemistry and physiology of mitochondrial fatty acid beta-oxidation and its genetic disorders. *Annual*

Review of Physiology, *78*, 23–44. http://dx.doi.org/10.1146/annurev-physiol-021115-105045.

Ishihara, T., Kohno, H., & Ishihara, N. (2015). Physiological roles of mitochondrial fission in cultured cells and mouse development. *Annals of the New York Academy of Sciences*, *1350*, 77–81. http://dx.doi.org/10.1111/nyas.12848.

Jang, H., Yang, J., Lee, E., & Cheong, J. H. (2015). Metabolism in embryonic and cancer stemness. *Archives of Pharmacal Research*, *38*(3), 381–388. http://dx.doi.org/10.1007/s12272-015-0558-y.

Jansen, S., Cashman, K., Thompson, J. G., Pantaleon, M., & Kaye, P. L. (2009). Glucose deprivation, oxidative stress and peroxisome proliferator-activated receptor-alpha (PPARA) cause peroxisome proliferation in preimplantation mouse embryos. *Reproduction*, *138*(3), 493–505. http://dx.doi.org/10.1530/REP-09-0038.

Jansen, S., Esmaeilpour, T., Pantaleon, M., & Kaye, P. L. (2006). Glucose affects monocarboxylate cotransporter (MCT) 1 expression during mouse preimplantation development. *Reproduction*, *131*(3), 469–479. http://dx.doi.org/10.1530/rep.1.00953.

Jevtic, P., & Levy, D. L. (2015). Nuclear size scaling during Xenopus early development contributes to midblastula transition timing. *Current Biology*, *25*(1), 45–52. http://dx.doi.org/10.1016/j.cub.2014.10.051.

Jewell, J. L., Kim, Y. C., Russell, R. C., Yu, F. X., Park, H. W., Plouffe, S. W., … Guan, K. L. (2015). Metabolism. Differential regulation of mTORC1 by leucine and glutamine. *Science*, *347*(6218), 194–198. http://dx.doi.org/10.1126/science.1259472.

Jiang, P., Du, W., & Wu, M. (2014). Regulation of the pentose phosphate pathway in cancer. *Protein & Cell*, *5*(8), 592–602. http://dx.doi.org/10.1007/s13238-014-0082-8.

Jin, R., Dobry, C. J., McCown, P. J., & Kumar, A. (2008). Large-scale analysis of yeast filamentous growth by systematic gene disruption and overexpression. *Molecular Biology of the Cell*, *19*(1), 284–296. http://dx.doi.org/10.1091/mbc.E07-05-0519.

Johnson, M. T., Mahmood, S., & Patel, M. S. (2003). Intermediary metabolism and energetics during murine early embryogenesis. *The Journal of Biological Chemistry*, *278*(34), 31457–31460. http://dx.doi.org/10.1074/jbc.R300002200. R300002200 [pii].

Johnson, M. H., & Nasr-Esfahani, M. H. (1994). Radical solutions and cultural problems: Could free oxygen radicals be responsible for the impaired development of preimplantation mammalian embryos in vitro? *Bioessays*, *16*(1), 31–38. http://dx.doi.org/10.1002/bies.950160105.

Jungheim, E. S., & Moley, K. H. (2008). The impact of type 1 and type 2 diabetes mellitus on the oocyte and the preimplantation embryo. *Seminars in Reproductive Medicine*, *26*(2), 186–195. http://dx.doi.org/10.1055/s-2008-1042957.

Kane, M. T., & Buckley, N. J. (1977). The effects of inhibitors of energy metabolism on the growth of one-cell rabbit ova to blastocysts in vitro. *Journal of Reproduction and Fertility*, *49*(2), 261–266.

Kaneko, K. J., & DePamphilis, M. L. (2013). TEAD4 establishes the energy homeostasis essential for blastocoel formation. *Development*, *140*(17), 3680–3690. http://dx.doi.org/10.1242/dev.093799.

Kang, C. M., & Jiang, Y. W. (2005). Genome-wide survey of non-essential genes required for slowed DNA synthesis-induced filamentous growth in yeast. *Yeast*, *22*(2), 79–90. http://dx.doi.org/10.1002/yea.1195.

Katz, J., Landau, B. R., & Bartsch, G. E. (1966). The pentose cycle, triose phosphate isomerization, and lipogenesis in rat adipose tissue. *The Journal of Biological Chemistry*, *241*(3), 727–740.

Kayikci, O., & Nielsen, J. (2015). Glucose repression in Saccharomyces cerevisiae. *FEMS Yeast Research*, *15*(6), fov068. http://dx.doi.org/10.1093/femsyr/fov068.

Kelly, S. J. (1977). Studies of the developmental potential of 4- and 8-cell stage mouse blastomeres. *The Journal of Experimental Zoology*, *200*(3), 365–376. http://dx.doi.org/10.1002/jez.1402000307.

Keramari, M., Razavi, J., Ingman, K. A., Patsch, C., Edenhofer, F., Ward, C. M., & Kimber, S. J. (2010). Sox2 is essential for formation of trophectoderm in the preimplantation embryo. *PLoS One*, *5*(11), e13952. http://dx.doi.org/10.1371/journal.pone.0013952.

Kidder, G. M. (2002). Trophectoderm development and function: The roles of Na+/K(+)-ATPase subunit isoforms. *Canadian Journal of Physiology and Pharmacology*, *80*(2), 110–115.

Kidder, G. M., & Watson, A. J. (2005). Roles of Na, K-ATPase in early development and trophectoderm differentiation. *Seminars in Nephrology*, *25*(5), 352–355. http://dx.doi.org/10.1016/j.semnephrol.2005.03.011.

Kim, S. Y., & Kim, J. (2010). Roles of dihydrolipoamide dehydrogenase Lpd1 in Candida albicans filamentation. *Fungal Genetics and Biology*, *47*(9), 782–788. http://dx.doi.org/10.1016/j.fgb.2010.06.005.

Kim, J. W., Tchernyshyov, I., Semenza, G. L., & Dang, C. V. (2006). HIF-1-mediated expression of pyruvate dehydrogenase kinase: A metabolic switch required for cellular adaptation to hypoxia. *Cell Metabolism*, *3*(3), 177–185. http://dx.doi.org/10.1016/j.cmet.2006.02.002.

Kingsbury, J. M., Sen, N. D., & Cardenas, M. E. (2015). Branched-chain aminotransferases control TORC1 signaling in Saccharomyces cerevisiae. *PLoS Genetics*, *11*(12), e1005714. http://dx.doi.org/10.1371/journal.pgen.1005714.

Kohn, M. J., Kaneko, K. J., & DePamphilis, M. L. (2012). Regulation of gene expression at the beginning of mammalian development. In R. A. Meyers (Ed.), *Epigenetic regulation and epigenomics: Vol. 1.* (pp. 325–364). Weinheim, Germany: Wiley-Blackwell.

Kotiadis, V. N., Duchen, M. R., & Osellame, L. D. (2014). Mitochondrial quality control and communications with the nucleus are important in maintaining mitochondrial function and cell health. *Biochimica et Biophysica Acta*, *1840*(4), 1254–1265. http://dx.doi.org/10.1016/j.bbagen.2013.10.041.

Land, G. A., McDonald, W. C., Stjernholm, R. L., & Friedman, L. (1975). Factors affecting filamentation in Candida albicans: Changes in respiratory activity of Candida albicans during filamentation. *Infection and Immunity*, *12*(1), 119–127.

Lane, M., & Gardner, D. K. (1994). Increase in postimplantation development of cultured mouse embryos by amino acids and induction of fetal retardation and exencephaly by ammonium ions. *Journal of Reproduction and Fertility*, *102*(2), 305–312.

Lane, M., & Gardner, D. K. (1996). Selection of viable mouse blastocysts prior to transfer using a metabolic criterion. *Human Reproduction*, *11*(9), 1975–1978.

Lane, M., & Gardner, D. K. (1997). Differential regulation of mouse embryo development and viability by amino acids. *Journal of Reproduction and Fertility*, *109*(1), 153–164.

Lane, M., & Gardner, D. K. (1998). Amino acids and vitamins prevent culture-induced metabolic perturbations and associated loss of viability of mouse blastocysts. *Human Reproduction*, *13*(4), 991–997.

Lane, M., & Gardner, D. K. (2005). Mitochondrial malate-aspartate shuttle regulates mouse embryo nutrient consumption. *The Journal of Biological Chemistry*, *280*(18), 18361–18367. http://dx.doi.org/10.1074/jbc.M500174200.

Laplante, M., & Sabatini, D. M. (2012). mTOR signaling in growth control and disease. *Cell*, *149*(2), 274–293. http://dx.doi.org/10.1016/j.cell.2012.03.017.

Lawitts, J. A., & Biggers, J. D. (1993). Culture of preimplantation embryos. *Methods in Enzymology*, *225*, 153–164.

Lee, S. M., Koh, H. J., Park, D. C., Song, B. J., Huh, T. L., & Park, J. W. (2002). Cytosolic NADP(+)-dependent isocitrate dehydrogenase status modulates oxidative damage to cells. *Free Radical Biology & Medicine*, *32*(11), 1185–1196.

Leese, H. J., Conaghan, J., Martin, K. L., & Hardy, K. (1993). Early human embryo metabolism. *Bioessays*, *15*(4), 259–264. http://dx.doi.org/10.1002/bies.950150406.

Li, E., & Zhang, Y. (2014). DNA methylation in mammals. *Cold Spring Harbor Perspectives in Biology*, *6*(5), a019133. http://dx.doi.org/10.1101/cshperspect.a019133.

Li, L., Zheng, P., & Dean, J. (2010). Maternal control of early mouse development. *Development*, *137*(6), 859–870. http://dx.doi.org/10.1242/dev.039487.

Lorenz, M. C., Pan, X., Harashima, T., Cardenas, M. E., Xue, Y., Hirsch, J. P., & Heitman, J. (2000). The G protein-coupled receptor gpr1 is a nutrient sensor that regulates pseudohyphal differentiation in Saccharomyces cerevisiae. *Genetics*, *154*(2), 609–622.

Lu, C. L., Qin, L., Liu, H. C., Candas, D., Fan, M., & Li, J. J. (2015). Tumor cells switch to mitochondrial oxidative phosphorylation under radiation via mTOR-mediated hexokinase II inhibition—A Warburg-reversing effect. *PLoS One*, *10*(3), e0121046. http://dx.doi.org/10.1371/journal.pone.0121046.

Ma, W. G., Song, H., Das, S. K., Paria, B. C., & Dey, S. K. (2003). Estrogen is a critical determinant that specifies the duration of the window of uterine receptivity for implantation. *Proceedings of the National Academy of Sciences of the United States of America*, *100*(5), 2963–2968. http://dx.doi.org/10.1073/pnas.0530162100.

Madhani, H. D., & Fink, G. R. (1998). The control of filamentous differentiation and virulence in fungi. *Trends in Cell Biology*, *8*(9), 348–353.

Maldonado, E. N., & Lemasters, J. J. (2014). ATP/ADP ratio, the missed connection between mitochondria and the Warburg effect. *Mitochondrion*, *19*(Pt. A), 78–84.

Mandal, S., Lindgren, A. G., Srivastava, A. S., Clark, A. T., & Banerjee, U. (2011). Mitochondrial function controls proliferation and early differentiation potential of embryonic stem cells. *Stem Cells (Dayton, Ohio)*, *29*(3), 486–495.

Mannella, C. A. (2006). Structure and dynamics of the mitochondrial inner membrane cristae. *Biochimica et Biophysica Acta*, *1763*(5–6), 542–548. http://dx.doi.org/10.1016/j.bbamcr.2006.04.006.

Marchi, S., Patergnani, S., & Pinton, P. (2014). The endoplasmic reticulum-mitochondria connection: One touch, multiple functions. *Biochimica et Biophysica Acta*, *1837*(4), 461–469. http://dx.doi.org/10.1016/j.bbabio.2013.10.015.

Martin, K. L., & Leese, H. J. (1995). Role of glucose in mouse preimplantation embryo development. *Molecular Reproduction and Development*, *40*(4), 436–443. http://dx.doi.org/10.1002/mrd.1080400407.

Martin, K. L., & Leese, H. J. (1999). Role of developmental factors in the switch from pyruvate to glucose as the major exogenous energy substrate in the preimplantation mouse embryo. *Reproduction, Fertility, and Development*, *11*(7–8), 425–433.

McDonough, J. A., Bhattacherjee, V., Sadlon, T., & Hostetter, M. K. (2002). Involvement of Candida albicans NADH dehydrogenase complex I in filamentation. *Fungal Genetics and Biology*, *36*(2), 117–127. http://dx.doi.org/10.1016/S1087-1845(02)00007-5.

McGowen, M. R., Erez, O., Romero, R., & Wildman, D. E. (2014). The evolution of embryo implantation. *The International Journal of Developmental Biology*, *58*(2–4), 155–161. http://dx.doi.org/10.1387/ijdb.140020dw.

McKnight, S. L. (2010). On getting there from here. *Science (New York, NY)*, *330*(6009), 1338–1339.

Morita, M., Gravel, S. P., Chenard, V., Sikstrom, K., Zheng, L., Alain, T., … Sonenberg, N. (2013). mTORC1 controls mitochondrial activity and biogenesis through 4E-BP-dependent translational regulation. *Cell Metabolism*, *18*(5), 698–711. http://dx.doi.org/10.1016/j.cmet.2013.10.001.

Murphy, M. P. (2009). How mitochondria produce reactive oxygen species. *The Biochemical Journal, 417*(1), 1–13. http://dx.doi.org/10.1042/BJ20081386.
Nagy, A. (2003). *Manipulating the mouse embryo: A laboratory manual.* Cold Spring Harbor, NY: Cold Spring Harbor Laboratory Press.
Nagy, A., Gertsenstein, M., Vintersten, K., & Behringer, R. (2003). Summary of mouse development. In *Manipulating the mouse embryo* (pp. 31–139). Cold Spring Harbor, NY: Cold Spring Harbor Laboratory Press.
Nasution, O., Srinivasa, K., Kim, M., Kim, Y. J., Kim, W., Jeong, W., & Choi, W. (2008). Hydrogen peroxide induces hyphal differentiation in Candida albicans. *Eukaryotic Cell,* 7(11), 2008–2011. http://dx.doi.org/10.1128/EC.00105-08.
Nelson, D. L., Lehninger, A. L., & Cox, M. M. (2008). *Lehninger principles of biochemistry.* New York: W.H. Freeman.
Newport, J., & Kirschner, M. (1982). A major developmental transition in early Xenopus embryos: II. Control of the onset of transcription. *Cell, 30*(3), 687–696.
Nicklin, P., Bergman, P., Zhang, B., Triantafellow, E., Wang, H., Nyfeler, B., … Murphy, L. O. (2009). Bidirectional transport of amino acids regulates mTOR and autophagy. *Cell, 136*(3), 521–534. http://dx.doi.org/10.1016/j.cell.2008.11.044.
Nieder, G. L., & Weitlauf, H. M. (1984). Regulation of glycolysis in the mouse blastocyst during delayed implantation. *The Journal of Experimental Zoology, 231*(1), 121–129. http://dx.doi.org/10.1002/jez.1402310116.
Nishioka, N., Inoue, K., Adachi, K., Kiyonari, H., Ota, M., Ralston, A., … Sasaki, H. (2009). The Hippo signaling pathway components Lats and Yap pattern Tead4 activity to distinguish mouse trophectoderm from inner cell mass. *Developmental Cell, 16*(3), 398–410. http://dx.doi.org/10.1016/j.devcel.2009.02.003. S1534-5807(09)00077-X [pii].
Nishioka, N., Yamamoto, S., Kiyonari, H., Sato, H., Sawada, A., Ota, M., … Sasaki, H. (2008). Tead4 is required for specification of trophectoderm in pre-implantation mouse embryos. *Mechanisms of Development, 125*(3–4), 270–283. http://dx.doi.org/10.1016/j.mod.2007.11.002. S0925-4773(07)00197-9 [pii].
Nothias, J. Y., Majumder, S., Kaneko, K. J., & DePamphilis, M. L. (1995). Regulation of gene expression at the beginning of mammalian development. *The Journal of Biological Chemistry, 270*(38), 22077–22080.
Nunnari, J., & Suomalainen, A. (2012). Mitochondria: In sickness and in health. *Cell, 148*(6), 1145–1159. http://dx.doi.org/10.1016/j.cell.2012.02.035.
O'Fallon, J. V., & Wright, R. W., Jr. (1986). Quantitative determination of the pentose phosphate pathway in preimplantation mouse embryos. *Biology of Reproduction, 34*(1), 58–64.
Onjiko, R. M., Moody, S. A., & Nemes, P. (2015). Single-cell mass spectrometry reveals small molecules that affect cell fates in the 16-cell embryo. *Proceedings of the National Academy of Sciences of the United States of America, 112*(21), 6545–6550. http://dx.doi.org/10.1073/pnas.1423682112.
Orsi, N. M., & Leese, H. J. (2001). Protection against reactive oxygen species during mouse preimplantation embryo development: Role of EDTA, oxygen tension, catalase, superoxide dismutase and pyruvate. *Molecular Reproduction and Development, 59*(1), 44–53. http://dx.doi.org/10.1002/mrd.1006.
Osthus, R. C., Shim, H., Kim, S., Li, Q., Reddy, R., Mukherjee, M., … Dang, C. V. (2000). Deregulation of glucose transporter 1 and glycolytic gene expression by c-Myc. *The Journal of Biological Chemistry, 275*(29), 21797–21800. http://dx.doi.org/10.1074/jbc.C000023200.
Owen, O. E., Kalhan, S. C., & Hanson, R. W. (2002). The key role of anaplerosis and cataplerosis for citric acid cycle function. *The Journal of Biological Chemistry, 277*(34), 30409–30412. http://dx.doi.org/10.1074/jbc.R200006200.

Pan, X., Harashima, T., & Heitman, J. (2000). Signal transduction cascades regulating pseudohyphal differentiation of Saccharomyces cerevisiae. *Current Opinion in Microbiology, 3*(6), 567–572.

Pandolfi, P. P., Roth, M. E., Karis, A., Leonard, M. W., Dzierzak, E., Grosveld, F. G., … Lindenbaum, M. H. (1995). Targeted disruption of the GATA3 gene causes severe abnormalities in the nervous system and in fetal liver haematopoiesis. *Nature Genetics, 11*(1), 40–44. http://dx.doi.org/10.1038/ng0995-40.

Pantaleon, M., Scott, J., & Kaye, P. L. (2008). Nutrient sensing by the early mouse embryo: Hexosamine biosynthesis and glucose signaling during preimplantation development. *Biology of Reproduction, 78*(4), 595–600. http://dx.doi.org/10.1095/biolreprod.107.062877. biolreprod.107.062877 [pii].

Pantaleon, M., Tan, H. Y., Kafer, G. R., & Kaye, P. L. (2010). Toxic effects of hyperglycemia are mediated by the hexosamine signaling pathway and o-linked glycosylation in early mouse embryos. *Biology of Reproduction, 82*(4), 751–758. http://dx.doi.org/10.1095/biolreprod.109.076661.

Patra, K. C., & Hay, N. (2014). The pentose phosphate pathway and cancer. *Trends in Biochemical Sciences, 39*(8), 347–354. http://dx.doi.org/10.1016/j.tibs.2014.06.005.

Patten, D. A., Wong, J., Khacho, M., Soubannier, V., Mailloux, R. J., Pilon-Larose, K., … Slack, R. S. (2014). OPA1-dependent cristae modulation is essential for cellular adaptation to metabolic demand. *The EMBO Journal, 33*(22), 2676–2691. http://dx.doi.org/10.15252/embj.201488349.

Payne, J. L., & Wagner, A. (2015). Mechanisms of mutational robustness in transcriptional regulation. *Frontiers in Genetics, 6*, 322. http://dx.doi.org/10.3389/fgene.2015.00322.

Pike, L. S., Smift, A. L., Croteau, N. J., Ferrick, D. A., & Wu, M. (2011). Inhibition of fatty acid oxidation by etomoxir impairs NADPH production and increases reactive oxygen species resulting in ATP depletion and cell death in human glioblastoma cells. *Biochimica et Biophysica Acta, 1807*(6), 726–734.

Ralston, A., & Rossant, J. (2008). Cdx2 acts downstream of cell polarization to cell-autonomously promote trophectoderm fate in the early mouse embryo. *Developmental Biology, 313*(2), 614–629. http://dx.doi.org/10.1016/j.ydbio.2007.10.054.

Ramanathan, A., & Schreiber, S. L. (2009). Direct control of mitochondrial function by mTOR. *Proceedings of the National Academy of Sciences of the United States of America, 106*(52), 22229–22232. http://dx.doi.org/10.1073/pnas.0912074106.

Riganti, C., Gazzano, E., Polimeni, M., Aldieri, E., & Ghigo, D. (2012). The pentose phosphate pathway: An antioxidant defense and a crossroad in tumor cell fate. *Free Radical Biology & Medicine, 53*(3), 421–436. http://dx.doi.org/10.1016/j.freeradbiomed.2012.05.006.

Rinaudo, P. F., Giritharan, G., Talbi, S., Dobson, A. T., & Schultz, R. M. (2006). Effects of oxygen tension on gene expression in preimplantation mouse embryos. *Fertility and Sterility, 86*(4 Suppl.), 1252–1265. http://dx.doi.org/10.1016/j.fertnstert.2006.05.017. 1265 e1–36; S0015-0282(06)01306-9 [pii].

Roberts, R. L., & Fink, G. R. (1994). Elements of a single MAP kinase cascade in Saccharomyces cerevisiae mediate two developmental programs in the same cell type: Mating and invasive growth. *Genes & Development, 8*(24), 2974–2985.

Rodriguez-Enriquez, S., Juarez, O., Rodriguez-Zavala, J. S., & Moreno-Sanchez, R. (2001). Multisite control of the Crabtree effect in ascites hepatoma cells. *European Journal of Biochemistry, 268*(8), 2512–2519.

Rohde, J. R., Bastidas, R., Puria, R., & Cardenas, M. E. (2008). Nutritional control via Tor signaling in Saccharomyces cerevisiae. *Current Opinion in Microbiology, 11*(2), 153–160. http://dx.doi.org/10.1016/j.mib.2008.02.013.

Rosas-Lemus, M., Uribe-Alvarez, C., Chiquete-Felix, N., & Uribe-Carvajal, S. (2014). In Saccharomyces cerevisiae fructose-1,6-bisphosphate contributes to the Crabtree effect

through closure of the mitochondrial unspecific channel. *Archives of Biochemistry and Biophysics*, *555–556*, 66–70. http://dx.doi.org/10.1016/j.abb.2014.05.027.
Rossant, J., & Tam, P. P. (2009). Blastocyst lineage formation, early embryonic asymmetries and axis patterning in the mouse. *Development*, *136*(5), 701–713. http://dx.doi.org/10.1242/dev.017178.
Rua, D., Tobe, B. T., & Kron, S. J. (2001). Cell cycle control of yeast filamentous growth. *Current Opinion in Microbiology*, *4*(6), 720–727.
Ryan, M. T., & Hoogenraad, N. J. (2007). Mitochondrial-nuclear communications. *Annual Review of Biochemistry*, *76*, 701–722. http://dx.doi.org/10.1146/annurev.biochem.76.052305.091720.
Saha, A., Connelly, S., Jiang, J., Zhuang, S., Amador, D. T., Phan, T., … Boss, G. R. (2014). Akt phosphorylation and regulation of transketolase is a nodal point for amino acid control of purine synthesis. *Molecular Cell*, *55*(2), 264–276. http://dx.doi.org/10.1016/j.molcel.2014.05.028.
Sato, M., & Sato, K. (2013). Maternal inheritance of mitochondrial DNA by diverse mechanisms to eliminate paternal mitochondrial DNA. *Biochimica et Biophysica Acta*, *1833*(8), 1979–1984. http://dx.doi.org/10.1016/j.bbamcr.2013.03.010.
Schieke, S. M., Phillips, D., McCoy, J. P., Jr., Aponte, A. M., Shen, R. F., Balaban, R. S., & Finkel, T. (2006). The mammalian target of rapamycin (mTOR) pathway regulates mitochondrial oxygen consumption and oxidative capacity. *The Journal of Biological Chemistry*, *281*(37), 27643–27652. http://dx.doi.org/10.1074/jbc.M603536200.
Schneper, L., Duvel, K., & Broach, J. R. (2004). Sense and sensibility: Nutritional response and signal integration in yeast. *Current Opinion in Microbiology*, 7(6), 624–630. http://dx.doi.org/10.1016/j.mib.2004.10.002.
Schonenberger, M. J., & Kovacs, W. J. (2015). Hypoxia signaling pathways: Modulators of oxygen-related organelles. *Frontiers in Cell and Developmental Biology*, *3*, 42. http://dx.doi.org/10.3389/fcell.2015.00042.
Schultz, R. M. (2002). The molecular foundations of the maternal to zygotic transition in the preimplantation embryo. *Human Reproduction Update*, *8*(4), 323–331.
Schuster, S., Boley, D., Moller, P., Stark, H., & Kaleta, C. (2015). Mathematical models for explaining the Warburg effect: A review focussed on ATP and biomass production. *Biochemical Society Transactions*, *43*(6), 1187–1194. http://dx.doi.org/10.1042/BST20150153.
Semenza, G. L. (1999). Regulation of mammalian O2 homeostasis by hypoxia-inducible factor 1. *Annual Review of Cell and Developmental Biology*, *15*, 551–578. http://dx.doi.org/10.1146/annurev.cellbio.15.1.551.
Sengupta, S., Peterson, T. R., & Sabatini, D. M. (2010). Regulation of the mTOR complex 1 pathway by nutrients, growth factors, and stress. *Molecular Cell*, *40*(2), 310–322. http://dx.doi.org/10.1016/j.molcel.2010.09.026. S1097-2765(10)00754-9 [pii].
Shyh-Chang, N., Daley, G. Q., & Cantley, L. C. (2013). Stem cell metabolism in tissue development and aging. *Development*, *140*(12), 2535–2547. http://dx.doi.org/10.1242/dev.091777.
Simmons, D. G., & Cross, J. C. (2005). Determinants of trophoblast lineage and cell subtype specification in the mouse placenta. *Developmental Biology*, *284*(1), 12–24. http://dx.doi.org/10.1016/j.ydbio.2005.05.010.
Sorrells, T. R., & Johnson, A. D. (2015). Making sense of transcription networks. *Cell*, *161*(4), 714–723. http://dx.doi.org/10.1016/j.cell.2015.04.014.
St John, J. C., Facucho-Oliveira, J., Jiang, Y., Kelly, R., & Salah, R. (2010). Mitochondrial DNA transmission, replication and inheritance: A journey from the gamete through the embryo and into offspring and embryonic stem cells. *Human Reproduction Update*, *16*(5), 488–509. http://dx.doi.org/10.1093/humupd/dmq002.

Starovoytova, A. N., Sorokin, M. I., Sokolov, S. S., Severin, F. F., & Knorre, D. A. (2013). Mitochondrial signaling in Saccharomyces cerevisiae pseudohyphae formation induced by butanol. *FEMS Yeast Research*, *13*(4), 367–374. http://dx.doi.org/10.1111/1567-1364.12039.

Stern, S., Biggers, J. D., & Anderson, E. (1971). Mitochondria and early development of the mouse. *The Journal of Experimental Zoology*, *176*(2), 179–191. http://dx.doi.org/10.1002/jez.1401760206.

Stotland, A., & Gottlieb, R. A. (2015). Mitochondrial quality control: Easy come, easy go. *Biochimica et Biophysica Acta*, *1853*(10 Pt. B), 2802–2811. http://dx.doi.org/10.1016/j.bbamcr.2014.12.041.

Strange, K., Denton, J., & Nehrke, K. (2006). Ste20-type kinases: Evolutionarily conserved regulators of ion transport and cell volume. *Physiology (Bethesda)*, *21*, 61–68. http://dx.doi.org/10.1152/physiol.00139.2005.

Sudbery, P. E. (2011). Growth of Candida albicans hyphae. *Nature Reviews. Microbiology*, *9*(10), 737–748. http://dx.doi.org/10.1038/nrmicro2636.

Sullivan, L. B., Gui, D. Y., Hosios, A. M., Bush, L. N., Freinkman, E., & Vander Heiden, M. G. (2015). Supporting aspartate biosynthesis is an essential function of respiration in proliferating cells. *Cell*, *162*(3), 552–563. http://dx.doi.org/10.1016/j.cell.2015.07.017.

Summers, M. C., McGinnis, L. K., Lawitts, J. A., Raffin, M., & Biggers, J. D. (2000). IVF of mouse ova in a simplex optimized medium supplemented with amino acids. *Human Reproduction*, *15*(8), 1791–1801.

Suwinska, A., Czolowska, R., Ozdzenski, W., & Tarkowski, A. K. (2008). Blastomeres of the mouse embryo lose totipotency after the fifth cleavage division: Expression of Cdx2 and Oct4 and developmental potential of inner and outer blastomeres of 16- and 32-cell embryos. *Developmental Biology*, *322*(1), 133–144. http://dx.doi.org/10.1016/j.ydbio.2008.07.019.

Taanman, J. W. (1999). The mitochondrial genome: Structure, transcription, translation and replication. *Biochimica et Biophysica Acta*, *1410*(2), 103–123.

Takahashi, M. (2012). Oxidative stress and redox regulation on in vitro development of mammalian embryos. *The Journal of Reproduction and Development*, *58*(1), 1–9.

Tarkowski, A. K., & Wroblewska, J. (1967). Development of blastomeres of mouse eggs isolated at the 4- and 8-cell stage. *Journal of Embryology and Experimental Morphology*, *18*(1), 155–180.

Teslaa, T., & Teitell, M. A. (2015). Pluripotent stem cell energy metabolism: An update. *The EMBO Journal*, *34*(2), 138–153. http://dx.doi.org/10.15252/embj.201490446.

Thompson, J. G., Mitchell, M., & Kind, K. L. (2007). Embryo culture and long-term consequences. *Reproduction, Fertility, and Development*, *19*(1), 43–52. RD06129 [pii].

Thomson, J. L. (1967). Effect of inhibitors of carbohydrate metabolism on the development of preimplantation mouse embryos. *Experimental Cell Research*, *46*(2), 252–262.

Thouas, G. A., Trounson, A. O., Wolvetang, E. J., & Jones, G. M. (2004). Mitochondrial dysfunction in mouse oocytes results in preimplantation embryo arrest in vitro. *Biology of Reproduction*, *71*(6), 1936–1942. http://dx.doi.org/10.1095/biolreprod.104.033589.

Tokunaga, C., Yoshino, K., & Yonezawa, K. (2004). mTOR integrates amino acid- and energy-sensing pathways. *Biochemical and Biophysical Research Communications*, *313*(2), 443–446.

Trimarchi, J. R., Liu, L., Porterfield, D. M., Smith, P. J., & Keefe, D. L. (2000). Oxidative phosphorylation-dependent and -independent oxygen consumption by individual preimplantation mouse embryos. *Biology of Reproduction*, *62*(6), 1866–1874.

Turrens, J. F. (2003). Mitochondrial formation of reactive oxygen species. *The Journal of Physiology*, *552*(Pt. 2), 335–344. http://dx.doi.org/10.1113/jphysiol.2003.049478.

Tyra, H. M., Spitz, D. R., & Rutkowski, D. T. (2012). Inhibition of fatty acid oxidation enhances oxidative protein folding and protects hepatocytes from endoplasmic reticulum stress. *Molecular Biology of the Cell*, *23*(5), 811–819. http://dx.doi.org/10.1091/mbc.E11-12-1011.

Vadlakonda, L., Dash, A., Pasupuleti, M., Anil Kumar, K., & Reddanna, P. (2013). Did we get pasteur, warburg, and crabtree on a right note? *Frontiers in Oncology*, *3*, 186.

Van Blerkom, J. (2009). Mitochondria in early mammalian development. *Seminars in Cell & Developmental Biology*, *20*(3), 354–364. http://dx.doi.org/10.1016/j.semcdb.2008.12.005.

Van Winkle, L. J. (2001). Amino acid transport regulation and early embryo development. *Biology of Reproduction*, *64*(1), 1–12.

Vander Heiden, M. G., Cantley, L. C., & Thompson, C. B. (2009). Understanding the Warburg effect: The metabolic requirements of cell proliferation. *Science*, *324*(5930), 1029–1033. http://dx.doi.org/10.1126/science.1160809.

Vastag, L., Jorgensen, P., Peshkin, L., Wei, R., Rabinowitz, J. D., & Kirschner, M. W. (2011). Remodeling of the metabolome during early frog development. *PLoS One*, *6*(2), e16881. http://dx.doi.org/10.1371/journal.pone.0016881.

Vellucci, V. F., Gygax, S. E., & Hostetter, M. K. (2007). Involvement of Candida albicans pyruvate dehydrogenase complex protein X (Pdx1) in filamentation. *Fungal Genetics and Biology*, *44*(10), 979–990. http://dx.doi.org/10.1016/j.fgb.2006.12.003.

Villa-Cuesta, E., Holmbeck, M. A., & Rand, D. M. (2014). Rapamycin increases mitochondrial efficiency by mtDNA-dependent reprogramming of mitochondrial metabolism in Drosophila. *Journal of Cell Science*, *127*(Pt. 10), 2282–2290. http://dx.doi.org/10.1242/jcs.142026.

Wai, T., & Langer, T. (2016). Mitochondrial dynamics and metabolic regulation. *Trends in Endocrinology and Metabolism*, *27*, 105–117. http://dx.doi.org/10.1016/j.tem.2015.12.001.

Wakil, S. J. (1989). Fatty acid synthase, a proficient multifunctional enzyme. *Biochemistry*, *28*(11), 4523–4530.

Wale, P. L., & Gardner, D. K. (2012). Oxygen regulates amino acid turnover and carbohydrate uptake during the preimplantation period of mouse embryo development. *Biology of Reproduction*, *87*(1), 24. http://dx.doi.org/10.1095/biolreprod.112.100552. 21–28.

Wang, H., & Dey, S. K. (2006). Roadmap to embryo implantation: Clues from mouse models. *Nature Reviews. Genetics*, 7(3), 185–199. http://dx.doi.org/10.1038/nrg1808.

Warburg, O. (1956). On the origin of cancer cells. *Science*, *123*(3191), 309–314.

Watkins, A. J., Papenbrock, T., & Fleming, T. P. (2008). The preimplantation embryo: Handle with care. *Seminars in Reproductive Medicine*, *26*(2), 175–185. http://dx.doi.org/10.1055/s-2008-1042956.

Watson, A. J. (1992). The cell biology of blastocyst development. *Molecular Reproduction and Development*, *33*(4), 492–504. http://dx.doi.org/10.1002/mrd.1080330417.

Waypa, G. B., Smith, K. A., & Schumacker, P. T. (2016). O_2 sensing, mitochondria and ROS signaling: The fog is lifting. *Molecular Aspects of Medicine*, *47–48*, 76–89. http://dx.doi.org/10.1016/j.mam.2016.01.002.

Whelan, S. P., & Zuckerbraun, B. S. (2013). Mitochondrial signaling: Forwards, backwards, and in between. *Oxidative Medicine and Cellular Longevity*, *2013*, 351613. http://dx.doi.org/10.1155/2013/351613.

Whiteway, M., & Bachewich, C. (2007). Morphogenesis in Candida albicans. *Annual Review of Microbiology*, *61*, 529–553. http://dx.doi.org/10.1146/annurev.micro.61.080706.093341.

Whitten, W. K. (1956). Culture of tubal mouse ova. *Nature*, *177*(4498), 96.

Wicklow, E., Blij, S., Frum, T., Hirate, Y., Lang, R. A., Sasaki, H., & Ralston, A. (2014). HIPPO pathway members restrict SOX2 to the inner cell mass where it promotes ICM fates in the mouse blastocyst. *PLoS Genetics*, *10*(10), e1004618. http://dx.doi.org/10.1371/journal.pgen.1004618.

Wise, D. R., & Thompson, C. B. (2010). Glutamine addiction: A new therapeutic target in cancer. *Trends in Biochemical Sciences*, *35*(8), 427–433. http://dx.doi.org/10.1016/j.tibs.2010.05.003.

Wojtczak, L. (1996). The Crabtree effect: A new look at the old problem. *Acta Biochimica Polonica*, *43*(2), 361–368.

Wright, R. M., Repine, T., & Repine, J. E. (1993). Reversible pseudohyphal growth in haploid Saccharomyces cerevisiae is an aerobic process. *Current Genetics*, *23*(5–6), 388–391.

Wu, G., Han, D., Gong, Y., Sebastiano, V., Gentile, L., Singhal, N., … Scholer, H. R. (2013). Establishment of totipotency does not depend on Oct4A. *Nature Cell Biology*, *15*(9), 1089–1097. http://dx.doi.org/10.1038/ncb2816.

Xu, X., Duan, S., Yi, F., Ocampo, A., Liu, G. H., & Izpisua Belmonte, J. C. (2013). Mitochondrial regulation in pluripotent stem cells. *Cell Metabolism*, *18*(3), 325–332. http://dx.doi.org/10.1016/j.cmet.2013.06.005.

Xu, Z. P., Wawrousek, E. F., & Piatigorsky, J. (2002). Transketolase haploinsufficiency reduces adipose tissue and female fertility in mice. *Molecular and Cellular Biology*, *22*(17), 6142–6147.

Xue, Y., Batlle, M., & Hirsch, J. P. (1998). GPR1 encodes a putative G protein-coupled receptor that associates with the Gpa2p Galpha subunit and functions in a Ras-independent pathway. *The EMBO Journal*, *17*(7), 1996–2007. http://dx.doi.org/10.1093/emboj/17.7.1996.

Yagi, R., Kohn, M. J., Karavanova, I., Kaneko, K. J., Vullhorst, D., DePamphilis, M. L., & Buonanno, A. (2007). Transcription factor TEAD4 specifies the trophectoderm lineage at the beginning of mammalian development. *Development*, *134*(21), 3827–3836. http://dx.doi.org/10.1242/dev.010223. dev.010223 [pii].

Yamada, M., Takanashi, K., Hamatani, T., Hirayama, A., Akutsu, H., Fukunaga, T., … Tomita, M. (2012). A medium-chain fatty acid as an alternative energy source in mouse preimplantation development. *Scientific Reports*, *2*, 930. http://dx.doi.org/10.1038/srep00930.

Yamanaka, Y., Ralston, A., Stephenson, R. O., & Rossant, J. (2006). Cell and molecular regulation of the mouse blastocyst. *Developmental Dynamics*, *235*(9), 2301–2314. http://dx.doi.org/10.1002/dvdy.20844.

Yin, F., Sancheti, H., & Cadenas, E. (2012). Mitochondrial thiols in the regulation of cell death pathways. *Antioxidants & Redox Signaling*, *17*(12), 1714–1727. http://dx.doi.org/10.1089/ars.2012.4639.

Zaytseva, Y. Y., Harris, J. W., Mitov, M. I., Kim, J. T., Butterfield, D. A., Lee, E. Y., … Evers, B. M. (2015). Increased expression of fatty acid synthase provides a survival advantage to colorectal cancer cells via upregulation of cellular respiration. *Oncotarget*, *6*(22), 18891–18904. http://dx.doi.org/10.18632/oncotarget.3783.

Zernicka-Goetz, M., Morris, S. A., & Bruce, A. W. (2009). Making a firm decision: Multifaceted regulation of cell fate in the early mouse embryo. *Nature Reviews. Genetics*, *10*(7), 467–477. http://dx.doi.org/10.1038/nrg2564.

CHAPTER NINE

Epigenetic Control of Early Mouse Development

C.Y. Lim*, B.B. Knowles[†,‡], D. Solter[‡,§,1], D.M. Messerschmidt[¶,1]
*Institute of Medical Biology, A*STAR, Singapore, Singapore
†Emerita, The Jackson Laboratory, Bar Harbor, ME, United States
‡Siriraj Center of Excellence for Stem Cell Research, Mahidol University, Bangkok, Thailand
§Emeritus, Max-Planck Institute of Immunobiology and Epigenetics, Freiburg, Germany
¶Institute of Molecular and Cell Biology, A*STAR, Singapore, Singapore
[1]Corresponding authors: e-mail address: davorsolter@mac.com; danielm@imcb.a-star.edu.sg

Contents

Abstract

Although the genes sequentially transcribed in the mammalian embryo prior to implantation have been identified, understanding of the molecular processes ensuring this transcription is still in development. The genomes of the sperm and egg are hypermethylated, hence transcriptionally silent. Their union, in the prepared environment of the egg, initiates their epigenetic genomic reprogramming into a totipotent zygote,

Current Topics in Developmental Biology, Volume 120
ISSN 0070-2153
http://dx.doi.org/10.1016/bs.ctdb.2016.05.002

in which the genome gradually becomes transcriptionally activated. During gametogenesis, sex-specific processes result in sperm and eggs with disparate epigenomes, both of which require drastic reprogramming to establish the totipotent genome of the zygote and the pluripotent inner cell mass of the blastocyst. Herein, we describe the factors, DNA and histone modifications, activation and repression of retrotransposons, and cytoplasmic localizations, known to influence the activation of the mammalian genome at the initiation of new life.

1. INTRODUCTION

The epigenome of male and female gametes at the time of syngamy is a product of a long, complex reprogramming process starting with the naïve epigenome of the primordial germ cells (PGCs). For the purposes of this chapter, we deem the epigenome to be the conglomerate of all the information present in the gametes and embryo but not encoded by the DNA sequence. In this chapter, we will discuss the relevant modifications of the DNA bases, describe changes in other chromatin components, and consider the molecules and structures present in the nucleoplasm, in the cytoplasm, and in the nuclear and cell membranes of the gametes and embryo. DNA and chromatin modifications determine which parts of the genome are open for transcription or subsequent modification, and they also control long-distance intra- and interchromosomal interactions and establishment of chromosomal territories in the nucleus. The molecular constituents of the mature gametes and zygote and their spatial arrangement, at this time of transcriptional silence, regulate completion of meiosis and activation of the embryonic genome, and they may also contribute to the morphogenesis of the early preimplantation embryo.

Since somatic cell nuclear transfer (SCNT) can replace the male and female pronucleus and round spermatids or nuclei of androgenetic haploid embryonic stem (ES) cells can replace sperm, we have to ask whether the epigenome elaboration during gametogenesis is absolutely necessary for normal development? Obviously, neither somatic cell nuclei nor the nuclei of androgenetic haploid ES cells possess the epigenome of a zygote or mature sperm, respectively, yet they can function as if they do, to a degree. Does this mean that an absolutely correct epigenome is not necessary for development, or is the very low success rate of normal development following SCNT a clear indication that the epigenome as a product of gametogenesis is essentially irreplaceable? The answer most compatible with current knowledge and observed results is the molecular components of the egg or zygote cytoplasm and nucleoplasm can sometimes reestablish the epigenome of the somatic cell to a state compatible with development. Yet, due to the stochastic nature of this reprogramming, failure is a more likely outcome. Understanding the establishment of the gametic epigenomes and their remodeling into the zygotic epigenome is a sine qua non for understanding totipotency and the molecular basis of mammalian development.

2. DNA-METHYLATION DYNAMICS DURING EPIGENETIC REPROGRAMMING

Owing to its methodological accessibility the best-studied epigenetic marks during preimplantation development are covalent modifications of the DNA. DNA methylation at the 5th position of cytosine (5-methylcytosine or 5mC) is the most common modification found in mammals and often referred to as the "5th base" of the DNA code (Fig. 1). Though recent reports have shown a significant level of non-CpG (CpH) methylation in mouse oocytes, the functional role of this modification remains unknown (Ramsahoye et al., 2000; Smallwood et al., 2011; Ziller et al., 2011). 5mC on the other hand has been extensively studied at many developmental stages and in many systems.

DNA methylation is a vital epigenetic control mechanism in mammalian embryonic development. When cells differentiate toward diverse lineages, 5mC imposes barriers restricting and guiding differentiation and preventing cells from reverting into undifferentiated stages. Also sex chromosome dosage compensation in females relies on DNA methylation, balancing expression levels of X-chromosomal genes by inactivating one of the chromosomes. Further, it is 5mC that is partially instrumental in repressing retrotransposons, which threaten the genome by retrotransposition, ie, transcription and reintegration into potentially sensitive sites of the genome. Dense DNA methylation is also found at pericentromeric repeats preventing genomic rearrangements in their vicinity. Changing levels of DNA methylation at splice acceptor/donor sites have more recently been implicated in creating alternatively spliced mRNAs, and finally, DNA methylation is vital to the coordinate expression of imprinted genes.

2.1 DNA-Methylation Dynamics During Germ Cell Formation

Most DNA-methylation marks, which are characteristic for somatic cells fulfilling their above-mentioned functions, are globally removed at two time

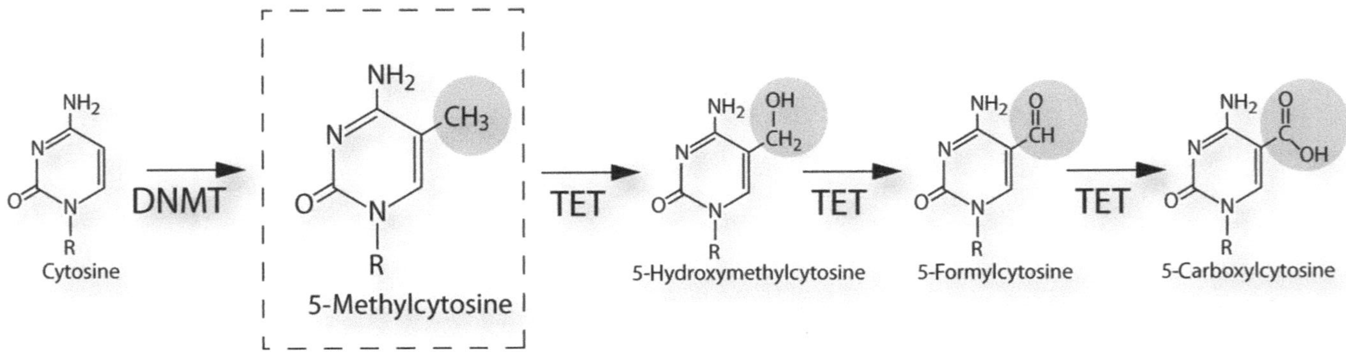

Fig. 1 Cytosine and its epigenetically relevant modifications. 5-Methylcytosine is considered the 5th base of the DNA. The methylation of the 5th carbon is achieved by DNA methyltransferases. 5-Methylcytosine can be oxidized iteratively to 5-hydroxy-, 5-formyl-, and 5-carboxy methyl cytosine by enzymes of the ten-eleven translocation (TET) family.

points during the mammalian life cycle, creating the specialized, hypomethylated epigenome of the PGCs, and the preimplantation embryo. The hypomethylation of the genome is an important step toward the toti- or pluripotent stage a new organism has to traverse for proper development. While they serve to erase the somatic epigenetic differentiation marks (and genomic imprints in germ cells), the mechanisms effecting global 5mC erasure also threaten regions better kept untouched and this severely complicates early embryonic development.

At the time of fertilization, the highly specialized, haploid oocyte and sperm give rise to the zygote, the first cell of the newly developing organism. In order to initiate development both gametes and their precursors themselves undergo an extensive differentiation and specialization process. PGCs originate from the embryonic epiblast of the E6.5 embryo. The cells in the epiblast, though of very early embryonic nature, have already undergone substantial differentiation toward a somatic fate. Germ cell and pluripotency genes have been robustly silenced, one X-chromosome in females has been inactivated, and proper genomic imprinting is established. The first step toward becoming a gamete therefore involves erasing these epigenetic marks. The extensive reprogramming and 5mC erasure is completed by E12.5–13.5 in both males and females with the genomes reaching their greatest extent of hypomethylation (~10% of all CpGs) (Guibert & Weber, 2012; Hackett et al., 2013; Kagiwada, Kurimoto, Hirota, Yamaji, & Saitou, 2013; Popp et al., 2010; Seisenberger et al., 2012; Yamaguchi et al., 2013). After this global erasure it is essential for germ cells to adopt their cell-fate-specific transcriptomes, acquire sex-specific imprints, and remethylate retrotransposons, which were derepressed during PGC reprogramming. This establishment of germ cell-specific methylation marks occurs at different times and in different cellular environments in males and females, ultimately leading to sperm- and oocyte-specific methylation patterns at the time of fertilization (Saitou & Yamaji, 2012; Sasaki & Matsui, 2008; reviewed in Messerschmidt, Knowles, & Solter, 2014).

The highly specialized state of the gametes is reflected in their epigenome, with its characteristic hypermethylation. In fact 80–90% of overall CpG methylation is found in mature sperm, the highest level of global DNA methylation of any cell in the mouse (Popp et al., 2010). The maternal genome shows lower, yet still significant, global methylation levels (~40%).

2.2 DNA-Methylation Dynamics in Early Embryos

The removal of the majority of 5mCs in both genomes is crucial to establish the pluripotency of the inner cell mass of the blastocyst and occurs during

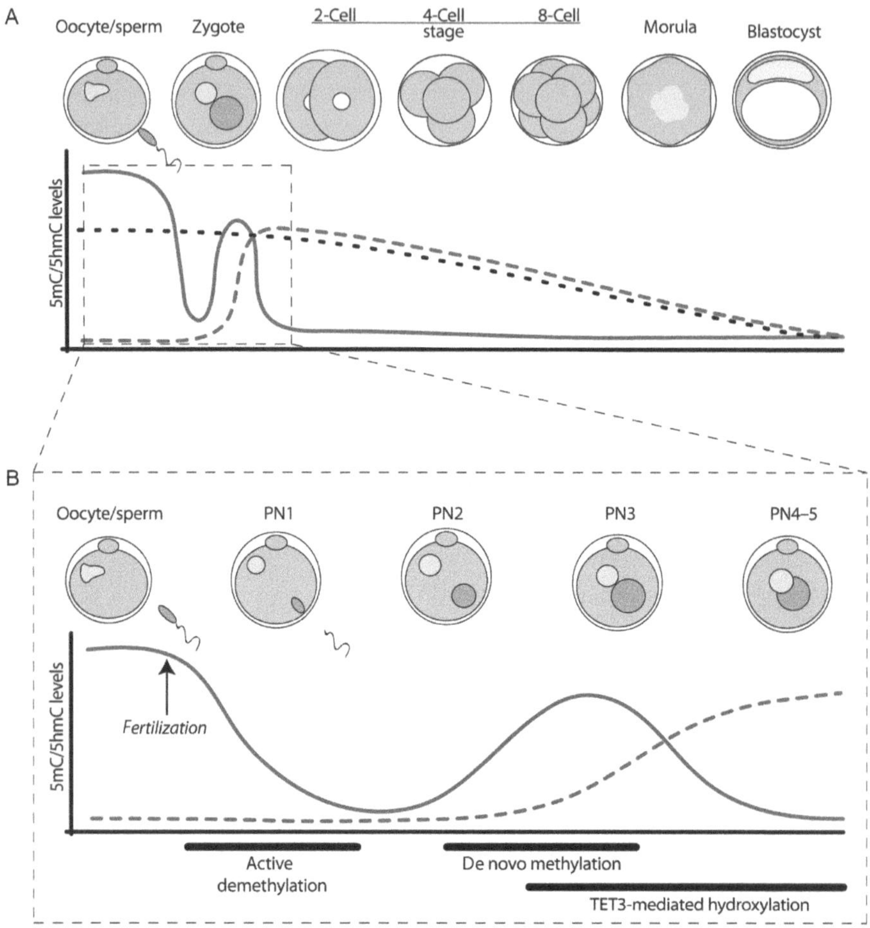

Fig. 2 DNA-demethylation dynamics in the preimplantation embryo. (A) DNA-methylation dynamics in the preimplantation embryo. The global paternal 5mC levels (*blue line*) decrease rapidly in the zygote, accompanied by an increase of 5hmC (*blue dotted line*). Maternal global 5mC levels decrease slowly over subsequent DNA-replication cycles. (B) Close-up illustration of global paternal DNA-methylation dynamics during zygotic progression from fertilization to pronuclear stage (PN) 5. 5mC is actively lost by an unknown mechanism (*red bar*), before de novo methylation activity (*green bar*) is counteracted by TET3-mediated hydroxylation (*pink bar*). (See the color plate.)

early embryogenesis (Fig. 2A). It begins in the zygote and it continues over subsequent cleavage stages. At the early blastocyst stage (32–64 cells) methylation levels are then again at a low, yet not as complete as seen in E14.5 PGCs. In contrast to PGCs DNA methylation of imprinted gene regions is preserved in embryos, enabling parent of origin-specific gene expression in somatic tissues. In addition, the imprinted paternal X-inactivation observed in early mouse embryos is not reversed until the late blastocyst.

The major peculiarity of the epigenetic reprogramming process in mammalian embryos, however, is the substantial epigenetic asymmetry of the genomes of maternal and paternal origin. Because of this asymmetry, or perhaps as a consequence of it, parental genomes are reprogrammed by divergent mechanisms and at distinct kinetics (Mayer, Niveleau, Walter, Fundele, & Haaf, 2000; Oswald et al., 2000; Santos, Hendrich, Reik, & Dean, 2002).

2.2.1 Passive DNA Demethylation

Traditionally it was assumed that the maternal genome only undergoes passive, replication-dependent DNA demethylation, although new evidence relativizes this assumption (Guo et al., 2014; Shen et al., 2014; Wang et al., 2014; see later). In order to allow dilution of 5mC it is necessary to disable the mechanisms maintaining DNA methylation in the first place. In mammals, the propagation of DNA methylation through S-phase is predominantly achieved by the DNA methyltransferase (DNMT) 1. DNMT1 displays high affinity to hemimethylated DNA originating at the replication fork (Bestor, 1988; Hitt, Wu, Cohen, & Linn, 1988; Pradhan, Bacolla, Wells, & Roberts, 1999; Ruchirawat, Noshari, & Lapeyre, 1987; Yoder, Soman, Verdine, & Bestor, 1997). Fittingly, expression of *Dnmt1* is cell cycle dependent and high in most mitotic cells (Kishikawa, Murata, Ugai, Yamazaki, & Yokoyama, 2003). The enzyme is attracted to replication foci by the nuclear antigen PCNA-interacting binding partner (NP95) where it converts hemi- to full-methylation (Arand et al., 2012; Bostick et al., 2007; Leonhardt, Page, Weier, & Bestor, 1992; Sharif et al., 2007) (Fig. 3A). The zygotic knockout of mouse *Dnmt1* is embryonic lethal at gastrulation, which is the consequence of substantial, global DNA hypomethylation (Arand et al., 2012; Hitt et al., 1988; Kurihara et al., 2008; Lei et al., 1996; Ruchirawat et al., 1987; Yoder et al., 1997). Deletion of NP95 has similar consequences (Sharif et al., 2007).

Though present, DNMT1 protein appears to be excluded from the nuclei of the early preimplantation embryo. This is an effective way to inhibit its DNA-methylation maintenance activity and realize the passive loss of 5mC (Branco, Oda, & Reik, 2008; Hirasawa et al., 2008; Howell et al., 2001; Ratnam et al., 2002). Despite the predominantly extranuclear localization, however, DNMT1 is still required to maintain DNA methylation at specialized regions (genomic imprints, retrotransposons, etc.) that retain 5mC marks throughout development. Only the complete (maternal

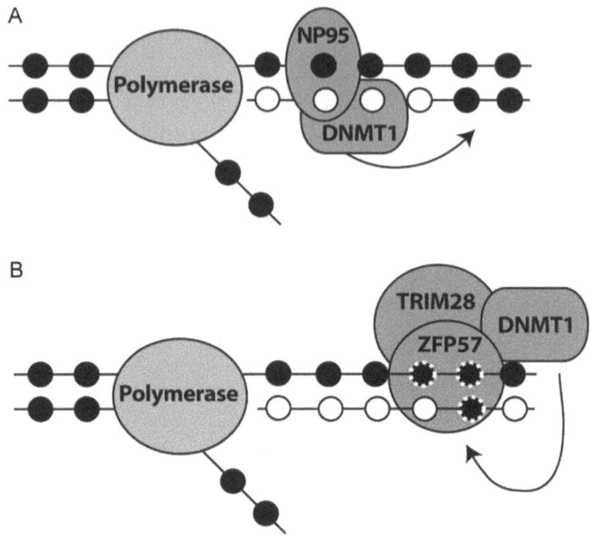

Fig. 3 Canonical and noncanonical DNMT1-mediated methylation maintenance modes. (A) In normal mitotic cells DNA methylation (*black circles*) of the newly synthesized strand is maintained by DNMT1, targeted to the replication fork by NP95. (B) In the early embryo DNMT1 levels are kept at low levels to allow the replication-mediated dilution of 5mC. Imprinted gene regions are recognized by ZFP57, which, through TRIM28, attracts DNMT1. This noncanonical targeting allows faithful maintenance of special DNA-methylation marks such as imprints (*red* (*black* in the print version) *circles*) in the light of global demethylation (*white circles*).

and zygotic), genetic removal of DNMT1 is able to abolish DNA methylation entirely (Branco et al., 2008; Cirio et al., 2008; Hirasawa et al., 2008; Kurihara et al., 2008). The DNMT1 requirement for DNA-methylation maintenance at specialized regions vs its drastically reduced, virtually undetectable nuclear levels, allows global demethylation, suggesting that a noncanonical targeting mechanism for DNMT1 is required.

2.2.2 Targeting of DNMT1 to Imprinted Gene Regions

The Krueppel-associated box (KRAB) domain zinc-finger protein 57 (ZFP57) is required for DNA-methylation maintenance, and its loss causes hypomethylation and misregulation of both paternally and maternally imprinted genes in embryos and ES cells (Li et al., 2008; Mackay et al., 2008; Quenneville et al., 2011; Zuo et al., 2012). It interacts with TRIM28, a component of a multifunctional epigenetic repressor complex. This complex includes, among others, the nucleosome-remodeling and histone deacetylation (NuRD) complex, the H3K9me3-catalyzing histone methyltransferase SETDB1, the heterochromatin protein 1 (HP1), and importantly the DNA methyltransferases DNMT1, DNMT3A, and DNMT3B (Iyengar & Farnham, 2011; Quenneville et al., 2011; Schultz, Ayyanathan, Negorev, Maul, & Rauscher, 2002; Schultz, Friedman, & Rauscher, 2001; Zuo et al., 2012). The loss of maternal *Trim28* alone is embryonic lethal (Messerschmidt et al., 2012). This is, or at least in part, the result of hypomethylation at several maternal and paternal ICRs (Lorthongpanich et al., 2013; Messerschmidt, 2012; Messerschmidt et al., 2012).

The complex is found at imprinted gene loci in mESCs (Quenneville et al., 2011) and embryos (Messerschmidt et al., 2012) and is therefore a prime candidate for a noncanonical targeting mechanism of DNMT1 (Fig. 3B). The analysis of ZFP57-binding sites, identified by ChIP-seq in mESCs, defined a ZFP57-binding motif (TGCCGC), which is found at most imprinted gene regions (Quenneville et al., 2011). Containing a CpG dinucleotide, this site is bound by ZFP57 with much higher affinity when methylated (Liu, Toh, Sasaki, Zhang, & Cheng, 2012; Quenneville et al., 2011). Hypomethylated sites are not bound by TRIM28 and cannot be restored by TRIM28 reexpression (Messerschmidt et al., 2012). Similarly ectopic reexpression of ZFP57 in *Zfp57*-deficient ESCs cannot restore lost imprinting (Zuo et al., 2012). In conclusion, DNMT1/NP95 appears to be targeted to imprinted regions in a noncanonical manner through interaction with TRIM28 and ZFP57. The latter's increased affinity to a 5mC in its DNA recognition sequence would allow for selective maintenance of the methylated vs the unmethylated allele. This targeting mode has been proposed to compensate for the observed reduction of nuclear DNMT1 levels in early embryos and thus enable DNA-methylation maintenance at imprinted regions. Only later, with increasing DNMT1 levels, canonical DNA-methylation maintenance ensues and ZFP57/TRIM28 complexes become dispensable for DNA-methylation maintenance in the imprinting context (Fig. 3).

2.2.3 Active DNA Demethylation

In contrast to the maternal genome, paternally inherited DNA is rapidly demethylated very shortly after zygote formation (Mayer et al., 2000; Oswald et al., 2000) (Fig. 2). This demethylation is credited to an active enzymatic process, as it completes before zygotic DNA replication has even begun. In any case, a single round of DNA replication could not account for the observed 5mC loss.

How then, is active DNA demethylation achieved? This question has been the subject of intense debate and has not been fully resolved today. A carbon–carbon bond-breaking enzyme with DNA demethylase activity remains elusive. Instead, several alternative, active, yet indirect, DNA demethylation pathways have been described (Messerschmidt et al., 2014; Wu & Zhang, 2010, 2014).

Most recently and noted, the discovery of the ten-eleven translocation (TET) family of dioxygenases and the existence of 5-hydroxymethylcytosine (5hmC) opened new possibilities of active DNA demethylation (Tahiliani et al., 2009). TET enzymes can achieve 5mC removal by catalyzing the iterative oxidation of 5mC to 5hmC to 5-formylcytosine (5fC) and finally to 5-carboxycytosine (5caC) (He et al., 2011; Inoue, Shen, Dai, He, & Zhang, 2011; Ito et al., 2011) (Fig. 1). TET-catalyzed conversion of 5mC might thus be is a plausible mechanism to allow cell cycle-independent removal of DNA methylation. In mice, the TET family comprises three members (*Tet1–3*), which are differentially expressed during development and in adult tissues (Tahiliani et al., 2009). *Tet1* and *Tet2* are detected in ESCs and PGCs. *Tet3* is expressed in oocytes, spermatozoa, and in preimplantation embryos. The loss of maternal *Tet3* can cause developmental arrest in some, but not all embryos (Gu et al., 2011). Importantly, 5hmC accumulation is prominent in the paternal pronucleus of the zygote, the site of active 5mC removal (Fig. 2). Indeed, TET3 localizes predominantly to the paternal pronucleus (Gu et al., 2011), where it was shown to be responsible for 5hmC appearance and in its absence 5hmC is not accumulated (Gu et al., 2011; Wossidlo et al., 2011). Lack of TET3 has been proposed to cause delayed activation of essential paternal alleles required for embryonic development (eg, *Nanog* and *Oct4*), which may cause the reduced fecundity and partial developmental failure of maternal TET3-null offspring (Gu et al., 2011). In more recent studies this reduced fertility was, however, attributed to a haploinsufficiency effect of the TET3 mutation that impacted embryonic development at later stages (Inoue, Shen, Matoba, & Zhang, 2015; Tsukada, Akiyama, & Nakayama, 2015). TET3-mediated hydroxylation may therefore not be essential, or compensatory mechanisms may be in place.

The bulk of 5mCs of the paternal genome is hydroxylated by the late zygote stage, but still remains modified (to 5hmC). It was shown that only a small subset of genomic loci do revert completely to unmodified cytosine before the first cleavage division (Gu et al., 2011; Guo et al., 2014; Shen et al., 2014; Wang et al., 2005). A direct decarboxylation process of 5caC could account for this loss (Schiesser et al., 2012). Alternatively, base excision repair (BER) activity had been proposed to mediate the removal of 5mC. XRCC1 (X-ray repair cross-complementing protein 1), a component of the BER pathway, is for instance found tightly bound to paternal, yet not maternal DNA (Hajkova et al., 2010; Wossidlo et al., 2010). Also a collaborative activity of the TET and BER pathways could be envisioned as 5hmC, 5fC, and 5caC can be targeted by thymidine DNA glycosylase (TDG) (He et al., 2011; Maiti & Drohat, 2011), though chemical inhibition of TET3 does not change the XRCC1 accumulation in the paternal pronucleus or prevent the incorporation of isotope-labeled deoxycytidine in prereplicating zygotes (Amouroux et al., 2016). TDG activity itself, however, was ruled out by the observation that genetic deletion from the mouse zygote has no effect (Guo et al., 2014). Certainly, other enzymes with similar activity could perform glycosylation/BER and a new race to identify alternative enzymes and pathways has been triggered.

Nonetheless, previous and particularly recent findings suggest that active demethylation routes may have been overestimated and paternal 5mC, as well as TET3-dependent 5hmC and possibly 5fC and 5caC, undergo replication-mediated dilution in the zygotic S-phase (Guo et al., 2014; Inoue et al., 2011; Inoue & Zhang, 2011; Shen et al., 2014; Wang et al., 2005). As described earlier, DNMT1, the maintenance methyltransferase, is largely excluded from the nuclei of preimplantation embryos (Hirasawa et al., 2008; Howell et al., 2001) and has much reduced affinity for oxidized 5mC derivatives (Hashimoto et al., 2012), further favoring the process. Though perhaps less appealing this demethylation pathway is most efficient on a genome-wide scale.

In summary, it is well established that in the mammalian (mouse) zygote three roads lead to demethylation: (1) active demethylation of 5mC to yield unmodified cytosine by an unknown enzymatic activity; (2) replication-dependent dilution by "nonpropagation" of 5mC marks; and (3) 5hmC-replication-dependent dilution after TET3-mediated oxidation of 5mC (Guo et al., 2014; Shen et al., 2014; Wang et al., 2014). Surprisingly, and it is not clear how these findings can be reconciled with immunofluorescent analysis of maternal and paternal pronuclei demethylation, it has been shown by reduced representation bisulfite sequencing that the maternal genome is also targeted by TET3 hydroxylation and TET3-independent active demethylation (Guo et al., 2014; Shen et al., 2014; Wang et al., 2014). Presumably, though detectable in the maternal pronucleus, the extent and proportion of the paternal and maternal genome targeted by the three possible demethylation routes differ substantially. To what extent, and to what purpose the individual pathways contribute to the demethylation process, both in maternal and in paternal genomes in the zygote will require locus-specific, deep-sequencing analyses and the result remain to be seen.

The current process of postfertilization demethylation described here, does to a certain degree disregard the dynamics of the process in the zygote itself, with most data obtained at the later pronuclear stages. A new study by Amouroux et al. (2016) has carefully timed 5mC disappearance and 5hmC appearance throughout progression through the zygotic stages and astoundingly found both processes to be uncoupled (Messerschmidt, 2016) (Fig. 2B). Using both, immunofluorescence and mass spectrometry, a clear reduction of 5mC in the early paternal pronucleus (PN2/3), with no concomitant 5hmC increase, was documented. The early loss of the bulk 5mC still occurs in the absence/inhibition of TET3/dioxygenases and is replication independent, which agrees with some (Mayer et al., 2000; Oswald et al., 2000; Wang et al., 2014) but is at odds with other previous findings (Guo et al., 2014; Shen et al., 2014). Yet again, these observations call for an alternative replication-/TET3-independent demethylation pathway (see earlier).

The formation of 5hmC requires preexisting 5mC; however, if the paternal genome is demethylated at pronuclear stage 2/3 (PN2/3) before TET3-mediated 5hmC appearance at PN4/5, what is the TET3 substrate? As an explanation, a hitherto underappreciated DNMT3a/DNMT1-mediated de novo DNA-methylation activity in the zygote is a prerequisite for paternal 5hmC accumulation. In this scenario, TET3 acts as a safe guard, hydroxylating/marking de novo methylated CpGs, which might then be further oxidized by TET3 or may follow a 5hmC replication-dependent dilution (Amouroux et al., 2016; Messerschmidt, 2016) (Fig. 2B).

2.3 Multiplicity of Demethylation Pathways as a Safety Net

Be this as it may, it is apparent that the demethylation dynamics in the zygote, active, passive, or a combination of both, in the paternal and/or maternal pronuclei is highly complex and exceedingly well-orchestrated processes. There is no doubt that epigenetic reprogramming and DNA demethylation must take place in the preimplantation embryo. Why and if the paternal genome must undergo active demethylation is unclear. Possibly the process is a mere consequence of the specialized architecture on which the paternal genome is delivered, or possibly some paternal genes are and must be activated early on (Gu et al., 2011). It seems that the more we learn and the deeper and detailed our insights get, the harder it becomes to perceive the overall process and its complexity. This is all the more bewildering as both, round spermatids injection into oocytes, as well as SCNT can give rise to normal pups, despite omitting the global active DNA demethylation process altogether (Chan, Smith, Egli, Regev, & Meissner,

2012; Polanski, Motosugi, Tsurumi, Hiiragi, & Hoffmann, 2008). Both processes, however, are notably less efficient than normal, natural conception (Chan et al., 2012; Polanski et al., 2008). Passive DNA demethylation would appear to be the most reliable and efficient way to remove sperm- and oocyte-specific DNA-methylation marks. This path to global hypomethylation may indeed be sufficient but not particularly efficient; thus, several additional DNA-demethylation mechanisms may have been necessary and evolutionary conserved. The mouse embryo shows a remarkable developmental potential and success rate from fertilization to birth. This may very well be the key to the astonishing complexity and multitude of partially redundant demethylation mechanisms in the zygote and cleavage-stage embryos: an epigenetic reprogramming safety net.

3. CHROMATIN REMODELING AND DYNAMICS

DNA is packaged in the female and male germ cells differently to meet the needs and functions of these highly specialized cells. In the oocyte, the maternal genome is packaged with maternally deposited histones, while the paternal DNA in the sperm is densely packed with highly basic protamines and some testis-specific histones. In the zygote, these two distinct sets of parental epigenomes must be dynamically remodeled to facilitate both the appropriate erasure of parent-specific epigenetic information and the acquisition of totipotency.

3.1 Histone Modification Dynamics

In the initial hours after fertilization, the female and male genomes can be clearly distinguished by different histone modification and chromatin profiles (Table 1). The female chromatin is enriched for modifications associated with transcriptional repression such as H3K9me2/3, H3K27me3, and H4K20me3, while the paternal genome undergoes protamine–histone exchange and becomes packaged with hyperacetylated H3- and H4-histones (Adenot et al., 1997; Erhardt et al., 2003; Lepikhov & Walter, 2004; Santos et al., 2005; Sarmento et al., 2004). After pronuclei formation, the chromatin content of the parental genomes is further altered via the activities of histone modifying enzymes, as part of the epigenetic reprogramming process to establish the embryonic chromatin signature (Cho, Park, Kwon, & Kang, 2012; Ma et al., 2015; Nakamura et al., 2012; Oda et al., 2009; Santos & Dean, 2004; Shao et al., 2014; Thomas,

Table 1 Histone Modifications Dynamics in Early Preimplantation Embryos

	Histone Mark	Maternal Contribution	1-Cell	2-Cell	References
I	H3K9me2	Yes	High in fPN; Absent in mPN	Asymmetric	Liu, Kim, and Aoki (2004), Yeo, Lee, Han, and Kang (2005), and Probst, Santos, Reik, Almouzni, and Dean (2007)
	H3K9me3	Yes	High in fPN; Absent in mPN	Asymmetric	Sarmento et al. (2004), Probst et al. (2007), and Wongtawan, Taylor, Lawson, Wilmut, and Pennings (2011)
II	H3K27me1	Yes	High in fPN; Low in mPN	Yes	Santos, Peters, Otte, Reik, and Dean (2005)
	H3K27me3	Yes	High in fPN; Low in mPN	Yes	Santos et al. (2005) and Wongtawan et al. (2011)
	H3K64me3	Yes	High in fPN; Absent in mPN	No	Daujat et al. (2009)
	H4K20me3	Yes	Low in fPN; Absent in mPN	No	van der Heijden et al. (2005), Probst et al. (2007), and Wongtawan et al. (2011)
III	H3K4me1	Yes	Present in both	Yes	Lepikhov and Walter (2004) and van der Heijden et al. (2005)
	H3K4me3	Yes	Present in both	Yes	Lepikhov and Walter (2004) and Lepikhov et al. (2008)
	H3K9me1	Yes	Present in both	Yes	Santos et al. (2005) and Puschendorf et al. (2008)
	H4K20me1	Yes	Present in both	Yes	van der Heijden et al. (2005) and Wongtawan et al. (2011)

Table 1 Histone Modifications Dynamics in Early Preimplantation Embryos—cont'd

	Histone Mark	Maternal Contribution	1-Cell	2-Cell	References
	H2Bub1	No	Present in both	Yes	Ooga, Fulka, Hashimoto, Suzuki, and Aoki (2015)
	H3K9ac	Yes	Present in both	Yes	Santos et al. (2005)
	Ph(Ser1) H4/H2A	Yes	Present in both	Yes	Sarmento et al. (2004)
	H3R17me	Yes	Present in both	Yes	Sarmento et al. (2004)
IV	H4Kac	No	Low in fPN; High in mPN	Yes	Adenot, Mercier, Renard, and Thompson, 1997 and Sarmento et al. (2004)
V	H3K79me2	Yes	Low in both	No	Ooga et al. (2008)
	H3K79me3	Yes	Low in both	No	Ooga et al. (2008)
	H4K20me2	No	Absent in both	No	van der Heijden et al. (2005) and Wongtawan et al. (2011)

Five different enrichment patterns of chromatin marks are apparent at the 1- and 2-cell embryos.

Dixon, Kueh, & Voss, 2008). High levels of H3K9me2/3 and H3K27me3 are maintained in the female pronucleus but not in the male pronucleus, while modifications such as H3K9ac, H3K4me1, H4K20me1, and H3R17me are acquired on both sets of epigenomes. In addition, some maternally incorporated marks such as H4K20me2, H3K64me3, and H3K79me2/3 are not detected in either pronucleus at the late 1-cell stage, indicating the absence of maternal enzymes responsible for these modifications in the early zygote (Burton et al., 2013; Daujat et al., 2009; Ooga et al., 2008; Ooga, Suzuki, & Aoki, 2013; Puschendorf et al., 2008; van der Heijden et al., 2005; Wongtawan et al., 2011). Although the paternal and maternal chromatin is highly asymmetric in the zygote, most of the differences are gradually resolved after the first cellular division. An exception to this are the H3K9 di- and trimethylation marks, which remain associated with the maternal chromatin in the 2-cell embryos, and are only resolved by the 4- to 8-cell stage (Liu et al., 2004; Puschendorf et al., 2008;

Yeo et al., 2005). Some chromatin marks such as H4K20me2, H4K20me3, and H4K64me3 become detectable only after the 4- to 8-cell stage, indicating the acquisition of these marks are linked to late stages of preimplantation development (Daujat et al., 2009; Wongtawan et al., 2011). In addition to the well-characterized modifications on the N-termini tails, acetylation of lysine residues within the H3 histone core domain is also dynamically present in cleavage-stage embryos (Ziegler-Birling, Daujat, Schneider, & Torres-Padilla, 2015). The functional impact of these histone core modifications on embryonic development remains to be elucidated.

In the very early stages of development, therefore, the highly asymmetrical paternal and maternal genomes are indexed by a more permissive and repressive histone modification profile, respectively (Burton & Torres-Padilla, 2010). Consequently, the paternal proncleus has been shown to be more highly associated with transcription factors such as Sp1 and TBP, and more transcriptionally active during minor zygotic gene activation (ZGA) at the S/G2-phase (Adenot et al., 1997; Aoki, Worrad, & Schultz, 1997; Worrad, Ram, & Schultz, 1994). This is further supported by the observation that inhibition of paternal H3K4 mono- and trimethylation, by blocking the activities of histone methyltransferases in the early zygote, significantly reduced minor ZGA from the male pronucleus and the developmental competence of the embryo (Aoshima, Inoue, Sawa, & Okada, 2015). On the other hand, asymmetric enrichment of the repressive H3K9me2 mark on the maternal chromatin by G9a methyltransferase, in the early zygote, serves as a recruitment signal for PGC7 to mediate the maintenance of DNA methylation in the maternal genome (Ma et al., 2015; Nakamura et al., 2012). The importance of the chromatin asymmetry of parental genomes is further underscored by the severe effects on preimplantation development observed when the enzymatic activities of the histone modifiers, particularly of histone H3K4- and K9-methylation, are perturbed (Ancelin et al., 2016; Andreu-Vieyra et al., 2010; Ciccone et al., 2009; Erhardt et al., 2003; Huang et al., 2015; Ma et al., 2015; Matoba et al., 2014; Tachibana et al., 2002; Tardat et al., 2015; Thomas et al., 2008; Wasson et al., 2016). Interestingly, while complete absence of histone demethylases in oocyte is embryonic lethal (Ancelin et al., 2016; Ciccone et al., 2009; Wasson et al., 2016), their partial loss during oogenesis is associated with perinatal lethality and behavioral changes in adult mice (Wasson et al., 2016). Thus, the differential histone codes on the paternal and maternal pronuclei are not simply corollaries of the parental genome decondensation process in the newly fertilized embryo, but rather they have a direct and important impact on embryonic development.

3.2 Organization of Heterochromatin

The formation and organization of pericentric and centric heterochromatin is a critical process in cells to facilitate gene regulation, centromere function, and chromosome segregation processes. Heterochromatin organization is highly dynamic and transits through distinct forms at different stages of preimplantation development. Unlike the mature heterochromatin chromocenters, marked by H3K9me3, H4K20me3, and HP1 seen in somatic cells, pericentric DNA at the 1-cell stage is spatially organized around the periphery of nucleolar precursor bodies (NPBs) of the parental pronuclei (Martin et al., 2006; Probst et al., 2007; Wongtawan et al., 2011). In the female pronucleus, heterochromatin is intensely marked by H3K9me2/3 and HP1 with no or very low levels of H4K20me3 (Arney, Bao, Bannister, Kouzarides, & Surani, 2002; Martin et al., 2006; Probst et al., 2007). These repressive histone modifications are absent in the male pronucleus, and heterochromatin organization appears to be influenced by H3K9me1 and the histone variant H3.3, together with HP1 (Arakawa et al., 2015; De La Fuente, Baumann, & Viveiros, 2015; Martin et al., 2006; Santenard et al., 2010; Santos et al., 2005). At the 2-cell stage, the heterochromatin undergoes initial restructuring, forming chromocenters that are enriched in H3K9me3 and HP1 in the nucleoplasm away from the NPB periphery. The spatial reorganization of heterochromatin in 2-cell embryos was shown to be critically dependent on the transcription of major satellite DNA from the parental genomes (Casanova et al., 2013; Probst et al., 2010, 2007). By the blastocyst stage, all pericentric heterochromatin is localized into chromocenters, and unassociated with the nucleoli bodies. However, these chromocenters, while enriched for H3K9me3 and HP1, still lack the H4K20me3 marks and HP1 association that are typical of mature heterochromatin in somatic cells (Wongtawan et al., 2011), suggesting that further "maturation" of the heterochromatin occurs after implantation and cellular differentiation.

3.3 Histone Variant Dynamics and Functions

In addition to the regulation conferred by posttranslational covalent modifications on canonical core histones, a special class of histone variants is also known to modulate chromatin function in preimplantation embryos. Histone variants differ in sequence, expression, and usage to their canonical homologues and can influence chromatin-templated processes by changing the structural properties of the nucleosomes, chromatin compaction, or recruitment of distinct binding partners (Ahmad & Henikoff, 2002;

Hake & Allis, 2006; Mattiroli, D'Arcy, & Luger, 2015; Maze, Noh, Soshnev, & Allis, 2014; Sarma & Reinberg, 2005). The expression and chromatin loading of the variants, encoded by genes located outside of the major histone gene clusters, are independently regulated from that of the canonical histones. While the replicative histones are only expressed during S-phase to facilitate the packaging of newly synthesized DNA, histone variants are expressed throughout the cell cycle and can be incorporated into chromatin by specialized chaperone and remodeling proteins in a replication-independent manner. To date, replication-independent variants of H2A, H2B, and H3 have been identified (Maze et al., 2014; Sarma & Reinberg, 2005). Of these, the expression and activities of histone H2A variants, H2A.Z, H2A.X, and macroH2A, and H3 variant H3.3 are more extensively characterized in preimplantation embryos (Fig. 4).

3.3.1 Histone H3 Variants

In mice, the subclasses of H3 histones are classified into the canonical replicative histones H3.1 and H3.2, and the replacement variants H3.3, CenH3$^{CENP\text{-}A}$, and the testis-specific H3.4 (Filipescu, Muller, & Almouzni, 2014; Maze et al., 2014). CenH3 is a unique H3 variant specifically deposited at the centromeres by its companion chaperone Hjurp and is

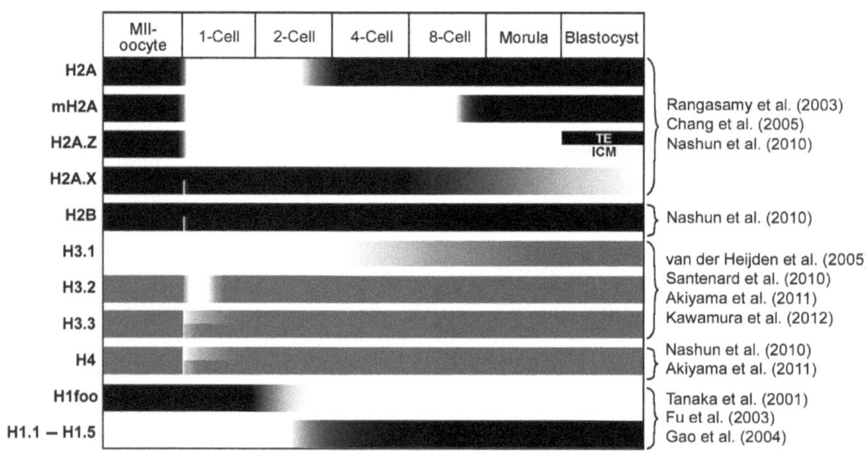

Fig. 4 Incorporation dynamics of linker and nucleosomal histones H1, H2A, H2B, H3, H4, and their variants in MII-oocytes and preimplantation embryos. Loss of maternally incorporated histones from the maternal pronuclei is triggered upon fertilization. Asymmetrical incorporation of H3.3–H4 dimers/tetramers into paternal, but not maternal, pronuclei within the first 4 h postfertilization. H2A.Z is present only in the trophectodermal cells at the blastocyst stage. *TE*, trophectoderm cells; *ICM*, inner cell mass.

implicated in the packaging of centromeric chromatin during centromere formation (Dunleavy et al., 2009; Foltz et al., 2009; Mattiroli et al., 2015). While a study has shown that CenH3 is essential for development and null embryos die at periimplantation stages (Howman et al., 2000), very little is known of the incorporation dynamics and function of CenH3 in early embryos.

H3.3 is expressed throughout the cell cycle from two coding genes, *H3f3a* and *H3f3b*, which are regulated independently. The amino acid sequence of H3.3 differs from that of replicative H3.1 and H3.2 at only 5 and 4 positions, respectively. However, these changes are sufficient to confer physical properties that lead to the instability of H3.3-containing nucleosomes, creating a more open chromatin structure at certain H3.3-enriched genomic loci (Jin et al., 2009). H3.3-specific chaperones, such as Hira, the Daxx–Atrx complex, and Dek, further mediate the selective incorporation of H3.3 into specialized genomic regions, which include active transcription sites, pericentric heterochromatin, and telomeres (Drane, Ouararhni, Depaux, Shuaib, & Hamiche, 2010; Goldberg et al., 2010; Ivanauskiene et al., 2014; Sawatsubashi et al., 2010; Wong et al., 2010). The distribution of H3.3 is additionally regulated by ATP-dependent chromatin-remodeling enzymes, such as Chd1, in a replication-independent manner, with important consequences on cellular differentiation and embryonic development (Gaspar-Maia et al., 2009; Guzman-Ayala et al., 2015; Konev et al., 2007; Piatti et al., 2015; Suzuki et al., 2015). More importantly, complete knockout of H3.3 by deleting both *H3f3a/b* genes was shown to result in early embryonic lethality, while a single knockout of either gene also led to gene-specific developmental phenotypes (Bush et al., 2013; Jang, Shibata, Starmer, Yee, & Magnuson, 2015; Tang et al., 2015). All of these studies point to H3.3 being an essential part of chromatin-mediated regulation of early mammalian embryonic development (Filipescu et al., 2014; Weber & Henikoff, 2014).

In zygotes, the distribution of H3.3 is shown, by studies that tracked the incorporation of N- or C-terminal epitope-tagged histones in preimplantation embryos, to be highly dynamic and asymmetrical (Akiyama, Suzuki, Matsuda, & Aoki, 2011; Kawamura, Akiyama, Tsukamoto, Suzuki, & Aoki, 2012; Santenard et al., 2010; Torres-Padilla, Bannister, Hurd, Kouzarides, & Zernicka-Goetz, 2006; van der Heijden et al., 2005). While maternal H3.3 is incorporated in the genome at euchromatic regions during oogenesis, this histone variant is actively removed, together with histone H4, from the maternal pronucleus upon fertilization

(Akiyama et al., 2011). The paternal genome, on the other hand, rapidly acquires histone H3.3, but not the replicative histone H3.1 or H3.2. At 4 h postfertilization, the newly formed male pronucleus appears to carry only H3.3-containing nucleosomes, while the female pronucleus exists in a relatively H3.3-free state (Akiyama et al., 2011; Santenard et al., 2010). Subsequently, H3.3 becomes deposited on both sets of genomes, though asymmetry is maintained in the male pronucleus by specific enrichment of H3.3 at the pericentric heterochromatin regions surrounding the NPBs in the late 1-cell embryo. The presence of H3.3 at both euchromatic and heterochromatic regions is detected in 2-cell-stage embryos, but the perinucleolar enrichment pattern is lost by the morula or blastocyst stages (Akiyama et al., 2011).

The asymmetrical H3.3 incorporation dynamics and localization in the male and female pronuclei at the 1-cell stage are implicated in regulation of early embryonic development. The initial wave of removing preexisting H3.3–H4 dimers/tetramers from the maternal pronucleus was proposed as a mechanism to erase the maternal epigenetic information and facilitating acquisition of totipotency in the zygote (Akiyama et al., 2011). In addition, the incorporation of H3.3 appears vital for the chromatinization of the paternal genome during protamine–histone exchange, failing of which led to the disruption of downstream chromatin-mediated processes culminating in developmental arrest. In 1-cell embryos depleted of maternally loaded H3.3-specific chaperone Hira, via Zp3cre-mediated gene ablation during oogenesis, the paternal genome was found to be devoid of nucleosomal histones (Lin, Koh, Wong, Conti, & Ramalho-Santos, 2014). Ablation of *H3f3b* during oogenesis also resulted in male pronuclei that are aberrantly smaller than the maternal pronuclei, consistent with abortive decondensation of the paternal genome (Tang et al., 2015). Consequently, the paternal genome could not be replicated, while replication of the maternal genome was significantly reduced in both Hira-depleted and *H3f3b*$^{-/-}$ zygotes. Loss of H3.3 incorporation into the paternal pronucleus also resulted in the loss of RNA polymerase I-mediated transcription of ribosomal RNA from both sets of genome during mid-to-late 1-cell stage (Lin et al., 2014). Consistent with aberrant rRNA transcription, the maternal pronuclei in *H3f3b*$^{-/-}$ embryos also exhibited dispersed NPBs instead of the one large NPB seen in normal embryos (Tang et al., 2015). H3.3 incorporation and the activities of its specific chaperones such as ATRX and DAXX were also shown to be essential and significantly impact centromeric and pericentromeric heterochromatin formation in the early preimplantation embryos (Arakawa et al., 2015; De La Fuente et al., 2015; Jang et al., 2015). Morpholino-mediated depletion of H3.3 or over-expression of a H3.3K27R mutant both resulted in mislocalization of the heterochromatic protein HP1, chromosomal condensation and segregation defects, and subsequent developmental arrest (Lin, Conti, & Ramalho-Santos, 2013; Santenard et al., 2010).

The importance of H3.3 function in the early zygotic stages is perhaps unsurprising. Deposition of the other H3 histones, H3.1 and H3.2, is tightly linked to S-phase, and hence these replicative histones are only incorporated as the parental genomes undergo their first replication (Kawamura et al., 2012; van der Heijden et al., 2005). H3.3 is, therefore, the only nucleosomal H3 that can be incorporated into the parental genomes prior to DNA replication to mediate key epigenetic events that occur upon fertilization. H3.3 is implicated in a wide array of chromatin-templated processes in these early embryos, and some of its functions have been linked to specific posttranslational modification marks on K4, K27, and K36 and its interaction with H1 linker histones and higher-order chromatin compaction (Aoshima et al., 2015; Lin et al., 2013; Santenard et al., 2010). It will be important to further dissect details of these correlations, as well as the activities of H3.3-specific chaperones and remodeling enzymes in the development of preimplantation embryos.

3.3.2 Histone H2A Variants

Among the nucleosomal histones H2A, H2B, H3, and H4, the H2A family contains the highest number of variants, which also exhibit the most sequence divergence from the canonical isoform (Mattiroli et al., 2015; Maze et al., 2014; Sarma & Reinberg, 2005). The H2A variants, including H2A.B, H2A.X, H2A.Z, and mH2A, typically carry a more conserved N-terminal histone-fold domain, while significantly divergent C-terminal domains (CTDs) mediate the differences in their chromatin distribution profiles, interacting partners, and function for each variant isoform.

In developing preimplantation embryos, loss of maternally loaded canonical H2A from the female pronucleus is triggered by fertilization (Nashun, Yukawa, Liu, Akiyama, & Aoki, 2010). The level of H2A incorporation remains low in 2-cell embryos before gradually increasing at the 4-cell stage to reach high levels in the blastocyst (Nashun et al., 2010). Diminished deposition of H2A in the 1-cell and 2-cell embryos suggests that the chromatin is loaded with one or more of the H2A variants. The incorporation dynamics of H2A.Z and mH2A were found to be similar to that of canonical H2A, with low or no deposition into the parental pronuclei in the zygote and 2-cell embryos before being detected at morula and blastocyst stages (Chang et al., 2005; Nashun et al., 2010). The H2A.X variant, however, was found to be actively incorporated into both male and female pronuclei soon after fertilization, and maintained in 2-cell and 4-cell embryos, before decreasing at the morula and blastocyst stages (Nashun et al., 2010). Hence, it appears that H2A.X is the predominant H2A isoform incorporated into the embryonic chromatin in early embryos, and is subsequently replaced with canonical H2A, H2A.Z, and mH2A as development proceeds.

Though H2A.Z and mH2A are both depleted in zygotes and 2-cell embryos, initial incorporation of mH2A is detected earlier at the morula stage, while H2A.Z is present only in the trophectodermal cells in the blastocysts (Chang et al., 2005; Costanzi, Stein, Worrad, Schultz, & Pehrson, 2000; Nashun et al., 2010). In trophectodermal cells, H2A.Z colocalizes with HP1 at pericentric heterochromatin, suggesting a role of H2A.Z in the lineage specification of TE cells (Rangasamy, Berven, Ridgway, & Tremethick, 2003). Interestingly, forced expression of H2A.Z at the 1- and 2-cell stages is detrimental to embryos, leading to delayed development (Nashun et al., 2010). Consistent with its incorporation and functional requirement at late preimplantation stages, H2A.Z-null embryos can develop normally to blastocysts but die soon after implantation (Faast et al., 2001).

The function of H2A.X in somatic cells is typically linked to its participation in DNA repair upon phosphorylation of S139 on its C-terminus. However, incorporation of H2A.X in early preimplantation embryos occurs in the absence of DNA damage, suggesting that H2A.X phosphorylation is uncoupled from DNA repair during preimplantation development (Ziegler-Birling, Helmrich, Tora, & Torres-Padilla, 2009). While H2A.X-null mice are viable, small litter sizes were obtained from mating H2A.$X^{-/-}$ females with wild-type males (Celeste et al., 2002). This may reflect partial lethality in embryos lacking maternal stores of H2A.X. A clearer picture of H2A.X requirement for preimplantation embryonic development would emerge from studies of *H2Afx* ablation during oogenesis.

3.4 Linker Histone H1 Dynamics and Functions

Chromatin structure in preimplantation embryos is additionally influenced by the H1 linker histones. Linker H1 histones play important roles in regulating the higher-order chromatin structure and gene expression. The CTD of histone H1 facilitates the interaction of H1 to the DNA and nucleosomal dyads, allowing for compaction and stabilization of the chromatin structure (Allan, Hartman, Crane-Robinson, & Aviles, 1980; Harshman, Young, Parthun, & Freitas, 2013). The CTD of linker histone H1 has also been shown to interact directly with the DNA methyltransferases, DNMT1 and DNMT3b, but not DNMT3a (Yang, Kim, Norwood Toro, & Skoultchi, 2013). Parallel enrichment of H1, DNMT1, and DNMT3b at the *H19*- and *Gtl2*-imprinted gene loci observed in mESCs also suggests that H1 histones may participate in gene silencing via compaction-independent mechanisms during preimplantation development (Yang et al., 2013). The H1 histones are also covalently modified by phosphorylation, methylation, citrullination, and acetylation at the N- and C-terminal tails, as well as within the globular domains (Annalisa & Robert, 2016). However, the function and effect of these H1 posttranslational modifications in preimplantation embryos has not been defined.

In the mouse, at least 11 histone H1 subtypes have been identified—5 somatic (H1.1, H1.2, H1.3, H1.4, and H1.5), 2 testis-specific (H1t, H1T2), 1 oocyte-specific (H1foo), and the replication-independent variants (H1.0 and H1X). The H1foo subtype is the predominant linker H1 present in oocytes and zygotes, though small amounts of H1.0 and H1.3 are also detectable (Adenot et al., 2000; Tanaka, Hennebold, Macfarlane, & Adashi, 2001). Upon fertilization, H1foo, and presumably some H1.0 and H1.3, is rapidly incorporated into the paternal genome as part of the decondensation process (Adenot et al., 2000; Fu et al., 2003; Gao et al., 2004; Lorthongpanich, Solter, & Lim, 2010; Tanaka et al., 2001). However, the level of H1foo decreases significantly as the embryo undergoes the first cleavage and becomes undetectable in 4-cell-stage embryos (Fu et al., 2003; Gao et al., 2004). By contrast, expression of the somatic subtypes (H1.1 to H1.5) is initiated at major ZGA. These somatic H1 subtypes are incorporated gradually into the embryonic chromatin from the late 2-cell stage and increased maximally by the late 4-cell stage (Fu et al., 2003; Gao et al., 2004).

Linker histone H1 subtypes are highly variable in sequence and expression patterns, and each H1 subtype may have a specific nonredundant role (Wang, Sirotkin, Buchold, Skoultchi, & Marzluff, 1997). However, single knockout of somatic linker histones H1.3 and H1.5 and the replication-independent H1.0 did not induce changes in the total H1-core histone ratio at the cellular level nor lead to any overt developmental phenotypes

(Fan, Sirotkin, Russell, Ayala, & Skoultchi, 2001). The importance of linker histone functions in embryonic development was ultimately revealed in a triple H1.3, H1.4, and H1.5 knockout mouse, which had a 50% reduction of total linker H1 content in the cells and exhibited mid-gestation lethality. The developmental phenotypes observed in the linker H1 triple-knockout mice are linked to a global reduction in nucleosomal spacing, consistent with recent findings that alteration of linker–nucleosomal histone ratios leads to disrupted gastrulation (Fan et al., 2003; Lim, Lorthongpanich, et al., 2013; Lim, Reversade, Knowles, & Solter, 2013). Excessive deposition of linker H1 in the absence of histone variant H3.3 is implicated in the chromosome condensation and segregation failures observed in mutant preimplantation embryos (Lin et al., 2013). These studies highlight the importance of linker histones and higher-order structures in the regulation of chromatin-templated processes during normal embryonic development.

4. RETROTRANSPOSONS AND NONCODING RNAs

4.1 The Viral Genome-Retention Dilemma

The epigenetic mechanisms controlling genome reprogramming, from gametes to soma and back again, are evolutionarily intertwined with those mitigating expression of the endogenized viral remnants present in the genomes of all organisms. Current estimates suggest half of the mouse and human genomes are made up of degenerate, partial, or complete genomes of DNA, reverse transcribed from long-terminal repeat (LTR)-containing and non-LTR RNA viruses and from degenerate DNA viruses. The completeness of the viral genomes, the ability of degenerate copies to replicate upon activation, and their capacity for reintegration into the host genome during the global DNA demethylation accompanying reprogramming pose a threat to host genome integrity. Viral reintegration into exons or introns can change host gene function, and reintegrated viral LTRs can change or coordinate host gene expression, points considered a treat, yet also potentially beneficial and a selective advantage during evolution (McClintock, 1956; Shapiro, 2005 for discussion).

4.2 Transcription of Viruses and Activation of the Embryonic Genome

Intracisternal A-type viral particles (IAP) were first noted in electron micrograms of 2-cell-stage mouse embryos, but not oocytes, leading to the suggestion that transcription of viral genomes is activated along with the host embryonic genome at the 2-cell stage (Biczysko, Pienkowski, Solter, & Koprowski, 1975; Calarco, 1975). Viral particles have also been detected by TEM in normal human blastocysts, shortly after transcriptional activation of the human embryonic genome, which occurs at the 4- to 8-cell stage (Grow et al., 2015).

Transcriptome analysis of mouse oocytes and the newly activated genome of the 2-cell-stage mouse embryo first revealed a striking level of stage-specific transcription from specific classes of retroviruses (Evisikov et al., 2006; Solter et al., 2004). Not only were retroviral elements themselves actively transcribed but LTR integration into the host genome created chimeric gene products that could change host gene function. The number of these products that had been evolutionarily preserved led to the suggestion that the viral LTR in chimeric constructs could coordinately control a network of viral LTR-promoted host gene transcripts, which could be of benefit to the host (Macfarlan et al., 2012; Peaston et al., 2004).

However, the preservation of repetitive elements in the genome probably requires more than just the handful of genes that have been identified to fall under the control of viral LTRs. Now, as we move into investigating a deeper level of gene expression control at the initiation of embryogenesis, a reassessment of the role of the considerable endogenized viral load in the genome is probably in order. A recent survey of the first genes expressed from the early human embryo demonstrated two groups of early transcripts, one from embryos prior to the 4-cell stage and another group from 4- to 8-cell-stage embryos (Tohonen et al., 2015). These two periods are analogous to the minor ZGA occurring in the mouse shortly after fertilization, and the major ZGA at the 2-cell stage. A search for regulatory elements in the promoters of the products of the early embryonic genes revealed significant overrepresentation of endogenous retroviral elements (Alu) in the far 5′ end of a novel set of PRD-like homeobox transcription factor genes, which themselves are proposed to function in the major embryonic genome activation. In this model endogenized viral elements may act indirectly by driving expression of genes that in turn regulate ZGA. In another study of early embryonic gene expression, a long noncoding RNA-containing repetitive elements representing SINES, LINEs, LTR/ERVs, and DNA transposons were found to be expressed in human blastocysts. Specific and control siRNA when injected into one of the blastomeres of 2-cell-stage embryo eliminated its contribution to the inner cell mass, suggesting long noncoding RNAs of repetitive element derivation could also be required for proper genome activation (Durruthy-Durruthy et al., 2015). Others working with mouse embryos have gathered evidence to suggest a proportion of zygotically activated long noncoding RNAs are required for the whole-scale activation of the mouse embryogenic genome at the 2-cell stage. They hypothesize that zygotically expressed noncoding RNAs bind to promoters of their cognate genomic loci to effect transcription at the 2-cell stage (Hamazaki, Uesaka, Nakashima, Agata, & Imamura, 2015).

Each of these approaches brings forward the concept that an early minor activation of the embryonic genome is required to prepare for the later, major genome activation. In each phase-specific recognition is required and one could question whether the endogenous viral elements in the host genome might add specificity by targeting the genes required for the major embryonic genome activation.

4.3 DNA and Histone Methylation Controlling Endogenized Retroviruses

When retroviruses were injected into mouse zygotes and embryos, they were subsequently found to be silenced by DNA methylation (Jahner et al., 1982) an effect that could spread to proximal host loci (Jahner & Jaenisch, 1985). The molecular basis for the specific methylation of endogenized viruses was found to be initiated by KRAB-containing zinc-finger transcription factors that interact with and localize the repressive TRIM28 complex to effect DNA methylation of the locus (Rowe et al., 2010; Wolf & Goff, 2007, 2009). It was assumed that the complex of TRIM28 and its associated molecules controls endogenous retrovirus expression in the same manner it controls imprinted gene expression.

The KRAB zinc-finger gene family members that effect recognition of specific DNA motifs in the viral genomes is one of the most rapidly expanding primate gene families, hypothetically because they play a role in the evolutionary race to control new, evolving transposable elements in the genome (Jacobs et al., 2014; Lukic, Nicolas, & Levine, 2014). Where the majority of endogenous retroviral elements controlled only by TRIM28/KRAB zinc-finger complexes the lack of maternal Trim28 should lead to their universal activation. However, transcripts from the LTR class II retroviral elements IAP and MT, not those of other LTR- and non-LTR-containing retroviral elements, are specifically upregulated in 2-cell-stage embryos lacking maternal TRIM28 (Leseva, Knowles, Messerschmidt, & Solter, 2015). This finding suggests that TRIM28-directed methylation is but one of the mechanisms important in endogenous retroviral control. In human embryos endogenous viral transcripts (HERVK) are also transcribed at the onset of embryonic genome activation, an activation correlating with direct demethylation of one of its LTRs (Grow et al., 2015). In this case the molecular basis for specific demethylation of this LTR unknown.

Although the TRIM28 molecular complex can modulate expression of some of the plethora of retrotransposons at the time of embryonic genome activation, other epigenetic mechanisms, which may or may not directly interact with the TRIM28 complex, are possible. As was the case for DNA-methylation control, the role of histone methylation in silencing endogenous retroviruses has been primarily investigated in vitro. From experiments primarily performed in ES cells the lysine-specific demethylase1 (*Kdm1A*) was reported to upregulate the mouse endogenous virus L and a series of host genes containing its LTRs. Embryogenesis in these mutant mice arrested at gastrulation (Macfarlan et al., 2011). This report and those of others correlating the histone methylation pathway with untimely viral activation and embryonic lethality (Hatanaka et al., 2015) are of interest in and of themselves, but is there more to be learned from control of endogenous viruses about the role of histone methylation during reprogramming and vice versa? In analogy to the TRIM28 maternal-null experiment cited earlier, lack of maternal KDM1A at the time of genomic reprogramming in the zygote leads to entirely different results from studies in ES cells or in conventional knockout mice. Lack of maternal KDM1A leads to expression of some LINE1 elements, not SINEs- or LTR-containing endogenous retroviral elements, aberrant expression of host genes and arrest of embryos at the 2-cell stage (Ancelin et al., 2016). These data draw our attention away from the explicit control of endogenous viral elements to the broader role of histone methylation in embryonic genome activation.

4.4 Histone Methylation Through the piRNA Pathway

The piRNA pathway is a major epigenetic program guarding the demethylated genome's integrity during meiosis (Huang et al., 2013). In *Drosophila*, long noncoding RNAs (~200 kb) originating from transposons, as well as from discrete genomic regions of the host that are predominantly composed of transposon remnants (piRNAs clusters), bind to Piwi-clade argonaute family proteins, which cleave and trim them giving rise to thousands of small piRNAs (24–32 nts) (Huang et al., 2013 for review). These piRNAs then act as guides, specifically interacting with the transposons, and any of the components of the piRNA clusters from which they arose, localizing Piwi proteins into specific genomic loci. These specific loci-bound Piwi proteins in turn attract histone methyltransferases, HP1A family molecules, and other members of the methylation complex to effectively silence transposon encoding and other loci (for extensive reviews, also see Iwasaki, Siomi, & Siomi, 2015; LeThomas, Toth, & Aravin, 2014).

This program has been described as a self-recognition component of the innate immune system to silence not only transposons but also cognate loci in the transposon-rich piRNA clusters by inducing a repressive chromatin state (Iwasaki et al., 2015). In *Drosophila*, the relationship between piRNA and one type of the noncoding RNA clusters from which they are derived (those which are bidirectionally transcribed with extended heterochromatic histone marks) suggests that piRNAs themselves are maternally inherited. Such transgenerationally inherited cytoplasmic factors and their Piwi-bound proteins interact with their homologous targets to silence transposons, elements within the piRNA clusters, and also to activate piRNA synthesis, perpetuating the process (see Le Thomas et al., 2016 for literature review and proof of principle). This interesting concept may be specific to *Drosophila* and remains to be tested in mammals.

Most studies of piRNAs in mammals have focused on transposon silencing in the male germline, ie, spermatozoa in the adult testis (eg, Aravin et al., 2006). Indeed, because female mice that contain mutations in components of the piRNA pathway are fertile, whereas their mutant male counterparts are sterile, it was only recently established that components of the piRNA pathway are indeed expressed in mammalian fetal oocytes (Lim, Lorthongpanich, et al., 2013), which are at an analogous stage of gametogenesis to the spermatogonia in the adult testis. Fetal human ovaries also contain piRNA sequences that map to transposable elements in the human genome and to the intergenomic piRNA loci from which they originated (Williams et al., 2015).

Although the mechanism for maintenance of genomic imprints is now mostly known, how these specific loci are chosen among all other DNA-nonmethylated loci remains a mystery. Perhaps the resolution to this puzzle will be made evident as the role of piRNAs, or other noncoding RNA molecules, in directing histone modifications to specific sites in the genome is established.

The piRNA pathway has evolved to protect the host genome from perturbations caused by many types of endogenized viruses in the host genome of many species. However, its mechanism of action may not be confined to quelling viral activity. Perhaps the piRNA recognition process in mammals has evolved to function in multiple ways in the mammalian genome acting not only to suppress endogenous retroviruses in the germline but also to support meiotic processes and to scaffold coordinate reprogramming of the embryonic genome. The tools to test this supposition in mammals are just now becoming available (Lim & Knowles, 2015; Williams et al., 2015).

5. EPIGENETICS BEYOND THE NUCLEUS

Epigenetic information in the form of cytoplasmic localization of specific mRNAs and proteins in eggs of nonmammalian species is absolutely essential for normal development of *Drosophila*, *Xenopus*, and the like. Though morphological asymmetries can be observed in mammalian eggs and zygotes, it is not clear if and how much they contribute to developmental regulation. Localization of specific molecules to definitive areas of mammalian eggs and zygotes has not been observed, which does not mean that it does not exist. However, orderly progression of early mammalian development appears to be based on small random differences, reinforced by positional clues, rather than determinative cytoplasmic localization of specific factors, as observed in nonmammalian species. The unique feature of mammalian development is the establishment of an early interaction between the mother and her embryo. Thus, the earliest goal of the mammalian embryo is to differentiate the cells necessary to provide this interaction, the trophectoderm of the mammalian blastocyst. As it is impossible to determine which part of the egg or zygote will become the embryo and which will become extraembryonic membranes, it is difficult to imagine that cytoplasmic localization directs this earliest cell-fate decision and even more difficult to visualize cytoplasmic localization to establish the axes of the early embryo.

5.1 Earliest Egg Asymmetry: Position of the Meiotic Spindle

Asymmetric cell divisions play an important part in development, and meiotic division in female gametes is the most extreme example of this type of cell division (for review, see Almonacid, Terret, & Velhac, 2014). The exact mechanism that propels the nucleus of the mouse oocyte to the periphery during meiosis I is not completely known, though several components have been identified. The actin network is required, and positioning of the meiotic spindle is the function of an activated F-actin connection between the spindle poles and the egg cortex. Due to the initial stochastic positioning of the spindle, one pole will always be closer to the egg cortex, thus the forces exerted by myosin will pull stronger in the direction closest to the egg surface (Schuh & Ellenberg, 2008). This increases the asymmetry so that the pulling forces in that direction will grow ever stronger until the spindle will finally be localized at the egg periphery. Though the exact biophysical process is more complicated with many additional components identified (Almonacid et al., 2014), the basic motif, which we will repeatedly encounter, is essentially very simple: the exploitation of small accidental irregularities to direct a path to final asymmetries.

Based on this simplified description, it is easy to see how meiotic asymmetry can be established but it is much harder to explain how, following fertilization, the pronuclei move toward the center of the zygote, which then divides symmetrically. We know that the cloud of F-actin surrounding the spindle is essential (Chew, Lorthongpanich, Ang, Knowles, & Solter, 2012) and that F-actin dynamics are regulated by the molecular components of subcortical maternal complex (Yu et al., 2014), but how the directionality of the movement of the pronuclei toward the center is established remains unknown. Obviously any process directing the movement of the egg nucleus toward the periphery during meiosis must be canceled, or directionality reversed, for the first zygotic division to take place in the center of the embryo. Is the unique presence of two separate pronuclei in any way instrumental? Considering that the molecular motors that first move the spindle toward the periphery and then the pronuclei toward the center are most likely the same, identifying the molecules and signals that provide directionality is a most fascinating epigenetic puzzle.

The presence of the meiotic chromosomes themselves appears to exert a direct and significant effect on the overlaying plasma membrane, ie, the elimination of microvilli, a reduction of cell surface glycoproteins, and a slight exvagination (Van Blerkom & Bell, 1986). Possibly these changes to the plasma membrane are transmitted to the cytoskeletal network and it in turn facilitates the pronuclear movement toward the cell center. Indeed, it has been suggested that the site of the meiotic division affects the spindle orientation during the first zygotic division (Plusa, Grabarek, Piotrowska, Glover, & Zernicka-Goetz, 2002).

5.2 First Division = First Asymmetry in the Mouse Embryo?

The mouse zygote is not absolutely symmetrical, one pole, the place of the previous meiotic divisions, is marked by the presence of the second polar body, and by convention this is called the animal pole, while the opposite side of the zygote is termed the vegetal pole. It has been suggested that the plane of the first division is predetermined and defined by the position of the second polar body and possibly by the sperm entry point and that the plane of the first division predicts the position of embryonic–abembryonic axis of the blastocyst (Gardner, 1996, 1997, 2001; Gardner & Davies, 2006; Piotrowska, Wianny, Pedersen, & Zernicka-Goetz, 2001; Piotrowska & Zernicka-Goetz, 2001; Plusa et al., 2002). These assertions of preformation have been challenged by proponents of the "epigenesis view" of mammalian development (Alarcón & Marikawa, 2008; Dard, Breuer, Maro, & Louvet-Vallée, 2008; Hiiragi & Solter, 2004; Kurotaki, Hatta, Nakao, Nabeshima, & Fujimori, 2007; Motosugi, Bauer, Polanski, Solter, & Hiiragi, 2005), suggesting that the plane of the first division is essentially random as regards the axis of the animal–vegetal pole. Proponents of this "epigenesis view" also suggest that the plane of the first division is not related to the embryonic–abembryonic axis of the blastocyst, but rather this axis is established by the mechanical forces exerted by the zona pellucida. This controversy has been extensively discussed recently (Solter, 2016) and is still unresolved.

For the "preformation view" to be seriously considered as a possible paradigm in mammalian development, some conditions have to be postulated. It is necessary to assume that the animal and vegetal half of the zygote (and the corresponding parts of the two blastomeres of the 2-cell-stage embryo) are different from each other in molecular composition and that these differences determine which part of the embryo (embryonic or abembryonic half of the blastocyst) each blastomere gives rise to. It has been repeatedly established that a single blastomere of the 2-cell-stage embryo can give rise to an entire organism. However, this does not necessarily exclude the possibility that the blastomeres are indeed different and have different developmental fates (Gardner, 1996) since only 20–25% of single blastomeres develop into a complete organism. Proponents of preformation would argue that the failed blastomeres of the 2-cell-stage embryo did not receive some essential, localized cytoplasmic component. Alternatively one could argue that these failures can be explained by technical problems with such a difficult experiment. Even development of twins from separated blastomeres of a single 2-cell-stage embryo does not exclude the possibility of localized essential determinants since they could in a rare case be accidentally equally divided between two sister blastomeres. Are the blastomeres of the 2-cell-stage embryo equal or different? This issue could not have been resolved experimentally in the past but technical advances now make it possible. When we examine the transcriptome and/or proteome of single blastomeres from a 2-cell-stage embryo, what will be found? Are there reproducible differences that will enable identification of a molecular profile that in every 2-cell-stage embryo distinguishes blastomere A from blastomere B, and if so, can we show each class of blastomeres has a different developmental fate?

Several recent publications have addressed these questions, but so far the results are somewhat inconclusive. When comparing the fragment of the oocyte containing the spindle with the rest of the M2 oocyte by microarray analysis, or comparing the second polar body with the zygote, transcriptome differences were observed (VerMilyea et al., 2011). In view of the previously noted morphological differences (Van Blerkom & Bell, 1986) this is probably not surprising. However, when individual blastomeres of 2- and 3-cell-stage embryos were examined, no differences were found (VerMilyea et al., 2011). In this study, technical issues inherent in the microarray approach limited the number of transcripts studied. Later studies using single-cell RNA sequencing have been more informative; however, examination of the transcriptome of individual blastomeres of the cleavage-stage embryos yielded mixed results. One report decidedly stated that reproducible differences in transcriptomes of sister blastomeres of the 2-cell-stage embryos can be found, although these differences were mostly restricted to the differences between low-abundance transcripts (Biase, Cao, & Zhong, 2014). However, four other groups found no differences between the sister blastomeres of the same 2-cell-stage embryo, while clear differences between blastomeres of different embryonic stages were observed (Deng, Ramsköld, Reinius, & Sandberg, 2014; Piras, Tomita, & Selvarajoo, 2014; Shi et al., 2015; Xue et al., 2013). Principal component analysis revealed a clear separation between blastomeres of the early, mid, and late 2-cell-stage embryos, but all individual blastomeres of the same stage (early, mid, and late) grouped together (Deng et al., 2014; Xue et al., 2013). These experiments and interpretation of the results are inherently very difficult due to the multiple technical limitations. A substantial number of RNA molecules (60% in one protocol) will be lost for technical reasons (Deng et al., 2014), and this loss will obviously have a greater impact on low-abundance RNAs. In a single report claiming reproducible differences between sister blastomeres of the 2-cell stage (Biase et al., 2014), the authors examined the expression level of differentially expressed genes at the RNA and protein level. Among the many genes found to be differentially expressed in sister blastomeres at the RNA level, four were studied at the protein level. Only one recapitulated difference observed at the RNA level also at the protein level (two blastomeres high and two low in a single 4-cell-stage embryo). Overall there are no obvious, reproducible differences between sister blastomeres of the early cleavage stages embryos detectable by transcriptome analysis, which does not exclude the possibility that such differences will be found with more advanced methods.

There are only two reports that describe differences in the developmental capacity of sister blastomeres of the 4-cell-stage embryo (Piotrowska-Nitsche, Perea-Gomez, Haraguchi, & Zernicka-Goetz, 2005; Piotrowska-Nitsche & Zernicka-Goetz, 2005). In cases where the embryo followed a specific cleavage pattern at the first two divisions, some blastomeres contributed predominantly to the trophectoderm. Chimeric embryos composed entirely of such blastomeres fail to develop (Piotrowska-Nitsche et al., 2005). Subsequently, blastomeres with this predominantly trophectodermal fate were found to express low level of the histone arginine methyltransferase CARM1 (Torres-Padilla, Parfitt, Kouzarides, & Zernicka-Goetz, 2007). Regional cytoplasmic localization if present/absent from one blastomere could indeed affect its developmental fate. However, for these described results to make sense, the putative element would have to have been localized in (or absent from) one half of the vegetal half of the zygote, the plane of the first division would have to separate these two halves rather cleanly, and the plane of the second division would have to cleanly separate the animal and vegetal part of the blastomere. These observations (Piotrowska-Nitsche et al., 2005) await independent confirmation, especially since attempts to repeat these results, showing developmental failure of embryos composed of only blastomeres with putative trophectoderm fate, did not succeed (Gardner, 2010; Marikawa & Alarcón, 2009). Furthermore, it is unfortunate that the only gene expression analysis of CARM1 in individual sister blastomeres of the 8-cell-stage embryos, where a low level of CARM1 in two blastomeres and a high level of CARM1 RNA in six blastomeres would be expected, does not include an observation of the cleavage pattern. Failure to observe the expected pattern of CARM1 distribution could be explained by a different cleavage sequence in all three embryos examined (Shi et al., 2015).

At present time, we can neither confirm nor eliminate the possibility that epigenetic information, in the form of cytoplasmic localization of fate-determining factors, occurs in the mammalian egg and zygote. However, the preponderance of data indicates the most parsimonious mechanism to create asymmetry in preimplantation mouse embryos is gradual amplification of small, random stochastic differences, found in any biological system, by signals derived from the position of each blastomere, from its increasing intercellular contacts and its environment.

5.3 Transgenerational Inheritance

It is becoming apparent that many factors and conditions can alter germ cells nongenetically (without altering the DNA sequence) and that these changes can be transmitted and affect the phenotype of the progeny (see Bohacek & Mansuy, 2015 for review). Mechanisms for this type of transgenerational nongenetic inheritance are not absolutely clear, although it has been suggested that diet, among other factors, can affect the sperm methylome, in turn influencing the phenotype of the progeny (Radford et al., 2014; Rando, 2012; Rando & Simmons, 2015). However, recent report (Shea et al., 2015; discussed by Whitelaw, 2015) implies that changes in the sperm methylome are not related to diet but rather are consequences of "epivariation" so that the observed effects of paternal diet on the offspring metabolism are probably mediated by other factors (see later). Nevertheless, interference with DNA methylome can have a transgenerational effect, as demonstrated by developmental defects in the grandprogeny of maternal grandparents whose folate metabolism had been disrupted by mutation of *Mtrr* (Padmanabhan et al., 2013). The methylome of the affected progeny, especially of the placenta, was significantly dysregulated. Developmental abnormalities were found even if mothers of the affected fetuses were wild type as long as either maternal grandparent had a *Mtrr* mutation. These results suggest that disruption of folate metabolism results in epigenetic changes in the germ cells, and these changes can be transmitted to the progeny for several generations (Padmanabhan et al., 2013). The most likely explanation of these results is that defective folate metabolism causes multiple methylation changes (epimutations), and subsequently these changes escape the epigenetic reprogramming that normally occurs in each generation in both PGCs and the zygote (see Section 2).

As already mentioned the diet most likely has no effect on the sperm methylome per se (Shea et al., 2015) yet diet does affect the phenotype of the offspring, especially their metabolism (Rando, 2012; Rando & Simmons, 2015). One possible mechanism has been suggested (Chen et al., 2016; Sharma et al., 2016). Manipulation of the diet (by feeding a high-fat or protein-restricted diet) was found to result in changes in sperm-associated small RNAs, mostly derived from tRNAs. Injection of a small

RNA fraction from the sperm of dietary-challenged males into normal zygotes resulted in mice displaying a typical metabolic disorder (Chen et al., 2016). Some of these small RNAs also repress genes regulated by retrotransposons in early embryos (Macfarlan et al., 2012; Peaston et al., 2004), suggesting a possible mechanism of how diet-induced changes in sperm-associated small RNAs can modulate metabolic response in the progeny (Sharma et al., 2016). Sperm born RNA has also been implicated in transgenerational nongenetic inheritance observed following loss of histone methylation in sperm (Siklenka et al., 2015). The connection between these tRNA-induced changes of the methylome and piRNA effects (see Section 4) has yet to be investigated.

Other environmental factors, such as the traumatic stress associated with postpartum and transient mother–litter separation, have also been reported to produce transgenerational effect on behavior and metabolism. In this case small sperm-associated RNAs as mediators also seem to be involved (Gapp et al., 2014). Mating with stressed males changes the methylome and behavior of the progeny, and the same effect can be achieved by injecting RNA isolated from sperm of these males into normal zygotes, thus establishing direct causal relationship between small sperm RNAs and a phenotypic effect (Gapp et al., 2014). Each of these results implicates a small sperm-associated RNAs as the mediators of transgenerational nongenetic inheritance, and this mechanism might also explain the previously puzzling observations of paramutation-like inheritance in mice (Rassoulzadegan et al., 2006). If altered sperm-associated small RNAs can mediate transgenerational nongenetic inheritance, one has to ask what is the function of the RNAs that are normally associated with sperm of unstressed individual. Is the low success rate of development following SCNT or fertilization with round spermatids the consequence of the absence of these small RNAs that are present in the sperm that passes through the epididymis.

Sperm-associated small RNAs may not be the only way to account for transgenerational nongenetic inheritance following stress. Cao-Lei et al. (2014) reported significant DNA-methylation changes in progeny of women exposed to a major stress during pregnancy (1998 Quebec ice storm). However, to use these observations to claim a transgenerational nongenetic inheritance, one would have to know whether these women transmitted a similar DNA-methylation change to their children, conceived years after the stress and more importantly to their grandchildren. This observation would distinguish between intrauterine stress to the fetus and permanent epigenetic changes to the germ cells.

Transgenerational nongenetic inheritance is a fascinating epigenetic phenomenon with many significant ramifications for human health. If it is shown to be widely present, the implication will be that our everyday behavior which we till now considered only relevant to our own health and well being may also irreversibly affect our progeny for generations to come. Dalgaard et al. (2016) described "epigenetic obesity" in mice and human caused by *Trim28* haploinsufficiency, or low level of TRIM28 levels, respectively. The effect is mediated by the effects of Trim28 on the expression of obesity-associated imprinted genes. Depending whether the expression of these genes is reduced or not, individuals with low TRIM28 expression levels will be obese or lean (Dalgaard et al., 2016). Such a multiplicity of phenotypes has been repeatedly observed in the absence of Trim28 (Messerschmidt et al., 2012), and it likely reflects which of the Trim28-targeted genes are not properly remethylated in absence or low level of Trim28. In the mice low level of Trim28 is the consequence of genetic mutation, while the reasons for a low level of Trim28 expression in humans are unknown. One could speculate that epigenetic changes themselves could permanently affect the level of expression from a given Trim28 allele and that such an epimutation can be inherited through generation causing obesity in some individuals with epimutation thus providing an example of transgenerational nongenetic inheritance that can significantly affect human health.

6. CODA

...counting the pebbles in a gravel-pit and describing their color is not geology...

This paraphrase of Darwin's letter to Fawcett neatly illustrates the current problem with the epigenetics and its role in mammalian development. In theory, we should be able to completely explain such a simple system and process restricted to an embryo composed of one-to-one hundred cells, as it develops over the course of 4 days, which is completely accessible in culture. Formerly, we lacked the proper technology to deal with the small amount of biological material available, but current advances have removed most of these technical obstacles. There is no denying that we now know much more about the role of epigenetic factors in regulating mouse development, but a significant amount of information is still unknown or controversial. How many unknown unknowns (to cite Donald Rumsfeld) like micro-RNAs are waiting in mammalian development? Are we asking the wrong questions? Are we buried under a sea of data we do not know how, or have no tools, to interpret? Is the system despite its small size too complicated to ever be completely understood? As the technology to intrude upon and change the genome of the human zygote has become not only available, but its use seriously contemplated, we must put in an extraordinary effort into really understanding the long-term transgenerational consequences of altering something we are now planning to improve.

REFERENCES

Adenot, P. G., Campion, E., Legouy, E., Allis, C. D., Dimitrov, S., Renard, J., et al. (2000). Somatic linker histone H1 is present throughout mouse embryogenesis and is not replaced by variant H1 degrees. *Journal of Cell Science*, *113*, 2897–2907.

Adenot, P. G., Mercier, Y., Renard, J. P., & Thompson, E. M. (1997). Differential H4 acetylation of paternal and maternal chromatin precedes DNA replication and differential transcriptional activity in pronuclei of 1-cell mouse embryos. *Development*, *124*, 4615–4625.

Ahmad, K., & Henikoff, S. (2002). Histone H3 variants specify modes of chromatin assembly. *Proceedings of the National Academy of Sciences of the United States of America*, *99*(Suppl. 4), 16477–16484.

Akiyama, T., Suzuki, O., Matsuda, J., & Aoki, F. (2011). Dynamic replacement of histone H3 variants reprograms epigenetic marks in early mouse embryos. *PLoS Genetics*, *7*, e1002279.

Alarcón, V. B., & Marikawa, Y. (2008). Spatial alignment of the mouse blastocyst axis across the first cleavage plane is caused by mechanical constraint rather than developmental bias among blastomeres. *Molecular Reproduction and Development*, *75*, 1143–1153.

Allan, J., Hartman, P. G., Crane-Robinson, C., & Aviles, F. X. (1980). The structure of histone H1 and its location in chromatin. *Nature*, *288*, 675–679.

Almonacid, M., Terret, M.-E., & Velhac, M.-H. (2014). Actin-based spindle positioning: New insights from female gametes. *Journal of Cell Science*, *127*, 477–483.

Amouroux, R., Nashun, B., Shirane, K., Nakagawa, S., Hill, P. W. S., D'Souza, Z., et al. (2016). De novo DNA methylation drives 5hmC accumulation in mouse zygotes. *Nature Cell Biology*, *18*, 225–233.

Ancelin, K., Syx, L., Borensztein, M., Ranisavljevic, N., Vassilev, I., Briseno-Roa, L., et al. (2016). Maternal LSD1/KDM1A is an essential regulator of chromatin and transcription landscapes during zygotic genome activation. *eLife*, *5*, e08851. http://dx.doi.org/10.7554/eLife.08851.

Andreu-Vieyra, C. V., Chen, R., Agno, J. E., Glaser, S., Anastassiadis, K., Stewart, A. F., et al. (2010). MLL2 is required in oocytes for bulk histone 3 lysine 4 trimethylation and transcriptional silencing. *PLoS Biology*, *8*, e1000453.

Annalisa, I., & Robert, S. (2016). The role of linker histone H1 modifications in the regulation of gene expression and chromatin dynamics. *Biochimica et Biophysica Acta*, *1859*, 486–495. http://dx.doi.org/10.1016/j.bbagrm.2015.09.003.

Aoki, F., Worrad, D. M., & Schultz, R. M. (1997). Regulation of transcriptional activity during the first and second cell cycles in the preimplantation mouse embryo. *Developmental Biology*, *181*, 296–307.

Aoshima, K., Inoue, E., Sawa, H., & Okada, Y. (2015). Paternal H3K4 methylation is required for minor zygotic gene activation and early mouse embryonic development. *EMBO Reports*, *16*, 803–812.

Arakawa, T., Nakatani, T., Oda, M., Kimura, Y., Sekita, Y., Kimura, T., et al. (2015). Stella controls chromocenter formation through regulation of Daxx expression in 2-cell embryos. *Biochemical and Biophysical Research Communications*, *466*, 60–65.

Arand, J., Spieler, D., Karius, T., Branco, M. R., Meilinger, D., Meissner, A., et al. (2012). In vivo control of CpG and non-CpG DNA methylation by DNA methyltransferases. *PLoS Genetics*, *8*, e1002750.

Aravin, A. A., Gaidatzis, D., Pfeffer, S., Lagos-Quintana, M., Landgraf, P., Iovina, N., et al. (2006). A novel class of small RNAs bind to MILI protein in mouse testes. *Nature*, *442*, 203–207.

Arney, K. L., Bao, S., Bannister, A. J., Kouzarides, T., & Surani, M. A. (2002). Histone methylation defines epigenetic asymmetry in the mouse zygote. *The International Journal of Developmental Biology*, *46*, 317–320.

Bestor, T. H. (1988). Cloning of a mammalian DNA methyltransferase. *Gene*, *74*(1), 9–12.

Biase, F. H., Cao, X., & Zhong, S. (2014). Cell fate inclination within 2-cell and 4-cell mouse embryo revealed by single-cell RNA sequencing. *Genome Research*, *24*, 1787–1796.

Biczysko, W., Pienkowski, M., Solter, D., & Koprowski, H. (1975). Virus particles in early mouse embryos. *Journal of the National Cancer Institute*, *51*, 1041–1050.

Bohacek, J., & Mansuy, I. M. (2015). Molecular insights into transgenerational non-genetic inheritance of acquired behaviours. *Nature Reviews. Genetics*, *16*, 641–652.

Bostick, M., Kim, J. K., Estève, P.-O., Clark, A., Pradhan, S., & Jacobsen, S. E. (2007). UHRF1 plays a role in maintaining DNA methylation in mammalian cells. *Science*, *317*, 1760–1764.

Branco, M. R., Oda, M., & Reik, W. (2008). Safeguarding parental identity: Dnmt1 maintains imprints during epigenetic reprogramming in early embryogenesis. *Genes & Development*, *22*, 1567–1571.

Burton, A., Muller, J., Tu, S., Padilla-Longoria, P., Guccione, E., & Torres-Padilla, M. E. (2013). Single-cell profiling of epigenetic modifiers identifies PRDM14 as an inducer of cell fate in the mammalian embryo. *Cell Reports*, *5*, 687–701.

Burton, A., & Torres-Padilla, M. E. (2010). Epigenetic reprogramming and development: A unique heterochromatin organization in the preimplantation mouse embryo. *Briefings in Functional Genomics*, *9*, 444–454.

Bush, K. M., Yuen, B. T., Barrilleaux, B. L., Riggs, J. W., O'Geen, H., Cotterman, R. F., et al. (2013). Endogenous mammalian histone H3.3 exhibits chromatin-related functions during development. *Epigenetics & Chromatin*, *6*, 7.

Calarco, P. (1975). Intracisternal A particle formation and inhibition in preimplantation mouse embryos. *Biology of Reproduction*, *12*, 448–454.

Cao-Lei, L., Massart, R., Suderman, M. J., Machnes, Z., Elgbeili, G., Laplante, D. P., et al. (2014). DNA methylation signatures triggered by prenatal maternal stress exposure to a natural disaster: Project Ice Storm. *PloS One*, *9*, e107653.

Casanova, M., Pasternak, M., El Marjou, F., Le Baccon, P., Probst, A. V., & Almouzni, G. (2013). Heterochromatin reorganization during early mouse development requires a single-stranded noncoding transcript. *Cell Reports*, *4*, 1156–1167.

Celeste, A., Petersen, S., Romanienko, P. J., Fernandez-Capetillo, O., Chen, H. T., Sedelnikova, O. A., et al. (2002). Genomic instability in mice lacking histone H2AX. *Science*, *296*, 922–927.

Chan, M. M., Smith, Z. D., Egli, D., Regev, A., & Meissner, A. (2012). Mouse ooplasm confers context-specific reprogramming capacity. *Nature Genetics*, *44*, 978–980.

Chang, C. C., Ma, Y., Jacobs, S., Tian, X. C., Yang, X., & Rasmussen, T. P. (2005). A maternal store of macroH2A is removed from pronuclei prior to onset of somatic macroH2A expression in preimplantation embryos. *Developmental Biology*, *278*, 367–380.

Chen, Q., Yan, M., Cao, Z., Li, X., Zhang, Y., Shi, J., et al. (2016). Sperm tsRNAs contribute to intergenerational inheritance of an acquired metabolic disorder. *Science*, *351*, 397–400.

Chew, T. G., Lorthongpanich, C., Ang, W. X., Knowles, B. B., & Solter, D. (2012). Symmetric cell division of the mouse zygote requires an actin network. *Cytoskeleton*, *69*, 1040–1046.

Cho, S., Park, J. S., Kwon, S., & Kang, Y. K. (2012). Dynamics of Setdb1 expression in early mouse development. *Gene Expression Patterns*, *12*, 213–218.

Ciccone, D. N., Su, H., Hevi, S., Gay, F., Lei, H., Bajko, J., et al. (2009). KDM1B is a histone H3K4 demethylase required to establish maternal genomic imprints. *Nature*, *461*, 415–418.

Cirio, M. C., Ratnam, S., Ding, F., Reinhart, B., Navara, C., & Chaillet, J. R. (2008). Preimplantation expression of the somatic form of Dnmt1 suggests a role in the inheritance of genomic imprints. *BMC Developmental Biology*, *8*, 9.

Costanzi, C., Stein, P., Worrad, D. M., Schultz, R. M., & Pehrson, J. R. (2000). Histone macroH2A1 is concentrated in the inactive X chromosome of female preimplantation mouse embryos. *Development*, *127*, 2283–2289.

Dalgaard, K., Landgraf, K., Heyne, S., Lempradi, A., Longinotto, J., Gossens, K., et al. (2016). Trim28 haploinsufficiency triggers bi-stable epigenetic obesity. *Cell*, *164*, 353–364.

Dard, N., Breuer, M., Maro, B., & Louvet-Vallée, S. (2008). Morphogenesis of the mammalian blastocyst. *Molecular and Cellular Endocrinology*, *282*, 70–77.

Daujat, S., Weiss, T., Mohn, F., Lange, U. C., Ziegler-Birling, C., Zeissler, U., et al. (2009). H3K64 trimethylation marks heterochromatin and is dynamically remodeled during developmental reprogramming. *Nature Structural & Molecular Biology*, *16*, 777–781.

De La Fuente, R., Baumann, C., & Viveiros, M. M. (2015). ATRX contributes to epigenetic asymmetry and silencing of major satellite transcripts in the maternal genome of the mouse embryo. *Development*, *142*, 1806–1817.

Deng, Q., Ramsköld, D., Reinius, B., & Sandberg, R. (2014). Single-cell RNA-seq reveals dynamic, random monoallelic gene expression in mammalian cells. *Science*, *343*, 193–196.

Drane, P., Ouararhni, K., Depaux, A., Shuaib, M., & Hamiche, A. (2010). The death-associated protein DAXX is a novel histone chaperone involved in the replication-independent deposition of H3.3. *Genes & Development*, *24*, 1253–1265.

Dunleavy, E. M., Roche, D., Tagami, H., Lacoste, N., Ray-Gallet, D., Nakamura, Y., et al. (2009). HJURP is a cell-cycle-dependent maintenance and deposition factor of CENP-A at centromeres. *Cell*, *137*, 485–497.

Durruthy-Durruthy, J., Sebastiano, V., Wossidlo, M., Depada, D., Cui, J., Grow, E. J., et al. (2015). The primate-specific noncoding RNA HPAT5 regulates pluripotency during human preimplantation development and nuclear reprogramming. *Nature Genetics*, *48*, 44–52.

Erhardt, S., Su, I. H., Schneider, R., Barton, S., Bannister, A. J., Perez-Burgos, L., et al. (2003). Consequences of the depletion of zygotic and embryonic enhancer of zeste 2 during preimplantation mouse development. *Development*, *130*, 4235–4248.

Evisikov, A. V., Graber, J. H., Brockman, J. M., Hampl, A., Holbrook, A. E., Singh, P., et al. (2006). Cracking the egg: Molecular dynamics and evolutionary aspects of the transition from the fully grown oocyte to embryo. *Genes & Development*, *20*, 2713–2727.

Faast, R., Thonglairoam, V., Schulz, T. C., Beall, J., Wells, J. R., Taylor, H., et al. (2001). Histone variant H2A.Z is required for early mammalian development. *Current Biology*, *11*, 1183–1187.

Fan, Y., Nikitina, T., Morin-Kensicki, E. M., Zhao, J., Magnuson, T. R., Woodcock, C. L., et al. (2003). H1 linker histones are essential for mouse development and affect nucleosome spacing in vivo. *Molecular and Cellular Biology*, *23*, 4559–4572.

Fan, Y., Sirotkin, A., Russell, R. G., Ayala, J., & Skoultchi, A. I. (2001). Individual somatic H1 subtypes are dispensable for mouse development even in mice lacking the H1(0) replacement subtype. *Molecular and Cellular Biology, 21*, 7933–7943.

Filipescu, D., Muller, S., & Almouzni, G. (2014). Histone H3 variants and their chaperones during development and disease: Contributing to epigenetic control. *Annual Review of Cell and Developmental Biology, 30*, 615–646.

Foltz, D. R., Jansen, L. E., Bailey, A. O., Yates, J. R., 3rd, Bassett, E. A., Wood, S., et al. (2009). Centromere-specific assembly of CENP-A nucleosomes is mediated by HJURP. *Cell, 137*, 472–484.

Fu, G., Ghadam, P., Sirotkin, A., Khochbin, S., Skoultchi, A. I., & Clarke, H. J. (2003). Mouse oocytes and early embryos express multiple histone H1 subtypes. *Biology of Reproduction, 68*, 1569–1576.

Gao, S., Chung, Y. G., Parseghian, M. H., King, G. J., Adashi, E. Y., & Latham, K. E. (2004). Rapid H1 linker histone transitions following fertilization or somatic cell nuclear transfer: Evidence for a uniform developmental program in mice. *Developmental Biology, 266*, 62–75.

Gapp, K., Jawaid, A., Sarkies, P., Bohacek, J., Pelczar, P., Prados, J., et al. (2014). Implication of sperm RNA in transgenerational inheritance of the effects of early trauma in mice. *Nature Neuroscience, 17*, 667–669.

Gardner, R. (1996). Can developmentally significant spatial patterning of the egg be discounted in mammals? *Human Reproduction Update, 2*, 3–27.

Gardner, R. (1997). The early blastocyst is bilaterally symmetrical and it axis of symmetry is aligned with the animal-vegetal axis of the zygote in the mouse. *Development, 124*, 289–301.

Gardner, R. (2001). Specification of embryonic axes begins before cleavage in normal mouse development. *Development, 128*, 839–847.

Gardner, R. (2010). Normal bias in the direction of fetal rotation depends on blastomere composition during early cleavage in the mouse. *PloS One, 5*, e9610.

Gardner, R., & Davies, T. J. (2006). An investigation of the origin and significance of bilateral symmetry of the pronuclear zygote in the mouse. *Human Reproduction, 21*, 492–502.

Gaspar-Maia, A., Alajem, A., Polesso, F., Sridharan, R., Mason, M. J., Heidersbach, A., et al. (2009). Chd1 regulates open chromatin and pluripotency of embryonic stem cells. *Nature, 460*, 863–868.

Goldberg, A. D., Banaszynski, L. A., Noh, K. M., Lewis, P. W., Elsaesser, S. J., Stadler, S., et al. (2010). Distinct factors control histone variant H3.3 localization at specific genomic regions. *Cell, 140*, 678–691.

Grow, E. J., Flynn, R. A., Chavez, S. L., Bayless, N. I., Wossidlo, M., Wesche, D. J., et al. (2015). Intrinsic retroviral reactivation in human preimplantation embryos and pluripotent cells. *Nature, 522*, 221–225.

Gu, T.-P., Guo, F., Yang, H., Wu, H.-P., Xu, G.-F., Liu, W., et al. (2011). The role of Tet3 DNA dioxygenase in epigenetic reprogramming by oocytes. *Nature, 477*, 606–610.

Guibert, S., & Weber, M. (2012). Epigenetics: Erase for a new start. *Nature, 492*, 363–364.

Guo, F., Li, X., Liang, D., Li, T., Zhu, P., Guo, H., et al. (2014). Active and passive demethylation of male and female pronuclear DNA in the mammalian zygote. *Cell Stem Cell, 15*, 447–458.

Guzman-Ayala, M., Sachs, M., Koh, F. M., Onodera, C., Bulut-Karslioglu, A., Lin, C. J., et al. (2015). Chd1 is essential for the high transcriptional output and rapid growth of the mouse epiblast. *Development, 142*, 118–127.

Hackett, J. A., Sengupta, R., Zylicz, J. J., Murakami, K., Lee, C., Down, T. A., et al. (2013). Germline DNA demethylation dynamics and imprint erasure through 5-hydroxymethylcytosine. *Science, 339*, 448–452.

Hajkova, P., Jeffries, S. J., Lee, C., Miller, N., Jackson, S. P., & Surani, M. A. (2010). Genome-wide reprogramming in the mouse germ line entails the base excision repair pathway. *Science*, *329*, 78–82.

Hake, S. B., & Allis, C. D. (2006). Histone H3 variants and their potential role in indexing mammalian genomes: The "H3 barcode hypothesis". *Proceedings of the National Academy of Sciences of the United States of America*, *103*, 6428–6435.

Hamazaki, N., Uesaka, M., Nakashima, K., Agata, K., & Imamura, T. (2015). Gene activation-associated long noncoding RNAs function in mouse preimplantation development. *Development*, *142*, 910–929.

Harshman, S. W., Young, N. L., Parthun, M. R., & Freitas, M. A. (2013). H1 histones: Current perspectives and challenges. *Nucleic Acids Research*, *41*, 9593–9609.

Hashimoto, H., Liu, Y., Upadhyay, A. K., Chang, Y., Howerton, S. B., Vertino, P. M., et al. (2012). Recognition and potential mechanisms for replication and erasure of cytosine hydroxymethylation. *Nucleic Acids Research*, *40*, 4841–4849.

Hatanaka, Y., Inoue, K., Oikawa, M., Kamimura, S., Ogonuki, N., Kodama, E. N., et al. (2015). Histone chaperone CAF-1 mediates repressive histone modifications to protect preimplantation mouse embryos from endogenous retrotransposons. *Proceedings of the National Academy of Sciences of the United States of America*, *112*, 14641–14646.

He, Y.-F., Li, B.-Z., Li, Z., Liu, P., Wang, Y., Tang, Q., et al. (2011). Tet-mediated formation of 5-carboxylcytosine and its excision by TDG in mammalian DNA. *Science*, *333*, 1303–1307.

Hiiragi, T., & Solter, D. (2004). First cleavage plane of the mouse egg is not predetermined but defined by the topology of the two apposing pronuclei. *Nature*, *430*, 360–364.

Hirasawa, R., Chiba, H., Kaneda, M., Tajima, S., Li, E., Jaenisch, R., et al. (2008). Maternal and zygotic Dnmt1 are necessary and sufficient for the maintenance of DNA methylation imprints during preimplantation development. *Genes & Development*, *22*, 1607–1616.

Hitt, M. M., Wu, T. L., Cohen, G., & Linn, S. (1988). De novo and maintenance DNA methylation by a mouse plasmacytoma cell DNA methyltransferase. *The Journal of Biological Chemistry*, *263*, 4392–4399.

Howell, C. Y., Bestor, T. H., Ding, F., Latham, K. E., Mertineit, C., Trasler, J. M., et al. (2001). Genomic imprinting disrupted by a maternal effect mutation in the Dnmt1 gene. *Cell*, *104*, 829–838.

Howman, E. V., Fowler, K. J., Newson, A. J., Redward, S., MacDonald, A. C., Kalitsis, P., et al. (2000). Early disruption of centromeric chromatin organization in centromere protein A (Cenpa) null mice. *Proceedings of the National Academy of Sciences of the United States of America*, *97*, 1148–1153.

Huang, X. J., Ma, X., Wang, X., Zhou, X., Li, J., Sun, S. C., et al. (2015). Involvement of G9A-like protein (GLP) in the development of mouse preimplantation embryos in vitro. *Reproduction, Fertility, and Development*. http://dx.doi.org/10.1071/RD14341.

Huang, X. A., Yin, H., Sweeney, S., Raha, B., Snyder, M., & Lin, H. (2013). A major epigenetic programming mechanism guided by piRNAs. *Developmental Cell*, *24*, 502–506.

Inoue, A., Shen, L., Dai, Q., He, C., & Zhang, Y. (2011). Generation and replication-dependent dilution of 5fC and 5caC during mouse preimplantation development. *Cell Research*, *21*, 1670–1676.

Inoue, A., Shen, L., Matoba, S., & Zhang, Y. (2015). Haploinsufficiency, but not defective paternal 5mC oxidation, accounts for the developmental defects of maternal Tet3 knockouts. *Cell Reports*, *10*, 463–470.

Inoue, A., & Zhang, Y. (2011). Replication-dependent loss of 5-hydroxymethylcytosine in mouse preimplantation embryos. *Science*, *334*, 194.

Ito, S., Shen, L., Dai, Q., Wu, S. C., Collins, L. B., Swenberg, J. A., et al. (2011). Tet proteins can convert 5-methylcytosine to 5-formylcytosine and 5-carboxylcytosine. *Science*, *333*, 1300–1303.

Ivanauskiene, K., Delbarre, E., McGhie, J. D., Kuntziger, T., Wong, L. H., & Collas, P. (2014). The PML-associated protein DEK regulates the balance of H3.3 loading on chromatin and is important for telomere integrity. *Genome Research*, *24*, 1584–1594.

Iwasaki, Y. W., Siomi, M. C., & Siomi, H. (2015). PIWI-interacting RNA: Its biogenesis and functions. *Annual Review of Biochemistry*, *84*, 405–433.

Iyengar, S., & Farnham, P. J. (2011). KAP1 protein: An enigmatic master regulator of the genome. *The Journal of Biological Chemistry*, *286*, 26267–26276.

Jacobs, F. M. J., Greenberg, D., Nguyen, N., Haeussler, M., Ewing, A. D., Katzman, S., et al. (2014). An evolutionary arms race between KRAB zinc-finger genes *ZNF91/93* and SVA/L1 retrotransposons. *Nature*, *516*, 242–245.

Jahner, D., & Jaenisch, R. (1985). Retrovirus-induced de novo methylation of flanking host sequences correlates with gene inactivity. *Nature*, *315*, 594–597.

Jahner, D., Stuhlmann, H., Stewart, C. L., Harbers, K., Lohler, J., Simon, I., et al. (1982). De novo methylation and expression of retroviral genomes during mouse development. *Nature*, *298*, 623–628.

Jang, C. W., Shibata, Y., Starmer, J., Yee, D., & Magnuson, T. (2015). Histone H3.3 maintains genome integrity during mammalian development. *Genes & Development*, *29*, 1377–1392.

Jin, C., Zang, C., Wei, G., Cui, K., Peng, W., Zhao, K., et al. (2009). H3.3/H2A.Z double variant-containing nucleosomes mark 'nucleosome-free regions' of active promoters and other regulatory regions. *Nature Genetics*, *41*, 941–945.

Kagiwada, S., Kurimoto, K., Hirota, T., Yamaji, M., & Saitou, M. (2013). Replication-coupled passive DNA demethylation for the erasure of genome imprints in mice. *The EMBO Journal*, *32*, 340–353.

Kawamura, M., Akiyama, T., Tsukamoto, S., Suzuki, M. G., & Aoki, F. (2012). The expression and nuclear deposition of histone H3.1 in murine oocytes and preimplantation embryos. *The Journal of Reproduction and Development*, *58*, 557–562.

Kishikawa, S., Murata, T., Ugai, H., Yamazaki, T., & Yokoyama, K. K. (2003). Control elements of Dnmt1 gene are regulated in cell-cycle dependent manner. *Nucleic Acids Research. Supplement (2001)*, *3*, 307–308.

Konev, A. Y., Tribus, M., Park, S. Y., Podhraski, V., Lim, C. Y., Emelyanov, A. V., et al. (2007). CHD1 motor protein is required for deposition of histone variant H3.3 into chromatin in vivo. *Science*, *317*, 1087–1090.

Kurihara, Y., Kawamura, Y., Uchijima, Y., Amamo, T., Kobayashi, H., Asano, T., et al. (2008). Maintenance of genomic methylation patterns during preimplantation development requires the somatic form of DNA methyltransferase 1. *Developmental Biology*, *313*, 335–346.

Kurotaki, Y., Hatta, K., Nakao, K., Nabeshima, Y., & Fujimori, T. (2007). Blastocyst axis is specified independently of early cell lineage but aligns with the ZP shape. *Science*, *316*, 1081–1092.

Le Thomas, A., Stuwe, E., Li, S., Du, D., Marinov, G., Rozhkov, N., et al. (2016). Trans-generationally inherited piRNAs trigger piRNA biogenesis by changing the chromatin of piRNA clusters and inducing precursor processing. *Genes & Development*, *28*, 1667–1680.

Lei, H., Oh, S. P., Okano, M., Jüttermann, R., Goss, K. A., Jaenisch, R., et al. (1996). De novo DNA cytosine methyltransferase activities in mouse embryonic stem cells. *Development*, *122*, 3195–3205.

Leonhardt, H., Page, A. W., Weier, H. U., & Bestor, T. H. (1992). A targeting sequence directs DNA methyltransferase to sites of DNA replication in mammalian nuclei. *Cell*, *71*, 865–873.

Lepikhov, K., & Walter, J. (2004). Differential dynamics of histone H3 methylation at positions K4 and K9 in the mouse zygote. *BMC Developmental Biology*, *4*, 12.

Lepikhov, L., Zakhartchenko, V., Hao, R., Yang, F., Wrenzycki, C., Niemann, H., Wolf, W., & Walter, J. (2008). Evidence for conserved DNA and histone H3 methylation reprogramming in mouse, bovine and rabbit zygote. *Epigenetics & Chromatin.* http://dx.doi.org/10.1186/1756-8935-1-8.

Leseva, M., Knowles, B. B., Messerschmidt, D. M., & Solter, D. (2015). Erase-maintain-establish: Natural reprogramming of the mammalian epigenome. *Cold Spring Harbor Symposia in Quantitative Biology.* http://dx.doi.org/10.1101/sqb.2015.80.027441.

LeThomas, A., Toth, K. F., & Aravin, A. A. (2014). To be or not to be a piRNA: Genomic origin and processing of piRNA. *Genome Biology, 15,* 204–217.

Li, X., Ito, M., Zhou, F., Youngson, N., Zuo, X., Leder, P., et al. (2008). A maternal-zygotic effect gene, Zfp57, maintains both maternal and paternal imprints. *Developmental Cell, 15,* 547–557.

Lim, A. K., & Knowles, B. B. (2015). Controlling endogenous retroviruses and their chimeric transcript during natural reprogramming in the oocyte. *The Journal of Infectious Diseases, 212,* s47–s51.

Lim, A. K., Lorthongpanich, C. C., Chew, T. G., Tan, C. W., Shue, Y. T., Kuramochi-Miyagawa, S., et al. (2013). The nuage mediates piRNA-dependent retrotransposon silencing in mouse primordial ovarian follicles. *Development, 140,* 3819–3825.

Lim, C. Y., Reversade, B., Knowles, B. B., & Solter, D. (2013). Optimal histone H3 to linker histone H1 chromatin ratio is vital for mesodermal competence in Xenopus. *Development, 140,* 853–860.

Lin, C. J., Conti, M., & Ramalho-Santos, M. (2013). Histone variant H3.3 maintains a decondensed chromatin state essential for mouse preimplantation development. *Development, 140,* 3624–3634.

Lin, C. J., Koh, F. M., Wong, P., Conti, M., & Ramalho-Santos, M. (2014). Hira-mediated H3.3 incorporation is required for DNA replication and ribosomal RNA transcription in the mouse zygote. *Developmental Cell, 30,* 268–279.

Liu, H., Kim, J. M., & Aoki, F. (2004). Regulation of histone H3 lysine 9 methylation in oocytes and early pre-implantation embryos. *Development, 131,* 2269–2280.

Liu, Y., Toh, H., Sasaki, H., Zhang, X., & Cheng, X. (2012). An atomic model of Zfp57 recognition of CpG methylation within a specific DNA sequence. *Genes & Development, 26,* 2374–2379.

Lorthongpanich, C., Cheow, L. F., Balu, S., Quake, S. R., Knowles, B. B., Burkholder, W. F., et al. (2013). Single-cell DNA-methylation analysis reveals epigenetic chimerism in preimplantation embryos. *Science, 341,* 1110–1112.

Lorthongpanich, C., Solter, D., & Lim, C. Y. (2010). Nuclear reprogramming in zygotes. *The International Journal of Developmental Biology, 54,* 1631–1640.

Lukic, S., Nicolas, J.-C., & Levine, A. J. (2014). The diversity of zinc-finger genes on human chromosome 19 provides an evolutionary mechanism for defense against inherited endogenous retroviruses. *Cell Death & Differentiation, 21,* 381–387.

Ma, X.-S., Chao, S.-B., Huang, X.-J., Lin, F., Qin, L., Wang, X.-G., et al. (2015). The dynamics and regulatory mechanism of pronuclear H3k9me2 asymmetry in mouse zygotes. *Scientific Reports, 5,* 17924.

Macfarlan, T. S., Gifford, W. D., Agarwal, S., Driscoll, S., Lettieri, K., Wang, J., et al. (2011). Endogenous retroviruses and neighboring genes are coordinately repressed by LSD1/KDM1A. *Genes and Development, 25,* 594–607.

Macfarlan, T. S., Gifford, W. D., Driscoll, S., Lettieri, K., Rowe, H., Bonanomi, D., et al. (2012). Embryonic stem cell potency fluctuates with endogenous retroviral activity. *Nature, 487,* 57–63.

Mackay, D. J. G., Callaway, J. L. A., Marks, S. M., White, H. E., Acerini, C. L., Boonen, S. E., et al. (2008). Hypomethylation of multiple imprinted loci in individuals with transient neonatal diabetes is associated with mutations in ZFP57. *Nature Genetics, 40,* 949–951.

Maiti, A., & Drohat, A. C. (2011). Thymine DNA glycosylase can rapidly excise 5-formylcytosine and 5-carboxylcytosine: Potential implications for active demethylation of CpG sites. *The Journal of Biological Chemistry*, *286*, 35334–35338.

Marikawa, Y., & Alarcón, V. B. (2009). Establishment of trophectoderm and inner cell mass lineages in the mouse embryo. *Molecular Reproduction and Development*, *76*, 1019–1032.

Martin, C., Beaujean, N., Brochard, V., Audouard, C., Zink, D., & Debey, P. (2006). Genome restructuring in mouse embryos during reprogramming and early development. *Developmental Biology*, *292*, 317–332.

Matoba, S., Liu, Y., Lu, F., Iwabuchi, K. A., Shen, L., Inoue, A., et al. (2014). Embryonic development following somatic cell nuclear transfer impeded by persisting histone methylation. *Cell*, *159*, 884–895.

Mattiroli, F., D'Arcy, S., & Luger, K. (2015). The right place at the right time: Chaperoning core histone variants. *EMBO Reports*, *16*, 1454–1466.

Mayer, W., Niveleau, A., Walter, J., Fundele, R., & Haaf, T. (2000). Demethylation of the zygotic paternal genome. *Nature*, *403*, 501–502.

Maze, I., Noh, K. M., Soshnev, A. A., & Allis, C. D. (2014). Every amino acid matters: Essential contributions of histone variants to mammalian development and disease. *Nature Reviews. Genetics*, *15*, 259–271.

McClintock, B. (1956). Controlling elements and the gene. *Cold Spring Harbor Symposia in Quantitative Biology*, *21*, 197–216.

Messerschmidt, D. M. (2012). Should I stay or should I go: Protection and maintenance of DNA methylation at imprinted genes. *Epigenetics*, *7*, 969–975.

Messerschmidt, D. M. (2016). A twist in zygotic reprogramming. *Nature Cell Biology*, *18*, 139–140.

Messerschmidt, D. M., de Vries, W., Ito, M., Solter, D., Ferguson-Smith, A., & Knowles, B. B. (2012). Trim28 is required for epigenetic stability during mouse oocyte to embryo transition. *Science*, *335*, 1499–1502.

Messerschmidt, D. M., Knowles, B. B., & Solter, D. (2014). DNA methylation dynamics during epigenetic reprogramming in the germline and preimplantation embryos. *Genes & Development*, *28*, 812–828.

Motosugi, N., Bauer, T., Polanski, Z., Solter, D., & Hiiragi, T. (2005). Polarity of the mouse embryo is established at blastocyst and is not pre-patterned. *Genes & Development*, *19*, 1081–1092.

Nakamura, T., Liu, Y. J., Nakashima, H., Umehara, H., Inoue, K., Matoba, S., et al. (2012). PGC7 binds histone H3K9me2 to protect against conversion of 5mC to 5hmC in early embryos. *Nature*, *486*, 415–419.

Nashun, B., Yukawa, M., Liu, H., Akiyama, T., & Aoki, F. (2010). Changes in the nuclear deposition of histone H2A variants during pre-implantation development in mice. *Development*, *137*, 3785–3794.

Oda, H., Okamoto, I., Murphy, N., Chu, J., Price, S. M., Shen, M. M., et al. (2009). Monomethylation of histone H4-lysine 20 is involved in chromosome structure and stability and is essential for mouse development. *Molecular and Cellular Biology*, *29*, 2278–2295.

Ooga, M., Fulka, H., Hashimoto, S., Suzuki, M. G., & Aoki, F. (2015). Analysis of chromatin structure in mouse preimplantation embryos by fluorescent recovery after photobleaching. *Epigenetics*. http://dx.doi.org/10.1080/15592294.2015.1136774.

Ooga, M., Inoue, A., Kageyama, S., Akiyama, T., Nagata, M., & Aoki, F. (2008). Changes in H3K79 methylation during preimplantation development in mice. *Biology of Reproduction*, *78*, 413–424.

Ooga, M., Suzuki, M. G., & Aoki, F. (2013). Involvement of DOT1L in the remodeling of heterochromatin configuration during early preimplantation development in mice. *Biology of Reproduction*, *89*, 145.

Oswald, J., Engemann, S., Lane, N., Mayer, W., Olek, A., Fundele, R., et al. (2000). Active demethylation of the paternal genome in the mouse zygote. *Current Biology: CB*, *10*, 475–478.

Padmanabhan, N., Jia, D., Geary-Joo, C., Wu, X., Fergusson-Smith, A. C., Fung, E., et al. (2013). Mutation in folate metabolism causes epigenetic instability and transgenerational effects on development. *Cell*, *155*, 81–93.

Peaston, A. E., Evsikov, A. V., Graber, J., de Vries, W. N., Solter, D., & Knowles, B. B. (2004). Retrotransposons regulate host genes in mouse oocytes and preimplantation embryos. *Developmental Cell*, *7*, 597–606.

Piatti, P., Lim, C. Y., Nat, R., Villunger, A., Geley, S., Shue, Y. T., et al. (2015). Embryonic stem cell differentiation requires full length Chd1. *Scientific Reports*, *5*, 8007.

Piotrowska, K., Wianny, F., Pedersen, R. A., & Zernicka-Goetz, M. (2001). Blastomeres arising from the first cleavage division have distinguishable fates in normal mouse development. *Development*, *128*, 3739–3748.

Piotrowska, K., & Zernicka-Goetz, M. (2001). Role for sperm in spatial patterning of the early mouse embryo. *Nature*, *409*, 517–521.

Piotrowska-Nitsche, K., Perea-Gomez, A., Haraguchi, S., & Zernicka-Goetz, M. (2005). Four-cell stage mouse blastomeres have different developmental properties. *Development*, *132*, 479–490.

Piotrowska-Nitsche, K., & Zernicka-Goetz, M. (2005). Spatial arrangement of individual 4-cell stage blastomeres and the order in which they are generated correlate with blastocyst pattern in the mouse embryo. *Mechanisms of Development*, *122*, 487–500.

Piras, V., Tomita, M., & Selvarajoo, K. (2014). Transcriptome-wide variability in single embryonic development cells. *Scientific Reports*, *4*, 7137.

Plusa, B., Grabarek, J. B., Piotrowska, K., Glover, D. M., & Zernicka-Goetz, M. (2002). Site of the previous meiotic division defines cleavage orientation in the mouse embryo. *Nature Cell Biology*, *4*, 811–815.

Polanski, Z., Motosugi, N., Tsurumi, C., Hiiragi, T., & Hoffmann, S. (2008). Hypomethylation of paternal DNA in the late mouse zygote is not essential for development. *The International Journal of Developmental Biology*, *52*, 295–298.

Popp, C., Dean, W., Feng, S., Cokus, S. J., Andrews, S., Pellegrini, M., et al. (2010). Genome-wide erasure of DNA methylation in mouse primordial germ cells is affected by AID deficiency. *Nature*, *463*, 1101–1105.

Pradhan, S., Bacolla, A., Wells, R. D., & Roberts, R. J. (1999). Recombinant human DNA (cytosine-5) methyltransferase. I. Expression, purification, and comparison of de novo and maintenance methylation. *The Journal of Biological Chemistry*, *274*, 33002–33010.

Probst, A. V., Okamoto, I., Casanova, M., El Marjou, F., Le Baccon, P., & Almouzni, G. (2010). A strand-specific burst in transcription of pericentric satellites is required for chromocenter formation and early mouse development. *Developmental Cell*, *19*, 625–638.

Probst, A. V., Santos, F., Reik, W., Almouzni, G., & Dean, W. (2007). Structural differences in centromeric heterochromatin are spatially reconciled on fertilisation in the mouse zygote. *Chromosoma*, *116*, 403–415.

Puschendorf, M., Terranova, R., Boutsma, E., Mao, X., Isono, K., Brykczynska, U., et al. (2008). PRC1 and Suv39h specify parental asymmetry at constitutive heterochromatin in early mouse embryos. *Nature Genetics*, *40*, 411–420.

Quenneville, S., Verde, G., Corsinotti, A., Kapopoulou, A., Jakobsson, J., Offner, S., et al. (2011). In embryonic stem cells, ZFP57/KAP1 recognize a methylated hexanucleotide to affect chromatin and DNA methylation of imprinting control regions. *Molecular Cell*, *44*, 361–372.

Radford, E. J., Ito, M., Shi, H., Corish, J. A., Yamazawa, K., Isganaitis, E., et al. (2014). In utero undernourishment perturbs the adult sperm methylome and intergenerational metabolism. *Science*, *345*, 785.

Ramsahoye, B. H., Biniszkiewicz, D., Lyko, F., Clark, V., Bird, A. P., & Jaenisch, R. (2000). Non-CpG methylation is prevalent in embryonic stem cells and may be mediated by DNA methyltransferase 3a. *Proceedings of the National Academy of Sciences of the United States of America*, *97*, 5237–5242.

Rando, O. J. (2012). Daddy issues: Paternal effects on phenotype. *Cell*, *151*, 702–708.

Rando, O. J., & Simmons, R. A. (2015). I'm eating for two: Parental dietary effects on offspring metabolism. *Cell*, *161*, 93–105.

Rangasamy, D., Berven, L., Ridgway, P., & Tremethick, D. J. (2003). Pericentric heterochromatin becomes enriched with H2A.Z during early mammalian development. *The EMBO Journal*, *22*, 1599–1607.

Rassoulzadegan, M., Grandjean, V., Gounon, P., Vincent, S., Gillot, I., & Cusin, P. (2006). RNA-mediated non-Mendelian inheritance of an epigenetic change in the mouse. *Nature*, *441*, 469–474.

Ratnam, S., Mertineit, C., Ding, F., Howell, C. Y., Clarke, H. J., Bestor, T. H., et al. (2002). Dynamics of Dnmt1 methyltransferase expression and intracellular localization during oogenesis and preimplantation development. *Developmental Biology*, *245*, 304–314.

Rowe, H. M., Jakobsson, J., Mesnard, D., Rougemont, J., Reynard, S., Aktas, T., et al. (2010). KAP1 controls endogenous retroviruses in embryonic stem cells. *Nature*, *463*, 237–240.

Ruchirawat, M., Noshari, J., & Lapeyre, J. N. (1987). Kinetic mechanisms and interaction of rat liver DNA methyltransferase with defined DNA substrates. *Molecular and Cellular Biochemistry*, *76*, 45–54.

Saitou, M., & Yamaji, M. (2012). Primordial germ cells in mice. *Cold Spring Harbor Perspectives in Biology*, *4*, a008375.

Santenard, A., Ziegler-Birling, C., Koch, M., Tora, L., Bannister, A. J., & Torres-Padilla, M. E. (2010). Heterochromatin formation in the mouse embryo requires critical residues of the histone variant H3.3. *Nature Cell Biology*, *12*, 853–862.

Santos, F., & Dean, W. (2004). Epigenetic reprogramming during early development in mammals. *Reproduction*, *127*, 643–651.

Santos, F., Hendrich, B., Reik, W., & Dean, W. (2002). Dynamic reprogramming of DNA methylation in the early mouse embryo. *Developmental Biology*, *241*(1), 172–182.

Santos, F., Peters, A. H., Otte, A. P., Reik, W., & Dean, W. (2005). Dynamic chromatin modifications characterise the first cell cycle in mouse embryos. *Developmental Biology*, *280*, 225–236.

Sarma, K., & Reinberg, D. (2005). Histone variants meet their match. *Nature Reviews. Molecular Cell Biology*, *6*, 139–149.

Sarmento, O. F., Digilio, L. C., Wang, Y., Perlin, J., Herr, J. C., Allis, C. D., et al. (2004). Dynamic alterations of specific histone modifications during early murine development. *Journal of Cell Science*, *117*, 4449–4459.

Sasaki, H., & Matsui, Y. (2008). Epigenetic events in mammalian germ-cell development: Reprogramming and beyond. *Nature Reviews. Genetics*, *2008*, 129–140.

Sawatsubashi, S., Murata, T., Lim, J., Fujiki, R., Ito, S., Suzuki, E., et al. (2010). A histone chaperone, DEK, transcriptionally coactivates a nuclear receptor. *Genes & Development*, *24*, 159–170.

Schiesser, S. S., Hackner, B. B., Pfaffeneder, T. T., Müller, M. M., Hagemeier, C. C., Truss, M. M., et al. (2012). Mechanism and stem-cell activity of 5-carboxycytosine decarboxylation determined by isotope tracing. *Angewandte Chemie, International Edition*, *51*, 6516–6520.

Schuh, M., & Ellenberg, J. (2008). A new model of asymmetric spindle positioning in mouse oocytes. *Current Biology*, *18*, 1986–1992.

Schultz, D. C., Ayyanathan, K., Negorev, D., Maul, G. G., & Rauscher, F. J. (2002). SETDB1: A novel KAP-1-associated histone H3, lysine 9-specific methyltransferase that contributes to HP1-mediated silencing of euchromatic genes by KRAB zinc-finger proteins. *Genes & Development, 16*, 919–932.

Schultz, D. C., Friedman, J. R., & Rauscher, F. J. (2001). Targeting histone deacetylase complexes via KRAB-zinc finger proteins: The PHD and bromodomains of KAP-1 form a cooperative unit that recruits a novel isoform of the Mi-2alpha subunit of NuRD. *Genes & Development, 15*, 428–443.

Seisenberger, S., Andrews, S., Krueger, F., Arand, J., Walter, J., Santos, F., et al. (2012). The dynamics of genome-wide DNA methylation reprogramming in mouse primordial germ cells. *Molecular Cell, 48*, 849–862.

Shao, G. B., Chen, J. C., Zhang, L. P., Huang, P., Lu, H. Y., Jin, J., et al. (2014). Dynamic patterns of histone H3 lysine 4 methyltransferases and demethylases during mouse preimplantation development. *In Vitro Cellular & Developmental Biology. Animal, 50*, 603–613.

Shapiro, J. A. (2005). Retrotransposons and regulatory suites. *BioEssays, 22*, 122–125.

Sharif, J., Muto, M., Takebayashi, S.-I., Suetake, I., Iwamatsu, A., Endo, T. A., et al. (2007). The SRA protein Np95 mediates epigenetic inheritance by recruiting Dnmt1 to methylated DNA. *Nature, 450*, 908–912.

Sharma, U., Conine, C. C., Shea, J. M., Boskovic, A., Derr, A. G., Bing, X. Y., et al. (2016). Biogenesis and function of tRNA fragments during sperm maturation and fertilization in mammals. *Science, 351*, 391–396.

Shea, J. M., Serra, R. W., Carone, B. R., Shulha, H. P., Kucukural, A., Ziller, M. J., et al. (2015). Genetic and epigenetic variation, but not diet, shape the sperm methylome. *Developmental Cell, 35*, 750–758.

Shen, L., Inoue, A., He, J., Liu, Y., Lu, F., & Zhang, Y. (2014). Tet3 and DNA replication mediate demethylation of both the maternal and paternal genomes in mouse zygotes. *Cell Stem Cell, 15*, 459–470.

Shi, J., Chen, Q., Li, X., Zheng, Y., Qiao, J., Tang, F., et al. (2015). Dynamic transcriptional symmetry-breaking in pre-implantation mammalian embryo development revealed by single-cell RNA-seq. *Development, 142*, 3468–3477.

Siklenka, K., Erkek, S., Godmann, M., Lambrot, R., McGraw, S., Lafleur, C., et al. (2015). Disruption of histone methylation in developing sperm impairs offspring health transgenerationally. *Science, 350*, 651.

Smallwood, S. A., Tomizawa, S.-I., Krueger, F., Ruf, N., Carli, N., Segonds-Pichon, A., et al. (2011). Dynamic CpG island methylation landscape in oocytes and preimplantation embryos. *Nature Publishing Group, 43*, 811–814.

Solter, D. (2016). Preformation versus epigenesist in early mammalian development. *Current Topics in Developmental Biology, 117*, 377–391. http://dx.doi.org/10.1016/bs.ctdb.2015.11.006.

Solter, D., Hiragi, T., Evsikov, A. V., Moyer, J., DeVries, W. N., Peaston, A. E., et al. (2004). Epigenetic mechanisms in early mammalian development. *Cold Spring Harbor Symposia in Quantitative Biology, 69*, 1–8.

Suzuki, S., Nozawa, Y., Tsukamoto, S., Kaneko, T., Manabe, I., Imai, H., et al. (2015). CHD1 acts via the Hmgpi pathway to regulate mouse early embryogenesis. *Development, 142*, 2375–2384.

Tachibana, M., Sugimoto, K., Nozaki, M., Ueda, J., Ohta, T., Ohki, M., et al. (2002). G9a histone methyltransferase plays a dominant role in euchromatic histone H3 lysine 9 methylation and is essential for early embryogenesis. *Genes & Development, 16*, 1779–1791.

Tahiliani, M., Koh, K. P., Shen, Y., Pastor, W. A., Bandukwala, H., Brudno, Y., et al. (2009). Conversion of 5-methylcytosine to 5-hydroxymethylcytosine in mammalian DNA by MLL partner TET1. *Science, 324*, 930–935.

Tanaka, M., Hennebold, J. D., Macfarlane, J., & Adashi, E. Y. (2001). A mammalian oocyte-specific linker histone gene H1oo: Homology with the genes for the oocyte-specific cleavage stage histone (cs-H1) of sea urchin and the B4/H1M histone of the frog. *Development*, *128*, 655–664.

Tang, M. C., Jacobs, S. A., Mattiske, D. M., Soh, Y. M., Graham, A. N., Tran, A., et al. (2015). Contribution of the two genes encoding histone variant h3.3 to viability and fertility in mice. *PLoS Genetics*, *11*, e1004964.

Tardat, M., Albert, M., Kunzmann, R., Liu, Z., Kaustov, L., Thierry, R., et al. (2015). Cbx2 targets PRC1 to constitutive heterochromatin in mouse zygotes in a parent-of-origin-dependent manner. *Molecular Cell*, *58*, 157–171.

Thomas, T., Dixon, M. P., Kueh, A. J., & Voss, A. K. (2008). Mof (MYST1 or KAT8) is essential for progression of embryonic development past the blastocyst stage and required for normal chromatin architecture. *Molecular and Cellular Biology*, *28*, 5093–5105.

Tohonen, V., Katayama, S., Vesterlund, L., Jouhilahti, E.-M., Sheikhi, M., Madissoon, E., et al. (2015). Novel PRD-like homeodomain transcription factors and retrotransposon elements in early human development. *Nature Communications*, *6*, 8207–8217.

Torres-Padilla, M. E., Bannister, A. J., Hurd, P. J., Kouzarides, T., & Zernicka-Goetz, M. (2006). Dynamic distribution of the replacement histone variant H3.3 in the mouse oocyte and preimplantation embryos. *The International Journal of Developmental Biology*, *50*, 455–461.

Torres-Padilla, M.-E., Parfitt, D.-E., Kouzarides, T., & Zernicka-Goetz, M. (2007). Histone arginine methylation regulates pluripotency in the early mouse embryo. *Nature*, *445*, 214–218.

Tsukada, Y.-I., Akiyama, T., & Nakayama, K. I. (2015). Maternal TET3 is dispensable for embryonic development but is required for neonatal growth. *Scientific Reports*, *5*, 15876.

Van Blerkom, J., & Bell, H. (1986). Regulation of development in the fully grown mouse oocyte: Chromosome-mediated temporal and spatial differentiation of the cytoplasm and plasma membrane. *Journal of Embryology and Experimental Morphology*, *93*, 213–238.

van der Heijden, G. W., Dieker, J. W., Derijck, A. A., Muller, S., Berden, J. H., Braat, D. D., et al. (2005). Asymmetry in histone H3 variants and lysine methylation between paternal and maternal chromatin of the early mouse zygote. *Mechanisms of Development*, *122*, 1008–1022.

VerMilyea, M. D., Maneck, M., Yoshida, N., Blochberger, I., Suzuki, E., Suzuki, T., et al. (2011). Transcriptome asymmetry within mouse zygotes but not between early embryonic sister blastomeres. *The EMBO Journal*, *30*, 1841–1851.

Wang, C., Ivanov, A., Chen, L., Fredericks, W. J., Seto, E., Rauscher, F. J., et al. (2005). MDM2 interaction with nuclear corepressor KAP1 contributes to p53 inactivation. *The EMBO Journal*, *24*, 3279–3290.

Wang, Z. F., Sirotkin, A. M., Buchold, G. M., Skoultchi, A. I., & Marzluff, W. F. (1997). The mouse histone H1 genes: Gene organization and differential regulation. *Journal of Molecular Biology*, *271*, 124–138.

Wang, L., Zhang, J., Duan, J., Gao, X., Zhu, W., Lu, X., et al. (2014). Programming and inheritance of parental DNA methylomes in mammals. *Cell*, *157*, 979–991.

Wasson, J. A., Simon, A. K., Myrick, D. A., Wolf, W., Driscoll, S., Pfaff, S. L., et al. (2016). Maternally provided LSD1/KDM1A enables the maternal-to-zygotic transition and prevents defects that manifest postnatally. *eLife*, *5*, e08848. http://dx.doi.org/10.7554/eLife.08848.

Weber, C. M., & Henikoff, S. (2014). Histone variants: Dynamic punctuation in transcription. *Genes & Development*, *28*, 672–682.

Whitelaw, E. (2015). Sperm DNA methylation: Not a vehicle for dietary reprogramming of offspring? *Developmental Cell, 21*, 668–669.

Williams, Z., Morozov, P., Mihailovic, A., Lin, C., Puvvula, P. K., Juranek, S., et al. (2015). Discovery and characterization of piRNAs in the human fetal ovary. *Cell Reports, 13*, 854–863.

Wolf, D., & Goff, S. P. (2007). Trim 28 mediates primer binding site-targeted leukemia virus in embryonic cells. *Cell, 131*, 46–57.

Wolf, D., & Goff, S. P. (2009). Embryonic stem cells use ZFP809 to silence retroviral DNAs. *Nature, 458*, 1201–1204.

Wong, L. H., McGhie, J. D., Sim, M., Anderson, M. A., Ahn, S., Hannan, R. D., et al. (2010). ATRX interacts with H3.3 in maintaining telomere structural integrity in pluripotent embryonic stem cells. *Genome Research, 20*, 351–360.

Wongtawan, T., Taylor, J. E., Lawson, K. A., Wilmut, I., & Pennings, S. (2011). Histone H4K20me3 and HP1alpha are late heterochromatin markers in development, but present in undifferentiated embryonic stem cells. *Journal of Cell Science, 124*, 1878–1890.

Worrad, D. M., Ram, P. T., & Schultz, R. M. (1994). Regulation of gene expression in the mouse oocyte and early preimplantation embryo: Developmental changes in Sp1 and TATA box-binding protein, TBP. *Development, 120*, 2347–2357.

Wossidlo, M., Arand, J., Sebastiano, V., Lepikhov, K., Boiani, M., Reinhardt, R., et al. (2010). Dynamic link of DNA demethylation, DNA strand breaks and repair in mouse zygotes. *The EMBO Journal, 29*, 1877–1888.

Wossidlo, M., Nakamura, T., Lepikhov, K., Marques, C. J., Zakhartchenko, V., Boiani, M., et al. (2011). 5-Hydroxymethylcytosine in the mammalian zygote is linked with epigenetic reprogramming. *Nature Communications, 2*, 241.

Wu, S. C., & Zhang, Y. (2010). Active DNA demethylation: Many roads lead to Rome. *Nature Reviews. Molecular Cell Biology, 11*, 607–620.

Wu, H., & Zhang, Y. (2014). Reversing DNA methylation: Mechanisms, genomics, and biological functions. *Cell, 156*, 45–68.

Xue, Z., Huang, K., Cai, C., Cai, L., Jiang, C.-Y., Feng, Y., et al. (2013). Genetic program in human and mouse early embryos revealed by single-cell RNA sequencing. *Nature, 500*, 593–597.

Yamaguchi, S., Hong, K., Liu, R., Inoue, A., Shen, L., Zhang, K., et al. (2013). Dynamics of 5-methylcytosine and 5-hydroxymethylcytosine during germ cell reprogramming. *Cell Research, 23*, 329–339.

Yang, S. M., Kim, B. J., Norwood Toro, L., & Skoultchi, A. I. (2013). H1 linker histone promotes epigenetic silencing by regulating both DNA methylation and histone H3 methylation. *Proceedings of the National Academy of Sciences of the United States of America, 110*, 1708–1713.

Yeo, S., Lee, K. K., Han, Y. M., & Kang, Y. K. (2005). Methylation changes of lysine 9 of histone H3 during preimplantation mouse development. *Molecules and Cells, 20*, 423–428.

Yoder, J. A., Soman, N. S., Verdine, G. L., & Bestor, T. H. (1997). DNA (cytosine-5)-methyltransferases in mouse cells and tissues. Studies with a mechanism-based probe. *Journal of Molecular Biology, 270*, 385–395.

Yu, X.-J., Yi, Z., Gao, Z., Qin, D., Zhai, Y., Chen, X., et al. (2014). The subcortical maternal complex controls symmetric division of mouse zygotes by regulating F-actin dynamics. *Nature Communications, 5*, 4887.

Ziegler-Birling, C., Daujat, S., Schneider, R., & Torres-Padilla, M. E. (2015). Dynamics of histone H3 acetylation in the nucleosome core during mouse pre-implantation development. *Epigenetics*. http://dx.doi.org/10.1080/15592294.2015.1103424.

Ziegler-Birling, C., Helmrich, A., Tora, L., & Torres-Padilla, M. E. (2009). Distribution of p53 binding protein 1 (53BP1) and phosphorylated H2A.X during mouse preimplantation development in the absence of DNA damage. *The International Journal of Developmental Biology*, *53*, 1003–1011.

Ziller, M. J., Müller, F., Liao, J., Zhang, Y., Gu, H., Bock, C., et al. (2011). Genomic distribution and inter-sample variation of non-CpG methylation across human cell types. *PLoS Genetics*, 7, e1002389.

Zuo, X., Sheng, J., Lau, H.-T., McDonald, C. M., Andrade, M., Cullen, D. E., et al. (2012). Zinc finger protein ZFP57 requires its co-factor to recruit DNA methyltransferases and maintains DNA methylation imprint in embryonic stem cells via its transcriptional repression domain. *The Journal of Biological Chemistry*, *287*, 2107–2118.

CHAPTER TEN

Capturing Identity and Fate Ex Vivo: Stem Cells from the Mouse Blastocyst

V. Garg*,†, S. Morgani*, A.-K. Hadjantonakis*,†,1
*Developmental Biology Program, Sloan Kettering Institute, Memorial Sloan Kettering Cancer Center, New York, NY, United States
†Biochemistry, Cell and Molecular Biology Program, Weill Cornell Graduate School of Medical Sciences, New York, NY, United States
[1]Corresponding author: e-mail address: hadj@mskcc.org

Contents

Current Topics in Developmental Biology, Volume 120
ISSN 0070-2153
http://dx.doi.org/10.1016/bs.ctdb.2016.04.007

Abstract

During mouse preimplantation development, three molecularly, morphologically, and spatially distinct lineages are formed, the embryonic epiblast, the extraembryonic primitive endoderm, and the trophectoderm. Stem cell lines representing each of these lineages have now been derived and can be indefinitely maintained and expanded in culture, providing an unlimited source of material to study the interplay of tissue-specific transcription factors and signaling pathways involved in these fundamental cell fate decisions. Here we outline our current understanding of the derivation, maintenance, and properties of these in vitro stem cell models representing the preimplantation embryonic lineages.

1. INTRODUCTION

The preimplantation mammalian embryo is a paradigm for a program of differentiation and self-organization. The totipotent zygote differentiates into the three distinct cell types of the blastocyst: the epiblast (EPI), primitive endoderm (PrE), and the trophectoderm (TE). The TE and PrE will form the extraembryonic supporting structures of the embryo, while the EPI will give rise to the embryo proper and extraembryonic mesoderm. The three blastocyst lineages arise in two sequential binary cell fate decisions: segregation of the TE from the inner cell mass (ICM), followed by the separation of the ICM into the EPI and the PrE (reviewed in Schrode et al., 2013) (Fig. 1). The zygote undergoes three rounds of cell division to form the 8-cell morula. During the next two rounds of cell division (8- to 16-cell and 16- to 32-cell stages), asymmetric cell division and cell rearrangements culminate in the segregation of TE and ICM (Johnson & Ziomek, 1981, 1983; McDole, Xiong, Iglesias, & Zheng, 2011; Watanabe, Biggins, Tannan, & Srinivas, 2014).

A combination of growth factor signaling and cell positioning drives the second cell fate decision. Until the 32-cell stage, ICM cells coexpress the lineage-associated transcription factors *Nanog* and *Gata6*, which later mark the EPI and the PrE, respectively (Chazaud, Yamanaka, Pawson, & Rossant, 2006; Guo et al., 2010; Kurimoto et al., 2006; Ohnishi et al., 2014; Plusa, Piliszek, Frankenberg, Artus, & Hadjantonakis, 2008). Thereafter, around the mid-blastocyst stage (~64 cells), the expression of these factors begins to resolve and becomes mutually exclusive, with EPI and PrE cells distributed in a "salt-and-pepper" fashion throughout the ICM (Chazaud et al., 2006). Prior to implantation into the uterus, EPI and PrE precursors sort into two distinct layers by a combination of cell migration, apoptosis, and downregulation of the PrE transcriptional program in internal ICM

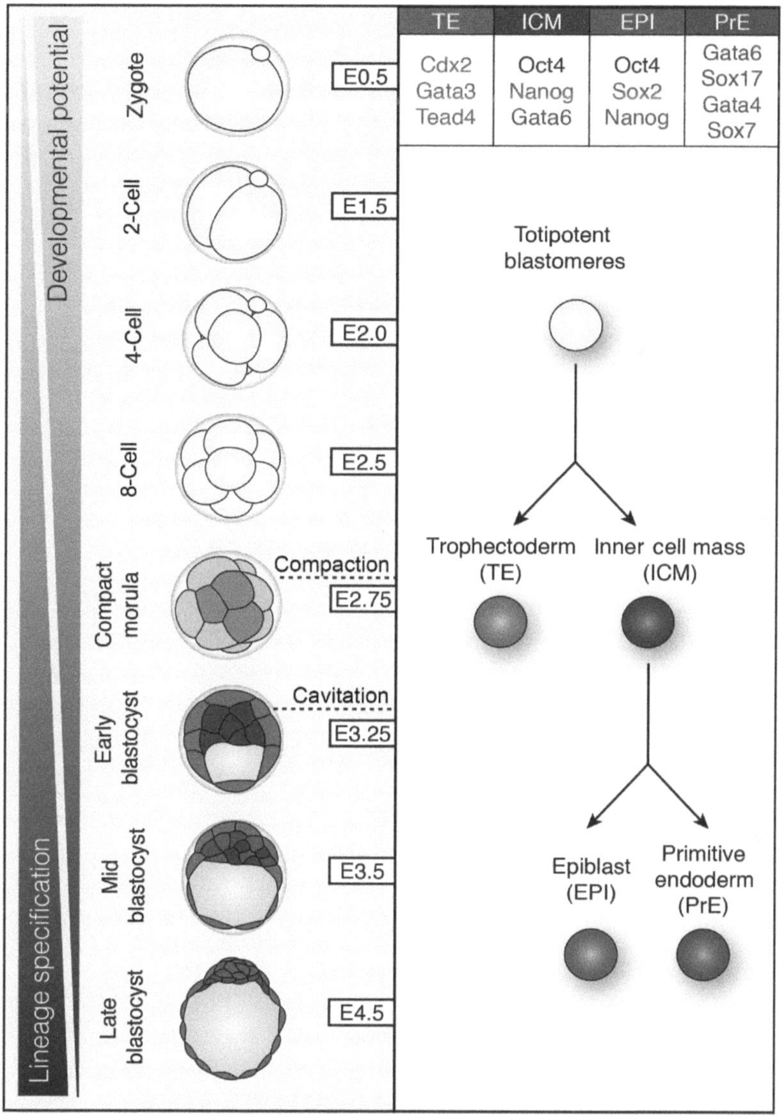

Fig. 1 Overview of mouse preimplantation development. Two cell fate decisions result in the specification of three lineages in the blastocyst prior to implantation. As lineages are specified developmental potential is restricted. Box on *top right* depicts lineage-specific transcription factors specifically expressed at each stage. *E*, embryonic day. (See the color plate.)

cells (Meilhac et al., 2009; Plusa et al., 2008; Saiz, Grabarek, Sabherwal, Papalopulu, & Plusa, 2013).

Although the preimplantation embryo is a good model to study cell fate decisions in vivo, at the late blastocyst stage the embryo comprises no more than 150 cells. However, in vitro stem cell counterparts of these early embryonic lineages can be derived and propagated indefinitely (Fig. 2). There are two kinds of stem cells representative of the EPI, embryonic stem (ES) cells, a naïve state of pluripotency derived from the preimplantation EPI (Evans & Kaufman, 1981; Martin, 1981), and epiblast stem cells (EpiSCs), a distinct primed state of pluripotency resembling the postimplantation EPI (Brons et al., 2007; Tesar et al., 2007). Stem cell lines representing the two extraembryonic lineages have also been isolated. Extraembryonic endoderm (XEN) stem cells resemble the PrE (Kunath et al., 2005), while trophoblast stem (TS) cells represent the TE (Tanaka, Kunath, Hadjantonakis, Nagy, & Rossant, 1998). As well as deriving these cell lines from the embryo, it is possible to generate these cell types by genetic manipulation or using specific culture conditions.

2. TS CELLS REPRESENT THE TE LINEAGE

The TE is an epithelial layer of cells encapsulating the ICM and the blastocoel. TE cells in contact with the ICM (polar TE) proliferate and differentiate into the extraembryonic ectoderm (ExE), the ectoplacental cone (EPC), and secondary giant cells, while the rest of the TE layer (mural TE) differentiates into nonproliferating primary trophoblast giant cells (TGCs), which facilitate implantation into the uterus (Gardner & Johnson, 1973; Gardner, Papaioannou, & Barton, 1973; Papaioannou, 1982). For a period of time, cells of the ExE self-renew, while also differentiating into other trophoblast derivatives. Additionally, the ExE relays paracrine signals to the EPI, such as BMP, necessary for primordial germ cell (PGC) formation (Ying, Liu, Marble, Lawson, & Zhao, 2000; Ying & Zhao, 2001), and activators of the Nodal pathway (Mesnard, Donnison, Fuerer, Pfeffer, & Constam, 2011). Later in development, it gives rise to the fetal portion of the placenta (Gardner, 1983), a supporting tissue of the embryo. TS cells are self-renewing cell lines representing the TE lineage.

2.1 Derivation and Culture

TS cells can be derived from mouse blastocysts or the ExE of postimplantation embryos (E6.5–E8.5) (Tanaka et al., 1998; Uy, Downs, & Gardner, 2002). They are similar to the TE lineage in terms of their gene expression profile, morphology, and developmental potential (Fig. 2). TS cells express markers of the early diploid ExE, including *Esrrb*, *Cdx2*, *Fgfr2*, and *Eomes*

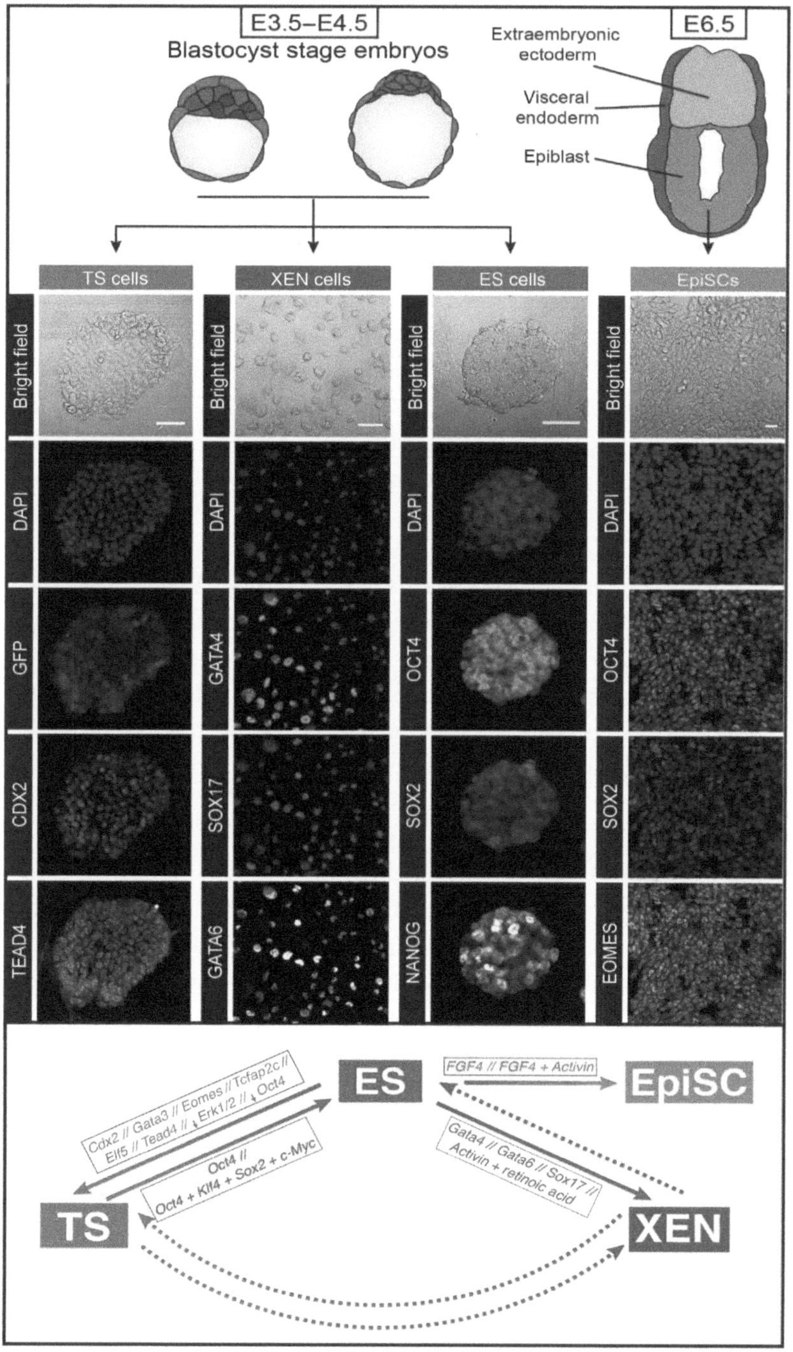

Fig. 2 See legend on next page.

(Beck, Erler, Russell, & James, 1995; Ciruna & Rossant, 1999; Luo et al., 1997; Orr-Urtreger et al., 1993). They have a tight epithelial-like morphology, which is in contrast to the enlarged and flattened morphology of differentiated TGCs (Tanaka et al., 1998).

In vivo, FGF4 secreted by EPI cells stimulates proliferation and maintenance of stem cell character in polar TE cells (Feldman, Poueymirou, Papaioannou, DeChiara, & Goldfarb, 1995; Gardner & Johnson, 1972; Haffner-Krausz, Gorivodsky, Chen, & Lonai, 1999; Niswander & Martin, 1992; Orr-Urtreger et al., 1993; Rossant & Ofer, 1977). Derivation protocols for TS cells routinely include FGF4, heparin (which prevents proteolytic cleavage of FGF and aids in signal transduction via FGF dimerization; Gospodarowicz & Cheng, 1986; Ornitz et al., 1992; Saksela, Moscatelli, Sommer, & Rifkin, 1988), and mouse embryonic fibroblast (MEF) feeders that provide key signaling cues for TS cell maintenance and proliferation (Erlebacher, Price, & Glimcher, 2004; Tanaka et al., 1998). Upon removal of FGF4 and MEFs, TS cell proliferation decreases, chromosome number increases, giant cell morphology is acquired, and giant cell and EPC markers are expressed (Faria, Ogren, Talamantes, Linzer, & Soares, 1991; Guillemot, Nagy, Auerbach, Rossant, & Joyner, 1994).

Studies of the signaling pathways involved in TE maintenance have uncovered a requirement for transforming growth factor β (TGFβ), as well as Activin and Nodal for promoting proliferation and blocking differentiation (Erlebacher et al., 2004; Guzman-Ayala, Ben-Haim, Beck, & Constam, 2004; Ma et al., 2001). While TGFβ is required for TS cell maintenance, its cytostatic effects via *c-myc* downregulation (Alexandrow & Moses, 1995) is prevented by FGF4 (Erlebacher et al., 2004). Thus, the combination of TGFβ and FGF4 maintains TS cell proliferation and self-renewal. Since TGFβ and/or Activin are known components of MEF-conditioned

Fig. 2 Lineage-specific stem cell types derived from preimplantation and postimplantation mouse embryos. TS cells, XEN cells, and ES cells represent the TE, PrE, and EPI of the blastocyst, respectively, and EpiSCs represent the postimplantation epiblast. Each stem cell type exhibits indefinite in vitro self-renewal, as well as the morphology, gene expression profile, and developmental potential of the cognate in vivo lineages they represent, and presumably are derived from. Immunofluorescence and bright field images depict morphology and common markers for each stem cell type. TS cells used in the analysis express GFP ubiquitously. (*Lower panel*) Interconversion of these stem cell types is possible by altering culture conditions and/or genetic manipulations. *Dashed lines* represent conversions that have yet to be demonstrated. (See the color plate.)

medium (MEF-CM) and Nodal (a TGFβ ligand) is expressed by EPI cells, these factors, along with FGF4, can effectively replace the use of undefined CM and feeder cells for TS cell culture (Erlebacher et al., 2004; Kubaczka et al., 2014; Schier, 2003).

A defined culture medium, "TX medium," comprised of standard TS cell medium supplemented with FGF4 and TGFβ, was recently formulated (Kubaczka et al., 2014). While culture in TX medium leads to a slower growth and higher apoptosis rates compared to standard TS medium, it reduces morphological differences between different TS cell lines and increases experimental consistency. Notably, maintenance of TS cells in TX medium requires Matrigel, an ill-defined, and growth factor-containing matrix (Hughes, Postovit, & Lajoie, 2010; Kleinman & Martin, 2005). Nevertheless, defined medium formulations will be useful to explore the mechanical and biochemical roles of ECM proteins in TS cell maintenance.

Although TS cells appear morphologically, transcriptionally, and epigenetically similar to in vivo TE cells (Kubaczka et al., 2014; Tanaka et al., 1998), the definitive test of their identity is their developmental potential in the embryo. TS cells injected into blastocysts are able to contribute to the TE and its derivatives, even after extended in vitro culture (Tanaka et al., 1998). However, the degree of contribution to different TE derivatives is variable, demonstrating that their developmental potential may be somewhat reduced or else limited by the particular assay used (Tanaka et al., 1998). Supporting this observation, TS cells demonstrate epigenetic differences compared to in vivo TE cells. TS cells have higher levels of 5-methylcytosine (5mC) (Senner, Krueger, Oxley, Andrews, & Hemberger, 2012) and reduced levels of 5-hydroxymethylation (5hmC) (Tahiliani et al., 2009) compared to the E7.5 TE, suggesting that TS cells may be epigenetically more restricted in their developmental potential.

2.2 Conversion to a TS Cell-Like State

The key transcriptional regulators of TE identity are now known (Latos & Hemberger, 2014). Many of these factors are required both for the establishment and maintenance of the TE and for the derivation, self-renewal, and proliferation of TS cells. A combination of TE-specific transcription factors, *Gata3*, *Eomes*, and *Tcfap2c* (GET) (Benchetrit et al., 2015) with *Ets2* (Kubaczka et al., 2015), can reprogram MEFs and fibroblasts into induced

TS cells. While the role of *Ets2* in the reprogramming process appears dispensable, it is required for TS cell proliferation and self-renewal (Odiatis & Georgiades, 2010; Wen et al., 2007).

2.3 Differentiation of TS Cells

TS cells differentiate into TE derivatives when deprived of FGF4 and Activin/Nodal in culture (Erlebacher et al., 2004; Tanaka et al., 1998). They primarily differentiate into TGCs (~50%; Hughes et al., 2004). However, specific fates can be achieved by genetic manipulation of transcription factors responsible for formation of specific cell types in vivo. The basic helix-loop-helix transcription factor (bHLH) *Mash2* is expressed in the EPC and down-regulated upon EPC differentiation into TGCs in vivo (Guillemot et al., 1994; Scott, Anson-Cartwright, Riley, Reda, & Cross, 2000). Embryos lacking *Mash2* produce excess TGCs, while its overexpression in TS cells promotes transient FGF4-independent proliferation before differentiation (Guillemot et al., 1994; Hughes et al., 2004). This suggests that *Mash2* suppresses TGC formation. In contrast, *Hand1*, another bHLH transcription factor and putative *Mash2* antagonist, promotes TGC differentiation, both in vivo and when ectopically expressed in TS cells (Cross et al., 1995; Hughes et al., 2004; Riley, Anson-Cartwright, & Cross, 1998; Scott et al., 2000). Like *Hand1*, treatment of TS cells with retinoic acid (RA) or overexpression of its downstream target *Stra13* can induce TGC differentiation (Hughes et al., 2004; Yan et al., 2001). Thus, the coordination of TGC-promoting and -antagonizing factors can be used to direct differentiation in vitro.

TS cell differentiation can also be enriched for syncytiotrophoblast (SynT) cells—a subtype of the labyrinth trophoblast within the innermost layer of the mouse placenta (Simmons & Cross, 2005). Overexpressing *Gcm1*, for example, promotes cell cycle exit and a restriction toward SynT fate, while blocking TGC formation (Hughes et al., 2004). Similarly, activation of PPARγ (Parast et al., 2009) or hypoxia-inducible factor (HIF) can mediate SynT vs TGC differentiation. The ECM composition can also direct TS cell differentiation via HIF toward SynTs or TGCs in a MAP2K1/2-dependent manner (Choi et al., 2013). TS cells maintained on CELLstart™ ECM substrate (Invitrogen), primarily consisting of fibronectin (Hughes, Radan, Betts, Postovit, & Lajoie, 2011), differentiate toward SynT, which can be reversed by activating MAP2K1.

2.4 Drawing Parallels

Derivation of TS-like (TSL) cells from species other than mouse has been problematic. Cells derived from rhesus monkey blastocysts express TE markers and show invasive behavior characteristic of TE cell types (Vandevoort, Thirkill, & Douglas, 2007). However, these cells were maintained without growth factors and progressively differentiated, suggesting that further optimization of culture conditions may be required to obtain stable self-renewing lines. Although it has not been possible to derive TS cells from human blastocysts, a recent study reported the derivation of cell lines from single blastomeres of early, preblastocyst human embryos (Zdravkovic, Nazor, Larocque, et al., 2015). When these were cultured as embryoid bodies (EBs), an outer layer of *Cdx2*-positive cells formed which when isolated by manual dissection were capable of self-renewal as TSL cells.

Rabbit ES cells treated with BMP4 can also generate TSL cells (Tan et al., 2011) that self-renew and contribute to placental derivatives in chimeras. Similar to mouse TS cells, these cells are dependent on FGF and TGFβ signaling. However, treating rhesus monkey ES cells with BMP4 produced PrE-like cells, indicating that the interpretation of signaling pathways may not be conserved across species (Kobayashi, Takada, Takahashi, Noda, & Torii, 2008). TSL cells have been derived from pig fibroblasts by genetic manipulation (Ezashi, Matsuyama, Telugu, & Roberts, 2011). Surprisingly, these induced trophoblast cells arise during reprogramming of fibroblasts using the ES cell-associated transcription factors *Oct4*, *Sox2*, *Klf4*, and *c-Myc*. They display stem-like characteristics, express TE markers, and can self-renew indefinitely. However, their developmental potential has not been tested in chimeras.

3. XEN CELLS REPRESENT THE PrE LINEAGE

XEN cells are the in vitro counterpart of the PrE lineage which in vivo gives rise to the visceral endoderm (VE) and parietal endoderm (PE) (Enders, Given, & Schlafke, 1978; Gardner, 1982; Kunath et al., 2005) (Fig. 2). Although XEN cells express markers of both the PE and VE, upon introduction into blastocysts, they preferentially contribute to the PE (Fujikura et al., 2002; Kunath et al., 2005; Shimosato, Shiki, & Niwa, 2007). Indeed, XEN cell morphology more closely resembles that of PE cells in culture (Fowler, Mitrangas, & Dziadek, 1990). XEN cells are heterogeneous exhibiting two distinct morphologies: rounded and refractile, similar to PE, or epithelial like, similar to VE. When plated at low densities, XEN cells are dispersed and form few cell–cell contacts, but at higher densities acquire an epithelial-like morphology (Kunath et al., 2005; Shimosato et al., 2007).

3.1 Derivation and Culture

Mouse XEN cells can be derived and cultured using various conditions (Niakan, Schrode, Cho, & Hadjantonakis, 2013). These include derivation from blastocysts plated ex vivo and transdifferentiation of ES cells to XEN-like cells by transcription factor overexpression or growth factor stimulation (Cho et al., 2012; Fujikura et al., 2002). The first derivation and characterization of XEN cells used culture conditions similar to those for TS and ES cells (Evans & Kaufman, 1981; Kunath et al., 2005). ICMs or whole blastocysts were cultured on a MEF feeder layer in medium supplemented with FGF4 and leukemia inhibitory factor (LIF), or MEF-CM. In the embryo, FGF4 secreted from EPI cells directs the specification and maturation of the PrE lineage (Kang, Piliszek, Artus, & Hadjantonakis, 2013; Messerschmidt & Kemler, 2010; Yamanaka, Lanner, & Rossant, 2010). Although exogenous FGF4 is not always required for XEN cell self-renewal (Cho et al., 2012; Kunath et al., 2005), it must be provided in some capacity, either endogenously or exogenously for derivation. This is demonstrated by the fact that XEN cells can only be derived from *Fgf4*-null ES cells or embryos in the presence of exogenous FGF4 (Cho et al., 2012; Kang et al., 2013).

3.2 Conversion to a XEN-Like State

XEN-like cells can also be established by growth factor addition to ES cell cultures (termed "cXEN" cells; see Section 6) (Cho et al., 2012) or ectopic expression of transcription factors in ES cells (referred to as iXEN cells) (Fig. 2). Overexpression of the GATA factors *Gata4* and *Gata6* efficiently induces a XEN-like cell fate in ES cells (Fujikura et al., 2002; Shimosato et al., 2007). These cells resemble blastocyst-derived XEN cells, expressing PrE markers (Fujikura et al., 2002; Murray & Edgar, 2001), and early VE genes (Abe et al., 1996). This change of state is accompanied by downregulation of pluripotency-associated markers. Transient induction of exogenous GATA transgenes leads to upregulation of endogenous *Gata6* and *Gata4*, which implies cross- and self-activation between these two transcription factors. Furthermore, endogenous *Gata6* expression activates downstream PrE markers including *Gata4*, *Sox17*, and *Sox7* (Niakan et al., 2010; Shimosato et al., 2007).

Like *Gata4* and *Gata6*, the SRY-related HMG-box transcription factor *Sox17* can induce a XEN-like fate when transiently expressed in ES cells. During this process, SOX17 binds to and activates the XEN-related genes, *Gata6* and *Gata4*, thus activating the XEN gene regulatory network (GRN) (Niakan et al., 2010). Consistently, XEN cells cannot be derived from *Sox17*$^{-/-}$ embryos (Artus, Piliszek, & Hadjantonakis, 2011; Kanai-Azuma et al., 2002; Niakan et al., 2010; Shimoda et al., 2007) and depletion of *Sox7* and *Sox17* by RNAi is known to block XEN cell self-renewal (Lim et al., 2008), suggesting that *Sox17* is required for its maintenance. SOX17 also partially represses the pluripotent state by binding and repressing a subset of targets of the pluripotency-associated factors *Nanog*, *Oct4*, and *Sox2* (Niakan et al., 2010). Detailed analysis of *Sox17*-induced iXEN cells revealed that ES cells transition through several intermediate states, including a VE-like state, before achieving a stable XEN state (McDonald, Biechele, Rossant, & Stanford, 2014).

3.3 Plasticity Within XEN Cell Populations

Although XEN cells show similar signaling dependency and transcriptional profiles to the in vivo PrE and its derivatives, they are biased toward a PE fate in chimeras (Kunath et al., 2005; McDonald et al., 2014). Nevertheless, they retain a degree of plasticity and can contribute to both PE and VE under specific conditions. XEN cells cultured in BMP4-supplemented medium differentiate into a subtype of the VE—the extraembryonic VE (exVE) (Artus et al., 2012; Paca et al., 2012). PE cells freshly isolated from Reichert's membrane of E7.5 embryos can also acquire a similar state when treated with BMP4 (Paca et al., 2012). In addition, treatment of XEN cells with Nodal or Cripto activates SMAD2 causing the upregulation of VE and anterior visceral endoderm (AVE)-specific genes (Kruithof-de Julio et al., 2011). Under these conditions, XEN cells efficiently contribute to both embryonic and exVE and the AVE in chimeras. These studies demonstrate the plasticity of XEN cells and highlight the critical influence of culture conditions on cell identity. In addition, detailed gene expression analyses comparing XEN cells to VE-like and PE-like cells revealed that XEN cells promiscuously express a combination of distinct ExEn markers, suggesting that, although are a good model for the PrE, they may not be equivalent (Brown et al., 2010).

3.4 Drawing Parallels

XEN-like cells can be derived from rat blastocysts using mouse ES (mES) cell culture conditions. These cells express ES cell markers, like *Oct4* and *SSEA1*, as well as the PrE markers *Gata6* and *Gata4*, and the TE marker *Eomes* (Galat et al., 2009). In chimeras, these cells can contribute to PE, VE, and TGCs but not the EPI. Thus, unlike mouse XEN cells, rat XEN-like cells may represent a metastable, early extraembryonic precursor that can differentiate into multiple extraembryonic lineages (Debeb et al., 2009). However, BMP pathway stimulation in mouse XEN cells causes differentiation into VE cells (Artus et al., 2012; Paca et al., 2012), as does activation of the BMP and Wnt pathways in rat XEN cells (Chuykin, Schulz, Guan, & Bader, 2013). Furthermore, mouse and rat XEN cells can be directed to PE by activating the cAMP pathway with recombinant parathyroid hormone-related peptide (Chuykin et al., 2013; van de Stolpe et al., 1993; Verheijen et al., 1999). Hence, while rat and mouse XEN cells differ in terms of developmental potential, they exploit similar mechanisms during differentiation into VE and PE, underscoring similarities across rodents and possibly mammals in general.

4. NAÏVE ES CELLS REPRESENT THE EPI LINEAGE OF THE BLASTOCYST

The pluripotent EPI gives rise to the embryo proper including all three germ layers—ectoderm, endoderm, mesoderm, and the germ cells. In addition, the EPI contributes to some extraembryonic tissues, including the yolk sac mesoderm. ES cells are self-renewing cell lines derived from the embryo, represent the preimplantation EPI, and retain pluripotency even after prolonged culture in vitro (Beddington & Robertson, 1989; Boroviak, Loos, Bertone, Smith, & Nichols, 2014; Brook & Gardner, 1997; Evans & Kaufman, 1981; Martin, 1981) (Fig. 2). They produce teratocarcinomas consisting of all three germ layers, form multidifferentiated EBs, and contribute to the germ line in chimeric mice (Bradley, Evans, Kaufman, & Robertson, 1984). Although it is possible to derive ES cells from various stages of preimplantation development (Tesar, 2005), they appear to transit through a blastocyst-like stage during derivation (Boroviak et al., 2014).

4.1 Derivation and Culture

ES cells can be derived and maintained using a number of different conditions. The original conditions, using fibroblast feeder cells and serum-supplemented medium (Evans & Kaufman, 1981; Martin, 1981), were inefficient and highly dependent on genetic background. The search for factors within embryonal carcinoma cell CM (used in Martin, 1981) that support ES cell maintenance and self-renewal led to the identification of LIF, a signaling molecule that functions via a heteromeric receptor complex (LIFR and GP130) to activate the canonical Janus kinase/signal transducers and activators of transcription (Jak/Stat) pathway (Davis et al., 1993; Murakami et al., 1993; Narazaki et al., 1994; Smith et al., 1988). It has been shown that activation of STAT3 downstream of LIF is necessary and sufficient to maintain pluripotency, possibly via one of its target genes *c-Myc* (Cartwright et al., 2005; Matsuda et al., 1999; Niwa, Burdon, Chambers, & Smith, 1998). This is consistent with the fact that STAT3 inactivation or *c-Myc* depletion leads to differentiation of ES cells, while sustained expression of *c-Myc* promotes self-renewal in the absence of LIF (Cartwright et al., 2005).

In addition to activating STAT3, LIF stimulates the MAPK/ERK pathway, known to promote differentiation (Burdon, Stracey, Chambers, Nichols, & Smith, 1999; Kunath et al., 2007). Moreover, the canonical Wnt pathway has been shown to maintain self-renewal (Ogawa, Nishinakamura, Iwamatsu, Shimosato, & Niwa, 2006; Sato, Meijer, Skaltsounis, Greengard, & Brivanlou, 2004). Therefore, a two inhibitor (2i) cocktail was developed consisting of a MEK (upstream of ERK) inhibitor and a GSK-3 (a Wnt antagonist) inhibitor (Buehr et al., 2008; Ying et al., 2008). Using 2i alone promotes proliferation and self-renewal of ES cells in the absence of serum or feeder cells. This cocktail has also facilitated the derivation of ES cell lines from nonpermissive strains, like NOD (Hanna et al., 2009; Nichols, Jones, et al., 2009), CBA (Ying et al., 2008), and DBA (Reinholdt et al., 2012) and is sufficient to revert certain mouse EpiSC lines (see Section 5) (Bernemann et al., 2011) and rat EpiSCs (Buehr et al., 2008) to an ES cell state spontaneously in culture.

4.2 The Pluripotency Network

Since ES cells were initially derived, much progress has been made to identify genes involved in the maintenance of the self-renewing state. Some of these factors are major components of the GRN that regulates the EPI lineage in vivo. *Oct4*, *Nanog*, *Sox2*, *Klf4*, and *c-Myc* are important for ES cell self-renewal and can induce a self-renewing state from somatic cells (induced pluripotent stem, iPS cells, see Section 4.3) (Kaji et al., 2009; Takahashi & Yamanaka, 2006; Wernig, Meissner, Foreman, et al., 2007). This key network of transcription factors is thought to be critical in maintaining an embryonic over an extraembryonic fate in both the embryo and ES cell cultures. These factors interact synergistically with one another and have many overlapping targets, including their own promoter regions, which often need more than one of these factors for activation (Chen, Xu, Yuan, et al., 2008; Kim, Chu, Shen, Wang, & Orkin, 2008; Nakatake, Fukui, Iwamatsu, et al., 2006; Rodda, Chew, Lim, et al., 2005; Tomioka, Nishimoto, Miyagi, et al., 2002).

A role for many of these factors in ES cells was suggested by their preimplantation phenotypes in mutant embryos. *Oct4*-null embryos develop into normal blastocysts. However, when implantation is delayed, the blastocysts collapse and lack an ICM population (Frum et al., 2013; Le Bin et al., 2014; Nichols et al., 1998). During ES cell derivation, *Oct4*-null embryos give rise to only TE cells, suggesting that it is required to maintain pluripotency (Nichols et al., 1998). In addition, the precise level of *Oct4* is important for ES cell maintenance—increased levels promote endoderm and mesoderm differentiation while decreased levels promote TE differentiation (Niwa, Miyazaki, & Smith, 2000). Similarly, *Sox2*-null embryos show a normal ICM at the early blastocyst stage, which gradually differentiates into extraembryonic tissues, resulting in periimplantation lethality (Avilion et al., 2003). In ES cells expressing a dominant-negative version of *Sox2*, pluripotency is lost and cells differentiate toward a TE fate (Li et al., 2007). *Sox2* levels are also important in ES cells wherein a twofold increase results in neuroectoderm, mesoderm, and TE differentiation (Kopp, Ormsbee, Desler, & Rizzino, 2008). OCT4 and SOX2 also form a complex on the *Fgf4* enhancer to induce gene expression (Ambrosetti, Basilico, & Dailey, 1997; Dailey, Yuan, & Basilico, 1994; Yuan, Corbi, Basilico, & Dailey, 1995). *Fgf4* expression is a functional characteristic of EPI cells in vivo as they signal to neighboring PrE cells, expressing *Fgfr2*, to drive the segregation of the two populations in the ICM (Guo et al., 2010; Kang et al., 2013; Krawchuk, Honma-Yamanaka, Anani, & Yamanaka, 2013; Nichols, Silva, Roode, & Smith, 2009; Ohnishi et al., 2014; Yamanaka et al., 2010).

The variant homeodomain-containing transcription factor NANOG is another key factor in this GRN. *Nanog*-null embryos fail to implant and at periimplantation stages consist only of extraembryonic tissue without an EPI (Mitsui et al., 2003). In *Nanog*-null ES cells, growth is retarded and cells display morphologies reminiscent of XEN (Mitsui et al., 2003). Overexpressing *Nanog* facilitates ES cell self-renewal in the absence of LIF, BMP, or 2i (Chambers et al., 2003; Ying, Nichols, Chambers, &

Smith, 2003; Ying et al., 2008). Interestingly, *Nanog* deletion does not affect self-renewal, but cells are more prone to differentiation, indicating that NANOG may prevent the exit of cells from a pluripotent state and into a differentiation program (Chambers et al., 2007). Additionally, other "secondary" pluripotency markers like ESRRβ, TBX3, and KLF4 promote self-renewal collectively but are not required individually (Nichols & Smith, 2012). ESRRβ, for example, has been used for reprogramming along with OCT4 and SOX2 but is only required in the placenta of the developing embryo (Feng et al., 2009; Luo et al., 1997). KLF4 is a direct target of STAT3 downstream of LIF signaling, and its overexpression causes ES cells to be partially independent of LIF for self-renewal (Hall et al., 2009; Li et al., 2005; Niwa, Ogawa, Shimosato, & Adachi, 2009). Also, TBX3 overexpression enhances self-renewal. It activates *Nanog* downstream of the PI3K/Akt pathway; however, it is not required for normal blastocyst and ICM development or for ES cell derivation (Davenport, Jerome-Majewska, & Papaioannou, 2003; Niwa et al., 2009; Waghray, Saiz, Jayaprakash, et al., 2015). These findings imply that OCT4 and SOX2 form a vital core of the GRN regulating pluripotency and self-renewal in ES cells, while the secondary factors are accessories to this function and possibly have overlapping roles and are required in different combinations with each other and OCT4 and SOX2 (Nichols & Smith, 2012).

Although these transcription factors have all been implicated in the maintenance of an ES cell state, they do not easily fit into a rigid framework with a singular ES cell-associated role. In development, these factors display very different expression patterns often associated with multiple lineages. *Nanog* is expressed in the EPI, *Oct4* in both EPI and PrE and *Sox2* in both EPI and TE. Additionally, *Nanog* and *Oct4* mutant embryos show defects in both embryonic and extraembryonic tissues (Frum et al., 2013; Messerschmidt & Kemler, 2010; Silva et al., 2009). This indicates that they contribute to the development of other tissues and are likely to influence the ES cell state via distinct mechanisms. Hence, in contrast to the well-established pluripotency network model, it has been proposed that pluripotency results from the antagonistic effect of multiple lineage-specific transcription factors, whose contradictory effects cancel each other out (Loh & Lim, 2011).

4.3 Conversion to a Pluripotent State

A clear demonstration of the potential of OCT4 and SOX2 to establish the ES cell GRN and maintain pluripotency comes from their ability to reprogram terminally differentiated somatic cells into self-renewing iPS cells along with KLF4 and c-MYC (Takahashi & Yamanaka, 2006). The same factors have also been used to reprogram human-derived terminally differentiated cells into iPS cells, thus displaying a conserved role in the pluripotency network across species (Park et al., 2008; Takahashi et al., 2007; Yu et al., 2007). Interestingly, Yu and colleagues were able to induce pluripotency by exchanging KLF4 and c-MYC for NANOG and LIN28, highlighting the core requirement for OCT4 and SOX2 in establishing pluripotency (Yu et al., 2007). It was also recently shown that lineage-specific markers such as *Gata6* are able to reprogram somatic cells to an ES cell fate (Shu et al., 2015).

4.4 Heterogeneity and Plasticity of ES Cells

In an effort to trace the developmental potential of cultured ES cells, a landmark study by Beddington and Robertson found that, although ES cells predominantly contributed to somatic lineages, they also exhibited a rare contribution to extraembryonic lineages like the TE, PE, and VE. Thus, ES cells might resemble an early ICM cell type that retains properties of all three lineages of the blastocyst (Beddington & Robertson, 1989). While this mimics the plasticity of the in vivo ICM cells, it may also indicate an inherent heterogeneity within ES cell cultures.

Recent single-cell analyses of ES cells have revealed that a number of transcription factors and signaling molecules are heterogeneously expressed, marking distinct subpopulations (Canham, Sharov, Ko, & Brickman, 2010; Chambers et al., 2007; Dietrich & Hiiragi, 2007; Hayashi, Lopes, Tang, & Surani, 2008; Kobayashi et al., 2009; Morgani et al., 2013; Singh, Hamazaki, Hankowski, & Terada, 2007; Toyooka, Shimosato, Murakami, Takahashi, & Niwa, 2008). In addition to transcriptional differences, these subpopulations exhibit distinct functional properties. Canham and colleagues identified a PrE-primed ES cell subpopulation coexpressing endoderm markers including *Hhex*, *Gata4*, *Gata6*, and *Sox7*, alongside pluripotency markers (Canham et al., 2010; Morgani et al., 2013). Genes specific to the 2-cell stage embryo, like *Zscan4*, a telomere maintenance gene (Falco et al., 2007), and the murine endogenous retrovirus, *MuERV-L* (Macfarlan et al., 2012), have also been detected in ES cell subpopulations. These ES cells are transcriptionally and functionally primed toward extraembryonic fates while retaining the capacity to interconvert with a pluripotent state. Therefore, it has been suggested that a small fraction of ES cells may exhibit features of the early totipotent cells of the embryo, coexpressing specific markers of all three lineages of the preimplantation embryo.

From an in vivo perspective, the heterogeneity of ES cells in culture may reflect heterogeneity within the ICM wherein cells oscillate between a naïve pluripotent state that is self-renewing, and an alternative pluripotent state that is primed for differentiation. It has been hypothesized that these primed cells are cells that are highly responsive to external signaling cues from the environment and are thus easily pushed to differentiate when presented with the appropriate signal (Canham et al., 2010). However, during ES cell derivation, certain genes that are homogeneously expressed within the ICM become heterogeneously expressed and vice versa (Tang et al., 2010). Surely, the methylation state and heterochromatin structure of the EPI cells are dynamic and undergo remodeling during the derivation process, leaving the heterochromatin structure of ES cells to be relatively fixed (Ueda et al., 2014). Additionally, while in vitro these subpopulations have been shown to dynamically interconvert, this may not be the case in vivo (Xenopoulos, Kang, Puliafito, Di Talia, & Hadjantonakis, 2015). Live imaging of a transcriptional reporter of *Nanog* did not show heterogeneity of expression within the blastocyst, possibly because the developmental timescale does not allow it. This highlights the importance of clonal and single-cell analyses to accurately characterize ES cell states and to determine whether they represent in vivo populations, different developmental stages, or states "trapped" ex vivo.

4.5 Drawing Parallels

Unlike the derivation of murine ES cells, it had been problematic to obtain ES cells from other mammals. Rabbit ES cells have been derived but cannot contribute to the germ line (Fang et al., 2006; Graves & Moreadith, 1993; Honda, Hirose, & Ogura, 2009; Intawicha et al., 2009). Indeed, these cells more closely resemble human ES (hES) cells and mouse EpiSCs (see Section 5) that lack the "naïve" characteristic of mES cells (Brons et al., 2007; Tesar et al., 2007; Thomson et al., 1998; Vallier, Alexander, & Pedersen, 2005). Similarly, attempts to derive ES cells from rat blastocysts generated unstable lines that underwent extensive spontaneous differentiation and were also unable to contribute to the germline (Buehr et al., 2003; Ouhibi et al., 1995; Vassilieva, Guan, Pich, & Wobus, 2000). However, using 2i conditions of mES cell culture, stable, and germline competent rat ES cell lines (rES cell) can be derived, suggesting the conservation of pluripotency maintenance pathways between the two rodent species (Buehr et al., 2008; Chen, Blair, & Smith, 2013; Li et al., 2008).

Although several nonhuman primate (NHP) ES cell lines have been established, they resemble hES cells more closely than mES cells (Sasaki et al., 2005; Suemori et al., 2001; Thomson et al., 1995, 1996). Generating chimeric monkeys using NHP ES cell lines to test their pluripotent potential has only recently been successful using naïve NHP pluripotent stem cell culture with both bFGF and 2i/LIF (Chen et al., 2015; Fang et al., 2014). This highlights the similarities and differences in culture conditions for maintaining a naïve state between mouse, human, and monkey. Like mES cells, NHP ES cells also require LIF/STAT3 signaling for the naïve pluripotency network. However, unlike mES cells, but similar to hES cells, they require bFGF signaling for self-renewal (Fang et al., 2014).

Until recently, there was no human equivalent to naïve mES cells. hES cells derived using standard protocols are generally considered to be more similar to the later, primed pluripotent EpiSC state (see Section 5) (Brons et al., 2007; Tesar et al., 2007; Yan et al., 2013), and 2i alone is not sufficient to maintain self-renewal. Several studies now report capturing this earlier state of human development in vitro (Chan et al., 2013; Gafni et al., 2013; Hanna et al., 2010; Takashima et al., 2014; Theunissen et al., 2014; Ware et al., 2014). However, these protocols use various cocktails of small-molecule inhibitors and it is not clear whether these conditions or cells are relevant to the human embryo. Recent data suggest that, although these conditions promote a state closer to the human preimplantation EPI, there are still significant differences. For example, *Esrrb* and *Klf2*, mouse pluripotency genes that are not associated with the human EPI, were induced under these conditions (Blakeley et al., 2015). Arguably, a naïve state similar to that found in the mouse may not occur or else be transient in human development.

5. EpiSCs REPRESENT THE EPI LINEAGE OF THE POSTIMPLANTATION EMBRYO

EpiSCs are another pluripotent stem cell type derived from the embryo resembling the postimplantation EPI (Brons et al., 2007; Tesar et al., 2007) (Fig. 2). Along with the pluripotency markers *Oct4*, *Sox2*, and *Nanog*, EpiSCs also express primitive streak (PS) markers (*Brachyury*, *Lefty2*, *Foxa2*, and *Eomes*), markers of the anterior mesendoderm and definitive endoderm (*Lefty1*, *Cer1*, and *Sox17*), and *Fgf5* and *Nodal* which are markers of the postimplantation EPI (Brons et al., 2007; Kojima et al., 2014). Global gene expression data recently revealed that rather than representing the early postimplantation EPI, EpiSCs in fact resemble the late-gastrulation EPI and are biased toward the anterior PS (Hayashi, Ohta, Kurimoto, Aramaki, & Saitou, 2011; Kojima et al., 2014). Unlike mES cells, EpiSCs do not grow in rounded compact colonies, but rather as single layered, flat, epithelial colonies. EpiSCs and the postimplantation EPI also contain primary cilia unlike the other stem cell types derived from the preimplantation embryo, including ES cells (Bangs, Schrode, Hadjantonakis, & Anderson, 2015).

5.1 Derivation and Culture

EpiSCs can be derived from the EPI of the postimplantation mouse embryo just before the onset of gastrulation (5.5 days postcoitum (dpc)) (Brons et al., 2007; Tesar et al., 2007). More recently, in an effort to track the persistence of pluripotency in the postimplantation embryo, Osorno et al. demonstrated that EpiSC lines can be efficiently established from EPI explants of embryos up to presomite stages (E7.5–E8.0) (Osorno et al., 2012). EpiSCs can also be derived from blastocyst-stage embryos in either ES cell or EpiSC conditions (Najm et al., 2011). Blastocyst-derived EpiSCs display similar gene expression profiles to EpiSCs derived from the postimplantation EPI, and distinct from that of ES cells. Both blastocyst-derived and postimplantation EPI-derived EpiSCs are responsive to Activin/Nodal signaling and form teratomas.

Among in vitro pluripotent cell types, EpiSCs are most similar to hES cells. They both depend on Activin/Nodal signaling for prolonged culture in vitro (Camus, Perea-Gomez, Moreau, & Collignon, 2006; James, Levine, Besser, & Hemmati-Brivanlou, 2005; Johansson & Wiles, 1995; Vallier, Reynolds, & Pedersen, 2004). In addition, while FGF2 is not required for maintenance, it aids Activin signaling to improve EpiSC cultures by sustaining expression of pluripotency markers (Vallier et al., 2005). Additionally, unlike ES cells the efficiency of deriving EpiSCs from pre- and postimplantation embryos is not heavily strain dependent.

Neither hES cells nor mouse EpiSCs respond to conventional mES cell culture conditions. In particular, LIF signaling fails to support self-renewal in hES cells and mouse EpiSCs (Brons et al., 2007; Dahéron et al., 2004; Tesar et al., 2007). While the disparity between mES cells and hES cells could be attributed to species-specific differences, the derivation and characterization of mouse EpiSCs would suggest a temporal difference. That is, hES cells likely represent a postimplantation EPI population much like EpiSCs despite being derived from preimplantation embryos. Interestingly, female hES cell lines have both X chromosomes active, while female mouse EpiSCs have a randomly inactivated X chromosome like in vivo postimplantation EPI cells (Monk & Harper, 1979; Tesar et al., 2007; Thomson et al., 1998). Thus, EpiSCs and hES cells exhibit differences that are not fully understood.

While they display characteristics of pluripotent cells, under standard conditions EpiSCs cannot contribute to chimeras when introduced into preimplantation (morula or blastocyst) stage embryos. However, when introduced into the postimplantation EPI at early-streak to late-streak stages (gastrulation stage) EpiSCs form chimeras efficiently (Huang, Osorno, Tsakiridis, & Wilson, 2012). Overexpressing E-cadherin prior to injecting EpiSCs into blastocysts facilitates the incorporation of these cells into the ICM and results in higher efficiencies of chimera contribution (Ohtsuka, Nishikawa-Torikai, & Niwa, 2012). These cells contribute to all three germ layers of the embryo and also to the PGC population demonstrating their *bona fide* pluripotent character and their functional similarity to the postimplantation EPI. Importantly, EpiSCs also do not form chimeras when injected into later stage, postsomitogenesis embryos, indicating that their ability to acquire the characteristics of the host tissue is limited to their compatibility with the environment rather than their lack of pluripotency (Brons et al., 2007; Tesar et al., 2007). This suggests to test their developmental potential cells must be grafted orthochronically.

5.2 Conversion to EpiSCs

As already mentioned, embryo-derived EpiSCs have been suggested to be more similar to late-gastrulation stage embryos (Kojima et al., 2014) and therefore do not offer a good unbiased model to study germ layer specification. However, a number of protocols have been developed which allow the conversion of ES cells to a transient EPI-like state reminiscent of the early prestreak embryo. These involve culture of ES cells in the presence of FGF or FGF and Activin for a 2-day period and generate cells that appear similar to those found in E5.75 embryos (Gouti et al., 2014; Hayashi et al., 2011). Furthermore, ES cells can be stably converted to an EpiSC state through selective long-term culture in serum-free medium with FGF and Activin (Guo et al., 2009). However, further analysis is necessary to determine what stage of development these cells are representative of and if they are the same as embryo-derived EpiSCs.

5.3 Drawing Parallels

The functional similarity of hES cells to mouse EpiSCs was demonstrated by interspecies chimeras where hES cells engrafted onto mouse postimplantation embryos differentiated into all three germ layers (Mascetti & Pedersen, 2016; Wu et al., 2015). Furthermore, single-cell transcriptome analysis of human preimplantation EPI cells and hES cells has highlighted the distinctive identities of the two groups (Yan et al., 2013). The majority of differences arise during the derivation process, in agreement with the finding that human EPI cells transit through a post-ICM intermediate during ES cell derivation, which is more similar to in vitro cultured hES cells than in vivo EPI cells (O'Leary et al., 2012). However, a detailed transcriptome analysis of this intermediate state is lacking.

6. INTERCONVERSION OF EMBRYO-DERIVED STEM CELL LINES

By the late blastocyst stage in vivo, the EPI, PrE, and TE are irreversibly transcriptionally and functionally segregated. However, their in vitro counterparts maintain a level of functional plasticity by genetic manipulations and modification of culture conditions (Fig. 2). This highlights the tractable feature of the in vitro systems to probe in vivo mechanisms.

While ES cells do not contribute efficiently to extraembryonic lineages, under certain conditions, mES cells can be driven toward a TE or PrE state. BMP4 functions via the Smad pathway to regulate *Cdx2* and direct differentiation of mES cells into TSL cells (Bernardo et al., 2011; Hayashi et al., 2010). However, these cells also coexpress *Brachyury*, unlike the in vivo TE and hence may in fact represent a mesodermal cell type. While mES cells grown in defined medium conditions containing BMP4 and LIF on gelatin or type I collagen self-renew, culturing on laminin or fibronectin in the absence of LIF induces upregulation of various TE-specific genes (Hayashi et al., 2010; Qi et al., 2004; Ying et al., 2003). In another study, type IV collagen was implicated in directing CDX2-dependent TS cell differentiation from mES cells precultured on MEF feeders (Schenke-Layland et al., 2007). Since differentiation to TS cells under these conditions was not seen in ES cells grown on fibronectin or laminin, it is specific to type IV collagen and is also dependent on FGF4 in medium without LIF. These studies highlight a pivotal role for the culture substrate on the maintenance and differentiation potential of ES cells.

In addition to modulating culture conditions, mES cells can acquire a TSL identity by genetic manipulation. For example, *Oct4* repression converts mES cells to TSL cells. A 1.5-fold increased expression of *Oct4* induces endodermal and mesodermal differentiation, while its repression leads to a TSL identity (Niwa et al., 2000). Similarly, SOX2 can promote differentiation toward a TE-like identity when deleted (Masui et al., 2007). Furthermore, genes implicated in driving TE differentiation and maintenance, such as *Cdx2*, *Eomes*, *Gata3*, and *Tead4*, have been shown to induce TE differentiation when overexpressed in mES cells (Nishioka, Inoue, Adachi, et al., 2009; Niwa, Toyooka, Shimosato, et al., 2005; Ralston, Cox, Nishioka, et al., 2010; Tolkunova et al., 2006).

While the individual transcription factors, *Tcfap2c* (Kuckenberg, Buhl, Woynecki, et al., 2010), *Cdx2* (Niwa et al., 2005), *Eomes* (Niwa et al., 2005), *Gata3* (Ralston et al., 2010), *Elf5* (Ng, Dean, Dawson, et al., 2008), and *Tead4* (Nishioka et al., 2009), have been used to differentiate mES cells to a TSL state, these cells do not fully acquire a TS cell identity. They vary in terms of their self-renewal capacity and in vivo developmental potential and demonstrate insufficient epigenetic reprogramming at key TS cell-specific gene promoters (Cambuli et al., 2014). These regions mark a cohort of so-called gatekeeper genes that are hypermethylated in ES cells, hypomethylated in TS cells but variably and incompletely demethylated during TSL state induction, suggesting an important lineage-specific epigenetic barrier separating the embryonic and extraembryonic lineages. In support of this idea, genome-wide DNA methylation profiles of the extraembryonic TS and XEN cells are vastly different from that of ES cells (Cambuli et al., 2014; Farthing et al., 2008; Senner et al., 2012). Thus, ES cells and TS cells are locked in distinct epigenetic states, such that factors able to induce a TS cell identity from MEFs are not able to successfully reprogram ES cells (Benchetrit et al., 2015; Kubaczka et al., 2015). This highlights the limited plasticity of ES cells after the first cell fate decision and establishment of the initial extraembryonic–embryonic boundary.

Several studies have also observed cells resembling the ExEn spontaneously emerge in mES cell populations cultured in the presence of LIF (Bradley et al., 1984; Canham et al., 2010; Doetschman, Eistetter, Katz, Schmidt, & Kemler, 1985; Niakan et al., 2010) or on the surface of EBs (Abe et al., 1996; Shen & Leder, 1992). Indeed targets of LIF include extraembryonic-specific genes including *Gata3*, *Eomes*, and *Gata6* (Kidder, Yang, & Palmer, 2008). Additionally, cells emerge in ES cell cultures that express high levels of *Sox17*. However, they are exclusively committed to a PrE fate and only contributing to the ExEn lineages in chimeric embryos (Niakan et al., 2010). ES cells from *Sox17* mutant blastocysts are also unable to form EBs with a differentiated epithelialized layer on the periphery (Niakan et al., 2010).

RA stimulates the differentiation of mES cells and F9 embryonal carcinoma cell lines to ExEn lineages (Artus, Panthier, & Hadjantonakis, 2010; Capo-Chichi et al., 2005; Soprano, Teets, & Soprano, 2007; Strickland, Smith, & Marotti, 1980). Also, Nodal is expressed in the PrE and drives mES cells toward a definitive endoderm fate (Kubo et al., 2004; Mesnard, Guzman-Ayala, & Constam, 2006). Since these factors lead to terminal differentiation of mES cells, a moderate dose of each was recently used to convert mES cells to stable XEN-like cells (termed "cXEN") under LIF and serum or 2i conditions (Cho et al., 2012; Niakan et al., 2013) (Fig. 2). Like their transgene-induced counterparts, cXEN cells are identical to embryo-derived XEN cells both morphologically and in their gene expression profile. In response to BMP4, both embryo-derived XEN cells and cXEN cells differentiate to a VE subtype (Artus et al., 2012; Cho et al., 2012; Paca et al., 2012).

Mouse pluripotent cells can also be generated from TS cells using the MEF reprogramming strategy for iPS cells (termed TSC-iPSCs) (Kuckenberg et al., 2011). Given that TS and ES cells display vastly different transcriptomes and epigenetic landscapes (Cambuli et al., 2014), this suggests that the epigenetic and lineage barrier separating these two cell types is surpassed more easily going from TS cells to iPS cells, rather than from ES cells to TS cells (see Section 6) (Benchetrit et al., 2015; Kubaczka et al., 2015), and/or these reprogramming factors are able to provide the necessary energy of activation to accomplish this switch. More importantly however, overexpression of *Oct4 alone* is also able to convert TS cells to iPS cells (termed, "OiPS cells"; Wu et al., 2011). Thus far, reprogramming XEN cells to iPS cells has not been reported.

7. CONCLUDING REMARKS

Establishing stem cell lines from, and representing, the cell types present in the preimplantation and postimplantation embryo has been instrumental in understanding the molecular mechanisms governing cell identity and lineage decisions. Utilizing a combination of in vitro and in vivo experiments, key players responsible for specifying and maintaining the cell lineages of the early mammalian embryo have been identified. Interconversion of cell types using in vitro manipulations, and conservation of signaling and transcription factor networks between mammalian species, has reinforced our understanding of these mechanisms. However, we still lack a complete knowledge of the factors involved in early in vivo development. Further examination of the in vitro cell types should help close this gap.

In parallel, an important open question to consider is how accurately these stem cell lines represent their corresponding in vivo lineages. Specific culture conditions can significantly alter the gene expression and epigenetic characteristics of these stem cells. In recent years the advancement and feasibility of whole-genome expression analyses in whole populations and, more recently, in single cells have allowed in depth studies of the heterogeneity inherent in stem cell cultures. Comparison with in vivo cells will help determine if the heterogeneity reflects particular developmental stages, or if it is merely an in vitro adaptation. These findings could then be supported with chimeras—the definitive test of developmental potential. As highlighted by several recent studies (Mascetti & Pedersen, 2016; Wu et al., 2015) the interpretation of chimera experiments must consider the stage and engrafted region of the host embryo and introduced cells.

ACKNOWLEDGMENTS

We thank members of our lab for stimulating discussions and comments on this manuscript. Our work on the early mammalian embryo and the stem cell types that can be derived from it is funded by NYSTEM (N13G-236) and by the National Institutes of Health (R01-DK084391 and P30-CA008748). S.M. is a Sir Henry Wellcome Postdoctoral Fellow working with the labs of Dr. Jenny Nichols (Wellcome Trust Stem Cell Institute, Cambridge, UK) and A.K.H.

REFERENCES

Abe, K., Niwa, H., Iwase, K., Takiguchi, M., Mori, M., Abé, S. I., & Yamamura, K. I. (1996). Endoderm-specific gene expression in embryonic stem cells differentiated to embryoid bodies. *Experimental Cell Research*, *229*(1), 27–34. http://doi.org/10.1006/excr.1996.0340.

Alexandrow, M. G., & Moses, H. L. (1995). Transforming growth factor beta and cell cycle regulation. *Cancer Research*, *55*(7), 1452–1457. http://doi.org/10.1006/excr.1996.0340.

Ambrosetti, D. C., Basilico, C., & Dailey, L. (1997). Synergistic activation of the fibroblast growth factor 4 enhancer by Sox2 and Oct-3 depends on protein-protein interactions facilitated by a specific spatial arrangement of factor binding sites. *Molecular and Cellular Biology*, *17*(11), 6321–6329. Retrieved from, http://www.pubmedcentral.nih.gov/articlerender.fcgi?artid=232483&tool=pmcentrez&rendertype=abstract.

Artus, J., Douvaras, P., Piliszek, A., Isern, J., Baron, M. H., & Hadjantonakis, A.-K. (2012). BMP4 signaling directs primitive endoderm-derived XEN cells to an extraembryonic visceral endoderm identity. *Developmental Biology*, *361*(2), 245–262. http://doi.org/10.1016/j.ydbio.2011.10.015.

Artus, J., Panthier, J.-J., & Hadjantonakis, A.-K. (2010). A role for PDGF signaling in expansion of the extra-embryonic endoderm lineage of the mouse blastocyst. *Development (Cambridge, England)*, *137*(20), 3361–3372. http://doi.org/10.1242/dev.050864.

Artus, J., Piliszek, A., & Hadjantonakis, A.-K. (2011). The primitive endoderm lineage of the mouse blastocyst: Sequential transcription factor activation and regulation of differentiation by Sox17. *Developmental Biology*, *350*(2), 393–404. http://doi.org/10.1016/j.ydbio.2010.12.007.

Avilion, A. A., Nicolis, S. K., Pevny, L. H., Perez, L., Vivian, N., & Lovell-Badge, R. (2003). Multipotent cell lineages in early mouse development depend on SOX2 function. *Genes & Development*, *17*(1), 126–140. http://doi.org/10.1101/gad.224503.

Bangs, F. K., Schrode, N., Hadjantonakis, A.-K., & Anderson, K. V. (2015). Lineage specificity of primary cilia in the mouse embryo. *Nature Cell Biology*, *17*(2), 113–122. http://doi.org/10.1038/ncb3091.

Beck, F., Erler, T., Russell, A., & James, R. (1995). Expression of Cdx-2 in the mouse embryo and placenta: Possible role in patterning of the extra-embryonic membranes. *Developmental Dynamics: An Official Publication of the American Association of Anatomists*, *204*(3), 219–227. http://doi.org/10.1002/aja.1002040302.

Beddington, R. S., & Robertson, E. J. (1989). An assessment of the developmental potential of embryonic stem cells in the midgestation mouse embryo. *Development (Cambridge, England)*, *105*(4), 733–737. Retrieved from, http://www.ncbi.nlm.nih.gov/pubmed/2598811.

Benchetrit, H., Herman, S., van Wietmarschen, N., Wu, T., Makedonski, K., Maoz, N., … Buganim, Y. (2015). Extensive nuclear reprogramming underlies lineage conversion into functional trophoblast stem-like cells. *Cell Stem Cell*, *17*(5), 543–556. http://doi.org/10.1016/j.stem.2015.08.006.

Bernardo, A. S., Faial, T., Gardner, L., Niakan, K. K., Ortmann, D., Senner, C. E., … Pedersen, R. A. (2011). BRACHYURY and CDX2 mediate BMP-induced differentiation of human and mouse pluripotent stem cells into embryonic and extraembryonic lineages. *Cell Stem Cell*, *9*(2), 144–155. http://doi.org/10.1016/j.stem.2011.06.015.

Bernemann, C., Greber, B., Ko, K., Sterneckert, J., Han, D. W., Araúzo-Bravo, M. J., & Schöler, H. R. (2011). Distinct developmental ground states of epiblast stem cell lines determine different pluripotency features. *Stem Cells (Dayton, Ohio)*, *29*(10), 1496–1503. http://doi.org/10.1002/stem.709.

Blakeley, P., Fogarty, N. M. E., del Valle, I., Wamaitha, S. E., Hu, T. X., Elder, K., … Niakan, K. K. (2015). Defining the three cell lineages of the human blastocyst by single-cell RNA-seq. *Development (Cambridge, England)*, *142*(18), 3151–3165. http://doi.org/10.1242/dev.123547.

Boroviak, T., Loos, R., Bertone, P., Smith, A., & Nichols, J. (2014). The ability of inner-cell-mass cells to self-renew as embryonic stem cells is acquired following epiblast specification. *Nature Cell Biology*, *16*(6), 516–528. http://doi.org/10.1038/ncb2965.

Bradley, A., Evans, M., Kaufman, M. H., & Robertson, E. (1984). Formation of germ-line chimaeras from embryo-derived teratocarcinoma cell lines. *Nature*, *309*(5965), 255–256. Retrieved from, http://www.ncbi.nlm.nih.gov/pubmed/6717601.

Brons, I. G. M., Smithers, L. E., Trotter, M. W. B., Rugg-Gunn, P., Sun, B., Chuva de Sousa Lopes, S. M., … Vallier, L. (2007). Derivation of pluripotent epiblast stem cells from mammalian embryos. *Nature*, *448*(7150), 191–195. http://doi.org/10.1038/nature05950.

Brook, F. A., & Gardner, R. L. (1997). The origin and efficient derivation of embryonic stem cells in the mouse. *Proceedings of the National Academy of Sciences of the United States of America*, *94*(11), 5709–5712. Retrieved from, http://www.pubmedcentral.nih.gov/articlerender.fcgi?artid=20843&tool=pmcentrez&rendertype=abstract.

Brown, K., Legros, S., Artus, J., Doss, M. X., Khanin, R., Hadjantonakis, A.-K., & Foley, A. (2010). A comparative analysis of extra-embryonic endoderm cell lines. *PloS One*, *5*(8), e12016. http://doi.org/10.1371/journal.pone.0012016.

Buehr, M., Meek, S., Blair, K., Yang, J., Ure, J., Silva, J., … Smith, A. (2008). Capture of authentic embryonic stem cells from rat blastocysts. *Cell*, *135*(7), 1287–1298. http://doi.org/10.1016/j.cell.2008.12.007.

Buehr, M., Nichols, J., Stenhouse, F., Mountford, P., Greenhalgh, C. J., Kantachuvesiri, S., … Smith, A. G. (2003). Rapid loss of Oct-4 and pluripotency in cultured rodent blastocysts and derivative cell lines. *Biology of Reproduction*, *68*(1), 222–229. http://doi.org/10.1095/biolreprod.102.006197.

Burdon, T., Stracey, C., Chambers, I., Nichols, J., & Smith, A. (1999). Suppression of SHP-2 and ERK signalling promotes self-renewal of mouse embryonic stem cells. *Developmental Biology*, *210*(1), 30–43. http://doi.org/10.1006/dbio.1999.9265.

Cambuli, F., Murray, A., Dean, W., Dudzinska, D., Krueger, F., Andrews, S., … Hemberger, M. (2014). Epigenetic memory of the first cell fate decision prevents complete ES cell reprogramming into trophoblast. *Nature Communications*, *5*, 5538. http://doi.org/10.1038/ncomms6538.

Camus, A., Perea-Gomez, A., Moreau, A., & Collignon, J. (2006). Absence of Nodal signaling promotes precocious neural differentiation in the mouse embryo. *Developmental Biology*, *295*(2), 743–755. http://doi.org/10.1016/j.ydbio.2006.03.047.

Canham, M. A., Sharov, A. A., Ko, M. S. H., & Brickman, J. M. (2010). Functional heterogeneity of embryonic stem cells revealed through translational amplification of an early endodermal transcript. *PLoS Biology*, *8*(5), e1000379. http://doi.org/10.1371/journal.pbio.1000379.

Capo-Chichi, C. D., Rula, M. E., Smedberg, J. L., Vanderveer, L., Parmacek, M. S., Morrisey, E. E., … Xu, X.-X. (2005). Perception of differentiation cues by GATA factors in primitive endoderm lineage determination of mouse embryonic stem cells. *Developmental Biology*, *286*(2), 574–586. http://doi.org/10.1016/j.ydbio.2005.07.037.

Cartwright, P., McLean, C., Sheppard, A., Rivett, D., Jones, K., & Dalton, S. (2005). LIF/STAT3 controls ES cell self-renewal and pluripotency by a Myc-dependent mechanism. *Development (Cambridge, England)*, *132*(5), 885–896. http://doi.org/10.1242/dev.01670.

Chambers, I., Colby, D., Robertson, M., Nichols, J., Lee, S., Tweedie, S., & Smith, A. (2003). Functional expression cloning of Nanog, a pluripotency sustaining factor in embryonic stem cells. *Cell*, *113*(5), 643–655. http://doi.org/10.1016/S0092-8674(03)00392-1.

Chambers, I., Silva, J., Colby, D., Nichols, J., Nijmeijer, B., Robertson, M., … Smith, A. (2007). Nanog safeguards pluripotency and mediates germline development. *Nature*, *450*(7173), 1230–1234. http://doi.org/10.1038/nature06403.

Chan, Y.-S., Göke, J., Ng, J.-H., Lu, X., Gonzales, K. A. U., Tan, C.-P., … Ng, H.-H. (2013). Induction of a human pluripotent state with distinct regulatory circuitry that resembles preimplantation epiblast. *Cell Stem Cell*, *13*(6), 663–675. http://doi.org/10.1016/j.stem.2013.11.015.

Chazaud, C., Yamanaka, Y., Pawson, T., & Rossant, J. (2006). Early lineage segregation between epiblast and primitive endoderm in mouse blastocysts through the Grb2-MAPK pathway. *Developmental Cell*, *10*(5), 615–624. http://doi.org/10.1016/j.devcel.2006.02.020.

Chen, Y., Blair, K., & Smith, A. (2013). Robust self-renewal of rat embryonic stem cells requires fine-tuning of glycogen synthase kinase-3 inhibition. *Stem Cell Reports*, *1*(3), 209–217. http://doi.org/10.1016/j.stemcr.2013.07.003.

Chen, Y., Niu, Y., Li, Y., Ai, Z., Kang, Y., Shi, H., … Li, T. (2015). Generation of cynomolgus monkey chimeric fetuses using embryonic stem cells. *Cell Stem Cell*, *17*(1), 116–124. http://doi.org/10.1016/j.stem.2015.06.004.

Chen, X., Xu, H., Yuan, P., et al. (2008). Integration of external signaling pathways with the core transcriptional network in embryonic stem cells. *Cell*, *133*(6), 1106–1117. http://doi.org/10.1016/j.cell.2008.04.043.

Cho, L. T. Y., Wamaitha, S. E., Tsai, I. J., Artus, J., Sherwood, R. I., Pedersen, R. A., ... Niakan, K. K. (2012). Conversion from mouse embryonic to extra-embryonic endoderm stem cells reveals distinct differentiation capacities of pluripotent stem cell states. *Development (Cambridge, England)*, *139*(16), 2866–2877. http://doi.org/10.1242/dev.078519.

Choi, H. J., Sanders, T. A., Tormos, K. V., Ameri, K., Tsai, J. D., Park, A. M., ... Maltepe, E. (2013). ECM-dependent HIF induction directs trophoblast stem cell fate via LIMK1-mediated cytoskeletal rearrangement. *PloS One*, *8*(2), e56949. http://doi.org/10.1371/journal.pone.0056949.

Chuykin, I., Schulz, H., Guan, K., & Bader, M. (2013). Activation of the PTHRP/adenylate cyclase pathway promotes differentiation of rat XEN cells into parietal endoderm, whereas Wnt/β-catenin signaling promotes differentiation into visceral endoderm. *Journal of Cell Science*, *126*(Pt. 1), 128–138. http://doi.org/10.1242/jcs.110239.

Ciruna, B. G., & Rossant, J. (1999). Expression of the T-box gene Eomesodermin during early mouse development. *Mechanisms of Development*, *81*(1–2), 199–203. http://doi.org/10.1016/S0925-4773(98)00243-3.

Cross, J. C., Flannery, M. L., Blanar, M. A., Steingrimsson, E., Jenkins, N. A., Copeland, N. G., ... Werb, Z. (1995). Hxt encodes a basic helix-loop-helix transcription factor that regulates trophoblast cell development. *Development (Cambridge, England)*, *121*(8), 2513–2523. Retrieved from, http://www.ncbi.nlm.nih.gov/entrez/query.fcgi?cmd=Retrieve&db=PubMed&dopt=Citation&list_uids=7671815.

Dahéron, L., Opitz, S. L., Zaehres, H., Lensch, M. W., Lensch, W. M., Andrews, P. W., ... Daley, G. Q. (2004). LIF/STAT3 signaling fails to maintain self-renewal of human embryonic stem cells. *Stem Cells (Dayton, Ohio)*, *22*(5), 770–778. http://doi.org/10.1634/stemcells.22-5-770.

Dailey, L., Yuan, H., & Basilico, C. (1994). Interaction between a novel F9-specific factor and octamer-binding proteins is required for cell-type-restricted activity of the fibroblast growth factor 4 enhancer. *Molecular and Cellular Biology*, *14*(12), 7758–7769. http://doi.org/10.1128/MCB.14.12.7758.Updated.

Davenport, T. G., Jerome-Majewska, L. A., & Papaioannou, V. E. (2003). Mammary gland, limb and yolk sac defects in mice lacking Tbx3, the gene mutated in human ulnar mammary syndrome. *Development (Cambridge, England)*, *130*, 2263–2273. http://doi.org/10.1242/dev.00431.

Davis, S., Aldrich, T. H., Stahl, N., Pan, L., Taga, T., Kishimoto, T., ... Yancopoulos, G. D. (1993). LIFR beta and gp130 as heterodimerizing signal transducers of the tripartite CNTF receptor. *Science (New York, N.Y.)*, *260*(5115), 1805–1808. Retrieved from, http://www.ncbi.nlm.nih.gov/pubmed/8390097.

Debeb, B. G., Galat, V., Epple-Farmer, J., Iannaccone, S., Woodward, W. A., Bader, M., ... Binas, B. (2009). Isolation of Oct4-expressing extraembryonic endoderm precursor cell lines. *PloS One*, *4*(9), e7216. http://doi.org/10.1371/journal.pone.0007216.

Dietrich, J.-E., & Hiiragi, T. (2007). Stochastic patterning in the mouse pre-implantation embryo. *Development (Cambridge, England)*, *134*(23), 4219–4231. http://doi.org/10.1242/dev.003798.

Doetschman, T. C., Eistetter, H., Katz, M., Schmidt, W., & Kemler, R. (1985). The in vitro development of blastocyst-derived embryonic stem cell lines: Formation of visceral yolk sac, blood islands and myocardium. *Journal of Embryology and Experimental Morphology*, *87*, 27–45. Retrieved from, http://www.ncbi.nlm.nih.gov/pubmed/3897439.

Enders, A. C., Given, R. L., & Schlafke, S. (1978). Differentiation and migration of endoderm in the rat and mouse at implantation. *The Anatomical Record*, *190*(1), 65–77. http://doi.org/10.1002/ar.1091900107.

Erlebacher, A., Price, K. A., & Glimcher, L. H. (2004). Maintenance of mouse trophoblast stem cell proliferation by TGF-beta/activin. *Developmental Biology*, *275*(1), 158–169. http://doi.org/10.1016/j.ydbio.2004.07.032.

Evans, M. J., & Kaufman, M. H. (1981). Establishment in culture of pluripotential cells from mouse embryos. *Nature*, *292*(5819), 154–156. http://doi.org/10.1038/292154a0.

Ezashi, T., Matsuyama, H., Telugu, B. P. V. L., & Roberts, R. M. (2011). Generation of colonies of induced trophoblast cells during standard reprogramming of porcine fibroblasts to induced pluripotent stem cells. *Biology of Reproduction*, *85*(4), 779–787. http://doi.org/10.1095/biolreprod.111.092809.

Falco, G., Lee, S. L., Stanghellini, I., Bassey, U. C., Hamatani, T., & Ko, M. S. H. (2007). Zscan4: A novel gene expressed exclusively in late 2-cell embryos and embryonic stem cells. *Developmental Biology*, *307*(2), 539–550. http://doi.org/10.1016/j.ydbio.2007.05.003.

Fang, Z. F., Gai, H., Huang, Y. Z., Li, S. G., Chen, X. J., Shi, J. J., … Sheng, H. Z. (2006). Rabbit embryonic stem cell lines derived from fertilized, parthenogenetic or somatic cell nuclear transfer embryos. *Experimental Cell Research*, *312*(18), 3669–3682. http://doi.org/10.1016/j.yexcr.2006.08.013.

Fang, R., Liu, K., Zhao, Y., Li, H., Zhu, D., Du, Y., … Deng, H. (2014). Generation of naive induced pluripotent stem cells from rhesus monkey fibroblasts. *Cell Stem Cell*, *15*(4), 488–496. http://doi.org/10.1016/j.stem.2014.09.004.

Faria, T. N., Ogren, L., Talamantes, F., Linzer, D. I., & Soares, M. J. (1991). Localization of placental lactogen-I in trophoblast giant cells of the mouse placenta. *Biology of Reproduction*, *44*(2), 327–331. Retrieved from, http://www.ncbi.nlm.nih.gov/pubmed/2009333.

Farthing, C. R., Ficz, G., Ng, R. K., Chan, C.-F., Andrews, S., Dean, W., … Reik, W. (2008). Global mapping of DNA methylation in mouse promoters reveals epigenetic reprogramming of pluripotency genes. *PLoS Genetics*, *4*(6), e1000116. http://doi.org/10.1371/journal.pgen.1000116.

Feldman, B., Poueymirou, W., Papaioannou, V. E., DeChiara, T. M., & Goldfarb, M. (1995). Requirement of FGF-4 for postimplantation mouse development. *Science (New York, N.Y.)*, *267*(5195), 246–249. Retrieved from, http://www.ncbi.nlm.nih.gov/pubmed/7809630.

Feng, B., Jiang, J., Kraus, P., Ng, J.-H., Heng, J.-C. D., Chan, Y.-S., … Ng, H.-H. (2009). Reprogramming of fibroblasts into induced pluripotent stem cells with orphan nuclear receptor Esrrb. *Nature Cell Biology*, *11*(2), 197–203. http://doi.org/10.1038/ncb1827.

Fowler, K. J., Mitrangas, K., & Dziadek, M. (1990). In vitro production of Reichert's membrane by mouse embryo-derived parietal endoderm cell lines. *Experimental Cell Research*, *191*(2), 194–203. Retrieved from, http://www.ncbi.nlm.nih.gov/pubmed/2124186.

Frum, T., Halbisen, M. A., Wang, C., Amiri, H., Robson, P., & Ralston, A. (2013). Oct4 cell-autonomously promotes primitive endoderm development in the mouse blastocyst. *Developmental Cell*, *25*(6), 610–622. http://doi.org/10.1016/j.devcel.2013.05.004.

Fujikura, J., Yamato, E., Yonemura, S., Hosoda, K., Masui, S., Nakao, K., … Niwa, H. (2002). Differentiation of embryonic stem cells is induced by GATA factors. *Genes & Development*, *16*(7), 784–789. http://doi.org/10.1101/gad.968802.

Gafni, O., Weinberger, L., Mansour, A. A., Manor, Y. S., Chomsky, E., Ben-Yosef, D., … Hanna, J. H. (2013). Derivation of novel human ground state naive pluripotent stem cells. *Nature*, *504*(7479), 282–286. http://doi.org/10.1038/nature12745.

Galat, V., Binas, B., Iannaccone, S., Postovit, L.-M., Debeb, B. G., & Iannaccone, P. (2009). Developmental potential of rat extraembryonic stem cells. *Stem Cells and Development*, *18*(9), 1309–1318. http://doi.org/10.1089/scd.2009.0115.

Gardner, R. L. (1982). Investigation of cell lineage and differentiation in the extraembryonic endoderm of the mouse embryo. *Journal of Embryology and Experimental Morphology*, *68*, 175–198. Retrieved from, http://www.ncbi.nlm.nih.gov/pubmed/7108421.

Gardner, R. L. (1983). Origin and differentiation of extraembryonic tissues in the mouse. *International Review of Experimental Pathology*, *24*, 63–133. Retrieved from, http://www.ncbi.nlm.nih.gov/pubmed/6302028.

Gardner, R. L., & Johnson, M. H. (1972). An investigation of inner cell mass and trophoblast tissues following their isolation from the mouse blastocyst. *Journal of Embryology and Experimental Morphology*, *28*(2), 279–312. Retrieved from, http://www.ncbi.nlm.nih.gov/pubmed/4672104.

Gardner, R. L., & Johnson, M. H. (1973). Investigation of early mammalian development using interspecific chimaeras between rat and mouse. *Nature: New Biology*, *246*(151), 86–89. Retrieved from, http://www.ncbi.nlm.nih.gov/pubmed/4586447.

Gardner, R. L., Papaioannou, V. E., & Barton, S. C. (1973). Origin of the ectoplacental cone and secondary giant cells in mouse blastocysts reconstituted from isolated trophoblast and inner cell mass. *Journal of Embryology and Experimental Morphology*, *30*(3), 561–572. Retrieved from, http://dev.biologists.org/content/30/3/561.full.pdf+html.

Gospodarowicz, D., & Cheng, J. (1986). Heparin protects basic and acidic FGF from inactivation. *Journal of Cellular Physiology*, *128*(3), 475–484. http://doi.org/10.1002/jcp.1041280317.

Gouti, M., Tsakiridis, A., Wymeersch, F. J., Huang, Y., Kleinjung, J., Wilson, V., & Briscoe, J. (2014). In vitro generation of neuromesodermal progenitors reveals distinct roles for wnt signalling in the specification of spinal cord and paraxial mesoderm identity. *PLoS Biology*, *12*(8), e1001937. http://doi.org/10.1371/journal.pbio.1001937.

Graves, K. H., & Moreadith, R. W. (1993). Derivation and characterization of putative pluripotential embryonic stem cells from preimplantation rabbit embryos. *Molecular Reproduction and Development*, *36*(4), 424–433. http://doi.org/10.1002/mrd.1080360404.

Guillemot, F., Nagy, A., Auerbach, A., Rossant, J., & Joyner, A. L. (1994). Essential role of Mash-2 in extraembryonic development. *Nature*, *371*(6495), 333–336. http://doi.org/10.1038/371333a0.

Guo, G., Huss, M., Tong, G. Q., Wang, C., Li Sun, L., Clarke, N. D., & Robson, P. (2010). Resolution of cell fate decisions revealed by single-cell gene expression analysis from zygote to blastocyst. *Developmental Cell*, *18*(4), 675–685. http://doi.org/10.1016/j.devcel.2010.02.012.

Guo, G., Yang, J., Nichols, J., Hall, J. S., Eyres, I., Mansfield, W., & Smith, A. (2009). Klf4 reverts developmentally programmed restriction of ground state pluripotency. *Development (Cambridge, England)*, *136*(7), 1063–1069. http://doi.org/10.1242/dev.030957.

Guzman-Ayala, M., Ben-Haim, N., Beck, S., & Constam, D. B. (2004). Nodal protein processing and fibroblast growth factor 4 synergize to maintain a trophoblast stem cell microenvironment. *Proceedings of the National Academy of Sciences of the United States of America*, *101*(44), 15656–15660. http://doi.org/10.1073/pnas.0405429101.

Haffner-Krausz, R., Gorivodsky, M., Chen, Y., & Lonai, P. (1999). Expression of Fgfr2 in the early mouse embryo indicates its involvement in preimplantation development. *Mechanisms of Development*, *85*(1–2), 167–172. http://doi.org/10.1016/S0925-4773(99)00082-9.

Hall, J., Guo, G., Wray, J., Eyres, I., Nichols, J., Grotewold, L., … Smith, A. (2009). Oct4 and LIF/Stat3 additively induce Krüppel factors to sustain embryonic stem cell self-renewal. *Cell Stem Cell*, *5*(6), 597–609. http://doi.org/10.1016/j.stem.2009.11.003.

Hanna, J., Cheng, A. W., Saha, K., Kim, J., Lengner, C. J., Soldner, F., … Jaenisch, R. (2010). Human embryonic stem cells with biological and epigenetic characteristics similar to those of mouse ESCs. *Proceedings of the National Academy of Sciences of the United States of America*, *107*(20), 9222–9227. http://doi.org/10.1073/pnas.1004584107.

Hanna, J., Markoulaki, S., Mitalipova, M., Cheng, A. W., Cassady, J. P., Staerk, J., … Jaenisch, R. (2009). Metastable pluripotent states in NOD-mouse-derived ESCs. *Cell Stem Cell*, *4*(6), 513–524. http://doi.org/10.1016/j.stem.2009.04.015.

Hayashi, Y., Furue, M. K., Tanaka, S., Hirose, M., Wakisaka, N., Danno, H., … Asashima, M. (2010). BMP4 induction of trophoblast from mouse embryonic stem cells in defined culture conditions on laminin. *In Vitro Cellular & Developmental Biology. Animal*, *46*(5), 416–430. http://doi.org/10.1007/s11626-009-9266-6.

Hayashi, K., Lopes, S. M. C. D. S., Tang, F., & Surani, M. A. (2008). Dynamic equilibrium and heterogeneity of mouse pluripotent stem cells with distinct functional and epigenetic states. *Cell Stem Cell*, *3*(4), 391–401. http://doi.org/10.1016/j.stem.2008.07.027.

Hayashi, K., Ohta, H., Kurimoto, K., Aramaki, S., & Saitou, M. (2011). Reconstitution of the mouse germ cell specification pathway in culture by pluripotent stem cells. *Cell*, *146*(4), 519–532. http://doi.org/10.1016/j.cell.2011.06.052.

Honda, A., Hirose, M., & Ogura, A. (2009). Basic FGF and Activin/Nodal but not LIF signaling sustain undifferentiated status of rabbit embryonic stem cells. *Experimental Cell Research*, *315*(12), 2033–2042. http://doi.org/10.1016/j.yexcr.2009.01.024.

Huang, Y., Osorno, R., Tsakiridis, A., & Wilson, V. (2012). In vivo differentiation potential of epiblast stem cells revealed by chimeric embryo formation. *Cell Reports*, *2*(6), 1571–1578. http://doi.org/10.1016/j.celrep.2012.10.022.

Hughes, M., Dobric, N., Scott, I. C., Su, L., Starovic, M., St-Pierre, B., … Cross, J. C. (2004). The Hand1, Stra13 and Gcm1 transcription factors override FGF signaling to promote terminal differentiation of trophoblast stem cells. *Developmental Biology*, *271*(1), 26–37. http://doi.org/10.1016/j.ydbio.2004.03.029.

Hughes, C. S., Postovit, L. M., & Lajoie, G. A. (2010). Matrigel: A complex protein mixture required for optimal growth of cell culture. *Proteomics*, *10*(9), 1886–1890. http://doi.org/10.1002/pmic.200900758.

Hughes, C. S., Radan, L., Betts, D., Postovit, L. M., & Lajoie, G. A. (2011). Proteomic analysis of extracellular matrices used in stem cell culture. *Proteomics*, *11*(20), 3983–3991. http://doi.org/10.1002/pmic.201100030.

Intawicha, P., Ou, Y.-W., Lo, N., Zhang, S., Chen, Y.-Z., Lin, T.-A., … Ju, J. (2009). Characterization of embryonic stem cell lines derived from New Zealand white rabbit embryos. *Cloning and Stem Cells*, *11*(1), 27–38. http://doi.org/10.1089/clo.2008.0040.

James, D., Levine, A. J., Besser, D., & Hemmati-Brivanlou, A. (2005). TGFbeta/activin/nodal signaling is necessary for the maintenance of pluripotency in human embryonic stem cells. *Development (Cambridge, England)*, *132*(6), 1273–1282. http://doi.org/10.1242/dev.01706.

Johansson, B. M., & Wiles, M. V. (1995). Evidence for involvement of activin A and bone morphogenetic protein 4 in mammalian mesoderm and hematopoietic development. *Molecular and Cellular Biology*, *15*(1), 141–151. Retrieved from, http://www.pubmedcentral.nih.gov/articlerender.fcgi?artid=231923&tool=pmcentrez&rendertype=abstract.

Johnson, M. H., & Ziomek, C. A. (1981). The foundation of two distinct cell lineages within the mouse morula. *Cell*, *24*(1), 71–80. Retrieved from, http://www.ncbi.nlm.nih.gov/pubmed/7237545.

Johnson, M. H., & Ziomek, C. A. (1983). Cell interactions influence the fate of mouse blastomeres undergoing the transition from the 16- to the 32-cell stage. *Developmental Biology*, *95*, 211–218. http://doi.org/10.1016/0012-1606(83)90019-2.

Kaji, K., Norrby, K., Paca, A., Mileikovsky, M., Mohseni, P., & Woltjen, K. (2009). Virus-free induction of pluripotency and subsequent excision of reprogramming factors. *Nature*, *458*(7239), 771–775. http://doi.org/10.1038/nature07864.

Kanai-Azuma, M., Kanai, Y., Gad, J. M., Tajima, Y., Taya, C., Kurohmaru, M., … Hayashi, Y. (2002). Depletion of definitive gut endoderm in Sox17-null mutant mice. *Development (Cambridge, England)*, *129*(10), 2367–2379. Retrieved from, http://dev.biologists.org/content/129/10/2367.short.

Kang, M., Piliszek, A., Artus, J., & Hadjantonakis, A.-K. (2013). FGF4 is required for lineage restriction and salt-and-pepper distribution of primitive endoderm factors but not their initial expression in the mouse. *Development (Cambridge, England)*, *140*(2), 267–279. http://doi.org/10.1242/dev.084996.

Kidder, B. L., Yang, J., & Palmer, S. (2008). Stat3 and c-Myc genome-wide promoter occupancy in embryonic stem cells. *PloS One*, *3*(12), 1–14. http://doi.org/10.1371/journal.pone.0003932.

Kim, J., Chu, J., Shen, X., Wang, J., & Orkin, S. H. (2008). An extended transcriptional network for pluripotency of embryonic stem cells. *Cell*, *132*(6), 1049–1061. http://doi.org/10.1016/j.cell.2008.02.039.

Kleinman, H. K., & Martin, G. R. (2005). Matrigel: Basement membrane matrix with biological activity. *Seminars in Cancer Biology*, *15*(5), 378–386. http://doi.org/10.1016/j.semcancer.2005.05.004.

Kobayashi, T., Mizuno, H., Imayoshi, I., Furusawa, C., Shirahige, K., & Kageyama, R. (2009). The cyclic gene Hes1 contributes to diverse differentiation responses of embryonic stem cells. *Genes & Development*, *23*(16), 1870–1875. http://doi.org/10.1101/gad.1823109.

Kobayashi, M., Takada, T., Takahashi, K., Noda, Y., & Torii, R. (2008). BMP4 induces primitive endoderm but not trophectoderm in monkey embryonic stem cells. *Cloning and Stem Cells*, *10*(4), 495–502. http://doi.org/10.1089/clo.2008.0030.

Kojima, Y., Kaufman-Francis, K., Studdert, J. B., Steiner, K. A., Power, M. D., Loebel, D. A. F., … Tam, P. P. L. (2014). The transcriptional and functional properties of mouse epiblast stem cells resemble the anterior primitive streak. *Cell Stem Cell*, *14*(1), 107–120. http://doi.org/10.1016/j.stem.2013.09.014.

Kopp, J. L., Ormsbee, B. D., Desler, M., & Rizzino, A. (2008). Small increases in the level of Sox2 trigger the differentiation of mouse embryonic stem cells. *Stem Cells (Dayton, Ohio)*, *26*(4), 903–911. http://doi.org/10.1634/stemcells.2007-0951.

Krawchuk, D., Honma-Yamanaka, N., Anani, S., & Yamanaka, Y. (2013). FGF4 is a limiting factor controlling the proportions of primitive endoderm and epiblast in the ICM of the mouse blastocyst. *Developmental Biology*, *384*(1), 65–71. http://doi.org/10.1016/j.ydbio.2013.09.023.

Kruithof-de Julio, M., Alvarez, M. J., Galli, A., Chu, J., Price, S. M., Califano, A., & Shen, M. M. (2011). Regulation of extra-embryonic endoderm stem cell differentiation by Nodal and Cripto signaling. *Development (Cambridge, England)*, *138*(18), 3885–3895. http://doi.org/10.1242/dev.065656.

Kubaczka, C., Senner, C., Araúzo-Bravo, M. J., Sharma, N., Kuckenberg, P., Becker, A., … Schorle, H. (2014). Derivation and maintenance of murine trophoblast stem cells under defined conditions. *Stem Cell Reports*, *2*(2), 232–242. http://doi.org/10.1016/j.stemcr.2013.12.013.

Kubaczka, C., Senner, C. E., Cierlitza, M., et al. (2015). Direct induction of trophoblast stem cells from murine fibroblasts. *Cell Stem Cell*, *17*(5), 557–568. http://doi.org/10.1016/j.stem.2015.08.005.

Kubo, A., Shinozaki, K., Shannon, J. M., Kouskoff, V., Kennedy, M., Woo, S., … Keller, G. (2004). Development of definitive endoderm from embryonic stem cells in culture. *Development (Cambridge, England)*, *131*(7), 1651–1662. http://doi.org/10.1242/dev.01044.

Kuckenberg, P., Buhl, S., Woynecki, T., et al. (2010). The transcription factor TCFAP2C/AP-2gamma cooperates with CDX2 to maintain trophectoderm formation. *Molecular and Cellular Biology*, *30*(13), 3310–3320. http://doi.org/10.1128/MCB.01215-09.

Kuckenberg, P., Peitz, M., Kubaczka, C., Becker, A., Egert, A., Wardelmann, E., … Schorle, H. (2011). Lineage conversion of murine extraembryonic trophoblast stem cells to pluripotent stem cells. *Molecular and Cellular Biology*, *31*(8), 1748–1756. http://doi.org/10.1128/MCB.01047-10.

Kunath, T., Arnaud, D., Uy, G. D., Okamoto, I., Chureau, C., Yamanaka, Y., … Rossant, J. (2005). Imprinted X-inactivation in extra-embryonic endoderm cell lines from mouse blastocysts. *Development (Cambridge, England)*, *132*(7), 1649–1661. http://doi.org/10.1242/dev.01715.

Kunath, T., Saba-El-Leil, M. K., Almousailleakh, M., Wray, J., Meloche, S., & Smith, A. (2007). FGF stimulation of the Erk1/2 signalling cascade triggers transition of pluripotent embryonic stem cells from self-renewal to lineage commitment. *Development (Cambridge, England)*, *134*(16), 2895–2902. http://doi.org/10.1242/dev.02880.

Kurimoto, K., Yabuta, Y., Ohinata, Y., Ono, Y., Uno, K. D., Yamada, R. G., … Saitou, M. (2006). An improved single-cell cDNA amplification method for efficient high-density oligonucleotide microarray analysis. *Nucleic Acids Research*, *34*(5), e42. http://doi.org/10.1093/nar/gkl050.

Latos, P. A., & Hemberger, M. (2014). Review: The transcriptional and signalling networks of mouse trophoblast stem cells. *Placenta*, *35*(Suppl.), S81–S85. http://doi.org/10.1016/j.placenta.2013.10.013.

Le Bin, G. C., Muñoz-Descalzo, S., Kurowski, A., Leitch, H., Lou, X., Mansfield, W., … Nichols, J. (2014). Oct4 is required for lineage priming in the developing inner cell mass of the mouse blastocyst. *Development (Cambridge, England)*, *141*(5), 1001–1010. http://doi.org/10.1242/dev.096875.

Li, Y., McClintick, J., Zhong, L., Edenberg, H. J., Yoder, M. C., & Chan, R. J. (2005). Murine embryonic stem cell differentiation is promoted by SOCS-3 and inhibited by the zinc finger transcription factor Klf4. *Blood*, *105*(2), 635–637. http://doi.org/10.1182/blood-2004-07-2681.

Li, J., Pan, G., Cui, K., Liu, Y., Xu, S., & Pei, D. (2007). A dominant-negative form of mouse SOX2 induces trophectoderm differentiation and progressive polyploidy in mouse embryonic stem cells. *The Journal of Biological Chemistry*, *282*(27), 19481–19492. http://doi.org/10.1074/jbc.M702056200.

Li, P., Tong, C., Mehrian-Shai, R., Jia, L., Wu, N., Yan, Y., … Ying, Q.-L. (2008). Germline competent embryonic stem cells derived from rat blastocysts. *Cell*, *135*(7), 1299–1310. http://doi.org/10.1016/j.cell.2008.12.006.

Lim, C. Y., Tam, W.-L., Zhang, J., Ang, H. S., Jia, H., Lipovich, L., … Lim, B. (2008). Sall4 regulates distinct transcription circuitries in different blastocyst-derived stem cell lineages. *Cell Stem Cell*, *3*(5), 543–554. http://doi.org/10.1016/j.stem.2008.08.004.

Loh, K. M., & Lim, B. (2011). A precarious balance: Pluripotency factors as lineage specifiers. *Cell Stem Cell*, *8*(4), 363–369. http://doi.org/10.1016/j.stem.2011.03.013.

Luo, J., Sladek, R., Bader, J. A., Matthyssen, A., Rossant, J., & Giguère, V. (1997). Placental abnormalities in mouse embryos lacking the orphan nuclear receptor ERR-beta. *Nature*, *388*(6644), 778–782. http://doi.org/10.1038/42022.

Ma, G. T., Soloveva, V., Tzeng, S. J., Lowe, L. A., Pfendler, K. C., Iannaccone, P. M., … Linzer, D. I. (2001). Nodal regulates trophoblast differentiation and placental development. *Developmental Biology*, *236*(1), 124–135. http://doi.org/10.1006/dbio.2001.0334.

Macfarlan, T. S., Gifford, W. D., Driscoll, S., Lettieri, K., Rowe, H. M., Bonanomi, D., … Pfaff, S. L. (2012). Embryonic stem cell potency fluctuates with endogenous retrovirus activity. *Nature*, *487*(7405), 57–63. http://doi.org/10.1038/nature11244.

Martin, G. (1981). Isolation of a pluripotent cell line from early mouse embryos cultured in medium conditioned by teratocarcinoma stem cells. *Proceedings of the National Academy of Sciences of the United States of America*, *78*(12), 7634–7638. Retrieved from, http://www.ncbi.nlm.nih.gov/pmc/articles/PMC349323/.

Mascetti, V. L., & Pedersen, R. A. (2016). Human-mouse chimerism validates human stem cell pluripotency. *Cell Stem Cell*, *18*(1), 67–72. http://doi.org/10.1016/j.stem.2015.11.017.

Masui, S., Nakatake, Y., Toyooka, Y., Shimosato, D., Yagi, R., Takahashi, K., ... Niwa, H. (2007). Pluripotency governed by Sox2 via regulation of Oct3/4 expression in mouse embryonic stem cells. *Nature Cell Biology*, *9*(6), 625–635. http://doi.org/10.1038/ncb1589.

Matsuda, T., Nakamura, T., Nakao, K., Arai, T., Katsuki, M., Heike, T., & Yokota, T. (1999). STAT3 activation is sufficient to maintain an undifferentiated state of mouse embryonic stem cells. *The EMBO Journal*, *18*(15), 4261–4269. http://doi.org/10.1093/emboj/18.15.4261.

McDole, K., Xiong, Y., Iglesias, P. A., & Zheng, Y. (2011). Lineage mapping the pre-implantation mouse embryo by two-photon microscopy, new insights into the segregation of cell fates. *Developmental Biology*, *355*(2), 239–249. http://doi.org/10.1016/j.ydbio.2011.04.024.

McDonald, A. C. H., Biechele, S., Rossant, J., & Stanford, W. L. (2014). Sox17-mediated XEN cell conversion identifies dynamic networks controlling cell-fate decisions in embryo-derived stem cells. *Cell Reports*, *9*(2), 780–793. http://doi.org/10.1016/j.celrep.2014.09.026.

Meilhac, S. M., Adams, R. J., Morris, S. A., Danckaert, A., Le Garrec, J. F., & Zernicka-Goetz, M. (2009). Active cell movements coupled to positional induction are involved in lineage segregation in the mouse blastocyst. *Developmental Biology*, *331*(2), 210–221. http://doi.org/10.1016/j.ydbio.2009.04.036.

Mesnard, D., Donnison, M., Fuerer, C., Pfeffer, P. L., & Constam, D. B. (2011). The micro-environment patterns the pluripotent mouse epiblast through paracrine Furin and Pace4 proteolytic activities. *Genes & Development*, *25*(17), 1871–1880. http://doi.org/10.1101/gad.16738711.

Mesnard, D., Guzman-Ayala, M., & Constam, D. B. (2006). Nodal specifies embryonic visceral endoderm and sustains pluripotent cells in the epiblast before overt axial patterning. *Development (Cambridge, England)*, *133*(13), 2497–2505. http://doi.org/10.1242/dev.02413.

Messerschmidt, D. M., & Kemler, R. (2010). Nanog is required for primitive endoderm formation through a non-cell autonomous mechanism. *Developmental Biology*, *344*(1), 129–137. http://doi.org/10.1016/j.ydbio.2010.04.020.

Mitsui, K., Tokuzawa, Y., Itoh, H., Segawa, K., Murakami, M., Takahashi, K., ... Yamanaka, S. (2003). The homeoprotein Nanog is required for maintenance of pluripotency in mouse epiblast and ES cells. *Cell*, *113*(5), 631–642. http://doi.org/10.1016/S0092-8674(03)00393-3.

Monk, M., & Harper, M. I. (1979). Sequential X chromosome inactivation coupled with cellular differentiation in early mouse embryos. *Nature*, *281*(5729), 311–313. Retrieved from, http://www.ncbi.nlm.nih.gov/pubmed/551278.

Morgani, S. M., Canham, M. A., Nichols, J., Sharov, A. A., Migueles, R. P., Ko, M. S. H., & Brickman, J. M. (2013). Totipotent embryonic stem cells arise in ground-state culture conditions. *Cell Reports*, *3*(6), 1945–1957. http://doi.org/10.1016/j.celrep.2013.04.034.

Murakami, M., Hibi, M., Nakagawa, N., Nakagawa, T., Yasukawa, K., Yamanishi, K., ... Kishimoto, T. (1993). IL-6-induced homodimerization of gp130 and associated activation of a tyrosine kinase. *Science (New York, N.Y.)*, *260*(5115), 1808–1810. Retrieved from, http://www.ncbi.nlm.nih.gov/pubmed/8511589.

Murray, P., & Edgar, D. (2001). Regulation of laminin and COUP-TF expression in extra-embryonic endodermal cells. *Mechanisms of Development*, *101*(1–2), 213–215. http://doi.org/10.1016/S0925-4773(00)00554-2.

Najm, F. J., Chenoweth, J. G., Anderson, P. D., Nadeau, J. H., Redline, R. W., McKay, R. D. G., & Tesar, P. J. (2011). Isolation of epiblast stem cells from preimplantation mouse embryos. *Cell Stem Cell*, *8*(3), 318–325. http://doi.org/10.1016/j.stem.2011.01.016.

Nakatake, Y., Fukui, N., Iwamatsu, Y., et al. (2006). Klf4 cooperates with Oct3/4 and Sox2 to activate the Lefty1 core promoter in embryonic stem cells. *Molecular and Cellular Biology, 26*(20), 7772–7782. http://doi.org/10.1128/MCB.00468-06.

Narazaki, M., Witthuhn, B. A., Yoshida, K., Silvennoinen, O., Yasukawa, K., Ihle, J. N., … Taga, T. (1994). Activation of JAK2 kinase mediated by the interleukin 6 signal transducer gp130. *Proceedings of the National Academy of Sciences of the United States of America, 91*(6), 2285–2289. Retrieved from, http://www.pubmedcentral.nih.gov/articlerender.fcgi?artid=43355&tool=pmcentrez&rendertype=abstract.

Ng, R. K., Dean, W., Dawson, C., et al. (2008). Epigenetic restriction of embryonic cell lineage fate by methylation of Elf5. *Nature Cell Biology, 10*(11), 1280–1290. http://doi.org/10.1038/ncb1786.

Niakan, K. K., Ji, H., Maehr, R., Vokes, S. A., Rodolfa, K. T., Sherwood, R. I., … Eggan, K. (2010). Sox17 promotes differentiation in mouse embryonic stem cells by directly regulating extraembryonic gene expression and indirectly antagonizing self-renewal. *Genes & Development, 24*(3), 312–326. http://doi.org/10.1101/gad.1833510.

Niakan, K. K., Schrode, N., Cho, L. T. Y., & Hadjantonakis, A.-K. (2013). Derivation of extraembryonic endoderm stem (XEN) cells from mouse embryos and embryonic stem cells. *Nature Protocols, 8*(6), 1028–1041. http://doi.org/10.1038/nprot.2013.049.

Nichols, J., Jones, K., Phillips, J. M., Newland, S. A., Roode, M., Mansfield, W., … Cooke, A. (2009). Validated germline-competent embryonic stem cell lines from non-obese diabetic mice. *Nature Medicine, 15*(7), 814–818. http://doi.org/10.1038/nm.1996.

Nichols, J., Silva, J., Roode, M., & Smith, A. (2009). Suppression of Erk signalling promotes ground state pluripotency in the mouse embryo. *Development (Cambridge, England), 136*(19), 3215–3222. http://doi.org/10.1242/dev.038893.

Nichols, J., & Smith, A. (2012). Pluripotency in the embryo and in culture. *Cold Spring Harbor Perspectives in Biology, 4*(8), a008128. http://doi.org/10.1101/cshperspect.a008128.

Nichols, J., Zevnik, B., Anastassiadis, K., Niwa, H., Klewe-Nebenius, D., Chambers, I., … Smith, A. (1998). Formation of pluripotent stem cells in the mammalian embryo depends on the POU transcription factor Oct4. *Cell, 95*(3), 379–391. http://doi.org/10.1016/S0092-8674(00)81769-9.

Nishioka, N., Inoue, K., Adachi, K., et al. (2009). The Hippo signaling pathway components Lats and Yap pattern Tead4 activity to distinguish mouse trophectoderm from inner cell mass. *Developmental Cell, 16*(3), 398–410. http://doi.org/10.1016/j.devcel.2009.02.003.

Niswander, L., & Martin, G. R. (1992). Fgf-4 expression during gastrulation, myogenesis, limb and tooth development in the mouse. *Development (Cambridge, England), 114*(3), 755–768. Retrieved from, http://www.ncbi.nlm.nih.gov/pubmed/1618140.

Niwa, H., Burdon, T., Chambers, I., & Smith, A. (1998). Self-renewal of pluripotent embryonic stem cells is mediated via activation of STAT3. *Genes & Development, 12*(13), 2048–2060. http://doi.org/10.1101/gad.12.13.2048.

Niwa, H., Miyazaki, J., & Smith, A. G. (2000). Quantitative expression of Oct-3/4 defines differentiation, dedifferentiation or self-renewal of ES cells. *Nature Genetics, 24*(4), 372–376. http://doi.org/10.1038/74199.

Niwa, H., Ogawa, K., Shimosato, D., & Adachi, K. (2009). A parallel circuit of LIF signalling pathways maintains pluripotency of mouse ES cells. *Nature, 460*(7251), 118–122. http://doi.org/10.1038/nature08113.

Niwa, H., Toyooka, Y., Shimosato, D., et al. (2005). Interaction between Oct3/4 and Cdx2 determines trophectoderm differentiation. *Cell, 123*(5), 917–929. http://doi.org/10.1016/j.cell.2005.08.040.

O'Leary, T., Heindryckx, B., Lierman, S., van Bruggen, D., Goeman, J. J., Vandewoestyne, M., … De Sutter, P. (2012). Tracking the progression of the human inner cell mass during embryonic stem cell derivation. *Nature Biotechnology, 30*(3), 278–282. http://doi.org/10.1038/nbt.2135.

Odiatis, C., & Georgiades, P. (2010). New insights for Ets2 function in trophoblast using lentivirus-mediated gene knockdown in trophoblast stem cells. *Placenta*, *31*(7), 630–640. http://doi.org/10.1016/j.placenta.2010.05.001.

Ogawa, K., Nishinakamura, R., Iwamatsu, Y., Shimosato, D., & Niwa, H. (2006). Synergistic action of Wnt and LIF in maintaining pluripotency of mouse ES cells. *Biochemical and Biophysical Research Communications*, *343*(1), 159–166. http://doi.org/10.1016/j.bbrc.2006.02.127.

Ohnishi, Y., Huber, W., Tsumura, A., Kang, M., Xenopoulos, P., Kurimoto, K., … Hiiragi, T. (2014). Cell-to-cell expression variability followed by signal reinforcement progressively segregates early mouse lineages. *Nature Cell Biology*, *16*(1), 27–37. http://doi.org/10.1038/ncb2881.

Ohtsuka, S., Nishikawa-Torikai, S., & Niwa, H. (2012). E-cadherin promotes incorporation of mouse epiblast stem cells into normal development. *PloS One*, 7(9), e45220. http://doi.org/10.1371/journal.pone.0045220.

Ornitz, D. M., Yayon, A., Flanagan, J. G., Svahn, C. M., Levi, E., & Leder, P. (1992). Heparin is required for cell-free binding of basic fibroblast growth factor to a soluble receptor and for mitogenesis in whole cells. *Molecular and Cellular Biology*, *12*(1), 240–247. Retrieved from, http://www.pubmedcentral.nih.gov/articlerender.fcgi?artid=364088&tool=pmcentrez&rendertype=abstract.

Orr-Urtreger, A., Bedford, M. T., Burakova, T., Arman, E., Zimmer, Y., Yayon, A., … Lonai, P. (1993). Developmental localization of the splicing alternatives of fibroblast growth factor receptor-2 (FGFR2). *Developmental Biology*, *158*(2), 475–486. http://doi.org/10.1006/dbio.1993.1205.

Osorno, R., Tsakiridis, A., Wong, F., Cambray, N., Economou, C., Wilkie, R., … Wilson, V. (2012). The developmental dismantling of pluripotency is reversed by ectopic Oct4 expression. *Development (Cambridge, England)*, *139*(13), 2288–2298. http://doi.org/10.1242/dev.078071.

Ouhibi, N., Sullivan, N. F., English, J., Colledge, W. H., Evans, M. J., & Clarke, N. J. (1995). Initial culture behaviour of rat blastocysts on selected feeder cell lines. *Molecular Reproduction and Development*, *40*(3), 311–324. http://doi.org/10.1002/mrd.1080400307.

Paca, A., Séguin, C. A., Clements, M., Ryczko, M., Rossant, J., Rodriguez, T. A., & Kunath, T. (2012). BMP signaling induces visceral endoderm differentiation of XEN cells and parietal endoderm. *Developmental Biology*, *361*(1), 90–102. http://doi.org/10.1016/j.ydbio.2011.10.013.

Papaioannou, V. E. (1982). Lineage analysis of inner cell mass and trophectoderm using microsurgically reconstituted mouse blastocysts. *Journal of Embryology and Experimental Morphology*, *68*(1979), 199–209. Retrieved from, http://www.ncbi.nlm.nih.gov/pubmed/7108422.

Parast, M. M., Yu, H., Ciric, A., Salata, M. W., Davis, V., & Milstone, D. S. (2009). PPARgamma regulates trophoblast proliferation and promotes labyrinthine trilineage differentiation. *PloS One*, *4*(11), e8055. http://doi.org/10.1371/journal.pone.0008055.

Park, I.-H., Zhao, R., West, J. A., Yabuuchi, A., Huo, H., Ince, T. A., … Daley, G. Q. (2008). Reprogramming of human somatic cells to pluripotency with defined factors. *Nature*, *451*(7175), 141–146. http://doi.org/10.1038/nature06534.

Plusa, B., Piliszek, A., Frankenberg, S., Artus, J., & Hadjantonakis, A.-K. (2008). Distinct sequential cell behaviours direct primitive endoderm formation in the mouse blastocyst. *Development (Cambridge, England)*, *135*(18), 3081–3091. http://doi.org/10.1242/dev.021519.

Qi, X., Li, T.-G., Hao, J., Hu, J., Wang, J., Simmons, H., … Zhao, G.-Q. (2004). BMP4 supports self-renewal of embryonic stem cells by inhibiting mitogen-activated protein kinase pathways. *Proceedings of the National Academy of Sciences of the United States of America*, *101*(16), 6027–6032. http://doi.org/10.1073/pnas.0401367101.

Ralston, A., Cox, B. J., Nishioka, N., et al. (2010). Gata3 regulates trophoblast development downstream of Tead4 and in parallel to Cdx2. *Development*, *137*(3), 395–403. http://doi.org/10.1242/dev.038828.

Reinholdt, L. G., Howell, G. R., Czechanski, A. M., Macalinao, D. G., Macnicoll, K. H., Lin, C.-S., ... John, S. W. M. (2012). Generating embryonic stem cells from the inbred mouse strain DBA/2J, a model of glaucoma and other complex diseases. *PloS One*, 7(11), e50081. http://doi.org/10.1371/journal.pone.0050081.

Riley, P., Anson-Cartwright, L., & Cross, J. C. (1998). The Hand1 bHLH transcription factor is essential for placentation and cardiac morphogenesis. *Nature Genetics*, *18*(3), 271–275. http://doi.org/10.1038/ng0398-271.

Rodda, D. J., Chew, J.-L., Lim, L.-H., et al. (2005). Transcriptional regulation of nanog by OCT4 and SOX2. *The Journal of Biological Chemistry*, *280*(26), 24731–24737. http://doi.org/10.1074/jbc.M502573200.

Rossant, J., & Ofer, L. (1977). Properties of extra-embryonic ectoderm isolated from postimplantation mouse embryos. *Journal of Embryology and Experimental Morphology*, *39*, 183–194. Retrieved from, http://www.ncbi.nlm.nih.gov/pubmed/560422.

Saiz, N., Grabarek, J. B., Sabherwal, N., Papalopulu, N., & Plusa, B. (2013). Atypical protein kinase C couples cell sorting with primitive endoderm maturation in the mouse blastocyst. *Development (Cambridge, England)*, *140*, 4311–4322. http://doi.org/10.1242/dev.093922.

Saksela, O., Moscatelli, D., Sommer, A., & Rifkin, D. B. (1988). Endothelial cell-derived heparan sulfate binds basic fibroblast growth factor and protects it from proteolytic degradation. *The Journal of Cell Biology*, *107*(2), 743–751. Retrieved from, http://www.pubmedcentral.nih.gov/articlerender.fcgi?artid=2115214&tool=pmcentrez&rendertype=abstract.

Sasaki, E., Hanazawa, K., Kurita, R., Akatsuka, A., Yoshizaki, T., Ishii, H., ... Tani, K. (2005). Establishment of novel embryonic stem cell lines derived from the common marmoset (Callithrix jacchus). *Stem Cells (Dayton, Ohio)*, *23*(9), 1304–1313. http://doi.org/10.1634/stemcells.2004-0366.

Sato, N., Meijer, L., Skaltsounis, L., Greengard, P., & Brivanlou, A. H. (2004). Maintenance of pluripotency in human and mouse embryonic stem cells through activation of Wnt signaling by a pharmacological GSK-3-specific inhibitor. *Nature Medicine*, *10*(1), 55–63. http://doi.org/10.1038/nm979.

Schenke-Layland, K., Angelis, E., Rhodes, K. E., Heydarkhan-Hagvall, S., Mikkola, H. K., & Maclellan, W. R. (2007). Collagen IV induces trophectoderm differentiation of mouse embryonic stem cells. *Stem Cells (Dayton, Ohio)*, *25*(6), 1529–1538. http://doi.org/10.1634/stemcells.2006-0729.

Schier, A. F. (2003). Nodal signaling in vertebrate development. *Annual Review of Cell and Developmental Biology*, *19*, 589–621. http://doi.org/10.1146/annurev.cellbio.19.041603.094522.

Schrode, N., Xenopoulos, P., Piliszek, A., Frankenberg, S., Plusa, B., & Hadjantonakis, A.-K. (2013). Anatomy of a blastocyst: Cell behaviors driving cell fate choice and morphogenesis in the early mouse embryo. *Genesis (New York, NY: 2000)*, *51*(4), 219–233. http://doi.org/10.1002/dvg.22368.

Scott, I. C., Anson-Cartwright, L., Riley, P., Reda, D., & Cross, J. C. (2000). The HAND1 basic helix-loop-helix transcription factor regulates trophoblast differentiation via multiple mechanisms. *Molecular and Cellular Biology*, *20*(2), 530–541. Retrieved from, http://mcb.asm.org/content/20/2/530.short.

Senner, C. E., Krueger, F., Oxley, D., Andrews, S., & Hemberger, M. (2012). DNA methylation profiles define stem cell identity and reveal a tight embryonic-extraembryonic lineage boundary. *Stem Cells (Dayton, Ohio)*, *30*(12), 2732–2745. http://doi.org/10.1002/stem.1249.

Shen, M. M., & Leder, P. (1992). Leukemia inhibitory factor is expressed by the preimplantation uterus and selectively blocks primitive ectoderm formation in vitro. *Proceedings of the National Academy of Sciences of the United States of America*, *89*(17), 8240–8244. Retrieved from, http://www.pubmedcentral.nih.gov/articlerender.fcgi?artid=49893&tool=pmcentrez&rendertype=abstract.

Shimoda, M., Kanai-Azuma, M., Hara, K., Miyazaki, S., Kanai, Y., Monden, M., & Miyazaki, J. (2007). Sox17 plays a substantial role in late-stage differentiation of the extra-embryonic endoderm in vitro. *Journal of Cell Science*, *120*(Pt. 21), 3859–3869. http://doi.org/10.1242/jcs.007856.

Shimosato, D., Shiki, M., & Niwa, H. (2007). Extra-embryonic endoderm cells derived from ES cells induced by GATA factors acquire the character of XEN cells. *BMC Developmental Biology*, 7(80), 1–12. http://doi.org/10.1186/1471-213X-7-80.

Shu, J., Zhang, K., Zhang, M., Yao, A., Shao, S., Du, F., … Deng, H. (2015). GATA family members as inducers for cellular reprogramming to pluripotency. *Cell Research*, *25*(2), 169–180. http://doi.org/10.1038/cr.2015.6.

Silva, J., Nichols, J., Theunissen, T. W., Guo, G., van Oosten, A. L., Barrandon, O., … Smith, A. (2009). Nanog is the gateway to the pluripotent ground state. *Cell*, *138*(4), 722–737. http://doi.org/10.1016/j.cell.2009.07.039.

Simmons, D. G., & Cross, J. C. (2005). Determinants of trophoblast lineage and cell subtype specification in the mouse placenta. *Developmental Biology*, *284*(1), 12–24. http://doi.org/10.1016/j.ydbio.2005.05.010.

Singh, A. M., Hamazaki, T., Hankowski, K. E., & Terada, N. (2007). A heterogeneous expression pattern for Nanog in embryonic stem cells. *Stem Cells (Dayton, Ohio)*, *25*(10), 2534–2542. http://doi.org/10.1634/stemcells.2007-0126.

Smith, A. G., Heath, J. K., Donaldson, D. D., Wong, G. G., Moreau, J., Stahl, M., & Rogers, D. (1988). Inhibition of pluripotential embryonic stem cell differentiation by purified polypeptides. *Nature*, *336*(6200), 688–690. http://doi.org/10.1038/336688a0.

Soprano, D. R., Teets, B. W., & Soprano, K. J. (2007). Role of retinoic acid in the differentiation of embryonal carcinoma and embryonic stem cells. *Vitamins and Hormones*, *75*(6), 69–95. http://doi.org/10.1016/S0083-6729(06)75003-8.

Strickland, S., Smith, K. K., & Marotti, K. R. (1980). Hormonal induction of differentiation in teratocarcinoma stem cells: Generation of parietal endoderm by retinoic acid and dibutyryl cAMP. *Cell*, *21*(2), 347–355. http://doi.org/10.1016/0092-8674(80)90471-7.

Suemori, H., Tada, T., Torii, R., Hosoi, Y., Kobayashi, K., Imahie, H., … Nakatsuji, N. (2001). Establishment of embryonic stem cell lines from cynomolgus monkey blastocysts produced by IVF or ICSI. *Developmental Dynamics: An Official Publication of the American Association of Anatomists*, *222*(2), 273–279. http://doi.org/10.1002/dvdy.1191.

Tahiliani, M., Koh, K. P., Shen, Y., Pastor, W. A., Bandukwala, H., Brudno, Y., … Rao, A. (2009). Conversion of 5-methylcytosine to 5-hydroxymethylcytosine in mammalian DNA by MLL partner TET1. *Science (New York, NY)*, *324*(5929), 930–935. http://doi.org/10.1126/science.1170116.

Takahashi, K., Tanabe, K., Ohnuki, M., Narita, M., Ichisaka, T., Tomoda, K., & Yamanaka, S. (2007). Induction of pluripotent stem cells from adult human fibroblasts by defined factors. *Cell*, *131*(5), 861–872. http://doi.org/10.1016/j.cell.2007.11.019.

Takahashi, K., & Yamanaka, S. (2006). Induction of pluripotent stem cells from mouse embryonic and adult fibroblast cultures by defined factors. *Cell*, *126*(4), 663–676. http://doi.org/10.1016/j.cell.2006.07.024.

Takashima, Y., Guo, G., Loos, R., Nichols, J., Ficz, G., Krueger, F., … Smith, A. (2014). Resetting transcription factor control circuitry toward ground-state pluripotency in human. *Cell*, *158*(6), 1254–1269. http://doi.org/10.1016/j.cell.2014.08.029.

Tan, T., Tang, X., Zhang, J., Niu, Y., Chen, H., Li, B., … Ji, W. (2011). Generation of trophoblast stem cells from rabbit embryonic stem cells with BMP4. *PloS One*, *6*(2), e17124. http://doi.org/10.1371/journal.pone.0017124.

Tanaka, S., Kunath, T., Hadjantonakis, A. K., Nagy, A., & Rossant, J. (1998). Promotion of trophoblast stem cell proliferation by FGF4. *Science (New York, NY)*, *282*(5396), 2072–2075. http://doi.org/10.1126/science.282.5396.2072.

Tang, F., Barbacioru, C., Bao, S., Lee, C., Nordman, E., Wang, X., … Surani, M. A. (2010). Tracing the derivation of embryonic stem cells from the inner cell mass by single-cell RNA-Seq analysis. *Cell Stem Cell*, *6*(5), 468–478. http://doi.org/10.1016/j.stem.2010.03.015.

Tesar, P. J. (2005). Derivation of germ-line-competent embryonic stem cell lines from preblastocyst mouse embryos. *Proceedings of the National Academy of Sciences of the United States of America*, *102*(23), 8239–8244. http://doi.org/10.1073/pnas.0503231102.

Tesar, P. J., Chenoweth, J. G., Brook, F. A., Davies, T. J., Evans, E. P., Mack, D. L., … McKay, R. D. G. (2007). New cell lines from mouse epiblast share defining features with human embryonic stem cells. *Nature*, *448*(7150), 196–199. http://doi.org/10.1038/nature05972.

Theunissen, T. W., Powell, B. E., Wang, H., Mitalipova, M., Faddah, D. A., Reddy, J., … Jaenisch, R. (2014). Systematic identification of culture conditions for induction and maintenance of naive human pluripotency. *Cell Stem Cell*, *15*(4), 471–487. http://doi.org/10.1016/j.stem.2014.07.002.

Thomson, J. A., Itskovitz-Eldor, J., Shapiro, S. S., Waknitz, M. A., Swiergiel, J. J., Marshall, V. S., & Jones, J. M. (1998). Embryonic stem cell lines derived from human blastocysts. *Science (New York, N.Y.)*, *282*(5391), 1145–1147. http://doi.org/10.1126/science.282.5391.1145.

Thomson, J. A., Kalishman, J., Golos, T. G., Durning, M., Harris, C. P., Becker, R. A., & Hearn, J. P. (1995). Isolation of a primate embryonic stem cell line. *Proceedings of the National Academy of Sciences of the United States of America*, *92*(17), 7844–7848. http://doi.org/PMC41242.

Thomson, J. A., Kalishman, J., Golos, T. G., Durning, M., Harris, C. P., & Hearn, J. P. (1996). Pluripotent cell lines derived from common marmoset (Callithrix jacchus) blastocysts. *Biology of Reproduction*, *55*(2), 254–259. http://doi.org/10.1172/JCI200521137.102.

Tolkunova, E., Cavaleri, F., Eckardt, S., Reinbold, R., Christenson, L. K., Schöler, H. R., & Tomilin, A. (2006). The caudal-related protein cdx2 promotes trophoblast differentiation of mouse embryonic stem cells. *Stem Cells (Dayton, Ohio)*, *24*(1), 139–144. http://doi.org/10.1634/stemcells.2005-0240.

Tomioka, M., Nishimoto, M., Miyagi, S., et al. (2002). Identification of Sox-2 regulatory region which is under the control of Oct-3/4-Sox-2 complex. *Nucleic Acids Research*, *30*(14), 3202–3213. http://doi.org/10.1093/nar/gkf435.

Toyooka, Y., Shimosato, D., Murakami, K., Takahashi, K., & Niwa, H. (2008). Identification and characterization of subpopulations in undifferentiated ES cell culture. *Development (Cambridge, England)*, *135*(5), 909–918. http://doi.org/10.1242/dev.017400.

Ueda, J., Maehara, K., Mashiko, D., Ichinose, T., Yao, T., Hori, M., … Yamagata, K. (2014). Heterochromatin dynamics during the differentiation process revealed by the DNA methylation reporter mouse, MethylRO. *Stem Cell Reports*, *2*(6), 910–924. http://doi.org/10.1016/j.stemcr.2014.05.008.

Uy, G. D., Downs, K. M., & Gardner, R. L. (2002). Inhibition of trophoblast stem cell potential in chorionic ectoderm coincides with occlusion of the ectoplacental cavity in the mouse. *Development (Cambridge, England)*, *129*(16), 3913–3924. Retrieved from, http://www.ncbi.nlm.nih.gov/pubmed/12135928.

Vallier, L., Alexander, M., & Pedersen, R. A. (2005). Activin/Nodal and FGF pathways cooperate to maintain pluripotency of human embryonic stem cells. *Journal of Cell Science*, *118*(Pt. 19), 4495–4509. http://doi.org/10.1242/jcs.02553.

Vallier, L., Reynolds, D., & Pedersen, R. A. (2004). Nodal inhibits differentiation of human embryonic stem cells along the neuroectodermal default pathway. *Developmental Biology, 275*(2), 403–421. http://doi.org/10.1016/j.ydbio.2004.08.031.

van de Stolpe, A., Karperien, M., Löwik, C. W., Jüppner, H., Segre, G. V., Abou-Samra, A. B., … Defize, L. H. (1993). Parathyroid hormone-related peptide as an endogenous inducer of parietal endoderm differentiation. *The Journal of Cell Biology, 120*(1), 235–243. Retrieved from, http://www.pubmedcentral.nih.gov/articlerender.fcgi?artid=2119495&tool=pmcentrez&rendertype=abstract.

Vandevoort, C. A., Thirkill, T. L., & Douglas, G. C. (2007). Blastocyst-derived trophoblast stem cells from the rhesus monkey. *Stem Cells and Development, 16*(5), 779–788. http://doi.org/10.1089/scd.2007.0020.

Vassilieva, S., Guan, K., Pich, U., & Wobus, A. M. (2000). Establishment of SSEA-1- and Oct-4-expressing rat embryonic stem-like cell lines and effects of cytokines of the IL-6 family on clonal growth. *Experimental Cell Research, 258*(2), 361–373. http://doi.org/10.1006/excr.2000.4940.

Verheijen, M. H., Karperien, M., Chung, U., van Wijuen, M., Heystek, H., Hendriks, J. A., … Defize, L. H. (1999). Parathyroid hormone-related peptide (PTHrP) induces parietal endoderm formation exclusively via the type I PTH/PTHrP receptor. *Mechanisms of Development, 81*(1–2), 151–161. Retrieved from, http://www.ncbi.nlm.nih.gov/pubmed/10330492.

Waghray, A., Saiz, N., Jayaprakash, A. D., et al. (2015). Tbx3 controls Dppa3 levels and exit from pluripotency toward mesoderm. *Stem Cell Reports, 5*(1), 97–110. http://doi.org/10.1016/j.stemcr.2015.05.009.

Ware, C. B., Nelson, A. M., Mecham, B., Hesson, J., Zhou, W., Jonlin, E. C., … Ruohola-Baker, H. (2014). Derivation of naive human embryonic stem cells. *Proceedings of the National Academy of Sciences of the United States of America, 111*(12), 4484–4489. http://doi.org/10.1073/pnas.1319738111.

Watanabe, T., Biggins, J. S., Tannan, N. B., & Srinivas, S. (2014). Limited predictive value of blastomere angle of division in trophectoderm and inner cell mass specification. *Development (Cambridge, England), 141*(11), 2279–2288. http://doi.org/10.1242/dev.103267.

Wen, F., Tynan, J. A., Cecena, G., Williams, R., Múnera, J., Mavrothalassitis, G., & Oshima, R. G. (2007). Ets2 is required for trophoblast stem cell self-renewal. *Developmental Biology, 312*(1), 284–299. http://doi.org/10.1016/j.ydbio.2007.09.024.

Wernig, M., Meissner, A., Foreman, R., et al. (2007). In vitro reprogramming of fibroblasts into a pluripotent ES-cell-like state. *Nature, 448*(7151), 318–324. http://doi.org/10.1038/nature05944.

Wu, J., Okamura, D., Li, M., Suzuki, K., Luo, C., Ma, L., … Izpisua Belmonte, J. C. (2015). An alternative pluripotent state confers interspecies chimaeric competency. *Nature, 521*(7552), 316–321. http://doi.org/10.1038/nature14413.

Wu, T., Wang, H., He, J., Kang, L., Jiang, Y., Liu, J., … Gao, S. (2011). Reprogramming of trophoblast stem cells into pluripotent stem cells by Oct4. *Stem Cells (Dayton, Ohio), 29*(5), 755–763. http://doi.org/10.1002/stem.617.

Xenopoulos, P., Kang, M., Puliafito, A., Di Talia, S., & Hadjantonakis, A.-K. (2015). Heterogeneities in Nanog expression drive stable commitment to pluripotency in the mouse blastocyst. *Cell Reports, 10*(9), 1508–1520. http://doi.org/10.1016/j.celrep.2015.02.010.

Yamanaka, Y., Lanner, F., & Rossant, J. (2010). FGF signal-dependent segregation of primitive endoderm and epiblast in the mouse blastocyst. *Development (Cambridge, England), 137*(5), 715–724. http://doi.org/10.1242/dev.043471.

Yan, J., Tanaka, S., Oda, M., Makino, T., Ohgane, J., & Shiota, K. (2001). Retinoic acid promotes differentiation of trophoblast stem cells to a giant cell fate. *Developmental Biology, 235*(2), 422–432. http://doi.org/10.1006/dbio.2001.0300.

Yan, L., Yang, M., Guo, H., Yang, L., Wu, J., Li, R., … Tang, F. (2013). Single-cell RNA-Seq profiling of human preimplantation embryos and embryonic stem cells. *Nature Structural & Molecular Biology*, *20*(9), 1131–1139. http://doi.org/10.1038/nsmb.2660.

Ying, Y., Liu, X. M., Marble, A., Lawson, K. A., & Zhao, G. Q. (2000). Requirement of Bmp8b for the generation of primordial germ cells in the mouse. *Molecular Endocrinology*, *14*(7), 1053–1063. http://doi.org/10.1210/mend.14.7.0479.

Ying, Q.-L., Nichols, J., Chambers, I., & Smith, A. (2003). BMP induction of Id proteins suppresses differentiation and sustains embryonic stem cell self-renewal in collaboration with STAT3. *Cell*, *115*(3), 281–292. http://doi.org/10.1016/S0092-8674(03)00847-X.

Ying, Q.-L., Wray, J., Nichols, J., Batlle-Morera, L., Doble, B., Woodgett, J., … Smith, A. (2008). The ground state of embryonic stem cell self-renewal. *Nature*, *453*(7194), 519–523. http://doi.org/10.1038/nature06968.

Ying, Y., & Zhao, G. Q. (2001). Cooperation of endoderm-derived BMP2 and extra-embryonic ectoderm-derived BMP4 in primordial germ cell generation in the mouse. *Developmental Biology*, *232*(2), 484–492. http://doi.org/10.1006/dbio.2001.0173.

Yu, J., Vodyanik, M. A., Smuga-Otto, K., Antosiewicz-Bourget, J., Frane, J. L., Tian, S., … Thomson, J. A. (2007). Induced pluripotent stem cell lines derived from human somatic cells. *Science (New York, NY)*, *318*(5858), 1917–1920. http://doi.org/10.1126/science.1151526.

Yuan, H., Corbi, N., Basilico, C., & Dailey, L. (1995). Developmental-specific activity of the FGF-4 enhancer requires the synergistic action of Sox2 and Oct-3. *Genes & Development*, *9*(21), 2635–2645. http://doi.org/10.1101/gad.9.21.2635.

Zdravkovic, T., Nazor, K. L., Larocque, N., et al. (2015). Human stem cells from single blastomeres reveal pathways of embryonic or trophoblast fate specification. *Development*, *142*(23), 4010–4025. http://doi.org/10.1242/dev.122846.

CHAPTER ELEVEN

Species-Specific Variation Among Mammals

R.A. Reijo Pera[*,1], **L. Prezzoto**[†]
[*]Montana State University, Bozeman, MT, United States
[†]Agricultural Research Centers, Montana State University, Bozeman, MT, United States
[1]Corresponding author: e-mail address: renee.reijopera@montana.edu

Contents

Abstract

Reproduction across mammalian species is conserved with a general pattern of fertilization followed by nascent embryo development in transcriptional silence for a variable length of time, a series of cleavage divisions that occur without growth in size of the embryo, compaction to form a morula, and production of a blastocyst. Following blastocyst formation, the embryo may implant immediately or after substantial differentiation of the epiblast and hypoblast layers. In this chapter, the shared and unique properties of several species, commonly used in studies of reproduction and embryology, are outlined.

Current Topics in Developmental Biology, Volume 120
ISSN 0070-2153
http://dx.doi.org/10.1016/bs.ctdb.2016.05.003

1. INTRODUCTION TO COMPARATIVE EMBRYOLOGY

Aristotle (384–322 BC) is often considered the "father of modern embryology." Although many others over the centuries preceding Aristotle's birth had pondered how human and animal development occurs, Aristotle used an experimental embryology approach to understanding development of the embryo (Thompson, 1910). In his famous work *On the Generation of Animals*, he reported the dissection of bird eggs over the course of their development and concluded: "Generation from the egg proceeds in an identical manner with all birds, but the full periods from conception to birth differ, as has been said. With the common hen after three days and three nights there is the first indication of the embryo; with larger birds the interval being longer, with smaller birds shorter." Aristotle then went on to describe timing of appearance of various organ systems and concluded that when the egg is 10 days old, all of "its parts are distinctly visible." As shown in Fig. 1, this work went on to be reproduced in many drawings and the

Fig. 1 Embryo chicks at different stages of development. *Historical figure from Fabricius ab Aquapendente, H. (1621).* De Formatine Ovi et Pulli. *Padua.*

chick became a premiere model organism. Remarkably, Aristotle implied that the core programs of development might be conserved across many species including humans. In essence, he eluted to the principle of universality that implies that important processes in development are conserved and laid the foundation for comparative embryology. Here, key characteristics of embryo development are compared beginning with an overview of the development of the chicken embryo and then focusing on mammalian species.

2. NONMAMMALIAN DEVELOPMENT: THE CHICKEN EMBRYO

Any discussion of species-specific differences in embryo development should include at least a brief discussion of the relevance of the chicken to our current methods and understanding. Given the ease of obtaining chicken embryos, the chick embryo has a unique status in studies of embryology. Major discoveries in the chicken include delineation of the function of arteries and veins (in 1628), identification of neurites and somites (1672) and the primary germ layers (1817–28), construction of the first genetic map (1936), establishment of a staging atlas (1951), and identification of the first oncogene (1976). Many researchers over the centuries, to the modern day, have explored the development of the chick embryo including Harvey, Malpighi, Rous, Waddington, Hutt, Varmus, Bishop, and Tabin (Stern, 2005). Advantages of the chicken embryo include accessibility, cost, ease of culture, development of precise methods for lineage tracing, ability to form chicken:quail chimeras, and visualization of early events. Disadvantages have been the absence of efficient transgenesis methods especially homologous recombination, inaccessibility to later stages of development (relative to species such as the zebrafish, for example), and difficulties in generating pluripotent stem cells for in vitro studies (Stern, 2005).

Nonetheless, the fundamental processes of chicken embryo development laid the groundwork for studies in other species with the production of the chicken staging atlas in 1951 (Hamburger & Hamilton, 1992). Development of the chicken embryo is apparent from the first day, when the egg is laid, as a thickening at the caudal end of the embryo that is termed the primitive streak. As the development moves to the second day, the blood islands form and the heart forms and begins to beat by approximately 44–48 h. By the end of the third day, the beak has begun to form and limb buds can be visualized. Remarkably by the fourth day, all of the major organ systems can be seen and through the next week, these organ systems are elaborated with formation of digits, enclosure of the heart in the thoracic cavity, and development of feather tracts and feathers. The chicken embryo will generally hatch on approximately day 21.

2.1 The Chicken Genome

Over the decades, many molecules and developmental determinants have been identified (Burt, 2005). Many, if not most, are conserved though timing of function and action may have diverged. The chicken genome sequence was first reported in 2004 (Project, International Chicken Genome Sequencing Consortium, 2004). Analysis indicates that there are approximately 20,000 genes with the majority having at least limited homology to the mouse and human (Burt, 2005). Although the genome is compact, consisting of approximately XMB, the chicken embryo like those of other species is populated by repetitive DNA arising from retroviral integration, duplications, and trimming of sequence, though only 11% of the chicken genome is repetitive (relative to >40–50% in mammalian species) (Project, International Chicken Genome Sequencing Consortium, 2004). The events of early development have been tracked at the whole-genome transcriptional level and reveal conserved and divergent pathways that have formed the basis for comparative studies with mammals.

3. RABBIT EMBRYO DEVELOPMENT

Following establishment of the chicken as a model for embryo development, the rabbit gained prominence with discoveries that included the identification of Hensen's node (also termed the primitive node) in 1876 that acts as an organizer of gastrulation (Viebahn, 2001) and in 1890 the first embryo transfer (ET) that resulted in birth of live young rabbits (Heape, 1891). That these discoveries of fundamental pathways and methods of reproduction were made in rabbits as a model was natural given key species-specific traits of rabbit embryology (Fischer, Chavatte-Palmer, Viebahn, Santos, & Duranthon, 2012). The unique combination of traits in the rabbit includes induction of ovulation with mating to allow precise timing of ovulation, a high number of cells in the blastocyst with the largest size of any mammal (~1 mm in size on day 6), embryonic genome activation that occurs predominantly over the course of the first few days of development (similar to humans), initiation of the early stages of gastrulation prior to implantation allowing observation and staging of gastrulation, and placentation that closely resembles that of humans in many aspects (Fischer et al., 2012). Notably, the rabbit–primate evolutionary distance is similar to that of the rodent–primate distance but is characterized by greater conservation of gene sequences (Graur, Duret, & Gouy, 1996).

Rabbit embryo development begins with ovulation at mating followed by fertilization approximately 10 h postcoitus. Cleavage divisions proceed for the first 2 days, followed by compaction late on day 2, and formation of the blastocyst on day 3. Remarkably in contrast to other species, the rabbit blastocyst is comprised of thousands of cells by day 5, increasing with the onset of gastrulation to 80,000–100,000 cells until implantation begins late on day 6. Pregnancy in rabbits is 31 days in length with >80% of embryos typically resulting in live birth (Adams, 1962).

3.1 Gene Expression

Over the last two decades, classical studies of rabbit embryology have been augmented with molecular genetic studies of development with important observations especially related to embryonic genome activation, the timing and mechanisms underlying X-chromosome inactivation (XCI) in the rabbit, and effects of environmental factors on gene expression profiles. Gene expression analysis indicates that there are seven discernable patterns of gene expression characterized as HLL (high–low–low indicating high expression at genome activation, low expression at the morula stage, and continued low expression in the blastocyst) and variations from high to low expression and vice versa through these stages of development (Leandri et al., 2009). The overall patterns suggest a gradual staged activation of the rabbit genome from early preimplantation development to the blastocyst stage with subsets of genes displaying "bursts" of expression throughout early development and others intermittently being expressed or maintained throughout this time period. The relatively late activation of the genome is in contrast to the mouse (where genome activation occurs predominantly at the 1- to 2-cell stage) and is similar to that of other mammalian species.

3.2 X-chromosome Inactivation

X-chromosome dosage is compensated between both sexes in mammals via inactivation of one of the two parental X chromosomes while maintaining active expression of a subset of genes on both chromosomes. Genes that escape X inactivation are generally those associated with Y-chromosome homologs. The process of XCI has been most thoroughly studied in mice

(Disteche & Berletch, 2015; Kamikawa & Donohoe, 2014). In mice, the paternal X chromosome is imprinted to be silenced during early embryo development via a mechanism whereby imprinted expression is established by expression of the noncoding RNA, *Xist*, that represses transcription from the paternal X chromosome (Okamoto et al., 2011). Subsequently the paternal X chromosome reactivated in the inner cell mass of the blastocyst and random XCI selection ensues as differentiated cell lineages form. In contrast, in other species including the rabbit and human, *Xist* is not imprinted and XCI begins later in development than occurs in mice and *Xist* is expressed from either the maternal or the paternal X chromosome in some cases resulting in transient inactivation of both X chromosomes (Okamoto et al., 2011). Subsequently in the rabbit as development ensues, the choice of which X chromosome will become inactive occurs downstream of *Xist* expression; furthering defining species-specific differences is the skewing of X inactivation in some tissues and the identity of the individual genes on both X chromosomes that escape XCI (Deng, Berletch, Nguyen, & Disteche, 2014).

4. BOVINE EMBRYO DEVELOPMENT

In vitro fertilization (IVF) and ET to produce bovine progeny have a long history beginning in the late 1940s with pioneering experiments on ET that led to ET and ultimately successful full-term pregnancy in the 1950s (Betteridge, 2003; Umbaugh, 1949; Willett, Black, Casida, Stone, & Buckner, 1951). Use of IVF and ET was spurred by the desire to maintain and optimize bovine genetic lines and to easily transport genetic stock between jurisdictions with minimal opportunity to transmit diseases (Seidel & Seidel, 1991). Today, more than 750,000 embryos are produced annually from superovulated donors and more than 450,000 embryos are produced via IVF (Betteridge, 2003; Mapletoft, 2013). The time course of bovine embryo development is similar to that of humans. The first cleavage division occurs on day 1, the major wave of genome activation occurs at the 4- to 8-cell stage, the morula is commonly observed on days 4–5, and the blastocyst stage appears on days 6 and 7, though some delay in these time points has been noted in some reports (Seidel & Seidel, 1991). In contrast to many other species, however, the bovine embryo then goes through a stage of elongation prior to attachment at approximately days 40–50 days following fertilization (Chang, 1952).

4.1 Bovine Embryo Loss and Large Offspring Syndrome

Bovine embryogenesis in vitro has been shown to differ from that in vivo in terms of some aspects of developmental timing, gene expression, and outcomes (Bó & Mapletoft, 2013). Indeed, bovine embryos provide a very attractive experimental model for the investigation of chromosome instability, culture-related embryo loss, and impacts of culture on long-term pre- and postnatal outcomes (Bó & Mapletoft, 2013; Saadeldin, Kim, Lee, & Jang, 2011). Routine observations indicate that more than 50% of bovine embryos are not able to produce viable pregnancies. Moreover, embryo fragmentation, disruptions of the spindle, chromosomal errors, and alterations of epigenetic programs are frequently associated with bovine embryos loss in vitro and in vivo (Bó & Mapletoft, 2013; Lazzari et al., 2006; Saadeldin et al., 2011).

Adverse long-term effects of bovine IVF include pregnancy loss, metabolic perturbations related to culture media, and an increase in frequency of large offspring syndrome (LOS) (Greve, Callesen, Hyttel, & Avery, 1995; Lazzari et al., 2006; Young, Sinclair, & Wilmut, 1998). LOS is observed primarily in bovine and ovine offspring following transfer of IVF-derived or cloned embryos and is characterized by diverse pathologies with an extended gestation length and increased birthweight as predominant features. Data suggest that LOS and other abnormalities of development and outcomes may be linked to the embryo culture conditions and media and increased length of culture; moreover, changes in gene expression, timing of development, and number of cells prior to the blastocyst stage are correlated with LOS (Lazzari et al., 2006; Rizos et al., 2003; Young et al., 1998).

5. OVERVIEW OF MOUSE EMBRYO DEVELOPMENT

Several chapters have already described mouse embryo development in detail. The mouse holds a prominent place in experimental embryology and developmental biology (Behringer, Gertsenstein, Nagy, & Nagy, 2014). This is in large part due to its relatively rapid life cycle, large number of progeny, ease of maintaining colonies, and genetic tractability especially in regard to isolation and construction of mutant lines. Moreover, it is clear that data from the mouse can be extrapolated to other species including data that detail the molecular and genetic determinants of cell fate, cell–cell interactions during embryogenesis, and developmental stages. Nonetheless, a combination of factors sets the mouse apart as a specialized mammal. These factors include the maternal centrosome generation, entry of embryos into diapause, early genome activation, and the ability to develop to fully mature and fertile adults with a single X chromosome (XO genotype). These are described briefly later.

5.1 Centrosome Generation

The centrosome is a cellular organelle that is composed of two centrioles surrounded by pericentriolar material and is involved in diverse cellular processes including sensory reception, locomotion, and embryogenesis (Clift & Schuh, 2013). In 1901, Bovari published his famous observations on chromosome segregation and the role of the centrosome in the roundworm *Ascaris* in which he observed that the inheritance of the centrosome was paternal, with the maternal centrosome degrading upon fertilization (Boveri, 1901). Over the last century, this observation has spurred many studies on the "inheritance of centrosomes" in which the replication of the centrosome has been thought to occur via a centrosomal replication cycle that requires existing centrosomes as a template for the generation of new centrosomes. However, research over the last decade has clearly indicated that centrosomes are derived de novo in most cases; moreover, it is clear that in mammalian species, the centrosome of the oocyte generally degrades prior to fertilization (Suthananthan et al., 1999). The mouse may be the exception in that there is reliance upon the maternal centrosome which does not degrade in the oocyte prior to or in the first cell cycle following fertilization; indeed, depletion of key components of the oocyte centrosome, the protein geminin, results in impaired embryo development and reduced fertility (Ma et al., 2016). In mammalian species, the centrosome is required for spermatogenesis and is passed to the zygote at fertilization (Suthananthan et al., 1999). Thus, historically, it has been suggested that the inheritance of the male centrosome in embryogenesis is a requirement (Ross & Normark, 2015). However, it is likely that in most species, the centrosome is generated de novo during embryogenesis (Ross & Normark, 2015).

5.2 Embryo Diapause

Embryonic diapause is defined as the temporary suspension of development of the embryo and occurs in more than 130 mammalian species including both the mouse and rat as well as a wide range of species of bears, seals, bats, and marsupials (Fenelon, Banerjee, & Murphy, 2014). Embryonic diapause occurs at the blastocyst stage for variable lengths of time and in response to diverse environmental conditions; notably, it does not occur in species such as the rabbit, cow, nonhuman primates, or humans (Fenelon et al., 2014). Presence or absence of diapause has biological consequences to reproduction under adverse environmental conditions and the ability to rapidly produce progency following lactation. Diapause may also be important to development of reproductive models in vitro. Indeed, derivation of embryonic stem cells may be linked to diapause. For example, in the mouse, entry into diapause may enhance derivation of embryonic stem cells; moreover, both the propagation of mouse embryonic stem cells and maintenance of pluripotency in diapause in vivo are dependent on LIF (leukemia inhibitory factor) (Fenelon et al., 2014). In contrast, in species that do not enter diapause, embryonic stem cells have been much more difficult to derive and generally do not require or benefit from LIF in culture (Fenelon et al., 2014). Thus, the presence or absence of embryo diapause, although widespread, is not universal and represents a specialized condition in commonly used reproductive models.

5.3 Genome Activation

Several studies have documented the onset of genome-wide gene expression in the newly formed mouse embryo following fertilization (Wang et al., 2004; Xue et al., 2013). The onset of gene expression in the mouse occurs at the 1- to 2-cell stage, thus prompting the term "zygotic gene activation." In general, in other species including rabbits, cows, nonhuman primates, and humans, the major wave of activation of the genome occurs at the 4- to 8-cell stage (Chavez et al., 2012; Dobson et al., 2004; Vera-Rodriguez, Chavez, Rubio, Reijo Pera, & Simon, 2015; Wong et al., 2010; Xue et al., 2013). Nonetheless, transcription of individual genes or gene classes, as in the mouse, may occur early in these other species as well. Thus, although historically, transcription is thought to occur earlier in the mouse than other species, activation of the embryonic genome may occur sequentially (Vassena et al., 2011). Genes that are activated early include those involved in ribosome biogenesis and assembly, protein synthesis, RNA metabolism, and transcription (Zeng & Schultz, 2005).

5.4 Mouse Development and the X Chromosome

A major difference across species is the requirement, or lack thereof, for the activity of two sex chromosomes. In humans, nonhuman primates, and horses, primarily, the absence of a second X chromosome or a

Y chromosome manifests as Turner syndrome (Reindollar, 2011). In humans, Turner syndrome occurs with complete or partial loss of the second sex chromosome (45,X) in 1–2% of all female conceptions. In more than 90% of cases, pregnancies are not carried to term. Diverse somatic characteristics are associated with surviving Turner syndrome females, including short stature and cardiovascular abnormalities. In addition, most Turner syndrome females are also infertile, establishing a link between the X chromosome and germ line formation and/or maintenance (Reindollar, 2011). Only the lack of a second sex chromosome results in infertility as females with an additional X chromosome (triple X syndrome) have normal fertility. Although some aspects of Turner syndrome, especially cognitive differences, can be modeled in the 39XO mouse, these mice are fertile and the 39XO genotype can be passed from one generation to the next (Lopes et al., 2010; Lynn & Davies, 2007). Thus, reproduction in mice and some other mammalian species is independent of a second X chromosome. In humans, genes that escape X inactivation, and thus are likely required in two copies, are implicated in Turner syndrome (Dominguez, Chiang, Sukhwani, Orwig, & Reijo Pera, 2014).

6. HUMAN AND NONHUMAN PRIMATE EMBRYO DEVELOPMENT

Human and nonhuman primate embryo development is remarkably similar in terms of embryo morphology, success rates of embryo development in development to the blastocyst stage, chromosomal stability and instability, and timing of embryo development (Burruel, Klooster, Barker, Reijo Pera, & Meyers, 2014). Here, the focus is on human development, in large part due to the advances in science linked to IVF. Nonetheless, references to studies in the nonhuman primate, *Rhesus macaque*, are included when substantially different.

Human development begins in relative transcriptional silence beginning with the fusion of the egg and sperm (Niakan, Han, Pedersen, Simon, & Reijo Pera, 2012). The oocyte to embryo transition then proceeds with a period of extraordinary epigenetic reprogramming of the gametic pronuclei. Concurrently, in both human and nonhuman primate embryogenesis, the first cleavage division begins, with minor waves of transcription, followed by the major wave of embryonic genome activation at approximately day 3 at the 8-cell stage. On day 4, compaction begins the process of distinguishing an outer layer of cells that will develop ultimately to form the trophectoderm and an inner cell mass that can be visually distinguished on day 5 with formation of the blastocyst. At approximately days 12–14, the embryo will begin the process of gastrulation, giving rise to the primary germ layers of the embryo, the endoderm, mesoderm, and ectoderm. Coincident with gastrulation, a small number of cells are set aside to form the primordial germ cells that will give rise to oocytes and sperm of the next generation (Niakan et al., 2012). Several aspects of human embryo development bear further discussion especially the relatively low success rates in reaching the blastocyst stage, frequency of chromosomal instability manifest as chromosomal and subchromosomal errors, and exquisite relationship to timing of the first cleavage divisions.

6.1 Low Success Rates and Chromosomal Errors

Abnormal human embryo development, leading to embryo arrest, is associated with diverse cell cycle behavior and gene expression profiles (Chen et al., 2009; Dobson et al., 2004; Eguizabal et al., 2011; Galan et al., 2010; Vanneste et al., 2009; Wong et al., 2010). In addition, studies that used whole-genome approaches have also demonstrated that aneuploidies (abnormalities in chromosome number) are remarkably common, occurring in as many as 50–90% of cleavage-stage human embryos, including those from fertile couples (Johnson et al., 2010; Vanneste et al., 2009). This is in sharp contrast to the frequency of aneuploidies in several other species, including the mouse, which typically exhibit less than 1% embryonic aneuploidy rates at this stage of development (Novak et al., 2006). The incidence of human embryonic aneuploidy observed in vitro may reflect that in vivo given that only about 30% of human conceptions result in live births and chromosomal abnormalities have been reported in more than 70% of spontaneous miscarriages (Causio, Fischetto, Sarcina, Geusa, & Tartagni, 2002; Lathi, Westphal, & Milki, 2008; Macklon, Geraedts, & Fauser, 2002; Spandorfer, Davis, Barmat, Chung, & Rosenwaks, 2004).

It is thought that chromosomal errors most commonly occur on the meiotic spindle during oogenesis, with other errors occurring less frequently on the mitotic spindle during embryonic cleavage divisions. Such errors could be perpetuated due to the apparent lack of cell cycle checkpoints in cleaving human embryos (Bishop et al., 2000). However, other studies have indicated that chromosomal abnormalities in human embryos at the cleavage stage can be complex and may be incongruent with this notion (Johnson et al., 2010; Vanneste et al., 2009), suggesting that alternative mechanisms distinct from meiotic and mitotic nondisjunction and/or spindle defects may contribute to complex alterations in chromosomal composition. In particular, the contribution of mitotic errors during the embryonic cleavage divisions has been documented and may contribute to embryo demise as often as meiotic errors (Chavez et al., 2012; Vera-Rodriguez et al., 2015).

6.2 Timing in Human Development

In 2010, a study tracked the development of 242 embryos from the zygote (day 1, 1 cell) to blastocyst stage and correlated imaging (movies) with molecular characteristics (Wong et al., 2010). Although several studies had reported the imaging of human embryos, these studies documented previously unrecognized aspects of human embryo development and correlated imaging with molecular programs and prospective prediction of success and failure to the blastocyst stage and beyond (Fig. 2). Multiple parameters were extracted and analyzed from time-lapse imaging, including blastomere size, thickness of the zona pellucida, degree of fragmentation, length of the first cell cycles, time intervals between the first few mitoses, and duration of the first cytokinesis (Wong et al., 2010). Three parameters that collectively predicted blastocyst formation, with 94.1% and 93.1% sensitivity and specificity, were identified: (1) duration of the first cytokinesis, (2) time interval between the end of the first cleavage and the initiation of the second, and (3) synchronicity of the blastomeres in the second cleavage division to yield a 4-cell embryo from a 2-cell embryo. It was observed that normal embryos that successfully developed into a blastocyst exhibited similar values in all three parameters, whereas arrested embryos were highly variable. This indicated that embryos that follow strict timing in cytokinesis and mitosis during the first two cleavage divisions are much more likely to successfully develop to blastocyst stage. Considering that embryonic genome activation does not occur until 1 and 2 days later (Braude, Bolton, & Moore, 1988), these data indicated that the fate of an embryo must be predetermined at least in large part by inherited factors rather than by genes activated in the nascent embryo.

6.3 Gene Expression and Cell Autonomy of Individual Blastomeres

In studies that paralleled imaging studies, gene expression analysis revealed several key findings (Wong et al., 2010). First, four unique embryonic stage-specific patterns of expression are observed. These patterns include

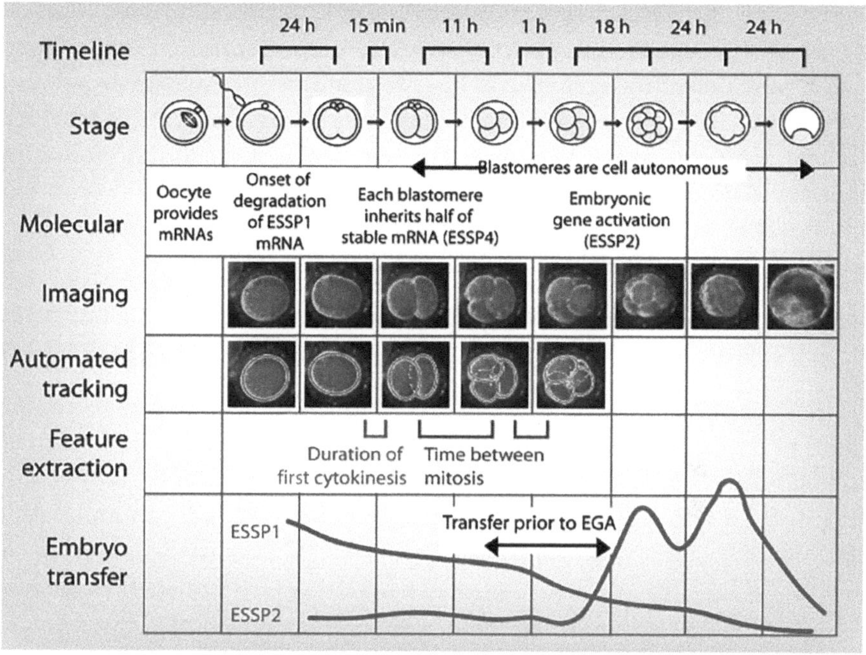

Fig. 2 Model for human embryo development based on correlated imaging and molecular analysis. Shown is timeline from zygote to blastocyst including critical brief times for prediction of successful development to blastocyst. Key molecular data indicate that human embryo development is characterized by four embryonic stage-specific patterns (ESSPs) of gene expression. During the cleavage divisions, blastomeres may arrest or progress independently. The outcome of cell autonomous development is that individual blastomeres may arrest or progress; as the embryo progresses to morula stage and beyond, blastocyst quality will be impacted by the number of cells that arrested or progressed beyond 8 cells. Imaging data demonstrate that critical periods of development predict success or failure: cytokinesis, second cleavage division, and synchronicity of appearance of third and fourth blastomeres. Data suggest improved ability to diagnose embryo outcome with key imaging predictors. *From Wong, C., Loewke, K., Bossert, N., Behr, B., DeJonge, C., Baer, T., et al. (2010). Non-invasive imaging of human embryos before embryonic genome activation predicts development to the blastocyst stage.* Nature Biotechnology, 28, *1115–1121.* (See the color plate.)

maternally inherited mRNAs destined for degradation, genes activated in either the minor or major waves of embryonic gene activation, later-activated genes expressed at the morula–blastocyst stage, and genes expressed stably across the entire preimplantation period. Notably, these studies have also indicated that degradation of maternal transcripts is not a spontaneous process that simply occurs through time. Instead, maternal degradation of RNA in human embryonic blastomeres must be an active process (that likely requires specific RNA degradation mechanisms) to target a specific subset of RNAs with a half-life of approximately 21 h. Furthermore, proper degradation of maternal transcripts may be a prerequisite for embryonic genome activation. Third, gene expression profiles of embryos that arrested in development appear to be as diverse and variable as their aberrant morphological phenotypes (Wong et al., 2010). These data indicated that the defects that we observed in the dynamic morphology of normal embryonic development reflected the intrinsic health of the embryo.

7. SPECIES-SPECIFIC GENES IN EMBRYO DEVELOPMENT

More recently, several studies have proposed that repetitive elements (REs) of retroviral origin may be enriched within the sequence of long noncoding RNAs (lncRNAs), might have driven evolution of lncRNAs, and thus may contribute to their biological function including in embryonic development (Kelley & Rinn, 2012; Ulitsky, Shkumatava, Jan, Sive, & Bartel, 2011). One specific class of REs, HERVH (Human Endogenous Retroviral Elements—H), has been shown to be specifically expressed in pluripotent stem cells (Carrieri et al., 2012; Gong & Maquat, 2011; Santoni, Guerra, & Luban, 2012; Ulitsky & Bartel, 2013).

In 2013, Au and colleagues combined second- and third-generation sequencing of transcripts in human embryonic stem cell (hESC) line, H1, with a custom-designed method for isoform identification and quantification (Au et al., 2013). More than 2000 gene isoforms were identified with transcripts derived from 216 novel, nonannotated gene loci (Au et al., 2013). The 23 most abundantly expressed transcripts were referred to as *HPAT1–23* (human pluripotency-associated transcripts 1–23, based on their abundance of expression) and in subsequent studies were further probed for function (Durruthy-Durruthy et al., 2015) (Fig. 3). Gene expression profiling of single cells of the embryo during nuclear reprogramming of somatic cells to pluripotent stem cells revealed that the *HPATs* (in particular, three members) were partners in the core regulatory network of OCT4, NANOG, SOX2, and SALL4 in stem cells. Moreover, reduced expression in human embryos changed the cell fate of pluripotent blastomeres so that progeny that lacked *HPAT* expression only contributed to trophectoderm (and not the pluripotent inner cell mass). These results demonstrated that a class of lncRNAs, comprised of human- and nonhuman primate-specific sequences, are critical for establishment and maintenance of pluripotency. Moreover, the mechanism by which the novel lncRNAs function was

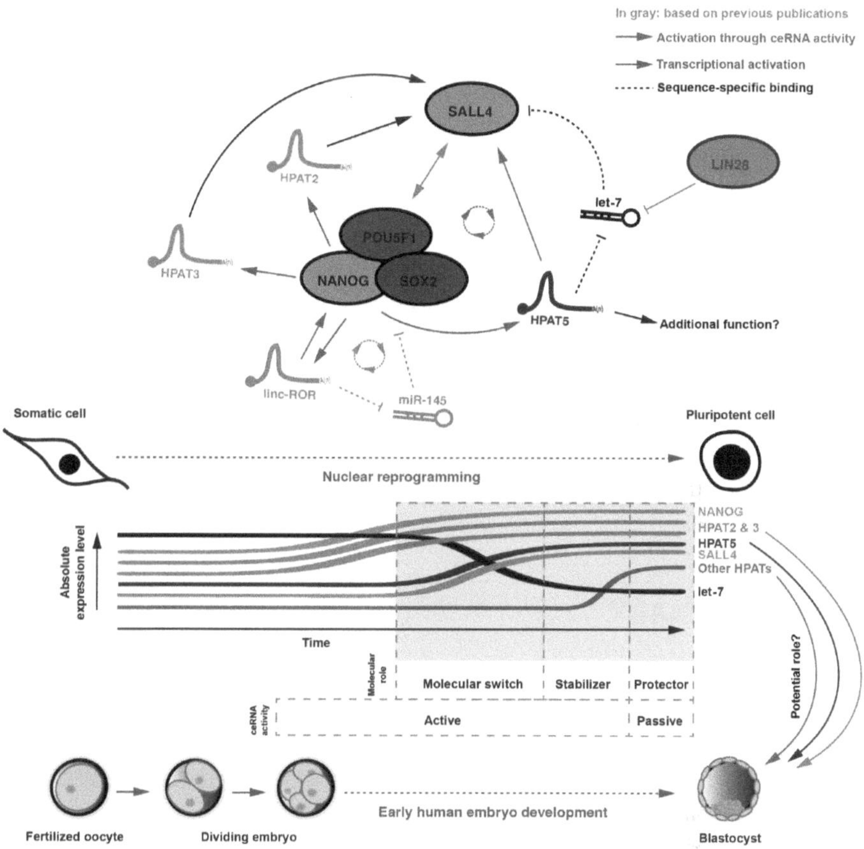

Fig. 3 Model of retroviral-derived long noncoding RNAs (lncRNAs) activity during embryo development, nuclear reprogramming, and maintenance of pluripotency in human stem cells. lncRNAs, enriched for HERVH and HERVK retroviral sequences (*HPATs, human pluripotency-associated transcripts*), exert its function as active and passive ceRNA to act as a molecular switch, stabilizer or protector during pluripotency establishment, and maintenance in human pluripotent cells. (See the color plate.)

dependent on *let*-7, a homolog of the first-identified miRNA in *Caenorhabditis elegans* that encodes a heterochronic (timing) regulator in development (Durruthy-Durruthy et al., 2015).

Additional functions of species-specific sequences in embryo development: In studies reported by Grow et al, the function of the most recently acquired human ERVs (endogenous retroviruses) was examined (Grow et al., 2015). HERVK repeatedly infected the primate lineage both before and after the divergence of the human and chimpanzee common ancestor and retained multiple copies of intact open reading frames encoding retroviral proteins (Grow et al., 2015). Grow and colleagues demonstrated that HERVK is transcribed during normal human embryogenesis, beginning at the 8-cell stage and continuing through the emergence of epiblast cells in preimplantation blastocysts (Grow et al., 2015). Expression ceases during hESC derivation from blastocyst outgrowths. Remarkably, both HERVK viral-like particles and Gag proteins are detected in vivo in human blastocysts, indicating that early human development proceeds in the presence of HERVK products. A mechanism for function suggests that HERVK can induce viral restriction pathways in early embryonic cells that may modulate the complex interactions between retroviral proteins and host factors in early human development. Thus, studies of human- and nonhuman primate-specific sequences are shown to play a role in species-specific aspects of development, in both determining developmental progression and timing through preimplantation development.

8. SUMMARY

Experimental embryology was introduced in experiments with the chick embryo, conducted by Aristotle in the 4th century BCE. These experiments set the stage for more than 2000 years with the concept of universality—that development across species is conserved. The universality of embryo development across mammalian species is evident in the major events following fertilization. Mammalian embryonic development begins in transcriptional silence followed by gene activation in the zygote and/or embryo. Cleavage divisions occur over the next 2–5 days followed by compaction of the embryo to form a morula and subsequent differentiation of distinct cell types with the formation of the blastocyst. Success in reaching blastocyst stage is variable with success rates that vary from 80–90% to less than 30–50% that may reflect meiotic and/or mitotic chromosomal errors. In addition to the chicken embryo, several common mammalian models have been developed and used throughout the last century, in particular the rabbit, cow, mouse, nonhuman primate, and humans (via IVF observations). Species-specific traits allow interrogation and comparisons of fundamental properties of embryo development. Recent data suggest that many aspects of species-specific embryo development, such as timing and immune modulation, may be manifest via expression of species-specific sequences derived from retroviral sequences introduced into the germ line. Unprecedented tools and methods to probe species-specific aspects of embryo development have been developed and will undoubtedly shed light on similarities and differences across species.

ACKNOWLEDGMENTS

The authors are grateful to members of their laboratories, past and present, who have contributed to the study of mammalian embryology. Funding for this project was obtained from the National Institutes for Health (NHLBI #U01HL100397) and as part of the Specialized Cooperative Centers Program in Reproduction and Infertility Research and National Institute for Child Health Development through Grant #P50 HD068158.

REFERENCES

Adams, C. (1962). Studies on prenatal mortality in the rabbit, *Oryctolagus cuniculus*: The effect of transferring varying numbers of eggs. *Journal of Endocrinology, 24*, 471–490.

Au, K., Sebastiano, V., Afshar, P., Durruthy-Durruthy, J., Lee, L., Williams, B., et al. (2013). Characterization of the human ESC transcriptome by hybrid sequencing. *Proceedings of the National Academy of Sciences of the United States of America, 110*, E4821–E4830.

Behringer, R., Gertsenstein, M., Nagy, K. V., & Nagy, A. (2014). *Manipulating the mouse embryo: A laboratory manual* (4th ed.). Cold Spring Harbor: Cold Spring Harbor Laboratory Press.

Betteridge, K. (2003). A history of farm animal embryo transfer and some associated techniques. *Animal Reproduction Science, 79*, 203–244.

Bishop, C. W., Qin, D. J., Agoulnik, Y., Harrison, A. I., Behringer, R., & Overbeek, P. A. (2000). A transgenic insertion upstream of sox9 is associated with dominant XX sex reversal in the mouse. *Nature Genetics, 26*, 490–494.

Bó, G., & Mapletoft, R. (2013). Evaluation and classification of bovine embryos. *Animal Reproduction, 10*, 344–348.

Boveri, T. (1901). *Zellen-Studien: uber die Natur der Centrosomen* (4th ed.). Jena: Fischer.

Braude, P., Bolton, V., & Moore, S. (1988). Human gene expression first occurs between the four- and eight-cell stages of preimplantation development. *Nature, 332*, 459–461.

Burruel, V., Klooster, K., Barker, C., Reijo Pera, R. A., & Meyers, S. (2014). Abnormal early cleavage events predict early embryo demise: Sperm oxidative stress and early abnormal cleavage. *Scientific Reports, 4*, 6598.

Burt, D. (2005). Chicken genome: Current status and future opportunities. *Genome Research, 15*, 1692–1698.

Carrieri, C., Cimatti, L., Biagioli, M., Beugnet, A., Zucchelli, S., Fedele, S., et al. (2012). Long non-coding antisense RNA controls Uchl1 translation through an embedded SINEB2 repeat. *Nature, 491*, 454–457.

Causio, F., Fischetto, R., Sarcina, E., Geusa, S., & Tartagni, M. (2002). Chromosome analysis of spontaneous abortions after in vitro fertilization (IVF) and intracytoplasmic sperm injection (ICSI). *European Journal of Obstetrics, Gynecology, and Reproductive Biology, 105*(1), 44–48.

Chang, M. (1952). Development of bovine blastocyst with a note on implantation. *The Anatomical Record, 113*, 143–161.

Chavez, S., Loewke, K., Han, J., Moussavi, F., Colls, P., Munne, S., et al. (2012). Dynamic blastomere behavior reflects human embryo ploidy by the four-cell stage. *Nature Communication, 3*, 1251.

Chen, A. E., Egli, D., Niakan, K., Deng, J., Akutsu, H., Yamaki, M., et al. (2009). Optimal timing of inner cell mass isolation increases the efficiency of human embryonic stem cell derivation and allows generation of sibling cell lines. *Cell Stem Cell*, *4*(2), 103–106. http://dx.doi.org/10.1016/j.stem.2008.12.001. S1934-5909(08)00618-8 [pii].

Clift, D., & Schuh, M. (2013). Restarting life: Fertilization and the transition from meiosis to mitosis. *Nature Reviews. Molecular Cell Biology*, *14*, 549–562.

Deng, X., Berletch, J., Nguyen, D., & Disteche, C. (2014). X chromosome regulation: Diverse patterns in development, tissues and disease. *Nature Reviews. Genetics*, *15*, 367–378.

Disteche, C., & Berletch, J. (2015). X-chromosome inactivation and escape. *Journal of Genetics*, *94*, 591–599.

Dobson, A. T., Raja, R., Abeyta, M. J., Taylor, T., Shen, S., Haqq, C., et al. (2004). The unique transcriptome through day 3 of human preimplantation development. *Human Molecular Genetics*, *13*(14), 1461–1470.

Dominguez, A., Chiang, H., Sukhwani, M., Orwig, K., & Reijo Pera, R. A. (2014). Human germ cell formation in xenotransplants of induced pluripotent stem cells carrying X chromosome aneuploidies. *Scientific Reports*, *4*, 6432.

Durruthy-Durruthy, J., Sebastiano, V., Wossidlo, M., Cepeda, D., Cui, J., Grow, E., et al. (2015). A novel primate-specific noncoding RNA modulates human embryo and pluripotent stem cell fate. *Nature*, *48*, 44–52.

Eguizabal, C., Montserrat, N., Vassena, R., Barragan, M., Garreta, E., Garcia-Quevedo, L., et al. (2011). Complete meiosis from human induced pluripotent stem cells. *Stem Cells*, *29*, 1186–1195.

Fenelon, J., Banerjee, A., & Murphy, B. (2014). Embryonic diapause: Development on hold. *The International Journal of Developmental Biology*, *58*, 163–174.

Fischer, B., Chavatte-Palmer, P., Viebahn, C., Santos, A. N., & Duranthon, V. (2012). Rabbit as a reproductive model for human health. *Reproduction*, *144*, 1–10.

Galan, A., Montaner, D., Poo, M., Valbuena, D., Ruiz, V., Aguilar, C., et al. (2010). Functional genomics of 5- to 8-cell stage human embryos by blastomere single-cell cDNA analysis. *PLoS One*, *5*, e13615.

Gong, C., & Maquat, L. (2011). lncRNAs transactivate STAU1-mediated mRNA decay by duplexing with 3′ UTRs via Alu elements. *Nature*, *470*, 284–288.

Graur, D., Duret, L., & Gouy, M. (1996). Phylogenetic position of the order Lagomorpha (rabbits, hares and allies). *Nature*, *379*, 333–335.

Greve, T., Callesen, H., Hyttel, P., & Avery, B. (1995). *From oocyte to calf: In vivo and in vitro*. Paris: Elsevier Biofutur.

Grow, E., Flynn, R., Chavez, S., Bayless, N., Wossidlo, M., Wesche, D., et al. (2015). Intrinsic retroviral reactivation in human preimplantation development and pluripotent cells. *Nature*, *522*, 221–225.

Hamburger, V., & Hamilton, H. (1992). A serioes of normal stages in the development of the chick embryo. 1951. *Developmental Dynamics*, *195*, 231–272.

Heape, W. (1891). Preliminary note on the transplantation and growth of mammalian ova within a uterine foster-mother. *Proceedings of the Royal Society B*, *48*, 457–458.

Johnson, D., Gemelos, G., Baner, J., Ryan, A., Cinnioglu, C., Banjevic, M., et al. (2010). Preclinical validation of a microarray method for full molecular karyotyping of blastomeres in a 24-h protocol. *Human Reproduction*, *25*, 1066–1075.

Kamikawa, Y., & Donohoe, M. (2014). The dynamics of X-chromosome inactivation in mouse development. *Molecular Reproduction and Development*, *81*, 141–147.

Kelley, D., & Rinn, J. (2012). Transposable elements reveal a stem cell-specific class of long noncoding RNAs. *Genome Biology*, *13*, R107.

Lathi, R., Westphal, L., & Milki, A. (2008). Aneuploidy in the miscarriages of infertile women and the potential benefit of preimplanation genetic diagnosis. *Fertility and Sterility*, *89*, 353–357.

Lazzari, G., Wrenzycki, C., Herrmann, D., Duchi, R., Kruip, T., Niemann, H., et al. (2006). Cellular and molecular deviations in bovine in vitro-produced embryos are related to the large offspring syndrome. *Biology of Reproduction, 67*, 767–775.

Leandri, R., Archilla, C., Bui, L., Peynot, N., Liu, Z., Cabau, C., et al. (2009). Revealing the dynamics of gene expression during embryonic genome activation and first differentiation in the rabbit embryo with a dedicated array screening. *Physiological Genomics, 36*, 98–113.

Lopes, A., Burgoyne, P., Ojarikre, A., Bauer, J., Sargent, C., Amorim, A., et al. (2010). Transcriptional changes in response to X chromosome dosage in the mouse: Implications for X inactivation and the molecular basis of Turner syndrome. *BMC Genomics, 11*, 82.

Lynn, P., & Davies, W. (2007). The 39, XO mouse as a model for the neurobiology of Turner syndrome and sex-biased neuropsychiatric disorders. *Behavioural Brain Research, 179*, 173–182.

Ma, X., Lin, F., Wang, Z., Hu, M., Huang, L., Meng, T., et al. (2016). Geminin deletion in mouse oocytes results in impaired embryo development and reduced fertility. *Molecular and Cell Biology, 27*, 768–775.

Macklon, N., Geraedts, J., & Fauser, B. (2002). Conception to ongoing pregnancy: The 'black box' of early pregnancy loss. *Human Reproduction Update, 8*, 333–343.

Mapletoft, R. (2013). History and perspectives on bovine embryo transfer. *Animal Reproduction, 10*, 168–173.

Niakan, K., Han, J., Pedersen, R., Simon, C., & Reijo Pera, R. A. (2012). Human pre-implantation embryo development. *Development, 139*, 829–841.

Novak, I., Lightfoot, D., Wang, H., Eriksson, A., Mahdy, E., & Hoog, C. (2006). Mouse embryonic stem cells form follicle-like ovarian structures but do not progress through meiosis. *Stem Cells, 24*, 1931–1936.

Okamoto, I., Patrat, C., Thépot, D., Peynot, N., Fauque, P., Daniel, N., et al. (2011). Eutherian mammals use diverse strategies to initiate X-chromosome inactivation during development. *Nature, 472*, 370–374.

Project, International Chicken Genome Sequencing Consortium. (2004). Sequence and comparative analysis of the chicken genome provide unique perspectives on vertebrate evolution. *Nature, 432*, 695–716.

Reindollar, R. (2011). Turner syndrome: Contemporary thoughts and reproductive issues. *Seminars in Reproductive Medicine, 29*, 342–352.

Rizos, D., Gutierrez-Adan, A., Perez-Garnelo, S., Fuente, J. D. l., Boland, M., & Lonergan, P. (2003). Bovine embryo culture in the presence or absence of serum: Implications for blastocyst development, cryotolerance, and messenger RNA expression. *Biology of Reproduction, 68*, 236–243.

Ross, L., & Normark, B. (2015). Evolutionary problems in centrosome and centriole biology. *Journal of Evolutionary Biology, 28*, 995–1004.

Saadeldin, M., Kim, B., Lee, B., & Jang, G. (2011). Effect of different culture media on the temporal gene expression in the bovine developing embryos. *Theriogenology, 75*, 995–1004.

Santoni, F., Guerra, J., & Luban, J. (2012). HERV-H RNA is abundant in human embryonic stem cells and a precise marker for pluripotency. *Retrovirology, 9*, 111.

Seidel, G. E., Jr., & Seidel, S. (1991). *Training manual for embryo transfer in cattle*. Rome: © FAO.

Spandorfer, S. D., Davis, O. K., Barmat, L. I., Chung, P. H., & Rosenwaks, Z. (2004). Relationship between maternal age and aneuploidy in in vitro fertilization pregnancy loss. *Fertility and Sterility, 81*, 1265–1269.

Stern, C. (2005). The chick: A great model system becomes even greater. *Developmental Cell, 8*, 9–17.

Suthananthan, A., Lyons, G., Charmawardena, V., Pushett, D., Lewis, I., & Trounson, A. (1999). Centriolar dynamics in the bovine embryo: Inheritance and perpetuation of the sperm centrosome during fertilization and development. *Protoplasma*, *206*, 263–269.

Thompson, D. (1910). *Aristotle: Historia animalium*. Oxford translation. Oxford: Clarendon Press.

Ulitsky, I., & Bartel, D. (2013). lincRNAs: Genomics, evolution, and mechanisms. *Cell*, *154*, 26–46.

Ulitsky, I., Shkumatava, A., Jan, C., Sive, H., & Bartel, D. (2011). Conserved function of lincRNAs in vertebrate embryonic development despite rapid sequence evolution. *Cell*, *147*, 1537–1550.

Umbaugh, R. (1949). Superovulation and ovum transfer in cattle. *American Journal of Veterinary Research*, *10*, 295–305.

Vanneste, E., Voet, T., Caignec, C. L., Ampe, M., Konings, P., Melotte, C., et al. (2009). Chromosome instability is common in human cleavage-stage embryos. *Nature Medicine*, *15*, 577–583.

Vassena, R., Boué, S., González-Roca, E., Aran, B., Auer, H., Veiga, A., et al. (2011). Waves of early transcriptional activation and pluripotency program initiation during human preimplantation development. *Development*, *138*, 3699–3709.

Vera-Rodriguez, M., Chavez, S., Rubio, C., Reijo Pera, R. A., & Simon, C. (2015). Prediction model for aneuploidy in early human embryo development revealed by single-cell analysis. *Nature Communications*, *6*, 7601.

Viebahn, C. (2001). Hensen's node. *Genesis*, *29*, 96–103.

Wang, Q. T., Piotrowska, K., Ciemerych, M. A., Milenkovic, L., Scott, M. P., Davis, R. W., et al. (2004). A genome-wide study of gene activity reveals developmental signaling pathways in the preimplantation mouse embryo. *Developmental Cell*, *6*, 133–144.

Willett, E., Black, W., Casida, L., Stone, W., & Buckner, P. (1951). Successful transplantation of a fertilized bovine ovum. *Science*, *113*, 247.

Wong, C., Loewke, K., Bossert, N., Behr, B., DeJonge, C., Baer, T., et al. (2010). Non-invasive imaging of human embryos before embryonic genome activation predicts development to the blastocyst stage. *Nature Biotechnology*, *28*, 1115–1121.

Xue, Z., Huang, K., Cai, C., Cai, L., Jiang, C.-Y., Feng, Y., et al. (2013). Genetic programs in human and mouse early embryos revealed by single-cell RNA sequencing. *Nature*, *500*, 593–597.

Young, L., Sinclair, K., & Wilmut, I. (1998). Large offspring syndrome in cattle and sheep. *Reviews of Reproduction*, *3*, 155–163.

Zeng, F., & Schultz, R. (2005). RNA transcript profiling during zygotic gene activation in the preimplantation mouse embryo. *Developmental Biology*, *283*, 40–57.

CHAPTER TWELVE

Genetic Analysis of Human Preimplantation Embryos

S. Garcia-Herrero, A. Cervero, E. Mateu, P. Mir, M.E. Póo, L. Rodrigo, M. Vera, C. Rubio[1]
Igenomix, Parc Científic Universitat de Valéncia, Valencia, Spain
[1]Corresponding author: e-mail address: Carmen.rubio@igenomix.com

Contents

Abstract

Preimplantation development comprises the initial stages of mammalian development, before the embryo implants into the mother's uterus. In normal conditions, after fertilization the embryo grows until reaching blastocyst stage. The blastocyst grows as the cells divide and the cavity expands, until it arrives at the uterus, where it "hatches" from the zona pellucida to implant into the uterine wall. Nevertheless, embryo quality and viability can be affected by chromosomal abnormalities, most of which occur during gametogenesis and early embryo development; human embryos produced in vitro are especially vulnerable. Therefore, the selection of chromosomally normal embryos for transfer in assisted reproduction can improve outcomes in poor-prognosis patients.

Current Topics in Developmental Biology, Volume 120
ISSN 0070-2153
http://dx.doi.org/10.1016/bs.ctdb.2016.04.009

Additionally, in couples with an inherited disorder, early diagnosis could prevent pregnancy with an affected child and would, thereby, avoid the therapeutic interruption of pregnancy. These concerns have prompted advancements in the use of preimplantation genetic diagnosis (PGD). Genetic testing is applied in two different scenarios: in couples with an inherited genetic disorder or carriers of a structural chromosomal abnormality, it is termed PGD; in infertile couples with increased risk of generating embryos with de novo chromosome abnormalities, it is termed preimplantation genetic screening, or PGS.

1. INTRODUCTION

The first instance of PGD was reported in 1989 for sex due to an X-linked disease. Since then, various approaches have been applied for single-gene disorders (Handyside, Kontogianni, Hardy, & Winston, 1990; Handyside et al., 1989) and for the assessment of aneuploidies (Munne et al., 2003). PGS initially incorporated fluorescence in situ hybridization (FISH) for a selected panel of chromosomes. However, several approaches were soon developed for 24-chromosome analysis, such as comparative genome hybridization (CGH), single-nucleotide polymorphism (SNP) arrays, and quantitative polymerase chain reaction (qPCR)-based techniques. A review by Simpson et al. proposed that array-CGH was the preferred diagnostic approach for assessing 24-chromosome aneuploidy (Simpson, 2012). More recently, next-generation sequencing (NGS) has been validated and successfully applied for PGS in single-cell and trophectoderm biopsies (Fiorentino et al., 2014; Vera-Rodriguez et al., 2015). Thus, a number of approaches are available for genetic testing in early embryos.

PGD is indicated for monogenic disorders including recessive disorders (such as cystic fibrosis or thalassemia); dominant disorders (ie, myotonic dystrophy, Huntington's disease, or achondroplasia); sex-linked disorders (hemophilia, Fragile X, and Duchenne muscular dystrophy); Rh blood group typing; and human leukocyte antigen (HLA) typing. PGD is also indicated for chromosomal disorders including carriers of balanced chromosomal rearrangements, such as reciprocal and Robertsonian translocations, inversions, and some cryptic deletion–duplication abnormalities.

A PGD/PGS cycle comprises the following steps: (1) ovarian superstimulation; (2) aspiration of ovarian follicles; (3) oocyte retrieval; (4) intracytoplasmic injection of oocytes with processed sperm; (5) in vitro culture of fertilized oocytes until day (D) 5–6, blastomere biopsy on D3 or trophectoderm biopsy on D5, D; (6) genetic testing; and (7) the transfer of a genetically normal embryo. Surplus embryos could be vitrified for a future transfer attempt.

2. BIOPSY STRATEGIES

There are different possible sources of genetic material for testing (Fig. 1): (1) the first and second polar bodies (PBs) before and after fertilization; (2) blastomere biopsies from cleavage embryo stage on D3; and (3) several trophoblast cells (usually 4–10) sampled from the blastocyst. Each of these stages has specific diagnostic advantages as well as critical limitations that relate to aneuploidy genesis during both meiosis and the preimplantation period of embryo development.

2.1 Polar-Body Biopsy

PB biopsy could be an option if only maternal mutations are evaluated or as a method to avoid ethical restrictions in some countries where the embryo genetic analysis is not allowed (Geraedts et al., 2010). Regarding the diagnostic capabilities with PBs, they are by-products of the meiotic division of

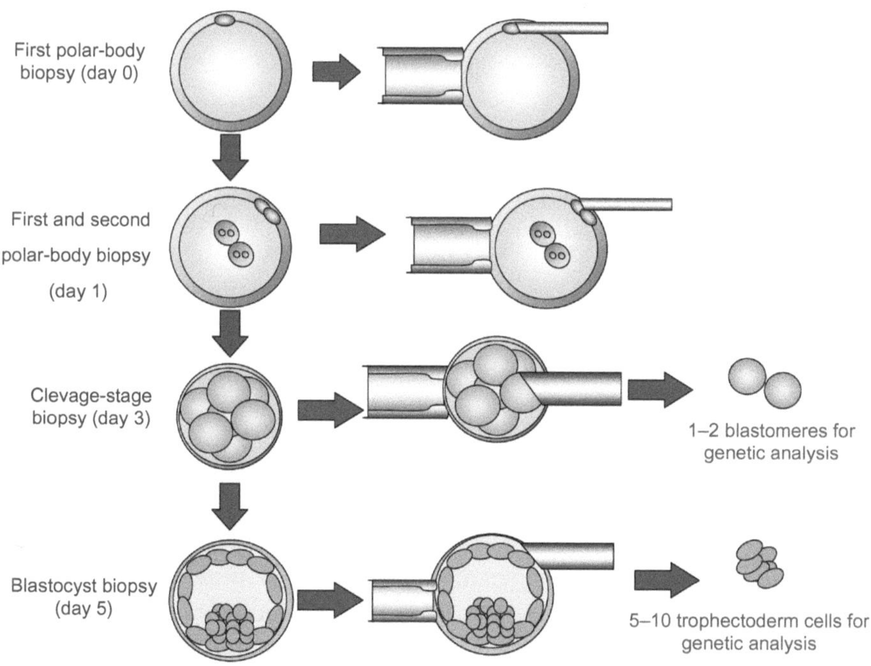

Fig. 1 Biopsy strategies: different possible sources of genetic material for testing.

the oocyte. Removal of the first and/or second PB is an indirect strategy to infer the genetic or chromosomal status of the oocyte. Both PBs are not required for successful fertilization or normal embryonic development.

Polar body biopsy is accepted as being less invasive and provides more time for analysis. Numerous reports have shown that PBs can be used for a large variety of investigations and diagnostic purposes. The range of procedures spans from genetic diseases to structural and numeric chromosome aberrations, however fails to capture as many as one in three embryonic aneuploidies. Furthermore, PBs provide information only on the female part, which is critical in the case of certain monogenetic diseases. In a recessive disease, PB analysis will only distinguish whether the oocyte carries the normal or the affected allele. The state of the corresponding embryo will be determined by the paternal allele and can be normal, unaffected (carrier), or affected. If only PB diagnosis is done, oocytes identified with the affected allele will not be used although one-half of them could result in an unaffected carrier. This can be seen as an ethical dilemma (Montag, Köster, Strowitzki, & Toth, 2013). Additionally, the inability to readily distinguish nondisjunction from premature separation of sister chromatids greatly limits the predictive value of the technique and may lead to an overdiagnosis of aneuploidy in as many as 45% of cases with first polar-body errors (Scott, Hong, & Scott, 2013).

2.2 Blastomere Biopsy

At cleavage stage, biopsy can be performed in the morning of D3, excluding poor quality embryos. At this stage, it is recommended to use Ca^{2+}/Mg^{2+}-free media because it allows the loosening of membrane adhesions between blastomeres making easier the cell removal and results in lower lysis and shorter procedure time. To remove cells from the embryo, it is necessary to pass through the zona pellucida (ZP). In the past, the most of the laboratories used acid tyrodes (AT) for zona hatching but nowadays, the use of laser drilling has spread worldwide and the use of AT tends to decrease significantly.

Two aspects are involved on the successful cleavage-stage biopsy. First, the cellular material removed from the embryo should be suitable for genetic analysis. This means that the cell should be intact and it should contain a single, clearly visible nucleus. Second, further development should not be impaired because of the biopsy procedure so the date record on blastocyst formation rates should be followed (De Vos & Van Steirteghem, 2001).

The main point of discussion on performing the genetic analysis on D3 is whether one cell is representative of the entire embryo due to the possible presence of mosaicism. Chromosomal mosaicism is defined as the presence of two or more chromosomally distinct cell lines within an embryo or an individual. The type of mosaicism and its clinical consequence depends on a variety of factors, including where and when the mosaicism is generated, making difficult to do a diagnosis of chromosomal competence of an embryo (Taylor et al., 2014).

The biopsy on D3 has a clear advantage over the PB biopsy and it is that chromosomal aberrations of both maternal and paternal origin can be detected. Possible disadvantages include the cellular mosaicism in embryos (Taylor et al., 2014), questioning whether the analyzed cell is representative for the rest of the embryo and the limited amount of tissue available for diagnosis.

2.3 Blastocyst Biopsy

Recently, it has been revealed that blastocyst biopsy does not compromise the developmental competence of embryos (Scott et al., 2013), and, in combination with comprehensive chromosome screening, is highly predictive of developmental potential of blastocysts (Scott et al., 2012). This upgraded form of PGS significantly improves implantation, both with fresh transfer or after vitrification (Schoolcraft & Katz-Jaffe, 2013; Scott et al., 2013).

Blastocyst-stage biopsy offers the advantage of removing several cells (4–10 cells) for analysis and if extra-embryonic trophectoderm cells are taken, the inner cell mass that will later form the fetus is left untouched. In this way, any risk of affecting fetal development is avoided. The analysis of more cells increases the probability to detect the presence of mosaicism in the embryo. The minimum ratio of aneuploidy to euploid cells that is needed to detect a chromosomal imbalance using aCGH has been determined (Mamas, Gordon, Brown, Harper, & Sengupta, 2012) demonstrating that it is possible to clearly show aneuploidy when at least 50% of cells or more in the biopsied sample carry the same aneuploidy. However, even when 25% of the cells are aneuploidy, a shift from normality can be clearly identified, indicating low-level mosaicism.

Few data are available regarding the concordance between TE cells, which will give origin to fetal membranes, and ICM, which will give origin to the fetus. Some authors have showed high accuracy of diagnosis and no major diagnostic impact of mosaicism at the blastocyst stage (Capalbo et al.,

2013), but in contrast there are some case reports that bring into discussion the possibility to a nonconcordance between ICM and TE (Haddad et al., 2013).

Despite these discrepancies, there are many advantages to perform PGD/PGS at blastocyst stage: the biopsy is less invasive, a larger number of cells can be removed and the cost reduction for couples due to the lower number of embryos to biopsy, which is associated to the self-selection of the embryos from D3 to D5/6.

3. APPLICATION TO MONOGENIC DISORDERS

3.1 Indications

Virtually any genetically inherited disease that is diagnosable in human beings can be identified in the embryo. As such, PGD can be applied for any genetic disease with a molecular diagnosis and/or defined marker linkage within a family.

Remarkably, in initial PGD the FISH procedure was applied for X-linked disorders to select female embryos (Harper et al., 1994). This approach has since been replaced, whenever possible, by specific diagnosis of the molecular defect. The advantage of a specific DNA diagnosis is that healthy male embryos can be selected and transferred, rather than discarded. Moreover, this approach allows detection of female carriers, which is important for X-linked dominant disorders (eg, fragile X syndrome) in which female carriers may manifest the disease.

PGD is also applicable in couples for whom one partner carriers a mutation predisposing to cancer or other late-onset diseases (Rechitsky et al., 2002). In these cases, PGD is an attractive alternative to prenatal diagnosis, circumventing the decision to terminate an established pregnancy. This procedure has been performed for several diseases, including the common syndromes of genetic predisposition to colon and breast cancer (Rechitsky et al., 2002; Spits et al., 2007). Despite the ethical and legal issues (Clancy, 2010), PGD for cancer predisposition syndromes is a reality and the number of cycles reported for this type of condition is increasing daily (De Rycke et al., 2015). Notably, for this indication, it is necessary to consider the legal issues, which vary by country.

Additionally, for human leukocyte antigen (HLA) matching, PGD is used to select a healthy and HLA-compatible embryo (Verlinsky et al., 2007) for an ill sibling in need of donor hematopoietic stem cells (Grewal, Kahn, MacMillan, Ramsay, & Wagner, 2004). HLA typing by

PGD is performed for acquired diseases, such as severe aplastic anemia or leukemia, or can be combined with diagnosis of a single-gene disorder to select an embryo free of the inherited condition and HLA-matched to an affected sibling (Verlinsky et al., 2007; Kuliev et al., 2005). This approach was applied for the first time in 2001 for Fanconi anemia (Verlinsky, Rechitsky, Schoolcraft, Strom, & Kuliev, 2001). In spite of ethical objections, including the instrumentalization of the future child (the new child becomes an instrument to cure another child), it is now an established clinical procedure with documented positive outcomes (Van de Velde et al., 2009). As in PGD for predisposition to cancer/disease, the laws regulating PGD for HLA matching vary by country.

3.2 Diagnostic Methods

- *Multiplex PCR*: Multiplex PCR uses targeted primers designed specifically for the mutation of interest combined with others for linked short tandem repeat (STR) markers. This has traditionally been the gold standard for PGD of monogenic disorders (Fiorentino et al., 2006). The use of polymorphic markers is applied during the pre-PGD workup, whereby analysis of the DNA samples of the patients and other relatives informs on the alleles to be expected in the embryos, as well as the markers that may co-segregate with the mutation. Such a combined study improves accuracy, minimizing potential errors caused by undetected allele drop-out (ADO) or contamination (Harton et al., 2013). ADO refers to the amplification failure of one of the two alleles, making a heterozygous locus appear homozygous, and potentially leading to misdiagnosis. Genotyping of the amplified products can be performed by means of different strategies such as amplification refractory mutation system (Moutou, Gardes, Nicod, & Viville, 2007), restriction enzyme digestion (Spits, De Rycke, et al., 2006), and real-time PCR and mini-sequencing (Fiorentino et al., 2003). Mini-sequencing encompasses a primer extension reaction that enables a quick and accurate detection of point mutations. The mini-sequencing primer is designed to anneal one base before the target site, and it will be elongated with only one dideoxynucleotide. The four different dideoxynucleotides are labeled with different fluorochromes, and the products can be analyzed on an automated DNA sequencing system. The use of multiplex PCR for markers has become widespread in PGD for monogenic disorders (Fiorentino et al., 2006) and HLA typing (Fiorentino et al., 2005).

The main advantage of multiplex PCR exclusively for linked markers is that the protocols can be used for several couples, independent of the mutation they carry. This saves time and resources in pre-PGD workups. However, the ability to use such indirect testing depends on the availability of appropriate family samples to determine the "at-risk haplotype."

- *Whole-genome amplification* (*WGA*): The creation of a robust and accurate multiplex protocol requires careful design, optimization, and validation before its clinical use. As a result, investment of time and resources is needed. In recent years, the use of WGA has been demonstrated as a practical and efficient alternative for performing PGD (Renwick, Trussler, Lashwood, Braude, & Ogilvie, 2010). WGA amplifies the entire genome, producing enough amplified DNA for several downstream applications. Multiple standard PCR assays may be performed for haplotyping and the direct analysis of mutations in monogenic diseases, thereby avoiding the need to optimize multiplex PCR protocols (Renwick et al., 2010). Moreover, WGA allows combining of PGD for single-gene disorders or HLA typing with array-CGH for the detection of chromosomal imbalances using the same sample (Rechitsky et al., 2015). These strategies are already clinically applied in some centers, with good clinical results (Rechitsky et al., 2015). However, WGA methods yield relatively high ADO rates (Spits, De Rycke, et al., 2006; Spits, Le Caignec, et al., 2006). This problem could be resolved with the application of a sufficient number of linked markers to avoid misdiagnosis and with the use of trophectoderm biopsies instead of single cells, given that the former gives lower ADO rates (Rechitsky et al., 2015).
- *Karyomapping*: This approach uses a high-density SNP array that allows evaluation of DNA haplotypes. With genotyping of the parents and close relatives with known disease status, the method identifies the parental haplotype linked to the mutation (Handyside et al., 2010). The main advantage of this platform is that it can be applied to a wide range of patients, eliminating the need for customized test development. A second advantage is that both single-gene disorders and aneuploidies are diagnosed on the same platform (Konstantinidis et al., 2015). However, the major limitation to this approach is the need to test relatives to identify the disease-linked parental haplotype. Therefore, karyomapping is not useful when family members are not available, when the "key" relative is recombinant, when the disease is caused by a de novo mutation, or in highly consanguineous couples.

- *Next-generation sequencing (NGS)*: NGS provides high-throughput and base pair-resolution data, permitting the analysis of multiple genetic loci and samples from different couples simultaneously. Moreover, like karyomapping, NGS allows the combined evaluation of aneuploidy and single-gene disorders from the same biopsy using a single platform. Several studies have been published showing the utility of NGS for testing single cells (Navin & Hicks, 2011; Wang et al., 2014). In 2013, Treff et al. published a specific protocol using NGS to test DNA from a trophectoderm biopsy, which was consistent with two conventional methodologies for PGD (Treff et al., 2013). The major concern relating to NGS is that an insufficient sequencing depth may result in false-positive or -negative identification of a mutation due to the presence of sequencing artifacts or ADO, respectively. Moreover, NGS has technical limitations in testing for dynamic mutations. Therefore, further studies are needed to evaluate this technology before its routine clinical use.

In summary, PGD provides screening for most genetic aberrations in embryos. Its utility is highlighted by the yearly increases in the number of PGD cases and indications tested. During the last two decades, PGD has reached a high level of accuracy and has enabled multiple diagnoses from the same sample. Diagnosis of a monogenic disease can be now combined with HLA typing and with the detection of chromosomal abnormalities.

4. CARRIERS OF STRUCTURAL ABNORMALITIES

4.1 Definitions

Balanced structural chromosome rearrangements are the most frequent chromosome abnormalities in the general population, with a prevalence of 0.4% in prenatal samples and 0.2% in newborns (Jacobs, Melville, Ratcliffe, Keay, & Syme, 1974; Van Dyke, Weiss, Roberson, & Babu, 1983). The most common structural chromosome rearrangements are translocations and inversions.

Translocations are structural chromosome abnormalities that occur after a double break in two different chromosomes and exchange of fragments between these chromosomes. Translocations can be of two types, reciprocal and Robertsonian. Reciprocal translocations are produced by breakage and exchange of distal segments between nonhomologous chromosomes (Fig. 2). The incidence of balanced reciprocal translocations in the general population and newborns is approximately 0.14% (Nielsen & Wohlert,

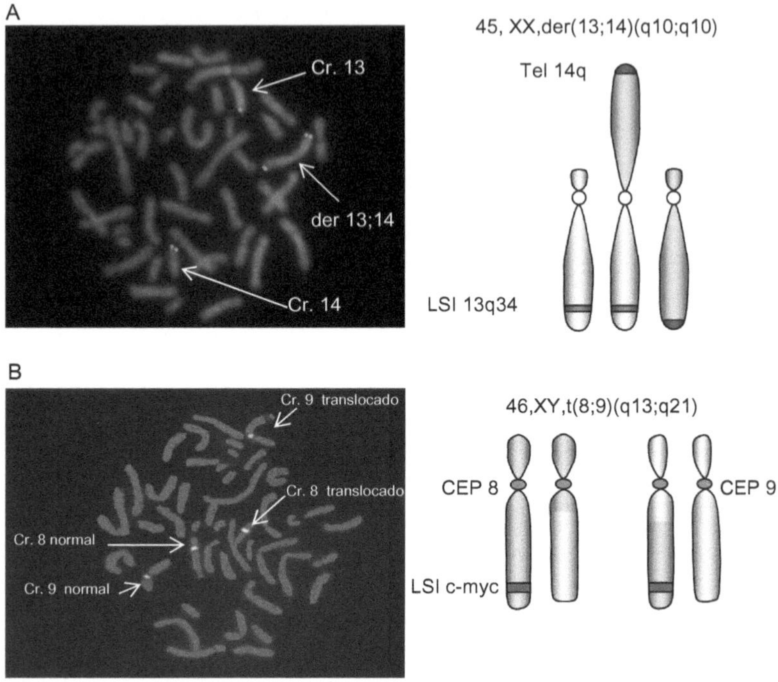

Fig. 2 An example of a Robertsonian (A) and reciprocal translocation (B). (See the color plate.)

1991; Schena, Shalon, Davis, & Brown, 1995). Robertsonian translocations arise by the fusion of two acrocentric chromosomes (chromosomes 13, 14, 15, 21, and 22) and loss of their short arms. Their incidence in the population is 0.12%, and the most common combination is the fusion of chromosomes 13 and 14 (Nielsen & Wohlert, 1991).

Inversions, on the other hand, are structural chromosome abnormalities that occur after a double intrachromosomal break, a 180 degree rotation of the fragment located between the two breakpoints, and subsequent reinsertion of the fragment into the chromosome. Inversions are classified according to the relative position of the centromere with respect to the inverted fragment and can be pericentric or paracentric inversions. In pericentric inversions, the centromere is within the inverted fragment; in paracentric inversions, both breakpoints are in the same arm of the chromosome (centromere not involved). Inverted chromosomes can establish conditions leading to the generation of recombinant gametes, some of which will have partial deletions and duplications.

When no loss of genetic material occurs, translocations, and inversions are balanced and the heterozygous phenotype is normal. However, the reproductive consequences for translocation and inversion carriers may vary depending on the chromosomes involved in the rearrangement, the size of the implicated fragments, and the locus in which breakage occurs. They could have an increased risk of having fertility problems or recurrent miscarriages, and producing offspring with congenital abnormalities and cognitive impairment. Because of this, the prevalence of such anomalies among infertile patients can be up to 25 times higher than in the general population (Campana, Serra, & Neri, 1986; Fryns & Van Buggenhout, 1998). These problems are mainly due to the production of unbalanced gametes during meiosis because of abnormal segregations in the translocation carriers or recombination events in the inversion carriers (Neri, Serra, Campana, & Tedeschi, 1983; Scriven, Handyside, & Ogilvie, 1998).

Unbalanced gametes will result in embryos with aneuploidies for chromosomes involved in the rearrangements. Some researchers have proposed that the chromosomes involved in rearrangements may interfere with the correct segregation of other chromosomes by disrupting chromosome alignment on the meiotic spindle. This phenomenon, known as the interchromosomal effect (ICE), was first described by Lejeune (1963), who observed an increased rate of carriers of balanced reciprocal translocations among the parents of children with trisomy 21 (Lejeune, 1963). However, conflicting results were reported. Some studies supported ICE (Blanco, Egozcue, Clusellas, & Vidal, 1998; Pang et al., 1999), and it seemed to be dependent on rearranged fragment size, patient, and chromosome (Boerke, Dieleman, & Gadella, 2007). Other studies did not find ICE (Estop et al., 2000; Guttenbach, Martinez-Exposito, Michelmann, Engel, & Schmid, 1997; Honda et al., 1999), or attributed the increased aneuploidy rate to other factors such as oligoasthenoteratozoospermia (OAT) syndrome, frequently observed within these patients (Moosani et al., 1995; Pang et al., 1999; Rives et al., 1999).

Nevertheless, PGD in patients carrying translocations and inversions improves their reproductive expectations, reducing the time to achieve a successful live birth from 4–6 years to less than 4 months, and decreases the incidence of miscarriage from more than 90% to less than 15% (Fischer, Colls, Escudero, & Munne, 2010; Munne et al., 2000; Verlinsky et al., 2005).

4.2 Diagnostic Methods

- *FISH*: Until recently FISH was the go-to technique for PGD in patients with structural chromosomal abnormalities in both polar body and interphase nuclei (Munne, Bahce, Schimmel, Sadowy, & Cohen, 1998; Munne, Scott, Sable, & Cohen, 1998). Although this technique improved the reproductive expectations of these couples, achieving clinical pregnancy rates of approximately 40% (Harper et al., 2012), FISH has some technical limitations. In particular, FISH is limited by the observation of split signals, cross-hybridization, chromosome polymorphisms, poor fixation quality, and loss of micronuclei or chromosomes during fixation (DeUgarte et al., 2008; Munne, 2002; Velilla, Escudero, & Munne, 2002). Additionally, this technique identifies only imbalances of the translocated chromosomes, and does not evaluate all 23 pairs of chromosomes for the presence of aneuploidy.
- *Array-CGH (aCGH)*: WGA approaches, including CGH (Fishel et al., 2010; Fragouli et al., 2010; Schoolcraft et al., 2010; Voullaire, Wilton, McBain, Callaghan, & Williamson, 2002; Wells, Alfarawati, & Fragouli, 2008; Wells et al., 2002; Wilton, Voullaire, Sargeant, Williamson, & McBain, 2003) and aCGH (Gutierrez-Mateo, Colls, Sanchez-Garcia, et al., 2011), have been successfully used for aneuploidy screening. With these technologies, aneuploidies, as well as imbalances for affected chromosomes, can be detected for all 24 chromosomes. Therefore, ICE can be ruled out.
- *SNP microarray*: Originally SNP arrays were designed to genotype human DNA at thousands of SNPs across the genome simultaneously; however, applications have expanded to include the detection and characterization of copy number variations. Biallelic SNPs, in which one of two bases is present, referred to generically as A and B, are valuable markers, and hundreds of thousands of SNPs can be genotyped simultaneously with the use of SNP arrays. Furthermore, for molecular cytogenetics, analysis of the ratio of the intensity of the B to the A alleles at heterozygous loci enables detection of duplications and deletions from whole chromosomes to small regions with high resolution. In the duplications, the B-allele ratio at heterozygous loci splits into two bands representing loci that are either AAB or ABB. In deletions, loss of heterozygosity (LOH) is detected by the absence of the heterozygous band. This technology has been applied to detect imbalances in carriers of structural chromosome abnormalities (Treff, Levy, et al., 2010; Treff,

Northrop, et al., 2011; Treff, Su, Tao, Levy, & Scott, 2010; Treff, Tao, et al., 2011; van Uum et al., 2012).
- *NGS*: Several NGS platforms have been validated for detecting whole chromosome aneuploidies, and few studies have shown also high concordance for segmental aneuploidies in carriers of translocations (Bono et al., 2015; Vera-Rodriguez et al., 2015).

5. ANEUPLOIDY SCREENING

5.1 Indications

PGS was introduced in clinical routine practice to improve pregnancy rates in subfertile couples, based on the assumption that high rates of chromosomal aneuploidy, frequently found in cleavage-stage embryos of these couples, were responsible for low pregnancy rates after ART. The main goals for most of the indications are not only to increase implantation and pregnancy rates, but also to decrease miscarriages, and the risk of aneuploid offspring, as well as to decrease the time to conceive. The following are the most common current indications:
- *Advanced maternal age* (*AMA*): AMA is the most common indication for PGS. Maternal age is a major factor in the prevalence of aneuploidy. A recent study reported that the rate of mis-segregation for the most clinically relevant aneuploidies (chromosomes 13, 16, 18, 21, 22) increased from 20% to 60% in women between the ages of 35 and 43 years (Kuliev, Zlatopolsky, Kirillova, Spivakova, & Cieslak Janzen, 2011).
- *Recurrent miscarriage* (*RM*): Although the definition of RM varies by country, generally speaking RM is the occurrence of 3 or more consecutive miscarriages with a gestational age up to 14 weeks (Rai & Regan, 2006). A leading cause of miscarriages is aneuploidy; indeed, aneuploidy is identified in 28–78% of products of conception in patients with recurrent pregnancy loss (Marquard, Westphal, Milki, & Lathi, 2010; Nybo Andersen, Wohlfahrt, Christens, Olsen, & Melbye, 2000).
- *Repetitive implantation failure* (*RIF*): RIF is defined as three or more failed IVF attempts or failed IVF treatments after the cumulative transfer of more than 10 good-quality embryos. Criteria to define RIF are not homogenous and an exhaustive and comprehensive definition has not yet been reached (Pehlivan et al., 2003; Rubio, Bellver, et al., 2013). RIF remains a big challenge to the clinician because its causes can be multiple and are still poorly defined (Garcia-Herrero et al., 2014).

Further, embryonic and endometrial factors can play an important role in this condition. Some authors argue that these couples produce more embryos with aneuploidies (Hodes-Wertz et al., 2012).

- *Severe male factor infertility* (*MF*): An increased incidence of chromosome abnormalities has been reported in sperm samples of infertile men with normal karyotypes by FISH (Rubio et al., 2001). For example Rubio et al. (2001) reported that OAT was associated with significant increases in sex chromosome disomies, disomy for chromosomes 18 and 21, and in the percentage of diploid sperm, particularly in those samples with markedly reduced sperm concentration ($<5 \times 10^6$/mL spermatozoa). Such conditions might, in part, explain the low implantation and high abortion rates observed in these patients.
- *Previous trisomic pregnancy* (*PTP*): Some studies suggest that a history of a trisomic pregnancy is associated with an increased risk of another aneuploid conception, so PGS could benefit this group of patients (Pagidas, Ying, & Keefe, 2008).

5.2 Diagnostic Methods

Several methods currently allow for the study of aneuploidies in human embryos, including FISH, SNP arrays, aCGH, and NGS. The evolution of the techniques for aneuploidy screening has provided more information about the genetic status of the embryo and faster results, which enables transfer of euploid embryos during the same cycle.

- FISH assay uses fluorescence nucleic acid probes complementary to DNA to visualize regions of interest. However, FISH cannot be performed for all chromosomes simultaneously (Abdelhadi, Colls, Sandalinas, Escudero, & Munne, 2003), therefore it has been used exclusively for those chromosomes for which aneuploidy is implicated in spontaneous miscarriages or for those compatible with live birth (Stephenson, Awartani, & Robinson, 2002), such as chromosomes 3, 15, 16, 17, 18, 21, 22, X, and Y. However, these chromosomes need to be assessed over multiple rounds of hybridization. Fortunately, new technologies have facilitated the transition from the study of a limited number of chromosomes by FISH, to the analysis of all 23 chromosome pairs simultaneously in a single cell.
- Array-based technologies provide results in less than 24 h and with less than 3% noninformative rate (Gutierrez-Mateo et al., 2011). One of the most used platforms is aCGH, for which the sample, a blastomere or trophectoderm biopsy, is placed in a sterile PCR tube for WGA.

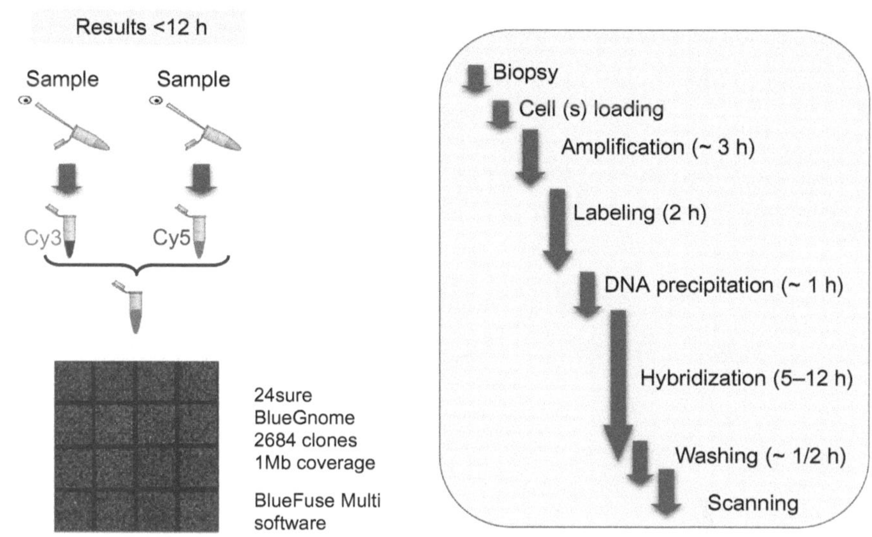

Fig. 3 CGH array flow chart. (See the color plate.)

Amplification quality needs to be ensured (eg, by gel electrophoresis), and then samples and control DNA are labeled with fluorophores (Rubio, Rodrigo, et al., 2013). Labeling mixes are combined and hybridized on the arrays (Fig. 3). Each probe in the array is specific to a different chromosomal region and occupies a discrete spot on a microscope slide. Chromosomal loss or gain is revealed by the color adopted by each spot after hybridization; this is because the technique involves the competitive hybridization of differentially labeled test and reference euploid DNA samples. Fluorescence intensity is detected with the use of a laser scanner, and specific software is used for data processing. Therefore, embryo transfer and vitrification of surplus euploid embryos can be scheduled on D5 when D3 biopsies are performed, or on D6 for trophectoderm biopsies (Rubio, Rodrigo, et al., 2013).

- NGS technology has been proposed as an effective technique for the analysis of copy number variation in single cells. The decrease in the cost of genome sequencing has position NGS as one of the most promising platforms for the study of not only aneuploidies, but also mitochondrial DNA or gene disorders in a simultaneous analysis (Yan et al., 2015). For NGS, most extended protocols share the first steps of the aCGH protocols, starting with WGA of a single cell. This is followed by a barcoding procedure, in which the different samples are labeled with unique sequences, in a way that they can be mixed later and sequenced as they can be individually identified (Fig. 4). This pooling step has significantly

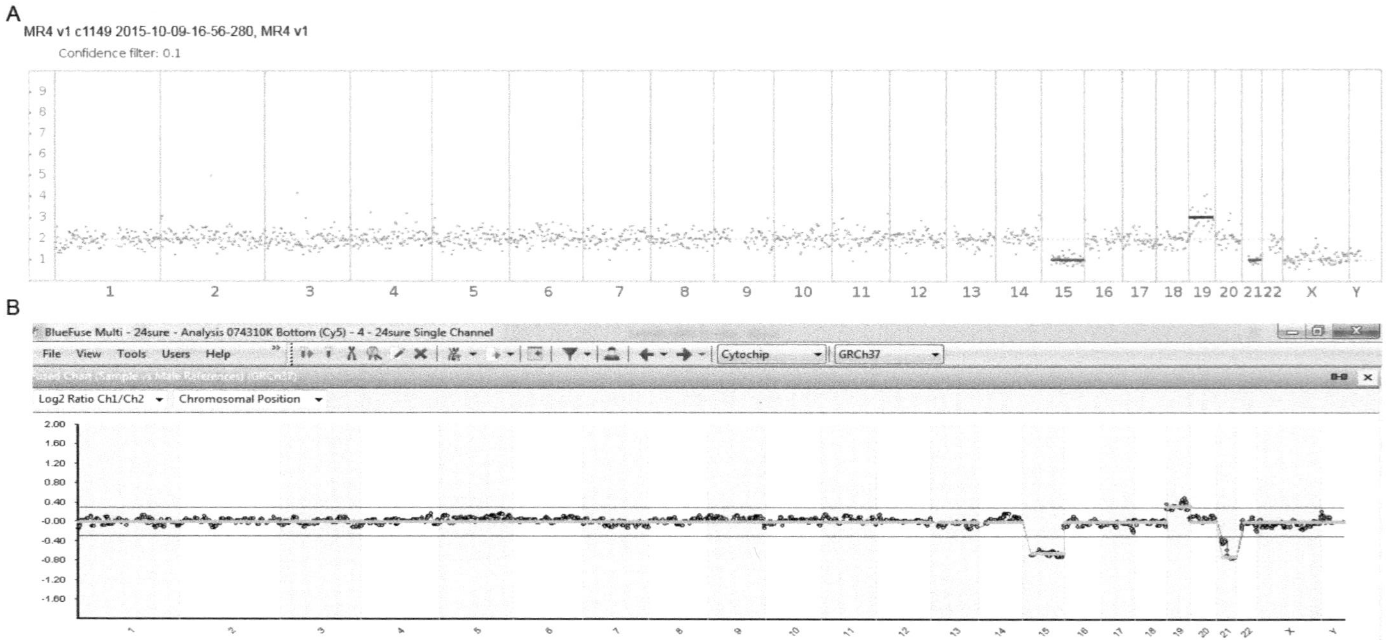

Fig. 4 NGS (A) and CGH arrays (B) profiles of a D5 embryo biopsy presenting two monosomies (15 and 21) and one trisomy (19). (See the color plate.)

reduced the cost of the technique and facilitated its transition into the clinical routine. The depth of sequencing is also an important aspect to consider, especially for the simultaneous study of aneuploidies and gene disorders, which would need high coverage on those regions of interest. An alternative is the parallel study of the mitochondrial DNA, which has been related to embryo quality (Diez-Juan et al., 2015). In addition, sequencing has been shown to detect lower mosaicism degrees in trophectoderm biopsies than previous technologies (Ruttanajit et al., 2016). Nevertheless, it would be important to consider the relevance of mosaicism in the trophectoderm, as there are no studies related to the effect of different percentages of aneuploidy cells in the blastocyst.

5.3 Accuracy and Results in Day-3 and Blastocyst Biopsies

D3 embryo biopsy has been highly criticized because of a high degree of mosaicism in cleavage-stage embryos and the possibility of self-correction of aneuploidies from cleavage to blastocyst stage; therefore, trophectoderm biopsy has been suggested as the most reliable option for PGS. Moreover, identical genetic constitution of both inner cell mass and trophectoderm from the same embryo has been demonstrated (Fragouli et al., 2008). Added to this, improvements in embryo culture conditions as well as in vitrification systems (Cobo et al., 2008) have paved the way for the current trend of trophectoderm embryo biopsies.

When comparing accuracy rates of D3 and trophectoderm embryo biopsies, all the studies using a 24-chromosome analysis, show comparable results in terms of high accuracy for both types of embryo biopsies. Those studies comparing the aCGH result of the day-3 biopsy to the results of the FISH technique applied on the remaining cells of the day-5 embryos showed low false-positive rates, from 1.9% to 2.7% (Gutierrez-Mateo et al., 2011; Mir et al., 2013). In Capalbo et al. (2013), isolated cells from the inner cell mass and from the trophectoderm were re-analyzed using FISH, showing a false-positive rate of 2.9% (2/70).

These results are consistent with another blinded study, in which 109 embryos were reanalyzed using the same aCGH technique used for the PGS analysis (PGS was performed on D3 in 50 of the embryos, and on trophectoderm biopsies in the remaining 59 embryos). High confirmation rates per individual chromosome in both types of embryo biopsy (98.8% for day-3 vs 98.5% for trophectoderm) were shown, and these rates are also reflected in high confirmation rates of the PGS diagnosis (98% for day-3 vs 97.6% for trophectoderm) (Mir et al., 2016).

The cases of false-positive results were diagnosed as chromosomally abnormal on the embryo biopsy but as chromosomally normal when reanalyzing the full blastocyst, therefore the most feasible explanation is that they were mosaic "euploid–aneuploid." It is challenging to calculate the real incidence of mosaicism in preimplantation embryos. First, because of technical limitations in aCGH mosaicism can be detected if it represents more than 25–40% of the cells (Treff, Su, et al., 2010). Second, the biopsy will always represent a small percentage of the total number of cells of the embryo. However, reviewing the clinical outcome of patients undergoing PGS with aCGH, which shows high pregnancy and implantation rates as well as reduced miscarriage rate (Gardner, Meseguer, Rubio, & Treff, 2015; Rodrigo et al., 2014; Rubio, Rodrigo, et al., 2013) and also a very low clinically recognizable error rate shown in PGS (Werner et al., 2014), it can be assumed that the number of "euploid–aneuploid" mosaic embryos in cleavage and blastocyst stage is not very high and should not have a major clinical impact.

5.4 Results and Randomized Controlled Trials in PGS

Several retrospective studies have described the clinical benefits obtained by the selection of euploid embryos (Gianaroli et al., 2002; Kahraman et al., 2000; Milan et al., 2010). However, a meta-analysis compiling nine published randomized controlled trials (RCTs) concluded that PGS should not be recommended (Mastenbroek, Twisk, van der Veen, & Repping, 2011). One RCT performed in RIF patients concluded no significant differences in clinical pregnancy rates between PGS and control groups (Blockeel et al., 2008). Four RCTs have been published for AMA patients; three of them indicated that PGS offered no benefit (Hardarson et al., 2008), but the fourth study described lower miscarriage rates and increased delivery rates (Schoolcraft, Katz-Jaffe, Stevens, Rawlins, & Munne, 2009). These studies have been criticized by several authors, who argued that some important pitfalls exist in study methodology, such as patient inclusion criteria, embryo biopsy procedures, embryo culture conditions, and type of genetic analyses performed (Cohen, Wells, & Munne, 2007; Mir et al., 2010; Rubio et al., 2009; Simpson, 2008).

Our own experience differs from the previously published studies. We conducted two prospective, randomized trials to evaluate the usefulness of

PGS in AMA patients between 41 and 44 years old and RIF patients <40 years of age. We observed a significant increase in live birth rates in the PGS group compared to the conventional blastocyst transfer group for the AMA study (32.3% vs 15.5%; $p=0.0099$), and a clear trend toward better live birth rates for the RIF study (47.9% vs 27.9%). We therefore concluded that PGS with classic FISH was beneficial for these two indications if proper blastomere biopsy procedures, good laboratory conditions, and validated and robust genetic analyses were applied (Rubio, Rodrigo, et al., 2013). Despite these results, there remained a need for a technique to analyze all chromosomes, while also producing reliable and accurate results in a short period of time.

With the application of aCGH in cleavage-stage biopsies, a retrospective study of 2858 cycles showed the clinical benefits of embryo selection based on aCGH analysis in poor reproductive-prognosis couples. The most frequent indication was AMA, followed by RIF, MF, and PTP. The percentage of informative embryos was 97.9%, with only 2.1% of samples noninformative mostly due to cell loss or damage during manipulation, resulting in defective amplification. Aneuploidy rates were 67.2% in women <38 years and 86.3% in women ≥38 years. Mean pregnancy rate per transfer and implantation rate in women <38 years were 59.0% and 49.0%, respectively. In women ≥38 years, pregnancy rates per transfer and implantation rates were as high as 51.1% and 46.2%, respectively. Miscarriage rates were also comparable in the <38 years and ≥38 years age groups (11.9% vs 9.9%) (Rodrigo et al., 2014). Another retrospective case–control study comparing the outcome in poor-prognosis patients who underwent cleavage-stage biopsies with fresh embryo transfer cycles and similar patients undergoing regular IVF cycles showed double implantation and ongoing pregnancy rate with a decrease in multiple pregnancy and miscarriage rates in PGS cases (Keltz et al., 2013).

Preliminary results of two RCTs comparing PGS with day-3 biopsies and blastocyst transfers without chromosomal analysis, performed in couples with AMA (38–41 years) and MF (≤2 million sperm/mL), have revealed promising results. In the AMA study, 86 cycles were assessed in the conventional blastocyst transfer group and 75 in the PGS group. Ongoing pregnancy rate per transfer and per cycle were significantly increased in the PGS group (26.5% vs 60.4%; $p=0.0001$ and 25.6% vs 42.7%; $p=0.0294$, respectively). Miscarriage rates were significantly higher in the blastocyst group than in the PGS group (43.6 vs 3.3; $p<0.0001$) (Rubio et al., 2014). In the MF study, in the group with conventional blastocyst transfer,

35 cycles were completed with 33 transfers and 15 ongoing pregnancies (45.4% ongoing pregnancy rate per transfer and 42.8% per cycle); in the PGS group, 33 cycles were performed with 31 transfers and 22 ongoing pregnancies (71.0% ongoing pregnancy rate per transfer and 66.7% per cycle). One-side chi-square test showed significant differences for ongoing pregnancy rates per transfer ($p = 0.0345$) and per cycle ($p < 0.0417$). In the blastocyst transfer group, miscarriage rate was 28.6%, whereas no miscarriage was observed in the PGS group (Rubio et al., 2015).

At the blastocyst stage, Wells et al. (2008) found that the probability of an individual analyzed by aCGH generating a pregnancy was 66.7% compared with 27.9% without aCGH testing. A multicenter retrospective study described an increase in the incidence of aneuploid embryos, which correlated with increased maternal age: similar implantation and ongoing pregnancy rates per transfer were observed after CCS in patients up to 42 years of age, after which these rates dramatically declined.

Moreover, PGS at blastocyst stage using aCGH has also been applied in good-prognosis patients with a high potential to increase overall pregnancy rates in IVF programs and to decrease multiple pregnancies when single-embryo transfer (SET) is performed. The first RCT comparing SET at blastocyst stage with and without aCGH in good-prognosis patients showed an aneuploidy rate of 44.9% among biopsied blastocysts, with a significantly higher clinical pregnancy rate in the CCS group (70.9% vs 45.8%, $p = 0.017$). There were no twin pregnancies. This study revealed the limitations of SET when conventional morphology was used alone, even in patients without an increased risk for aneuploidy, since the CGH arrays group implanted with greater efficiency and yielded a lower miscarriage rate than those selected without CGH arrays (Yang et al., 2012).

REFERENCES

Abdelhadi, I., Colls, P., Sandalinas, M., Escudero, T., & Munne, S. (2003). Preimplantation genetic diagnosis of numerical abnormalities for 13 chromosomes. *Reproductive Biomedicine Online*, *6*, 226–231.

Blanco, J., Egozcue, J., Clusellas, N., & Vidal, F. (1998). FISH on sperm heads allows the analysis of chromosome segregation and interchromosomal effects in carriers of structural rearrangements: Results in a translocation carrier, t(5;8)(q33;q13). *Cytogenetics and Cell Genetics*, *83*, 275–280.

Blockeel, C., Schutyser, V., De Vos, A., Staessen, C., Haentjens, P., Van der Elst, J., et al. (2008). Prospectively randomized controlled trial of PGS in IVF/ICSI patients with poor implantation. *Reproductive Biomedicine Online*, *17*, 848–854.

Boerke, A., Dieleman, S. J., & Gadella, B. M. (2007). A possible role for sperm RNA in early embryo development. *Theriogenology*, *68*(Suppl. 1), S147–S155.

Bono, S., Biricik, A., Spizzichino, L., Nuccitelli, A., Minasi, M., Greco, E., et al. (2015). Validation of a semiconductor next-generation sequencing-based protocol for preimplantation genetic diagnosis of reciprocal translocations. *Prenatal Diagnosis, 35*, 938–944.

Campana, M., Serra, A., & Neri, G. (1986). Role of chromosome aberrations in recurrent abortion: A study of 269 balanced translocations. *American Journal of Medical Genetics, 24*, 341–35.6.

Capalbo, A., Wright, G., Elliott, T., Ubaldi, F. M., Rienzi, L., & Nagy, Z. P. (2013). FISH reanalysis of inner cell mass and trophectoderm samples of previously array-CGH screened blastocysts shows high accuracy of diagnosis and no major diagnostic impact of mosaicism at the blastocyst stage. *Human Reproduction, 28*, 2298–2307.

Clancy, T. (2010). A clinical perspective on ethical arguments around prenatal diagnosis and preimplantation genetic diagnosis for later onset inherited cancer predispositions. *Familial Cancer, 9*, 9–14.

Cobo, A., Bellver, J., Domingo, J., Pérez, S., Crespo, J., Pellicer, A., et al. (2008). New options in assisted reproduction technology: The Cryotop method of oocyte vitrification. *Reproductive Biomedicine Online, 17*, 68–72.

Cohen, J., Wells, D., & Munne, S. (2007). Removal of 2 cells from cleavage stage embryos is likely to reduce the efficacy of chromosomal tests that are used to enhance implantation rates. *Fertility and Sterility, 87*, 496–503.

De Rycke, M., Belva, F., Goossens, V., Moutou, C., SenGuptam, S. B., Traeger-Synodinos, J., et al. (2015). ESHRE PGD Consortium data collection XIII: Cycles from January to December 2010 with pregnancy follow-up to October (2011). *Human Reproduction, 30*, 1763–1789.

De Vos, A., & Van Steirteghem, A. (2001). Aspects of biopsy procedures prior to preimplantation genetic diagnosis. *Prenatal Diagnosis, 21*, 767–780.

DeUgarte, C. M., Li, M., Surrey, M., Danzer, H., Hill, D., & DeCherney, A. H. (2008). Accuracy of FISH analysis in predicting chromosomal status in patients undergoing preimplantation genetic diagnosis. *Fertility and Sterility, 90*, 1049–1054.

Diez-Juan, A., Rubio, C., Marin, C., Martinez, S., Al-Asmar, N., Riboldi, M., et al. (2015). Mitochondrial DNA content as a viability score in human euploid embryos: Less is better. *Fertility and Sterility, 104*, 534–541.

Estop, A. M., Cieply, K., Munne, S., Surti, U., Wakim, A., & Feingold, E. (2000). Is there an interchromosomal effect in reciprocal translocation carriers? Sperm FISH studies. *Human Genetics, 106*, 517–524.

Fiorentino, F., Biricik, A., Nuccitelli, A., De Palma, R., Kahraman, S., Iacobelli, M., et al. (2006). Strategies and clinical outcome of 250 cycles of preimplantation genetic diagnosis for single gene disorders. *Human Reproduction, 21*, 670–684.

Fiorentino, F., Bono, S., Biricik, A., Nuccitelli, A., Cotroneo, E., Cottone, G., et al. (2014). Application of next-generation sequencing technology for comprehensive aneuploidy screening of blastocysts in clinical preimplantation genetic screening cycles. *Human Reproduction, 29*, 2802–2813.

Fiorentino, F., Kahraman, S., Karadayi, H., Biricik, A., Sertye, S., Karlikaya, G., et al. (2005). Short tandem repeats haplotyping of the HLA region in preimplantation HLA matching. *European Journal of Human Genetics, 13*, 953–958.

Fiorentino, F., Magli, M. C., Podini, D., Ferraretti, A. P., Nuccitelli, A., Vitale, N., et al. (2003). The minisequencing method: An alternative strategy for preimplantation genetic diagnosis of single gene disorders. *Molecular Human Reproduction, 9*, 399–410.

Fischer, J., Colls, P., Escudero, T., & Munne, S. (2010). Preimplantation genetic diagnosis (PGD) improves pregnancy outcome for translocation carriers with a history of recurrent losses. *Fertility and Sterility, 94*, 283–289.

Fishel, S., Gordon, A., Lynch, C., Dowell, K., Ndukwe, G., Kelada, E., et al. (2010). Live birth after polar body array comparative genomic hybridization prediction of embryo ploidy-the future of IVF? *Fertility and Sterility, 93*, 1006.e7–1006.e10.

Fragouli, E., Katz-Jaffe, M., Alfarawati, S., Stevens, J., Colls, P., Goodall, N. N., et al. (2010). Comprehensive chromosome screening of polar bodies and blastocysts from couples experiencing repeated implantation failure. *Fertility and Sterility, 94*, 875–887.

Fragouli, E., Lenzi, M., Ross, R., Katz-Jaffe, M., Schoolcraft, W. B., & Wells, D. (2008). Comprehensive molecular cytogenetic analysis of the human blastocyst stage. *Human Reproduction, 23*, 2596–2608.

Fryns, J. P., & Van Buggenhout, G. (1998). Structural chromosome rearrangements in couples with recurrent fetal wastage. *European Journal of Obstetrics, Gynecology, and Reproductive Biology, 81*, 171–176.

Garcia-Herrero, S., Rodrigo, L., Mateu, E., Peinado, V., Milán, M., Campos-Galindo, I., et al. (2014). Médecine thérapeutique/Médecine de la reproduction, gynécologie et endocrinologie. *John Libbey Eurotext, 14*(2), 112–119.

Gardner, D. K., Meseguer, M., Rubio, C., & Treff, N. R. (2015). Diagnosis of human preimplantation embryo viability. *Human Reproduction Update, 21*, 727–747.

Geraedts, J., Collins, J., Gianaroli, L., Goossens, V., Handyside, A., Harper, J., et al. (2010). What next for preimplantation genetic screening? A polar body approach. *Human Reproduction, 25*, 575–577.

Gianaroli, L., Magli, M. C., Ferraretti, A. P., Tabanelli, C., Trombetta, C., & Boudjema, E. (2002). The role of preimplantation diagnosis for aneuploidies. *Reproductive Biomedicine Online, 4*(Suppl. 3), 31–36.

Grewal, S. S., Kahn, J. P., MacMillan, M. L., Ramsay, N. K., & Wagner, J. E. (2004). Successful hematopoietic stem cell transplantation for Fanconi anemia from an unaffected HLA-genotype-identical sibling selected using preimplantation genetic diagnosis. *Blood, 103*, 1147–1151.

Gutierrez-Mateo, C., Colls, P., Sanchez-Garcia, J., et al. (2011). Validation of microarray comparative genomic hybridization for comprehensive chromosome analysis of embryos. *Fertility and Sterility, 95*, 953–958.

Guttenbach, M., Martinez-Exposito, M. J., Michelmann, H. W., Engel, W., & Schmid, M. (1997). Incidence of diploid and disomic sperm nuclei in 45 infertile men. *Human Reproduction, 12*, 468–473.

Haddad, G., He, W., Gill, J., Witz, C., Wang, C., Kaskar, K., et al. (2013). Mosaic pregnancy after transfer of a "euploid" blastocyst screened by DNA microarray. *Journal of Ovarian Research, 6*, 70.

Handyside, A. H., Harton, G. L., Mariani, B., Thornhill, A. R., Affara, N., Shaw, M. A., et al. (2010). Karyomapping: A universal method for genome wide analysis of genetic disease based on mapping crossovers between parental haplotypes. *Journal of Medical Genetics, 47*, 651–658.

Handyside, A. H., Kontogianni, E. H., Hardy, K., & Winston, R. M. (1990). Pregnancies from biopsied human preimplantation embryos sexed by Y-specific DNA amplification. *Nature, 344*, 768–770.

Handyside, A. H., Pattinson, J. K., Penketh, R. J., Delhanty, J. D., Winston, R. M., & Tuddenham, E. G. (1989). Biopsy of human preimplantation embryos and sexing by DNA amplification. *Lancet, 1*, 347–349.

Hardarson, T., Hanson, C., Lundin, K., Hillensjö, T., Nilsson, L., Stevic, J., et al. (2008). Preimplantation genetic screening in women of advanced maternal age caused a decrease in clinical pregnancy rate: A randomized controlled trial. *Human Reproduction, 23*, 2806–2812.

Harper, J. C., Coonen, E., Ramaekers, F. C., Delhanty, J. D., Handyside, A. H., Winston, R. M., et al. (1994). Identification of the sex of human preimplantation embryos in two hours using an improved spreading method and fluorescent in-situ hybridization (FISH) using directly labelled probes. *Human Reproduction, 9*, 721–724.

Harper, J. C., Wilton, L., Traeger-Synodinos, J., Goossens, V., Moutou, C., SenGupta, S. B., et al. (2012). The ESHRE PGD Consortium: 10 years of data collection. *Human Reproduction Update*, *18*, 234–247.

Harton, G. L., Munne, S., Surrey, M., Grifo, J., Kaplan, B., McCulloh, D. H., et al. (2013). Diminished effect of maternal age on implantation after preimplantation genetic diagnosis with array comparative genomic hybridization. *Fertility and Sterility*, *100*, 1695–1703.

Hodes-Wertz, B., Grifo, J., Ghadir, S., Ghadir, S., Kaplan, B., Laskin, C. A., et al. (2012). Idiopathic recurrent miscarriage is caused mostly by aneuploid embryos. *Fertility and Sterility*, *98*, 675–680.

Honda, H., Miharu, N., Ohashi, Y., Honda, N., Hara, T., & Ohama, K. (1999). Analysis of segregation and aneuploidy in two reciprocal translocation carriers, t(3;9)(q26.2;q32) and t(3;9)(p25;q32), by triple-color fluorescence in situ hybridization. *Human Genetics*, *105*, 428–436.

Jacobs, P. A., Melville, M., Ratcliffe, S., Keay, A. J., & Syme, J. (1974). A cytogenetic survey of 11,680 newborn infants. *Annals of Human Genetics*, *37*, 359–376.

Kahraman, S., Bahce, M., Samli, H., Imirzalioğlu, N., Yakisn, K., Cengiz, G., et al. (2000). Healthy births and ongoing pregnancies obtained by preimplantation genetic diagnosis in patients with advanced maternal age and recurrent implantation failure. *Human Reproduction*, *15*, 2003–2007.

Keltz, M. D., Vega, M., Sirota, I., Lederman, M., Moshier, E. L., Gonzales, E., et al. (2013). Preimplantation genetic screening (PGS) with comparative genomic hybridization (CGH) following day 3 single cell blastomere biopsy markedly improves IVF outcomes while lowering multiple pregnancies and miscarriages. *Journal of Assisted Reproduction and Genetics*, *30*, 1333–1339.

Konstantinidis, M., Prates, R., Goodall, N. N., Fischer, J., Tecson, V., Lemma, T., et al. (2015). Live births following Karyomapping of human blastocysts: Experience from clinical application of the method. *Reproductive Biomedicine Online*, *31*, 394–403.

Kuliev, A., Rechitsky, S., Verlinsky, O., Tur-Kaspa, I., Kalakoutis, G., Angastiniotis, M., et al. (2005). Preimplantation diagnosis and HLA typing for haemoglobin disorders. *Reproductive Biomedicine Online*, *11*, 362–370.

Kuliev, A., Zlatopolsky, Z., Kirillova, I., Spivakova, J., & Cieslak Janzen, J. (2011). Meiosis errors in over 20,000 oocytes studied in the practice of preimplantation aneuploidy testing. *Reproductive Biomedicine Online*, *22*, 2–8.

Lejeune, J. (1963). Autosomal disorders. *Pediatrics*, *32*, 326–337.

Mamas, T., Gordon, A., Brown, A., Harper, J., & Sengupta, S. (2012). Detection of aneuploidy byarray comparative genomic hybridization using cell lines to mimic a mosaic trophectoderm biopsy. *Fertility and Sterility*, *97*, 943–947.

Marquard, K., Westphal, L. M., Milki, A. A., & Lathi, R. B. (2010). Etiology of recurrent pregnancy loss in women over the age of 35 years. *Fertility and Sterility*, *94*, 1473–1477.

Mastenbroek, S., Twisk, M., van der Veen, F., & Repping, S. (2011). Preimplantation genetic screening: A systematic review and meta-analysis of RCTs. *Human Reproduction Update*, *17*, 454–466.

Milan, M., Cobo, A. C., Rodrigo, L., Mateu, E., Mercader, A., Buendía, P., et al. (2010). Redefining advanced maternal age as an indication for preimplantation genetic screening. *Reproductive Biomedicine Online*, *21*, 649–657.

Mir, P., Mateu, E., Mercader, A., Herrer, R., Rodrigo, L., Vera, M., et al. (2016). Confirmation rates of array-CGH in day-3 embryo and blastocyst biopsies for preimplantation genetic screening. *Journal of Assisted Reproduction and Genetics*, *33*, 59–66.

Mir, P., Rodrigo, L., Mateu, E., Peinado, V., Milán, M., Mercader, A., et al. (2010). Improving FISH diagnosis for preimplantation genetic aneuploidy screening. *Human Reproduction*, *25*, 1812–1817.

Mir, P., Rodrigo, L., Mercader, A., Buendía, P., Mateu, E., Milán-Sánchez, M., et al. (2013). False positive rate of an arrayCGH platform for single-cell preimplantation genetic screening and subsequent clinical application on day-3. *Journal of Assisted Reproduction and Genetics, 30*, 143–149.

Montag, M., Köster, M., Strowitzki, T., & Toth, B. (2013). Polar body biopsy. *FertSteril, 100*, 603–607.

Moosani, N., Pattinson, H. A., Carter, M. D., Cox, D. M., Rademaker, A. W., & Martin, R. H. (1995). Chromosomal analysis of sperm from men with idiopathic infertility using sperm karyotyping and fluorescence in situ hybridization. *Fertility and Sterility, 64*, 811–817.

Moutou, C., Gardes, N., Nicod, J. C., & Viville, S. (2007). Strategies and outcomes of PGD of familial adenomatous polyposis. *Molecular Human Reproduction, 13*, 95–101.

Munne, S. (2002). Preimplantation genetic diagnosis of numerical and structural chromosome abnormalities. *Reproductive Biomedicine Online, 4*, 183–196.

Munne, S., Bahce, M., Schimmel, T., Sadowy, S., & Cohen, J. (1998). Case report: Chromatid exchange and predivision of chromatids as other sources of abnormal oocytes detected by preimplantation genetic diagnosis of translocations. *Prenatal Diagnosis, 18*, 1450–1458.

Munne, S., Sandalinas, M., Escudero, T., Fung, J., Gianaroli, L., & Cohen, J. (2000). Outcome of preimplantation genetic diagnosis of translocations. *Fertility and Sterility, 73*, 1209–1218.

Munne, S., Sandalinas, M., Escudero, T., Velilla, E., Walmsley, R., Sadowy, S., et al. (2003). Improved implantation after preimplantation genetic diagnosis of aneuploidy. *Reproductive Biomedicine Online*, 7, 91–97.

Munne, S., Scott, R., Sable, D., & Cohen, J. (1998). First pregnancies after preconception diagnosis of translocations of maternal origin. *Fertility and Sterility, 69*, 675–681.

Navin, N., & Hicks, J. (2011). Future medical applications of single-cell sequencing in cancer. *Genome Medicine, 3*, 31.

Neri, G., Serra, A., Campana, M., & Tedeschi, B. (1983). Reproductive risks for translocation carriers: Cytogenetic study and analysis of pregnancy outcome in 58 families. *American Journal of Medical Genetics, 16*, 535–561.

Nielsen, J., & Wohlert, M. (1991). Chromosome abnormalities found among 34,910 newborn children: Results from a 13-year incidence study in Arhus, Denmark. *Human Genetics, 87*, 81–83.

Nybo Andersen, A. M., Wohlfahrt, J., Christens, P., Olsen, J., & Melbye, M. (2000). Maternal age and fetal loss: Population based register linkage study. *BMJ, 320*, 1708–1712.

Pagidas, K., Ying, Y., & Keefe, D. (2008). Predictive value of preimplantation genetic diagnosis for aneuploidy screening in repeated IVF-ET cycles among women with recurrent implantation failure. *Journal of Assisted Reproduction and Genetics, 25*, 103–106.

Pang, M. G., Hoegerman, S. F., Cuticchia, A. J., Moon, S. Y., Doncel, G. F., Acosta, A. A., et al. (1999). Detection of aneuploidy for chromosomes 4, 6, 7, 8, 9, 10, 11, 12, 13, 17, 18, 21, X and Y by fluorescence in-situ hybridization in spermatozoa from nine patients with oligoasthenoteratozoospermia undergoing intracytoplasmic sperm injection. *Human Reproduction, 14*, 1266–1273.

Pehlivan, T., Rubio, C., Rodrigo, L., Romero, J., Remohi, J., Simón, C., et al. (2003). Impact of preimplantation genetic diagnosis on IVF outcome in implantation failure patients. *Reproductive Biomedicine Online, 6*, 232–237.

Rai, R., & Regan, L. (2006). Recurrent miscarriage. *Lancet, 368*, 601–611.

Rechitsky, S., Pakhalchuk, T., San Ramos, G., Goodman, A., Zlatopolsky, Z., & Kuliev, A. (2015). First systematic experience of preimplantation genetic diagnosis for single-gene disorders, and/or preimplantation human leukocyte antigen typing, combined with 24-chromosome aneuploidy testing. *Fertility and Sterility, 103*, 503–512.

Rechitsky, S., Verlinsky, O., Chistokhina, A., Sharapova, T., Ozen, S., Masciangelo, C., et al. (2002). Preimplantation genetic diagnosis for cancer predisposition. *Reproductive Biomedicine Online*, *5*, 148–155.

Renwick, P., Trussler, J., Lashwood, A., Braude, P., & Ogilvie, C. M. (2010). Preimplantation genetic haplotyping: 127 diagnostic cycles demonstrating a robust, efficient alternative to direct mutation testing on single cells. *Reproductive Biomedicine Online*, *20*, 470–476.

Rives, N., Saint Clair, A., Mazurier, S., Sibert, L., Siméon, N., Joly, G., et al. (1999). Relationship between clinical phenotype, semen parameters and aneuploidy frequency in sperm nuclei of 50 infertile males. *Human Genetics*, *105*, 266–272.

Rodrigo, L., Mateu, E., Mercader, A., Cobo, A. C., Peinado, V., Milán, M., et al. (2014). New tools for embryo selection: Comprehensive chromosome screening by array comparative genomic hybridization. *BioMed Research International*, *2014*, 517125.

Rubio, C., Bellver, J., Rodrigo, L., Bosch, E., Mercader, A., Vidal, C., et al. (2013). Preimplantation genetic screening using fluorescence in situ hybridization in patients with repetitive implantation failure and advanced maternal age: Two randomized trials. *Fertility and Sterility*, *99*, 1400–1407.

Rubio, C., Castillón, G., Rodrigo, L., Bellver, J., Guillen, A., Remohí, J., et al. (2014). Improvement of clinical outcome in severe male factor infertility with embryo selection based on Array-CGH: A randomized controlled trial. *Fertility and Sterility*, *102*(Suppl.), e24–e25.

Rubio, C., Gil-Salom, M., Simon, C., Vidal, F., Rodrigo, L., Mínguez, Y., et al. (2001). Incidence of sperm chromosomal abnormalities in a risk population: Relationship with sperm quality and ICSI outcome. *Human Reproduction*, *16*, 2084–2092.

Rubio, C., Gimenez, C., Fernandez, E., Vendrell, X., Velilla, E., Parriego, M., et al. (2009). The importance of good practice in preimplantation genetic screening: Critical viewpoints. *Human Reproduction*, *24*, 2045–2047.

Rubio, C., Rodrigo, L., Bellver, J., Castillon, G., Guillen, A., Vidal, C., et al. (2015). Improved clinical outcome in women between 38 and 41 years of age by 24-chromosome aneuploidy screening. *Fertility and Sterility*, *104*(Suppl.), e58.

Rubio, C., Rodrigo, L., Mir, P., Mateu, E., Peinado, V., Milán, M., et al. (2013). Use of array comparative genomic hybridization (array-CGH) for embryo assessment: Clinical results. *Fertility and Sterility*, *99*, 1044–1048.

Ruttanajit, T., Chanchamroen, S., Cram, D. S., Sawakwongpra, K., Suksalak, W., Leng, X., et al. (2016). Detection and quantitation of chromosomal mosaicism in human blastocysts using copy number variation sequencing. *Prenatal Diagnosis*, *36*, 154–162.

Schena, M., Shalon, D., Davis, R. W., & Brown, P. O. (1995). Quantitative monitoring of gene expression patterns with a complementary DNA microarray. *Science*, *270*, 467–470.

Schoolcraft, W. B., Fragouli, E., Stevens, J., Munne, S., Katz-Jaffe, M. G., & Wells, D. (2010). Clinical application of comprehensive chromosomal screening at the blastocyst stage. *Fertility and Sterility*, *94*, 1700–1706.

Schoolcraft, W. B., & Katz-Jaffe, M. G. (2013). Comprehensive chromo- some screening of trophectoderm with vitrification facilitates elective single-embryo transfer for infertile women with advanced maternal age. *Fertility and Sterility*, *100*, 615–619.

Schoolcraft, W. B., Katz-Jaffe, M. G., Stevens, J., Rawlins, M., & Munne, S. (2009). Preimplantation aneuploidy testing for infertile patients of advanced maternal age: A randomized prospective trial. *Fertility and Sterility*, *92*, 157–162.

Scott, R. T., Jr., Ferry, K., Su, J., Tao, X., Scott, K., & Treff, N. R. (2012). Comprehensive chromosome screening is highly predictive of the reproductive potential of human embryos: Anumblinded, nonselection study. *Fertility and Sterility*, *97*, 870–875.

Scott, K. L., Hong, K. H., & Scott, R. T., Jr. (2013). Selecting the optimal time to perform biopsy for preimplantation genetic testing. *Fertility and Sterility*, *100*, 608–614.

Scriven, P. N., Handyside, A. H., & Ogilvie, C. M. (1998). Chromosome translocations: Segregation modes and strategies for preimplantation genetic diagnosis. *Prenatal Diagnosis, 18*, 1437–1449.

Simpson, J. L. (2008). What next for preimplantation genetic screening? Randomized clinical trial in assessing PGS: Necessary but not sufficient. *Human Reproduction, 23*, 2179–2181.

Simpson, J. L. (2012). Preimplantation genetic diagnosis to improve pregnancy outcomes in subfertility. *Best Practice & Research. Clinical Obstetrics & Gynaecology, 26*, 805–815.

Spits, C., De Rycke, M., Van Ranst, N., Verpoest, W., Lissens, W., Van Steirteghem, A., et al. (2007). Preimplantation genetic diagnosis for cancer predisposition syndromes. *Prenatal Diagnosis, 27*, 447–456.

Spits, C., De Rycke, M., Verpoest, W., Lissens, W., Van Steirteghem, A., Liebaers, I., et al. (2006). Preimplantation genetic diagnosis for Marfan syndrome. *Fertility and Sterility, 86*, 310–320.

Spits, C., Le Caignec, C., De Rycke, M., Van Haute, L., Van Steirteghem, A., Liebaers, I., et al. (2006). Optimization and evaluation of single-cell whole-genome multiple displacement amplification. *Human Mutation, 27*, 496–503.

Stephenson, M. D., Awartani, K. A., & Robinson, W. P. (2002). Cytogenetic analysis of miscarriages from couples with recurrent miscarriage: A case-control study. *Human Reproduction, 17*, 446–451.

Taylor, T. H., Gitlin, S. A., Patrick, J. L., Crain, J. L., Wilson, J. M., & Griffin, D. K. (2014). The origin, mechanisms, incidence and clinical consequences of chromosomal mosaicism in humans. *Human Reproduction Update, 20*, 571–581.

Treff, N. R., Fedick, A., Tao, X., Devkota, B., Taylor, D., & Scott, R. T., Jr. (2013). Evaluation of targeted next-generation sequencing-based preimplantation genetic diagnosis of monogenic disease. *Fertility and Sterility, 99*. 1377-1384.e6.

Treff, N. R., Levy, B., Su, J., Northrop, L. E., Tao, X., & Scott, R. T., Jr. (2010). SNP microarray-based 24 chromosome aneuploidy screening is significantly more consistent than FISH. *Molecular Human Reproduction, 16*, 583–589.

Treff, N. R., Northrop, L. E., Kasabwala, K., Su, J., Levy, B., & Scott, R. T., Jr. (2011). Single nucleotide polymorphism microarray-based concurrent screening of 24-chromosome aneuploidy and unbalanced translocations in preimplantation human embryos. *Fertility and Sterility, 95*. 1606-12.e1-2.

Treff, N. R., Su, J., Tao, X., Levy, B., & Scott, R. T., Jr. (2010). Accurate single cell 24 chromosome aneuploidy screening using whole genome amplification and single nucleotide polymorphism microarrays. *Fertility and Sterility, 94*, 2017–2021.

Treff, N. R., Tao, X., Schillings, W. J., Bergh, P. A., Scott, R. T., Jr., & Levy, B. (2011). Use of single nucleotide polymorphism microarrays to distinguish between balanced and normal chromosomes in embryos from a translocation carrier. *Fertility and Sterility, 96*, e58–e65.

Van de Velde, H., De Rycke, M., De Man, C., De Hauwere, K., Fiorentino, F., Kahraman, S., et al. (2009). The experience of two European preimplantation genetic diagnosis centres on human leukocyte antigen typing. *Human Reproduction, 24*, 732–740.

Van Dyke, D. L., Weiss, L., Roberson, J. R., & Babu, V. R. (1983). The frequency and mutation rate of balanced autosomal rearrangements in man estimated from prenatal genetic studies for advanced maternal age. *American Journal of Human Genetics, 35*, 301–308.

van Uum, C. M., Stevens, S. J., Dreesen, J. C., Drüsedau, M., Smeets, H. J., Hollanders-Crombach, B., et al. (2012). SNP array-based copy number and genotype analyses for preimplantation genetic diagnosis of human unbalanced translocations. *European Journal of Human Genetics, 20*, 938–944.

Velilla, E., Escudero, T., & Munne, S. (2002). Blastomere fixation techniques and risk of misdiagnosis for preimplantation genetic diagnosis of aneuploidy. *Reproductive Biomedicine Online, 4*, 210–217.

Vera-Rodriguez, M., Michel, C. E., Mercader, A., Bladon, A. J., Rodrigo, L., Kokocinski, F., et al. (2015). Distribution patterns of segmental aneuploidies in human blastocysts identified by next-generation sequencing. *Fertility and Sterility*, *105*(4), 1047–1055.

Verlinsky, Y., Rechitsky, S., Schoolcraft, W., Strom, C., & Kuliev, A. (2001). Preimplantation diagnosis for Fanconi anemia combined with HLA matching. *JAMA*, *285*, 3130–3133.

Verlinsky, Y., Rechitsky, S., Sharapova, T., Laziuk, K., Barsky, I., Verlinsky, O., et al. (2007). Preimplantation diagnosis for immunodeficiencies. *Reproductive Biomedicine Online*, *14*, 214–223.

Verlinsky, Y., Tur-Kaspa, I., Cieslak, J., Bernal, A., Morris, R., Taranissi, M., et al. (2005). Preimplantation testing for chromosomal disorders improves reproductive outcome of poor-prognosis patients. *Reproductive Biomedicine Online*, *11*, 219–225.

Voullaire, L., Wilton, L., McBain, J., Callaghan, T., & Williamson, R. (2002). Chromosome abnormalities identified by comparative genomic hybridization in embryos from women with repeated implantation failure. *Molecular Human Reproduction*, *8*, 1035–1041.

Wang, L., Cram, D. S., Shen, J., Wang, X., Zhang, J., Song, Z., et al. (2014). Validation of copy number variation sequencing for detecting chromosome imbalances in human preimplantation embryos. *Biology of Reproduction*, *91*, 37.

Wells, D., Alfarawati, S., & Fragouli, E. (2008). Use of comprehensive chromosomal screening for embryo assessment: Microarrays and CGH. *Molecular Human Reproduction*, *14*, 703–710.

Wells, D., Escudero, T., Levy, B., Hirschhorn, K., Delhanty, J. D., & Munne, S. (2002). First clinical application of comparative genomic hybridization and polar body testing for preimplantation genetic diagnosis of aneuploidy. *Fertility and Sterility*, *78*, 543–549.

Werner, M. D., Leondires, M. P., Schoolcraft, W. B., Miller, B. T., Copperman, A. B., Robins, E. D., et al. (2014). Clinically recognizable error rate after the transfer of comprehensive chromosomal screened euploid embryos is low. *Fertility and Sterility*, *102*, 1613–1618.

Wilton, L., Voullaire, L., Sargeant, P., Williamson, R., & McBain, J. (2003). Preimplantation aneuploidy screening using comparative genomic hybridization or fluorescence in situ hybridization of embryos from patients with recurrent implantation failure. *Fertility and Sterility*, *80*, 860–868.

Yan, L., Huang, L., Xu, L., Huang, J., Ma, F., Zhu, X., et al. (2015). Live births after simultaneous avoidance of monogenic diseases and chromosome abnormality by next-generation sequencing with linkage analyses. *Proceedings of the National academy of Sciences of the United States of America*, *112*, 15964–15969.

Yang, Z., Liu, J., Collins, G. S., Salem, S. A., Liu, X., Lyle, S. S., et al. (2012). Selection of single blastocysts for fresh transfer via standard morphology assessment alone and with array CGH for good prognosis IVF patients: Results from a randomized pilot study. *Mol. Cytogenet*, *5*. 24-8166-5-24.

INDEX

Note: Page numbers followed by "*f*" indicate figures, and "*t*" indicate tables.

R

Z

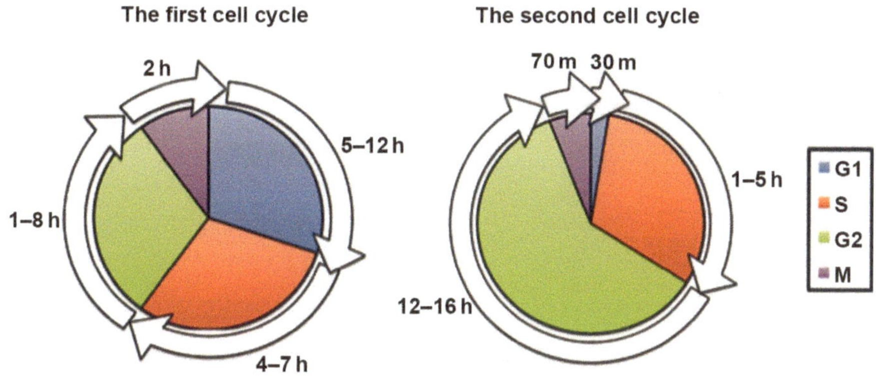

N. Palmer and P. Kaldis, Fig. 1 The estimated duration of each of the substages of interphase and mitosis during the first and second murine embryonic cell cycles.

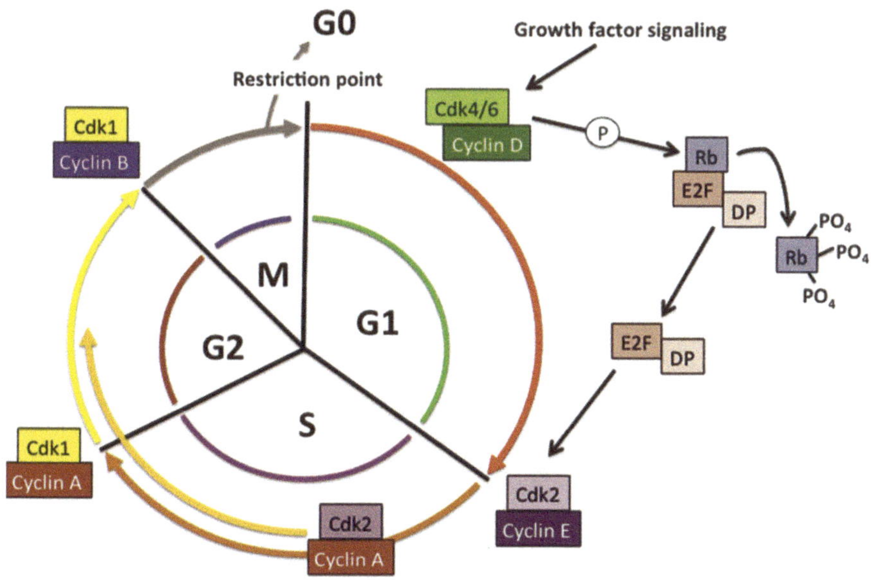

N. Palmer and P. Kaldis, Fig. 2 The role of the interphase cyclin-dependent kinases in a typical mammalian cell cycle.

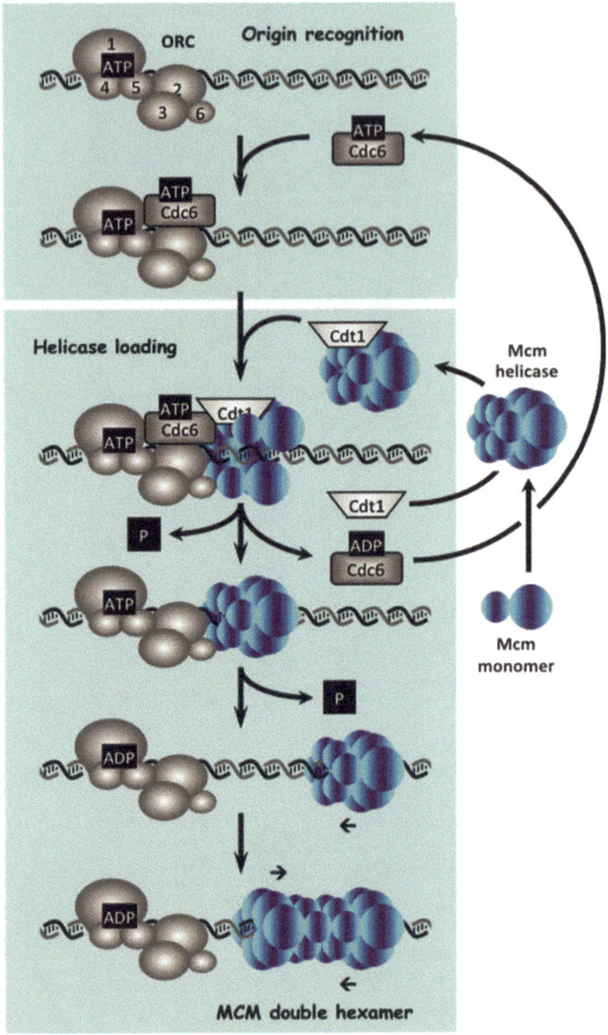

M.L. DePamphilis, Fig. 3 Assembly of prereplication complexes (preRCs) begins during mitotic exit. PreRC assembly begins with the binding of the six subunit origin recognition complex (ORC) to DNA. ORC contains adenosine triphosphate (ATP) bound to subunits 1, 4, and 5. This event allows Cdc6·ATP to form a stable complex with the ORC·DNA site. Cdt1 can then chaperone the six subunit MCM helicase then to the DNA·ORC·Cdc6·ATP site. Cdc6·ADP is released with concomitant loading of the MCM helicase onto the DNA. It is the interaction of Cdt1·MCM with a DNA·ORC·Cdc6·ATP site that causes the MCM hexamer to open and expose its central channel. DNA then associates with the central channel within the six Mcm subunits and activates the Cdc6 ATPase. Hydrolysis of the Cdc6-bound ATP releases Cdt1 and Cdc6·ADP to be recycled

(Continued)

M.L. DePamphilis, Fig. 3—Cont'd into Cdc6·ATP and Cdt1·MCM. Once loaded, the MCM helicase can slide along double-strand DNA without unwinding it, because MCM helicase activity is not activated until the preRC is converted into a preinitiation complex. Loading of the MCM helicase stimulates hydrolysis of ORC-bound ATP. This event allows the MCM helicase to migrate along the DNA. Exchange of ADP for ATP resets ORC for a new round of preRC assembly. In multicellular eukarya, this event may also trigger release of the Orc1 subunit from ORC·chromatin sites and its subsequent ubiquitin-dependent degradation. A second MCM helicase must be loaded in opposite orientation to the first in order to achieve bidirectional DNA replication. *See Bell, S. P., & Kaguni, J. M. (2013). Helicase loading at chromosomal origins of replication.* Cold Spring Harbor Perspectives in Biology *5; Coster, G., Frigola, J., Beuron, F., Morris, E. P., & Diffley, J. F. (2014). Origin licensing requires ATP binding and hydrolysis by the MCM replicative helicase.* Molecular Cell *55, 666–677.*

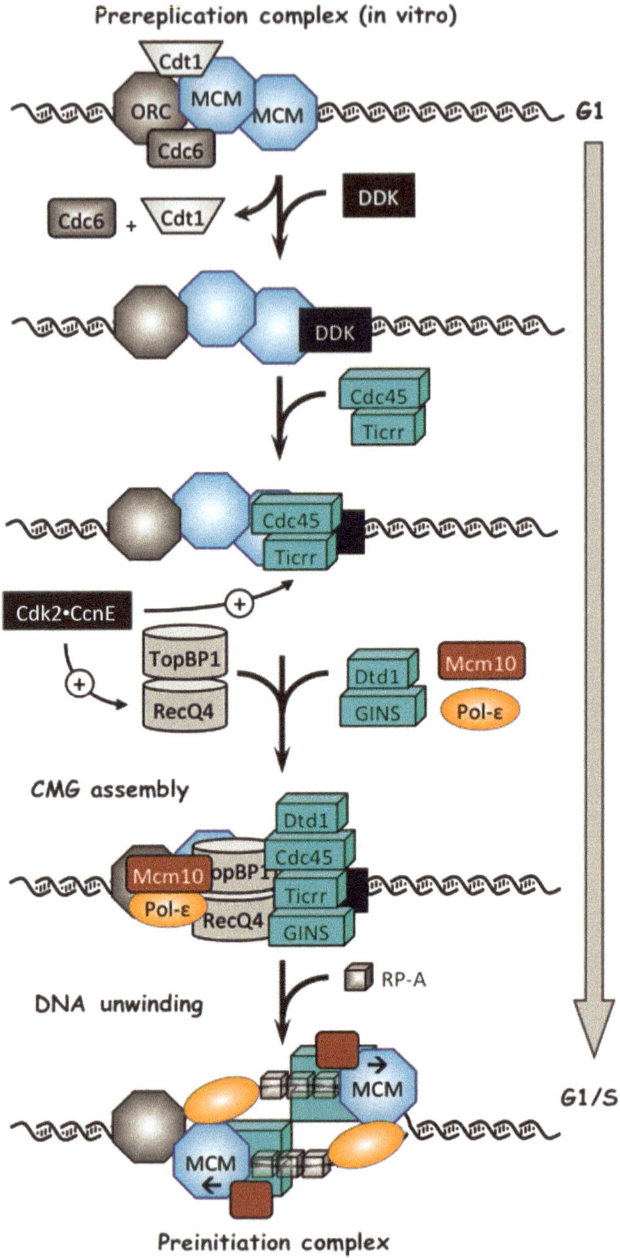

M.L. DePamphilis, Fig. 4 Conversion of prereplication complexes (preRCs) into preinitiation complexes (preICs) occurs at the end of G1 phase. PreICs are assembled at the G1 to S phase transition. Except for the two MCM helicases, the stoichiometry of other proteins is not indicated, only their order of addition. Ticrr is also known as Treslin or Sld3, and Dtd1 as DNA unwinding element-binding protein B or DUE-B. TopBP1 is a topoisomerase II binding protein. RecQ4 is a DNA helicase. GINS is a heterotypic four subunit complex that, together with Cdc45, is essential for efficient MCM helicase activity. CDK·cyclin is the only direct participant not bound to chromatin. *See Tanaka, S., & Araki, H. (2013). Helicase activation and establishment of replication forks at chromosomal origins of replication.* Cold Spring Harbor Perspectives in Biology *5, a010371.*

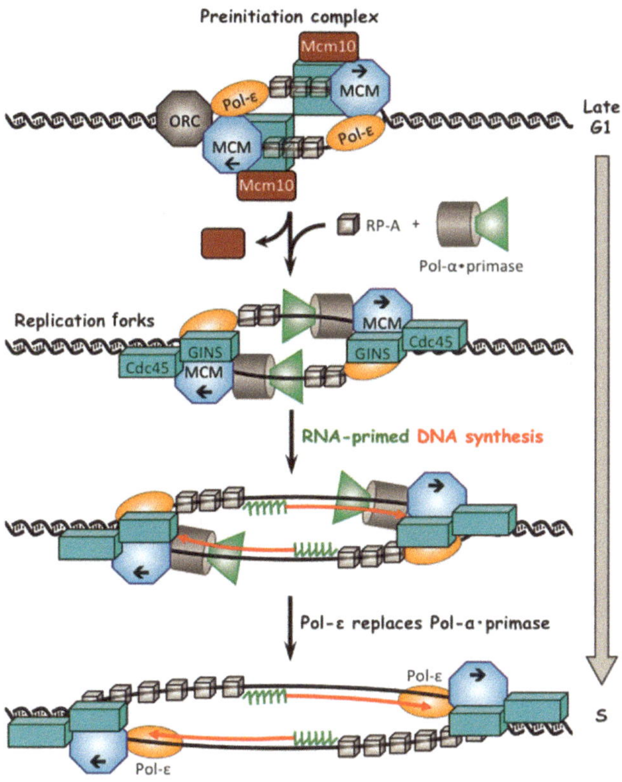

M.L. DePamphilis, Fig. 5 Conversion of preinitiation complexes into replication forks occurs at the beginning of S phase. RNA-primed DNA synthesis begins with the loading of the DNA polymerase-α·primase. DNA synthesis is linked to DNA unwinding by a physical interaction between the MCM–GINS–Cdc45 complex and DNA polymerase-α·primase. After synthesis of 100–200 nucleotides, DNA polymerase-ε replaces DNA polymerase-α·primase and DNA synthesis continues without interruption on the leading strand template. DNA polymerase-α·primase initiates RNA-primed DNA synthesis on the lagging strand template to produce an Okazaki fragment (Fig. 1).

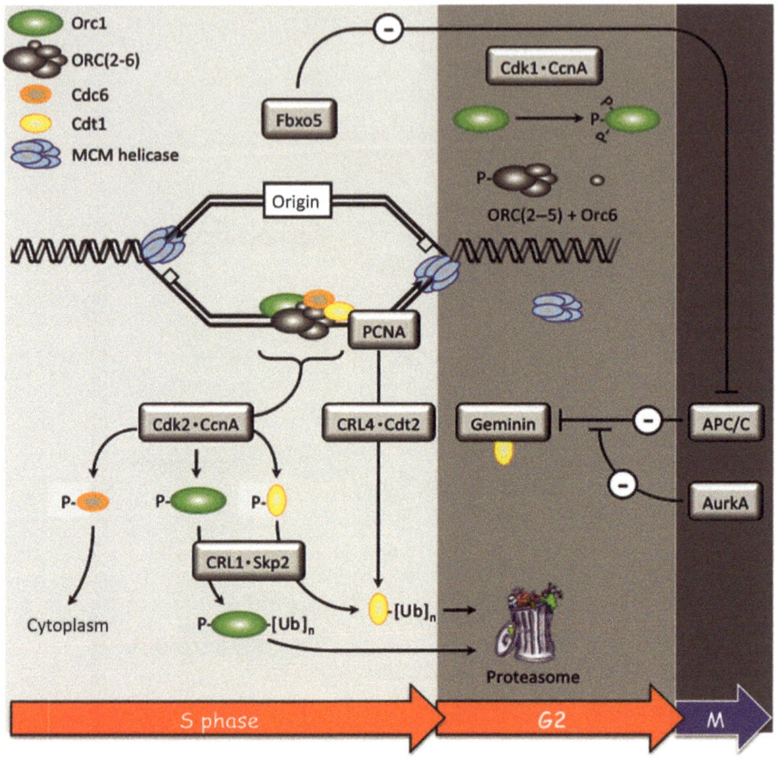

M.L. DePamphilis, Fig. 9 Multiple concerted pathways prevent DNA rereplication. When cells enter S phase, seven concerted pathways exist can inactivate the helicase loader, thereby preventing rereplication of DNA prior to mitosis. Cdk2·CcnA phosphorylates Orc1, Cdc6, and Cdt1, thereby inhibiting their ability to form helicase loaders. In addition, Cdc6-P is localized to the cytoplasm where it cannot facilitate origin licensing, and Orc1-P is ubiquitinated and targeted for destruction. Inactivation of Orc1 results in dissociation of ORC(2–5) and Orc6 from chromosomes. Cdt1-P is ubiquitinated by the CRL1·Skp2 ubiquitin ligase. Cdt1 that is not phosphorylated is ubiquitinated by the CRL4·Cdt2 ubiquitin ligase bound to PCNA and DNA. PCNA is the eukaryotic sliding clamp protein that facilitates DNA synthesis by DNA polymerases-δ and -ε. The ubiquitinated proteins are then degraded by the 26S proteasome. Geminin is a protein that binds specifically to Cdt1 and inhibits its activity. These activities are available from S through early M phase. As cells exit mitosis, geminin, and cyclin A are ubiquitinated by the anaphase-promoting complex (originally termed cyclosome, APC/C), an activity that is inhibited during S to early M phase by the APC/C specific inhibitor Fbxo5/Emi1. Reassembly of the helicase loader occurs by reassembly of Orc1, ORC(2–5), Orc6, and Cdc6 onto chromatin, an event that is triggered by the Orc1 subunit during the anaphase to G1 phase transition.

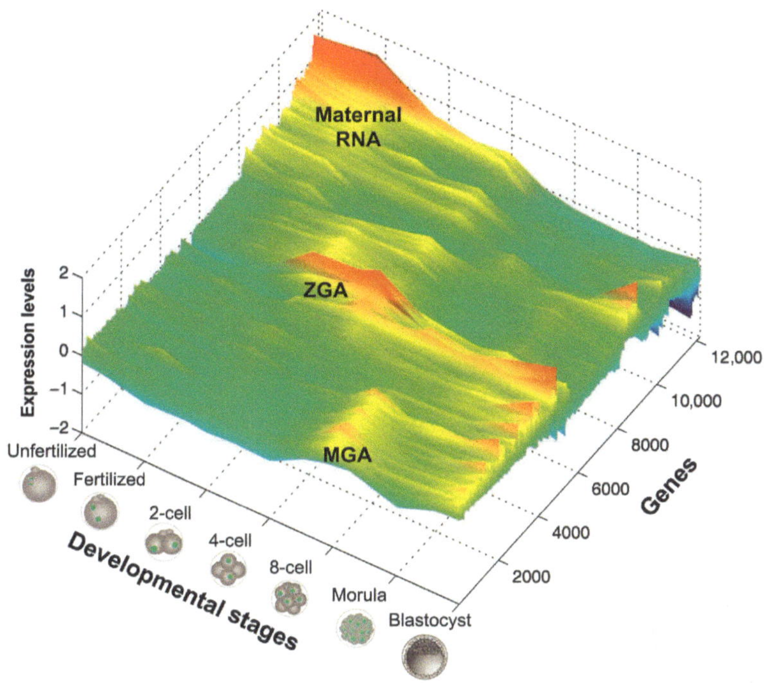

M.S.H. Ko, Fig. 1 A bird's eye view of gene expression patterns during mouse preimplantation development. The expression changes of 12,179 genes are shown. A *curve* represents the expression of each gene over the preimplantation period and is color coded by its expression levels (log-scale). Maternal RNA, RNAs stored in oocytes; MGA, mid-preimplantation gene activation; ZGA, zygotic genome activation. *Adapted from Ko, M. S. (2005). Molecular biology of preimplantation embryos: Primer for philosophical discussions.* Reproductive Biomedicine Online, 10*(Suppl. 1), 80–87.*

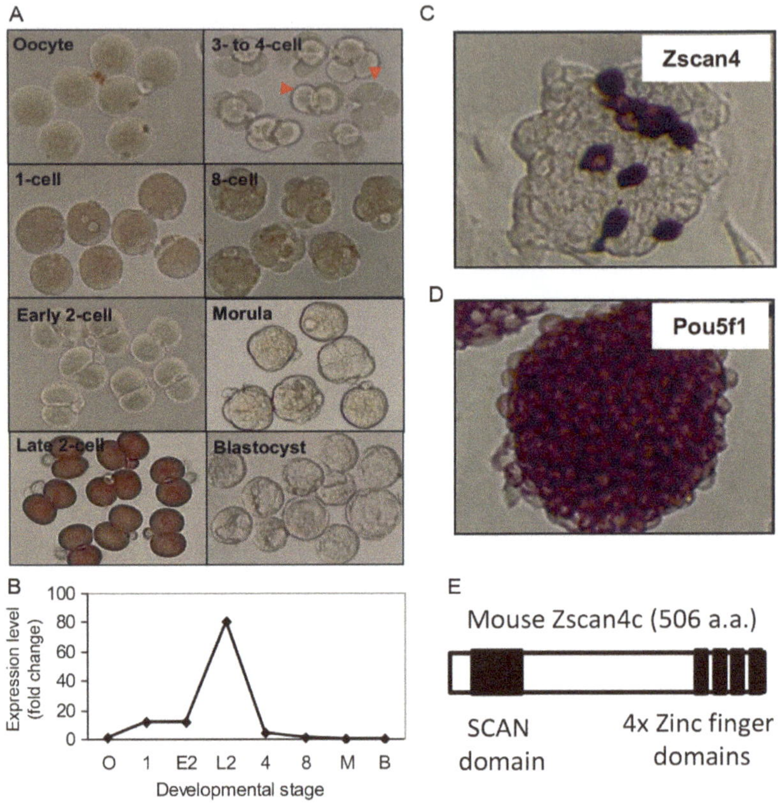

M.S.H. Ko, Fig. 2 Expression profiles of *Zscan4* in mouse preimplantation embryos and mouse ES cells. (A) Whole-mount RNA in situ hybridization of mouse preimplantation embryos with a *Zscan4* probe. (B) qRT-PCR analyses of *Zscan4* expression during preimplantation development. (C) Whole-mount RNA in situ hybridization of mouse ES cell colonies with a *Zscan4* probe. (D) As a control, whole-mount RNA in situ hybridization with *Pou5fa* (a.k.a. *Oct3/4* or *Oct4*). (E) Structure of mouse Zscan4 protein (Zscan4c). *Adapted from Falco, G., Lee, S. L., Stanghellini, I., Bassey, U. C., Hamatani, T., & Ko, M. S. (2007). Zscan4: A novel gene expressed exclusively in late 2-cell embryos and embryonic stem cells.* Developmental Biology, 307, *539–550.*

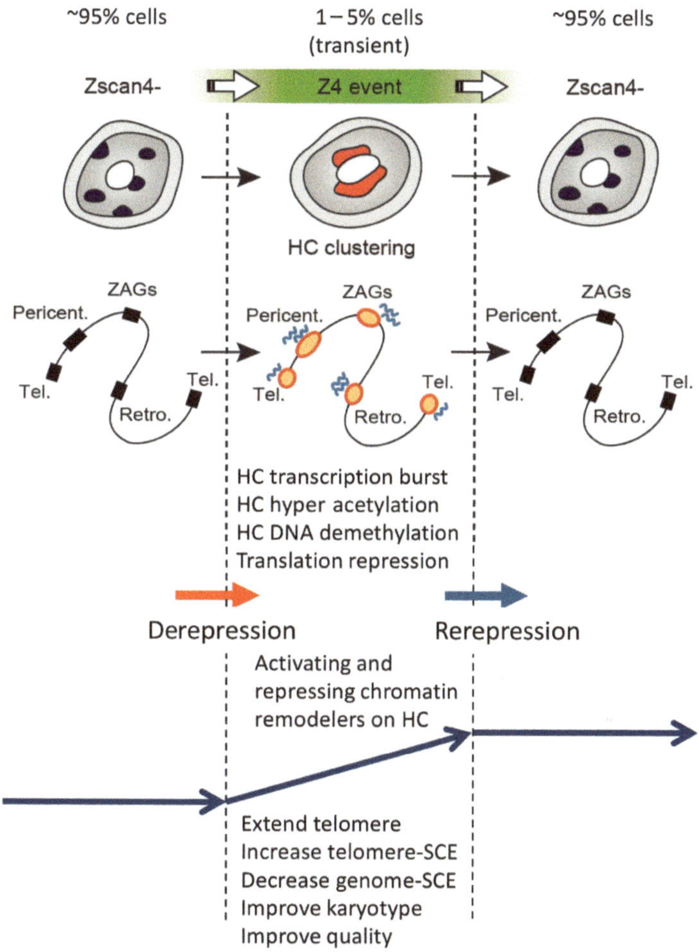

M.S.H. Ko, Fig. 3 A schematic summary of the Z4 event. Pericent., pericentomeres; Retro., retrotransposons; Tel., telomeres; ZAGs, Z4 event-associated genes. *Adapted from Akiyama, T., Xin, L., Oda, M., Sharov, A. A., Amano, M., Piao, Y., et al. (2015). Transient bursts of Zscan4 expression are accompanied by the rapid derepression of heterochromatin in mouse embryonic stem cells.* DNA Research, 22, *307–318.*

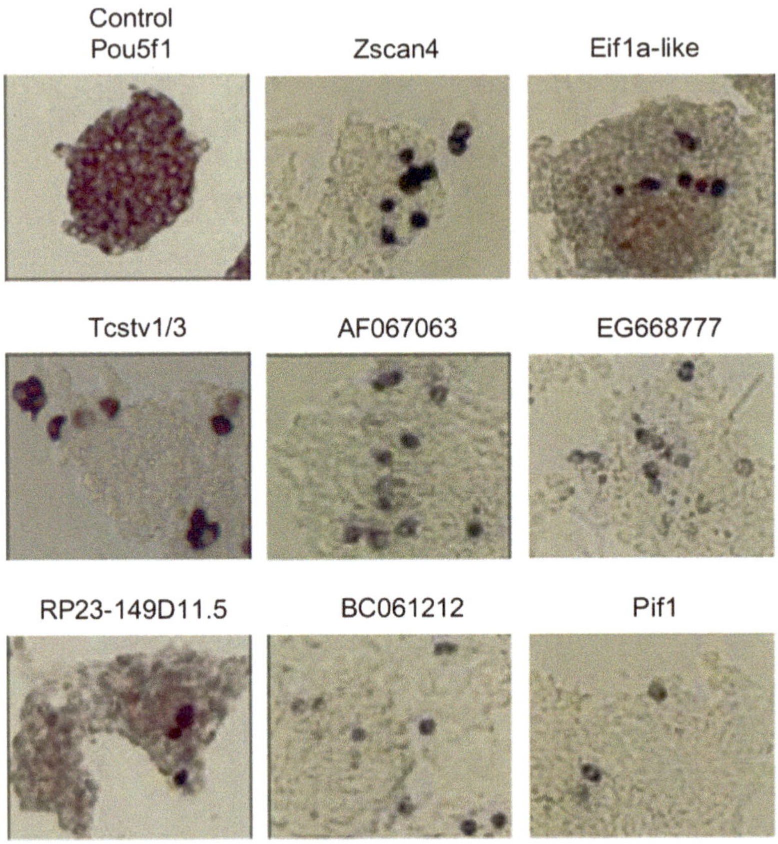

M.S.H. Ko, Fig. 4 Whole-mount RNA in situ hybridization of ZAGs in mouse ES cells. *Adapted from Amano, T., Hirata, T., Falco, G., Monti, M., Sharova, L. V., Amano, M., et al. (2013). Zscan4 restores the developmental potency of embryonic stem cells.* Nature Communications, 4, *1966.*

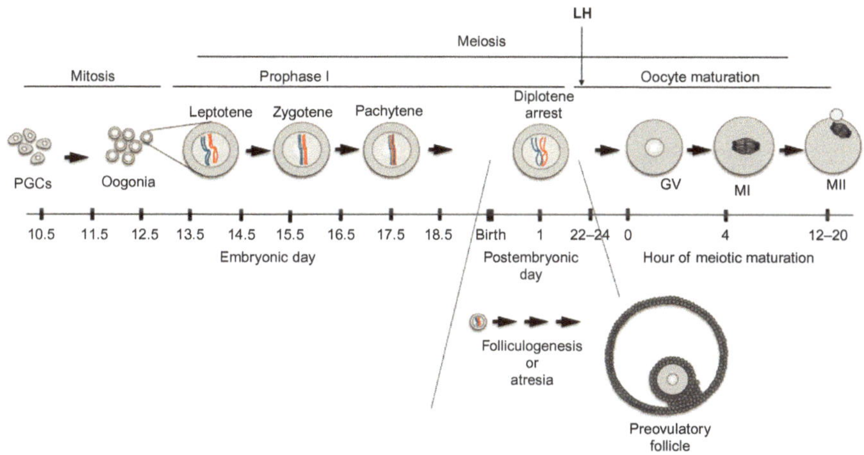

L. Bury *et al.*, Fig. 1 Stages of oogenesis and folliculogenesis in the mouse. Primordial germ cells (PGCs) divide mitotically giving rise to oogonia that populate the genital ridge. Around embryonic day 12.5–13.5, oocytes enter prophase I of meiosis, that includes the stages zyogotene, pachytene, and diplotene upon which they enter a resting state, dictyate. This arrest begins before birth and is characterized by oocyte growth and follicle formation. Within the large, preovulatory follicle, the oocyte and surrounding granulosa cells reside in the fluid filed antrum. After the onset of puberty, matured follicles can be activated by a periodic surge of luteinizing hormone (LH), which promotes maturation of the oocyte. As oocytes reenter meiosis I, they undergo breakdown of their germinal vesicle, congression of homologous chromosomes at metaphase I, followed by their segregation at anaphase I. Upon polar body extrusion, oocytes become arrested in meiosis II until fertilization.

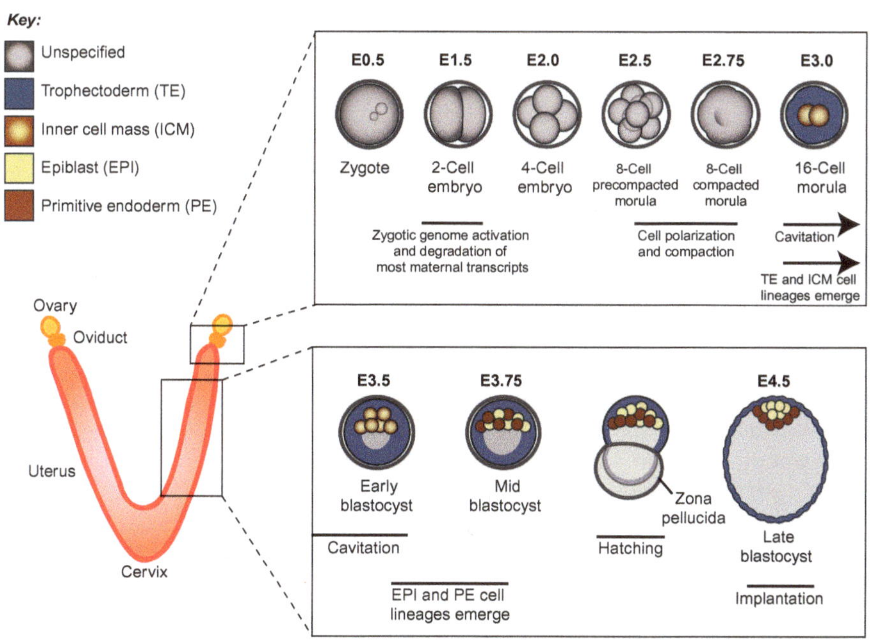

A.A. Lokken and A. Ralston, Fig. 2 Morphological events and embryo staging. Embryos proceed through stereotyped stages and are essentially synchronized within a litter. Fertilization occurs within the oviduct, and embryogenesis proceeds as embryos move toward the uterus, entering around E3.5.

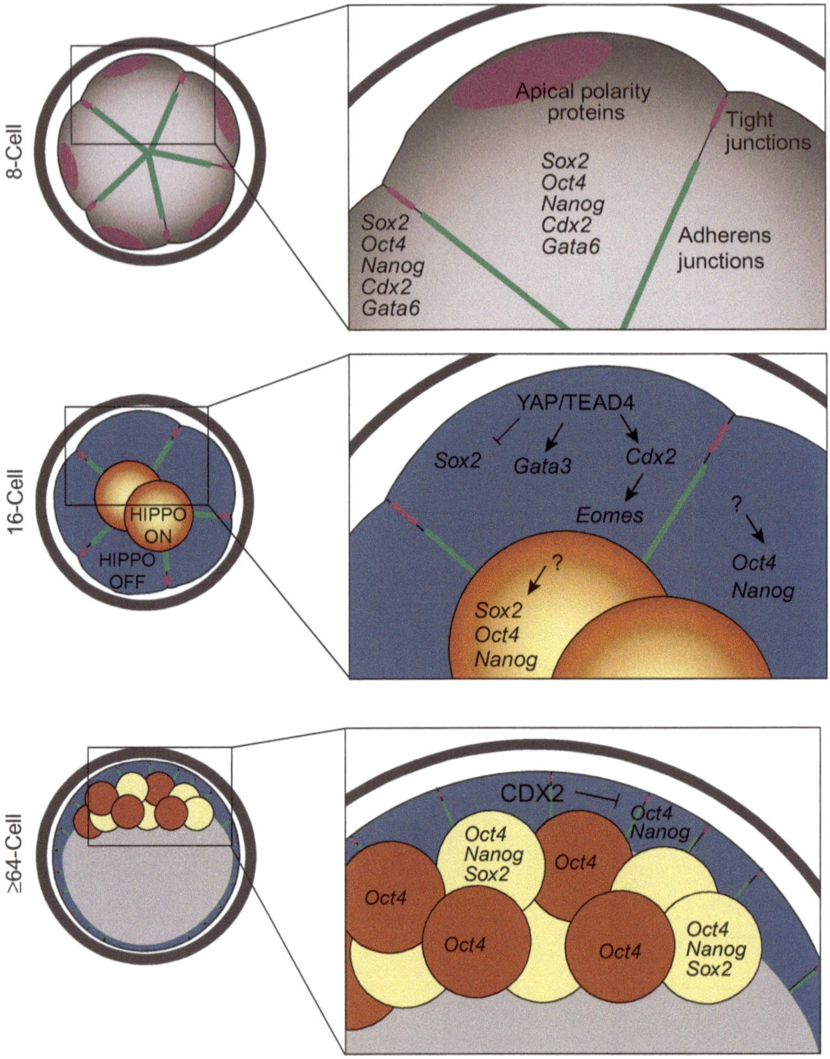

A.A. Lokken and A. Ralston, Fig. 3 Segregation of ICM and TE lineages. At the late 8-cell stage, cells polarize and the embryo compacts. At this stage, many genes that will later identify unique cell populations are expressed throughout the embryo. By the 16-cell stage, the HIPPO signaling pathway is active within inside cells, and inactivated by cell polarity components in outside cells. Thus proteins that are negatively regulated by HIPPO, such as YAP and WWTR1, are active in outside cells where they partner with TEAD4 to regulate expression of ICM and TE genes, *Sox2* and *Cdx2/Gata3*, respectively. At this stage, *Oct4* and *Nanog* are expressed ubiquitously, indicating that these genes are not regulated by HIPPO signaling. By the late blastocyst stage, CDX2 represses expression of *Oct4* and *Nanog* within TE cells, thereby restricting the expression of these genes to the ICM.

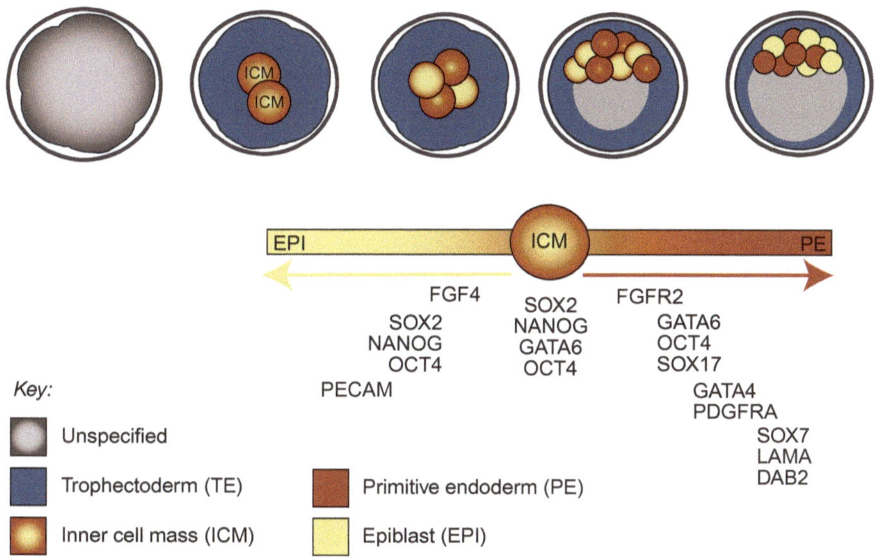

A.A. Lokken and A. Ralston, Fig. 4 The salt and pepper ICM. In the immature ICM, *Fgf4* and *Fgfr2* are the earliest to show complementary expression patterns in the ICM. Subsequently, EPI and PE genes become upregulated in *Fgf4*- and *Fgfr2*-expressing cells, respectively. Later, EPI and PE populations begin to sort into discrete layers and express markers of mature EPI and PE cells.

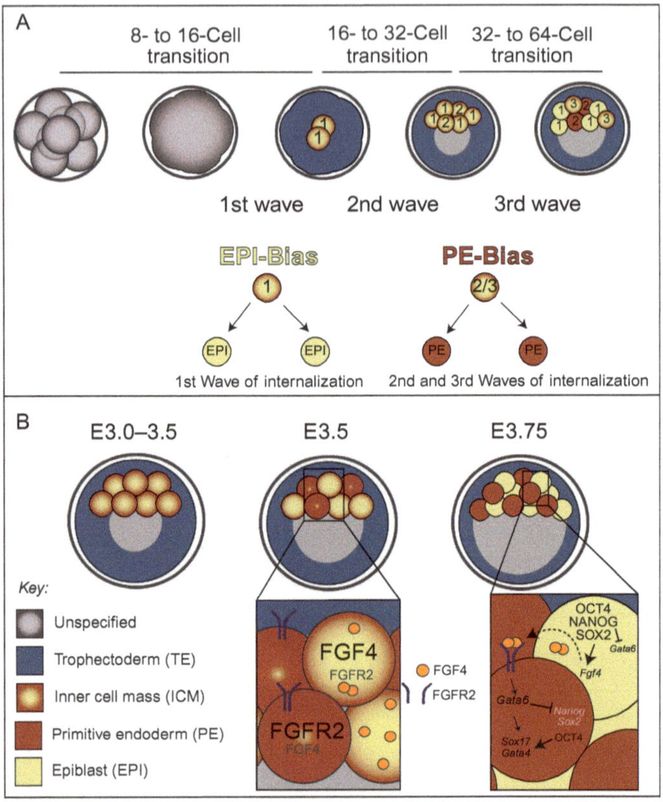

A.A. Lokken and A. Ralston, Fig. 5 The origins of ICM heterogeneity. (A) The cleavage history-dependent model of ICM cell heterogeneity posits that cells acquire biased developmental potential, depending on when and how they came to occupy inside space, with the first ICM cells more inclined to become EPI, and later-arising cells more likely to become PE, having descended from TE cells, and inherited PE determinants such as FGFR2. (B) The stochastic gene expression model of ICM cell heterogeneity posits that the initially identical cells of the ICM spontaneously express slightly different levels of FGF4 or FGFR2, and this signaling is amplified in the process of specifying EPI and PE progenitors.

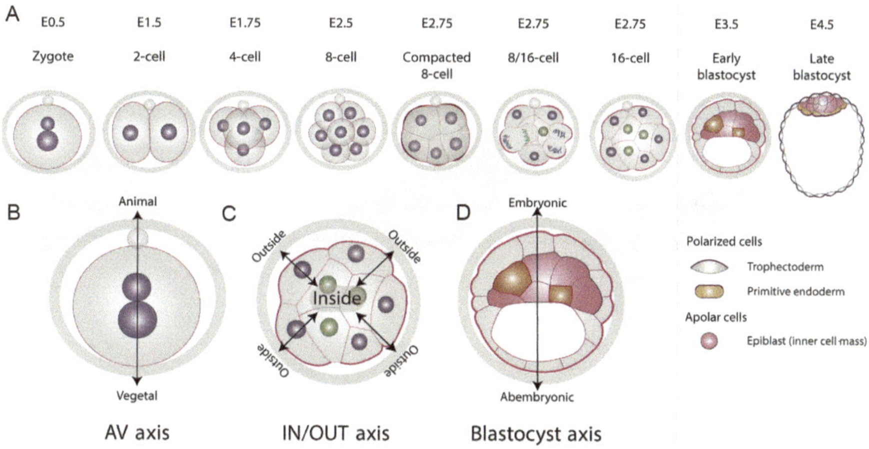

C.Y. Leung *et al.*, Fig. 1 An overview of the polarity axes of preimplantation development. (A) Schematic view of preimplantation development. (B) From the zygote stage onward the animal–vegetal axis is defined by the polar body marking the animal pole. (C) After compaction and asymmetric divisions, cells can divide along the IN/OUT axis into the "inside" of the embryo, making them "inside cells." (D) By the blastocyst stage an inner cell mass is formed which marks the embryonic pole of the embryo. There are two types of polarized cells in the E4.5 blastocyst, the trophectoderm and the primitive endoderm, while the epiblast cells that later form the embryo proper remain apolar.

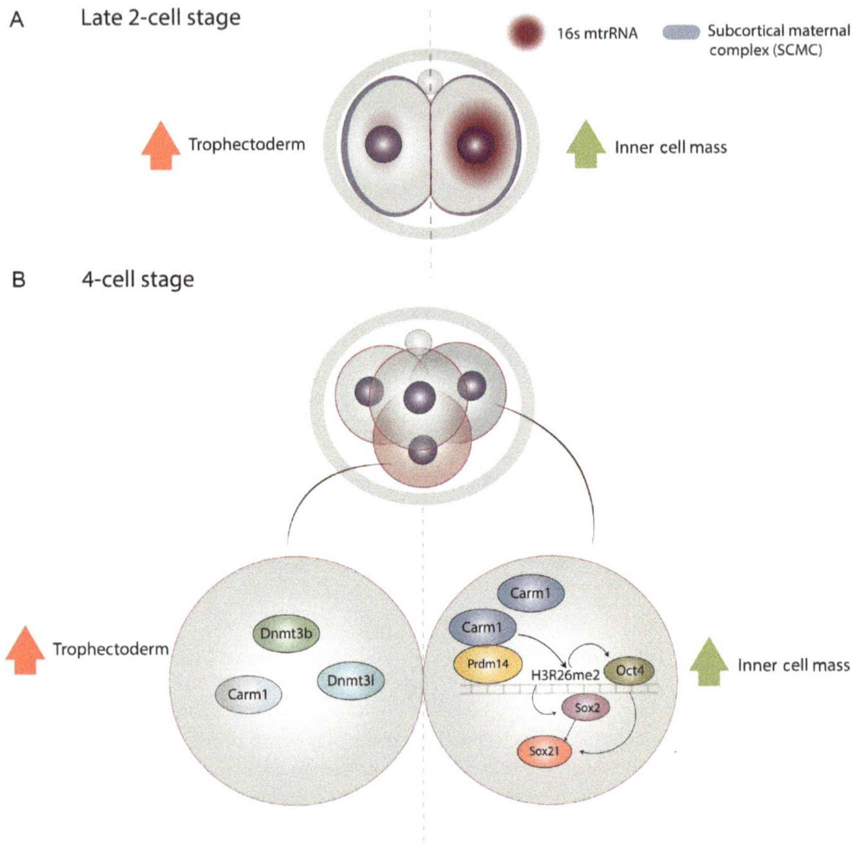

C.Y. Leung *et al.*, Fig. 2 Molecular hetereogeneities in the 2- and 4-cell embryo. (A) There are two molecular hetereogeneities in the 2-cell embryo, the subcortical maternal complex (SCMC) which is excluded from the cell–cell contact regions of the embryo, and the presence of 16S mtrRNA. 16S mtrRNA is the only heterogeneous molecule at the 2-cell stage known to be a bona fide cell-fate regulator. (B) A hypothetical model on the molecular differences of cell-fate regulators between 4-cell blastomeres. The defining difference in the levels of Prdm14 starts off a cascade that involves Carm1 and possibly Sox21 to promote the pluripotent lineage.

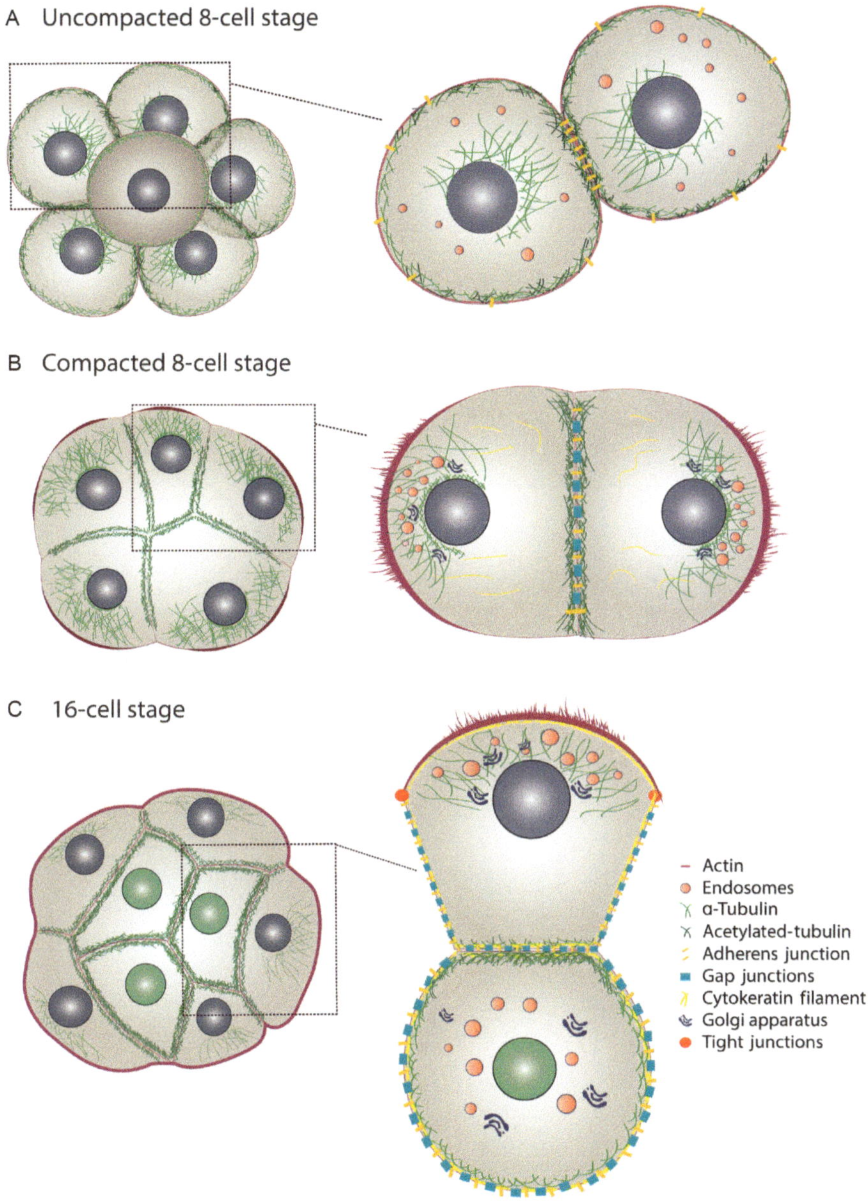

C.Y. Leung *et al.*, Fig. 3 The cytoplasmic and surface polarity of the (A) uncompacted 8-cell stage embryo, (B) compacted 8-cell stage embryo, and (C) the 8–16 cell stage embryo.

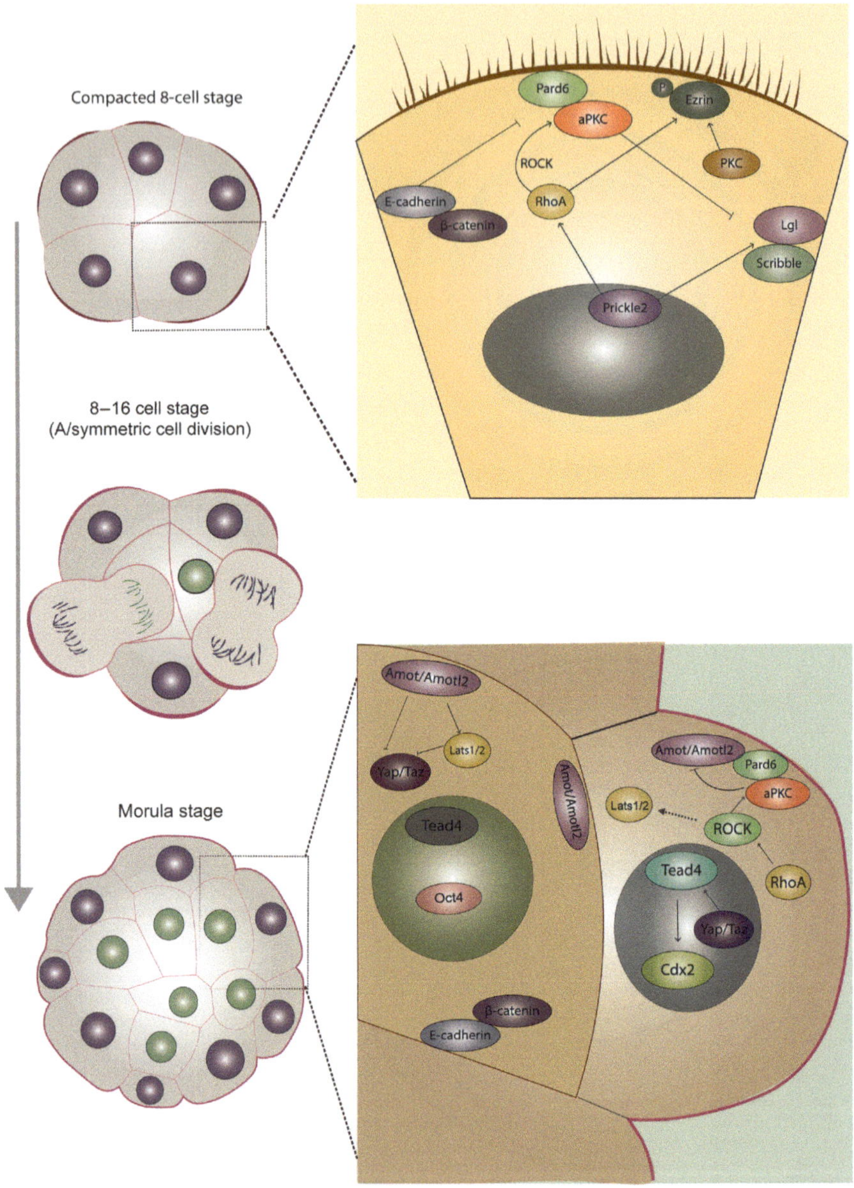

C.Y. Leung *et al.*, Fig. 4 The molecular landscape of 8-cell stage polarization formation and its role in preimplantation cell-fate decision. In the late 8-cell stage, Par complex and phosphorylated Ezrin enriched at the apical domain. The Par complex apical localization was antagonized by E-cadherin. RhoA activity is also required for the apical domain formation. Other conserved polarity regulators are also present in the mouse embryo. Polarity is essential in segregating trophectoderm (TE) and inner cell mass (ICM) lineage through regulating the activity of the core hippo pathway regulators.

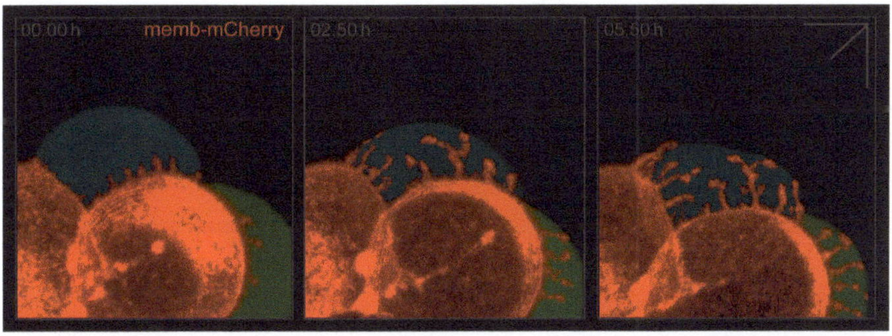

M.D. White *et al.*, Fig. 2 Filopodia extended by some blastomeres facilitate changes in cell shape that compact the entire embryo. One cell of the 2-cell embryo was injected with membrane-targeted mCherry (memb-mCherry) to label half of the embryo. This allows visualization of memb-mCherry positive filopodia that extend from some blastomeres onto their unlabeled neighbors (DIC image pseudo-colored *blue* and *green*) during compaction. Cells are initially relatively spherical (*left panel*) but both filopodia-extending cells and their neighbors flatten and draw closer together as filopodia are extended (*middle, right panel*). Scale bars: 10 μm in *x, y, z*.

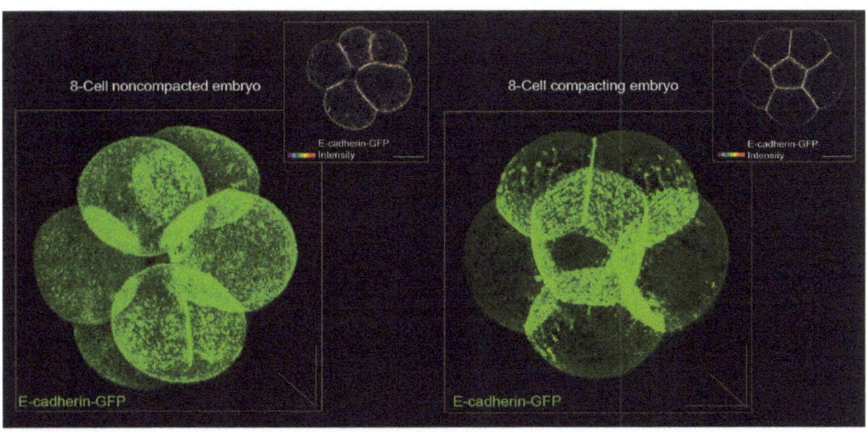

M.D. White *et al.*, Fig. 3 E-cadherin redistributes to basolateral cell membranes during compaction. E-cadherin-GFP expression in an embryo injected at the 1-cell stage. At the early 8-cell stage, prior to the onset of compaction, E-cadherin-GFP is distributed across the entire cell membrane (*left panel*). As compaction begins, E-cadherin-GFP becomes predominantly localized to the basolateral cell membrane and compaction-specific filopodia (*right panel*). Insets show intensity of E-cadherin-GFP expression in single 2D planes. Scale bars: 10 μm in *x, y* and *x, y, z*.

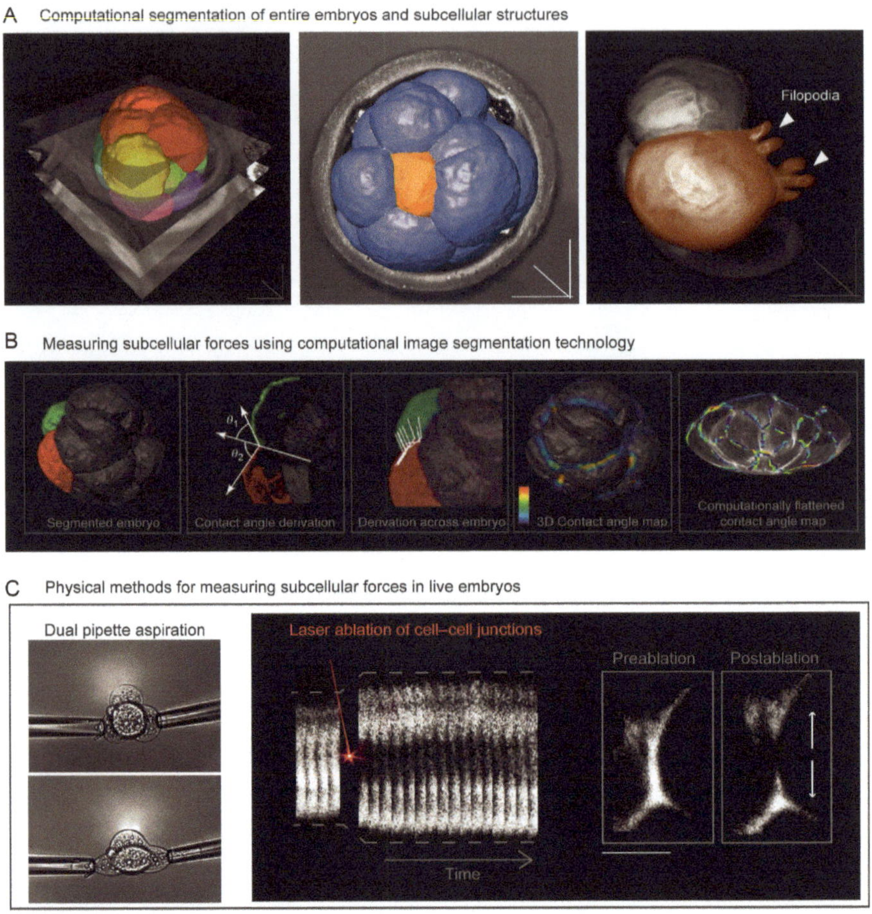

M.D. White *et al.*, Fig. 4 Emerging technologies for quantitative investigations of mechanisms controlling compaction. New quantitative imaging-based technologies are facilitating the investigation of compaction. (A) Computational segmentation of entire embryos (*left*, *middle panel*) or selected cells and filopodia (*arrowhead*, *right panel*). Embryos expressing fluorescently labeled proteins are imaged in 4D using two-photon microscopy, and individual features are computationally segmented. (B) Image segmentation technologies enable calculation of the subcellular distribution of biomechanical forces. Contact angles at cell–cell junctions result from the balance of subcellular forces and can be measured at multiple points along the junction. This provides a 3D map of contact angles, which can be projected to a 2D map of the apical surface of the embryo. (C) Dual pipette aspiration is a physical method for measuring subcellular forces in the embryo. An embryo is held between two micropipettes (*top left panel*) and aspiration is increased (*bottom left panel*) until the cells separate. Although this method provides useful information about the force of cell–cell adhesion, it is not compatible with ongoing development of the embryo. A less invasive physical technique for measuring subcellular forces is laser ablation. Targeting high laser energy into a localized region of cell cortex ablates cortical structure causing rapid recoil about the ablation that reflects the size and directionality of subcellular forces in that region. When cell–cell junctions are laser ablated (*right panel*) the degree by which adjacent vertices separate (*white arrows*, *right panel*) reveals the residual tensile forces. Scale bars: 5 μm in *x*, *y* and 10 μm in *x*, *y*, *z*.

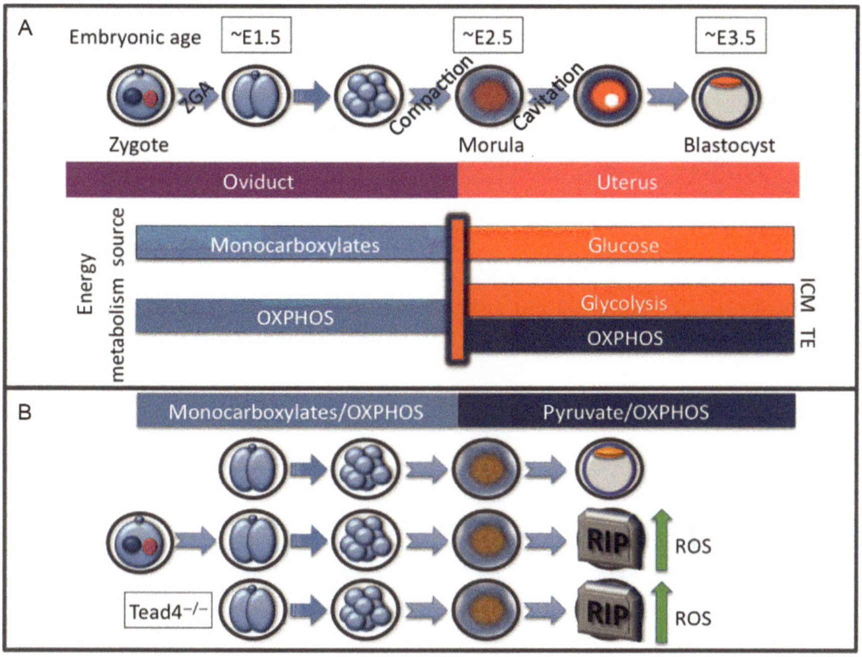

K.J. Kaneko, Fig. 1 Overview of mouse preimplantation development. (A) In the oviduct, transcriptionally silent egg is fertilized by the sperm to form the zygote (1-cell embryo), initially with two parental pronuclei, which are chromatin-modified and replicated prior to formation of a 2-cell embryo. Zygotic gene activation (ZGA) begins in the late 1-cell stage, but the major ZGA occurs after the 2-cell embryos are formed at ~E1.5. After the 8-cell stage, Cadherin1-mediated compaction occurs ~E2.5 to form the morula, which now enters the uterus. Outer cells of morula differentiate into trophectoderm (TE), whereas inner cells become the inner cell mass (ICM). Morula-to-blastocyst transition occurs during cavitation, where TE initiates and expands a fluid-filled cavity called the blastocoel. Mouse embryos use monocarboxylates, lactate, and pyruvate, for energy prior to the morula stage. At the morula stage, there is a switch to glucose for energy substrate with corresponding increase in O_2 consumption. Whereas glucose and monocarboxylates are present in both oviduct and the uterus, preimplantation embryos cannot use glucose for energy prior to the morula stage, whereas they preferentially use glucose for energy after the morula stage. Thus glycolytic activity is inhibited prior to the morula stage, presumably due to regulatory mechanisms involving glucose transporters and metabolic inhibition of hexokinase, PFK, and/or other enzymes involved in Embden-Meyerhof-Parnas pathway. Whereas monocarboxylates can only be metabolized as energy through oxidative phosphorylation (OXPHOS), glucose can be metabolized either by glycolysis or OXPHOS. As TE must use energy consuming Na^+/K^+ ATPase for blastocoel expansion, it relies on complete oxidation of glucose through OXPHOS in vivo. ICM, on the other hand, is concerned with replication, maintenance of pluripotency, and anticipation of rapid increase in biomass during early stages of implantation in a hypoxic environment, all suited for glycolytic energy metabolism.

(Continued)

K.J. Kaneko, Fig. 1—Cont'd (B) Mouse preimplantation embryos have the capacity for metabolic adaptation, as evidenced by the 2-cell embryos to develop to blastocysts in complete absence of glucose in culture (Jansen, Cashman, Thompson, Pantaleon, & Kaye, 2009; Martin & Leese, 1995). Wild-type embryos can undergo metabolic reprogramming at the morula stage to utilize pyruvate in the absence of glucose. Glucose, however, is essential for 1-cell embryos to develop to blastocysts, presumably because glucose metabolism is needed not for energy but for alleviation of oxidative stress (ROS) during blastocyst formation. Interestingly, *Tead4*$^{-/-}$ 2-cell embryos cannot develop to blastocysts in the absence of glucose, but can form blastocoel by addition of an antioxidant during glucose-free culture (Kaneko & DePamphilis, 2013).

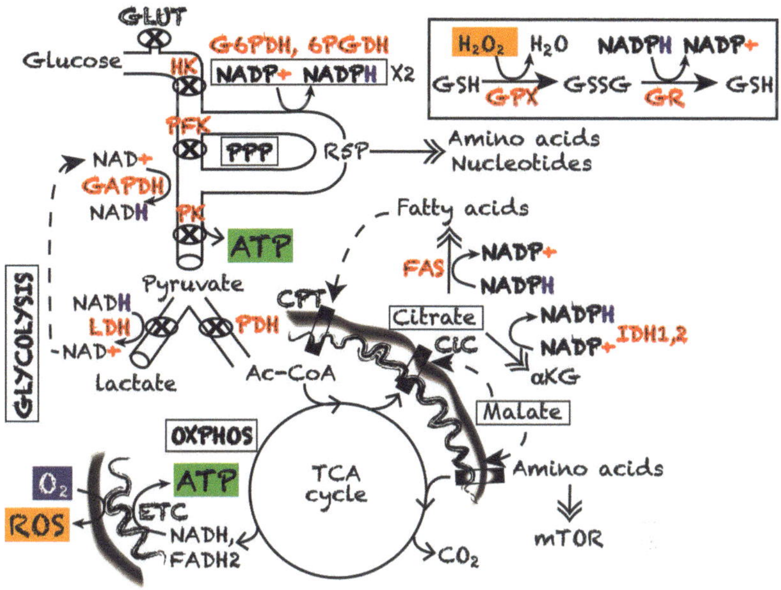

K.J. Kaneko, Fig. 2 Overview of relevant energy metabolism. Glucose can be metabolized in a number of ways, which is regulated at multiple points. Glucose entry into cells can be controlled by glucose transporters (GLUT), which is one target of Hypoxia Inducible Factor 1 (HIF1α). Glucose is then phosphorylated by hexose kinase (HK) to glucose 6-phosphate (G6P), which can be directed toward either Embden–Meyerhof–Parnas pathway (EMP) for ATP generation or diverted toward pentose phosphate pathway (PPP) to generate NADPH and ribose 5-phosphate (R5P). G6P is also a substrate for glycogen and glucosamine pathways (not shown). EMP is regulated at multiple points by phosphofructokinase (PFK), pyruvate kinase (PK), and glyceraldehyde phosphate dehydrogenase (GAPDH), which relies on the availability of NAD^+. Whether glucose is metabolized by glycolysis or oxidative phosphorylation (OXPHOS) is determined by the fate of pyruvate. Lactose dehydrogenase (LDH) produces lactate and regenerates NAD^+ for glycolysis, while for OXPHOS, pyruvate is transported to the mitochondria to generate acetyl-CoA (Ac-CoA), which is fed into the TCA cycle through the action of pyruvate dehydrogenase complex (PDH). TCA cycle generates reducing equivalents in the form of NADH and $FADH_2$, which is used by the electron transport chain (ETC) in the inner membrane to generate an electrochemical gradient that ultimately drives ATP synthesis. O_2 serves as the final electron acceptor to produce H_2O, but the chemical properties of O_2 make it a promiscuous electron acceptor, resulting in production of reactive oxygen species (ROS). Cells have multiple pathways to neutralize ROS, including superoxide dismutase (not shown) that reacts with oxygen radical to produce hydrogen peroxide (H_2O_2), which is then neutralized by either catalase (not shown) or glutathione peroxidase (GPX) to form H_2O. Reduction of H_2O_2 results in oxidation of glutathione (GSH) to glutathione disulfide (GSSG), which must be reduced by glutathione reductase (GR) and NADPH to regenerate GSH (*inset*). Amino acids can be metabolized for energy through pyruvate or Ac-CoA intermediates which feed into the TCA cycle and OXPHOS.

(Continued)

K.J. Kaneko, Fig. 2—Cont'd To regenerate TCA components that leave the cycle for biosynthetic pathways such as citrate, some amino acids (like glutamine) can be metabolized into TCA components such as α-ketoglutarate (αKG) by a process known as anaplerosis. Amino acids, especially leucine and glutamine, upregulate mTOR activity, leading to increase in OXPHOS activity. Fatty acids are also energy substrates, which enter the mitochondria through regulated transport systems, which include carnitine palmitoyltransferase 1 (CPT), a rate-limiting step that is irreversibly inhibited by etomoxir. Upon entering the mitochondria, fatty acids are metabolized to Ac-CoA units which feed into the TCA cycle. As FAs are synthesized in the cytoplasm by fatty acid synthase (FAS), which requires citrate as a substrate, glucose and amino acids can provide carbon sources for citrate synthesis in the mitochondria. However, citrate must be transported out of mitochondria by citrate carrier (CiC), which is an antiporter that requires dicarboxylates such as malate for exchange. FAS also requires NADPH, which is mostly provided by G6P dehydrogenase (G6PDH) and 6-phosphogluconate dehydrogenase (6PGDH) in the PPP, as well as by NADPH-dependent isocitrate dehydrogenases 1 and 2 (IDH1, 2) that oxidize isocitrate (which can be derived from cytoplasmic citrate) to αKG. Thus in addition to its role in fatty acid biosynthesis, cytoplasmic citrate has multiple roles in preimplantation embryos as it not only can inhibit PFK (thus regulate glycolysis) but also regenerate NADPH (Dumollard, Carroll, Duchen, Campbell, & Swann, 2009). Cytosolic malate is also an important regulator of redox balance as part of "malate-aspartate" shuttle, which is essential during preimplantation development (Lane & Gardner, 2005), as well as being a substrate for regenerating NADPH via malic enzyme (not shown).

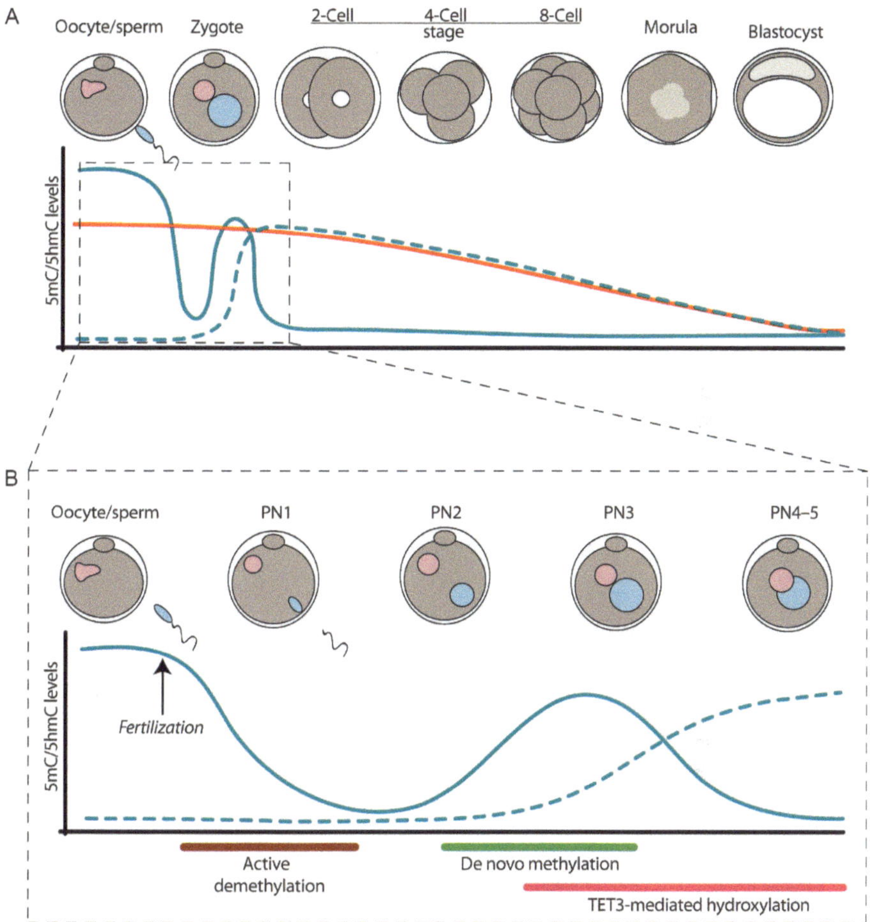

C.Y. Lim *et al.*, Fig. 2 DNA-demethylation dynamics in the preimplantation embryo. (A) DNA-methylation dynamics in the preimplantation embryo. The global paternal 5mC levels (*blue line*) decrease rapidly in the zygote, accompanied by an increase of 5hmC (*blue dotted line*). Maternal global 5mC levels decrease slowly over subsequent DNA-replication cycles. (B) Close-up illustration of global paternal DNA-methylation dynamics during zygotic progression from fertilization to pronuclear stage (PN) 5. 5mC is actively lost by an unknown mechanism (*red bar*), before de novo methylation activity (*green bar*) is counteracted by TET3-mediated hydroxylation (*pink bar*).

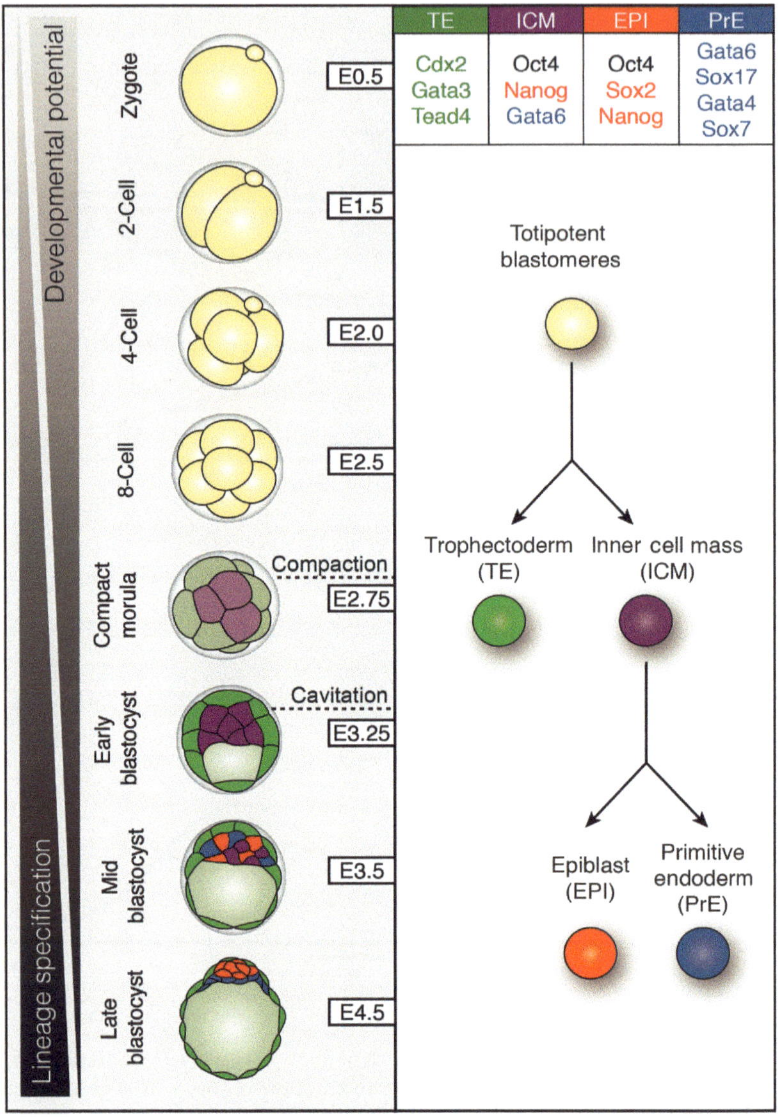

V. Garg *et al.*, Fig. 1 Overview of mouse preimplantation development. Two cell fate decisions result in the specification of three lineages in the blastocyst prior to implantation. As lineages are specified developmental potential is restricted. Box on *top right* depicts lineage-specific transcription factors specifically expressed at each stage. *E*, embryonic day.

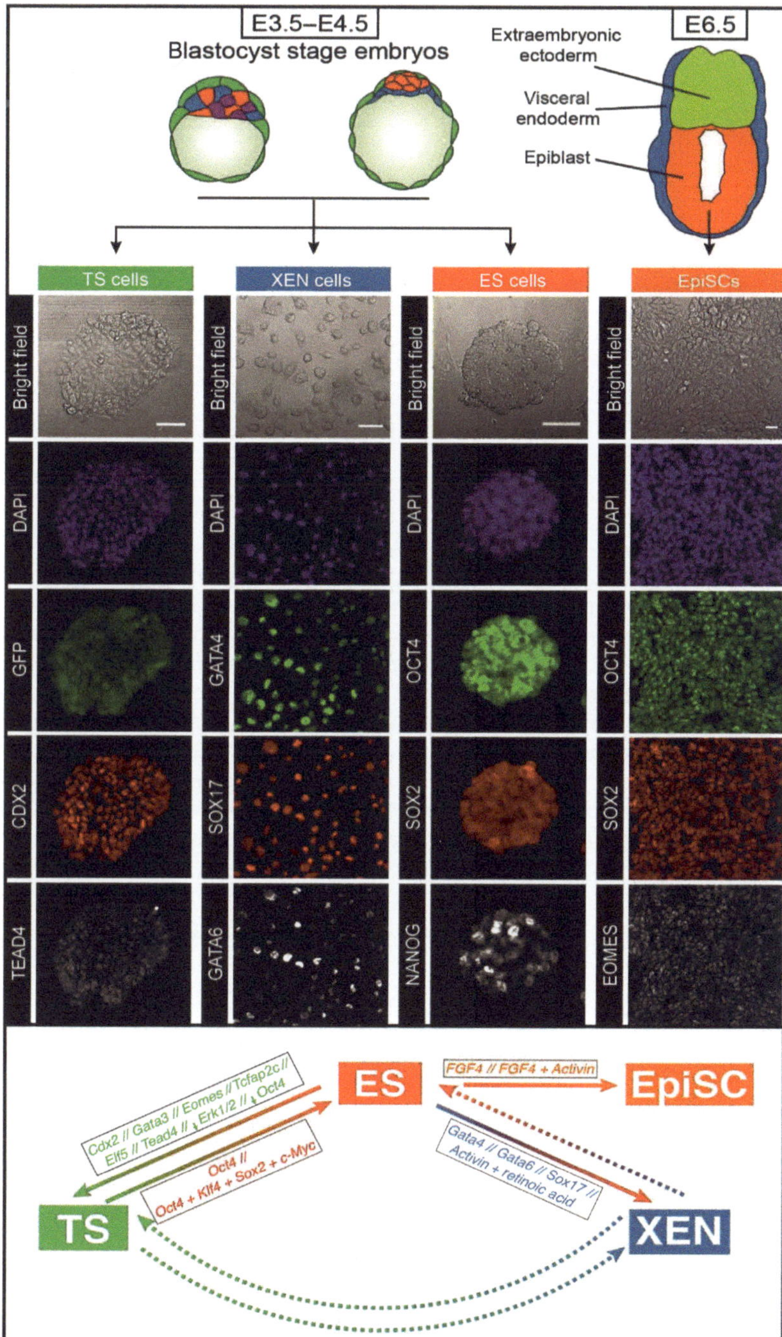

V. Garg *et al.*, Fig. 2 See legend on next page.

V. Garg *et al.*, Fig. 2 Lineage-specific stem cell types derived from preimplantation and postimplantation mouse embryos. TS cells, XEN cells, and ES cells represent the TE, PrE, and EPI of the blastocyst, respectively, and EpiSCs represent the postimplantation epiblast. Each stem cell type exhibits indefinite in vitro self-renewal, as well as the morphology, gene expression profile, and developmental potential of the cognate in vivo lineages they represent, and presumably are derived from. Immunofluorescence and bright field images depict morphology and common markers for each stem cell type. TS cells used in the analysis express GFP ubiquitously. (*Lower panel*) Interconversion of these stem cell types is possible by altering culture conditions and/or genetic manipulations. *Dashed lines* represent conversions that have yet to be demonstrated.

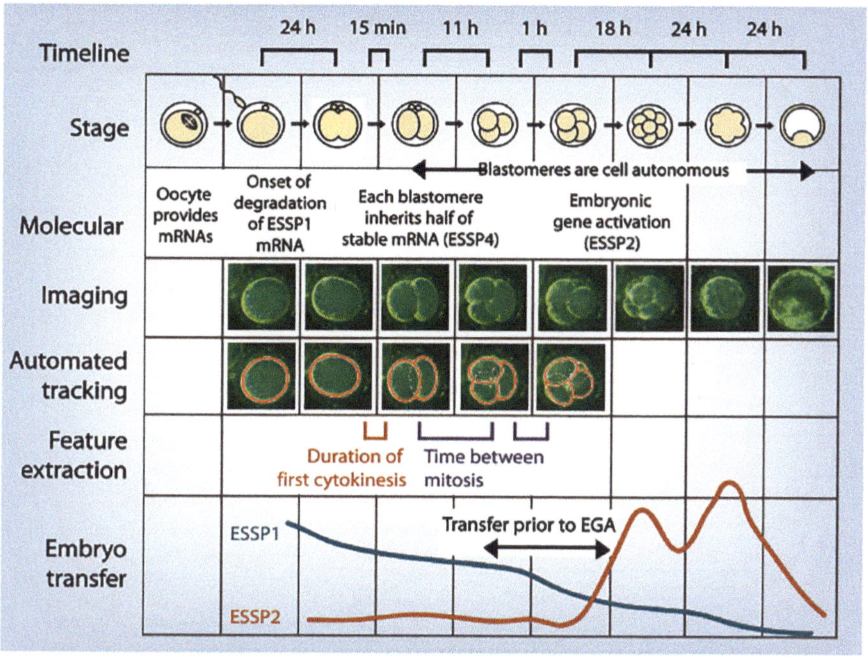

R.A. Reijo Pera and L. Prezzoto, Fig. 2 Model for human embryo development based on correlated imaging and molecular analysis. Shown is timeline from zygote to blastocyst including critical brief times for prediction of successful development to blastocyst. Key molecular data indicate that human embryo development is characterized by four embryonic stage-specific patterns (ESSPs) of gene expression. During the cleavage divisions, blastomeres may arrest or progress independently. The outcome of cell autonomous development is that individual blastomeres may arrest or progress; as the embryo progresses to morula stage and beyond, blastocyst quality will be impacted by the number of cells that arrested or progressed beyond 8 cells. Imaging data demonstrate that critical periods of development predict success or failure: cytokinesis, second cleavage division, and synchronicity of appearance of third and fourth blastomeres. Data suggest improved ability to diagnose embryo outcome with key imaging predictors. *From Wong, C., Loewke, K., Bossert, N., Behr, B., DeJonge, C., Baer, T., et al. (2010). Non-invasive imaging of human embryos before embryonic genome activation predicts development to the blastocyst stage.* Nature Biotechnology, 28, *1115–1121.*

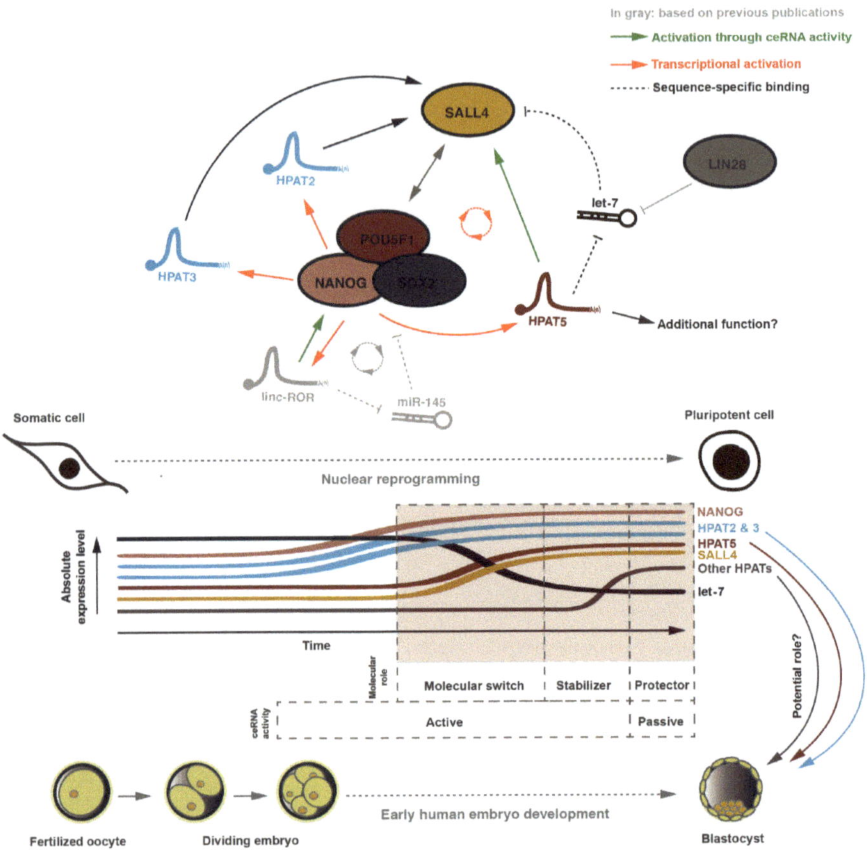

R.A. Reijo Pera and L. Prezzoto, Fig. 3 Model of retroviral-derived long noncoding RNAs (lncRNAs) activity during embryo development, nuclear reprogramming, and maintenance of pluripotency in human stem cells. lncRNAs, enriched for HERVH and HERVK retroviral sequences (*HPATs, human pluripotency-associated transcripts*), exert its function as active and passive ceRNA to act as a molecular switch, stabilizer or protector during pluripotency establishment, and maintenance in human pluripotent cells.

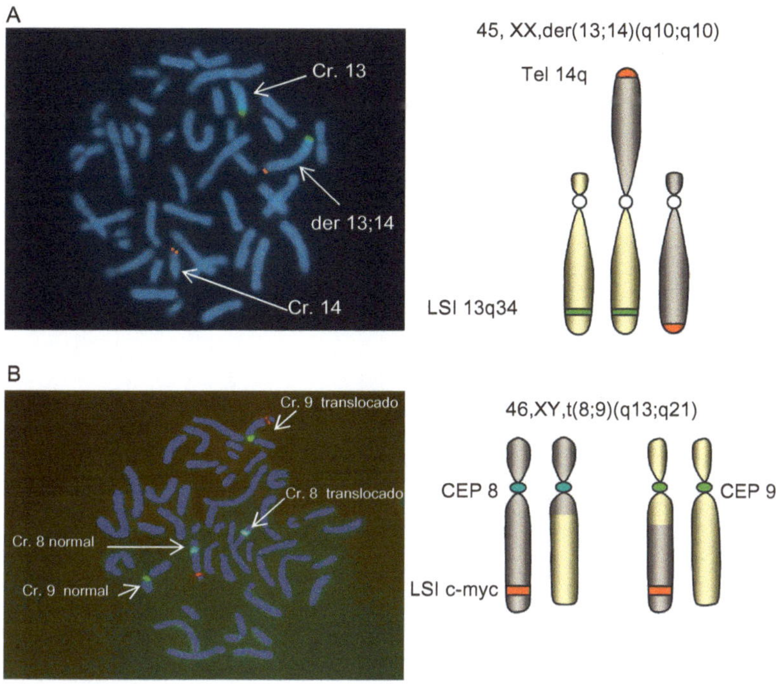

S. Garcia-Herrero *et al.*, Fig. 2 An example of a Robertsonian (A) and reciprocal translocation (B).

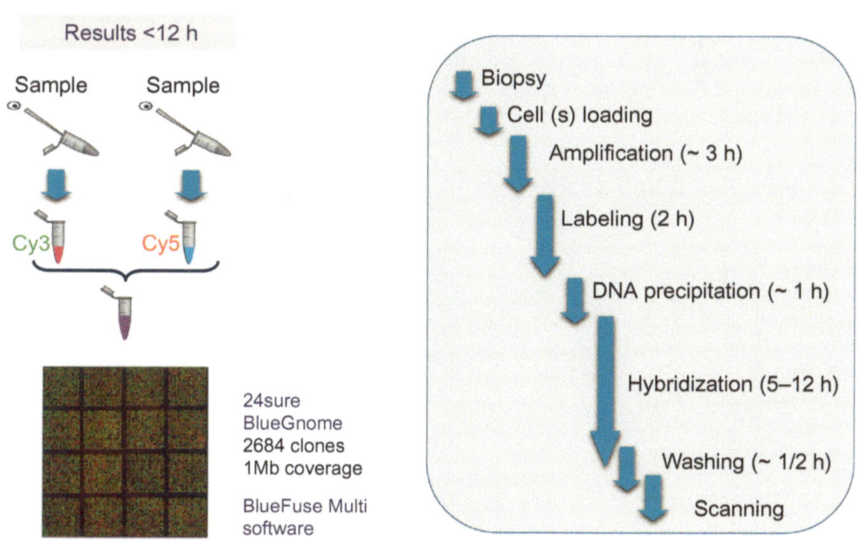

S. Garcia-Herrero *et al.*, Fig. 3 CGH array flow chart.

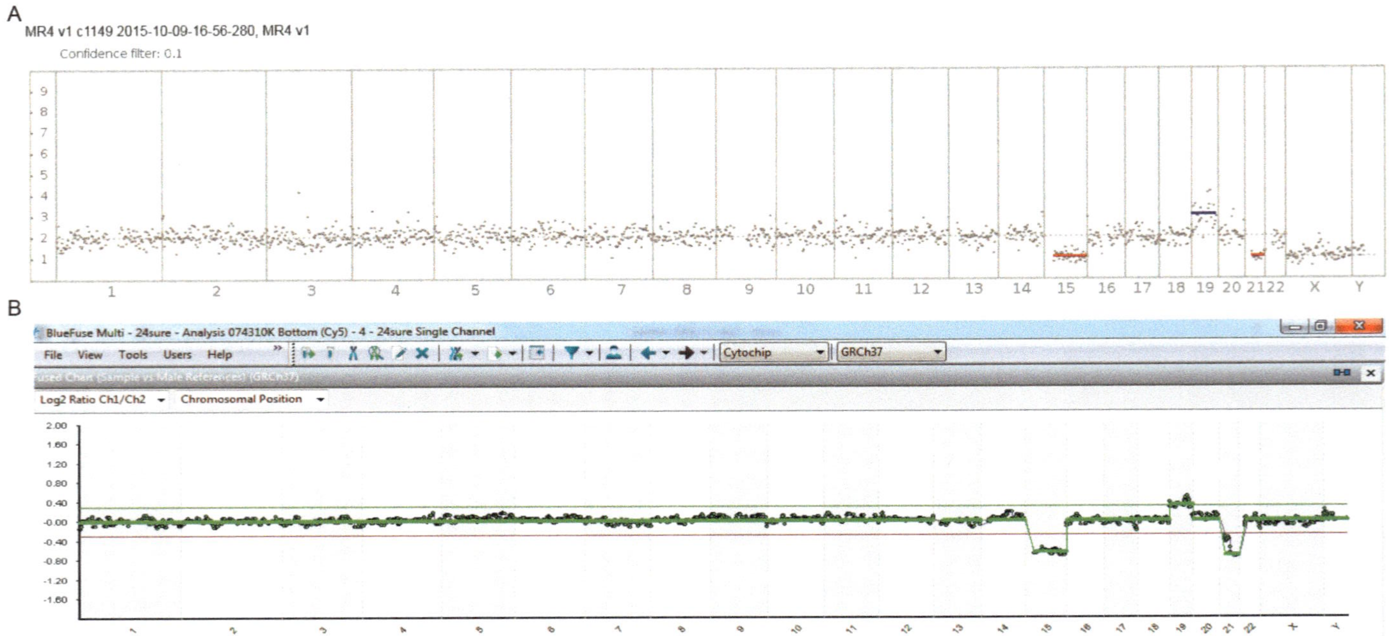

S. Garcia-Herrero *et al.*, Fig. 4 NGS (A) and CGH arrays (B) profiles of a D5 embryo biopsy presenting two monosomies (15 and 21) and one trisomy (19).

Kolibri 360 GmbH
Ferdinand-Jühlke-Straße 7
99095 Erfurt, Deutschland
produktsicherheit@kolibri360.de